Effective Knowledge Management Systems in Modern Society

Murray E. Jennex
San Diego State University, USA

A volume in the Advances in Knowledge Acquisition, Transfer, and Management (AKATM) Book Series

IGI Global
DISSEMINATOR OF KNOWLEDGE

Published in the United States of America by
IGI Global
Information Science Reference (an imprint of IGI Global)
701 E. Chocolate Avenue
Hershey PA, USA 17033
Tel: 717-533-8845
Fax: 717-533-8661
E-mail: cust@igi-global.com
Web site: http://www.igi-global.com

Library of Congress Cataloging-in-Publication Data

Names: Jennex, Murray E., 1956- editor.
Title: Effective knowledge management systems in modern society / Murray E. Jennex, editor.
Description: Hershey, PA : Information Science Reference, [2018]
Identifiers: LCCN 2017043236| ISBN 9781522554271 (hardcover) | ISBN 9781522554288 (ebook)
Subjects: LCSH: Knowledge management. | Organizational learning--Management.
Classification: LCC HD30.2 .E376 2018 | DDC 658.4/038011--dc23 LC record available at https://lccn.loc.gov/2017043236

This book is published in the IGI Global book series Advances in Knowledge Acquisition, Transfer, and Management (AKATM) (ISSN: 2326-7607; eISSN: 2326-7615)

British Cataloguing in Publication Data
A Cataloguing in Publication record for this book is available from the British Library.

For electronic access to this publication, please contact: eresources@igi-global.com.

Advances in Knowledge Acquisition, Transfer, and Management (AKATM) Book Series

Murray E. Jennex
San Diego State University, USA

ISSN:2326-7607
EISSN:2326-7615

Mission

Organizations and businesses continue to utilize knowledge management practices in order to streamline processes and procedures. The emergence of web technologies has provided new methods of information usage and knowledge sharing.

The **Advances in Knowledge Acquisition, Transfer, and Management (AKATM) Book Series** brings together research on emerging technologies and their effect on information systems as well as the knowledge society.**AKATM** will provide researchers, students, practitioners, and industry leaders with research highlights surrounding the knowledge management discipline, including technology support issues and knowledge representation.

Coverage

- Cognitive Theories
- Cultural Impacts
- Information and Communication Systems
- Knowledge acquisition and transfer processes
- Knowledge management strategy
- Knowledge Sharing
- Organizational Learning
- Organizational Memory
- Small and Medium Enterprises
- Virtual communities

IGI Global is currently accepting manuscripts for publication within this series. To submit a proposal for a volume in this series, please contact our Acquisition Editors at Acquisitions@igi-global.com or visit: http://www.igi-global.com/publish/.

The Advances in Knowledge Acquisition, Transfer, and Management (AKATM) Book Series (ISSN 2326-7607) is published by IGI Global, 701 E. Chocolate Avenue, Hershey, PA 17033-1240, USA, www.igi-global.com. This series is composed of titles available for purchase individually; each title is edited to be contextually exclusive from any other title within the series. For pricing and ordering information please visit http://www.igi-global.com/book-series/advances-knowledge-acquisition-transfer-management/37159. Postmaster: Send all address changes to above address.

Titles in this Series

For a list of additional titles in this series, please visit: www.igi-global.com/book-series

Contemporary Knowledge and Systems Science
W. B. Lee (Hong Kong Polytechnic University, China) and Farzad Sabetzadeh (Hong Kong Polytechnic University, China)
Information Science Reference • copyright 2018 • 340pp • H/C (ISBN: 9781522556558) • US $185.00 (our price)

Enhancing Knowledge Discovery and Innovation in the Digital Era
Miltiadis D. Lytras (The American College of Greece, Greece) Linda Daniela (University of Latvia, Latvia) and Anna Visvizi (The American College of Greece, Greece)
Information Science Reference • copyright 2018 • 363pp • H/C (ISBN: 9781522541912) • US $215.00 (our price)

N-ary Relations for Logical Analysis of Data and Knowledge
Boris Kulik (Russian Academy of Science, Russia) and Alexander Fridman (Russian Academy of Science, Russia)
Information Science Reference • copyright 2018 • 297pp • H/C (ISBN: 9781522527824) • US $225.00 (our price)

Radical Reorganization of Existing Work Structures Through Digitalization
Punita Duhan (Meera Bai Institute of Technology, India) Komal Singh (Meera Bai Institute of Technology, India) and Rahul Verma (Bhai Parmanand Institute of Business Studies, India)
Business Science Reference • copyright 2018 • 287pp • H/C (ISBN: 9781522531913) • US $235.00 (our price)

Maximizing Social Science Research Through Publicly Accessible Data Sets
S. Marshall Perry (Saint Mary's College of California, USA)
Information Science Reference • copyright 2018 • 328pp • H/C (ISBN: 9781522536161) • US $195.00 (our price)

Promoting Interdisciplinarity in Knowledge Generation and Problem Solving
Mohammed Nasser Al-Suqri (Sultan Qaboos University, Oman) Abdullah Khamis Al-Kindi (Sultan Qaboos University, Oman) Salim Said AlKindi (Sultan Qaboos University, Oman) and Naifa Eid Saleem (Sultan Qaboos University, Oman)
Information Science Reference • copyright 2018 • 324pp • H/C (ISBN: 9781522538783) • US $195.00 (our price)

Techniques for Coding Imagery and Multimedia Emerging Research and Opportunities
Shalin Hai-Jew (Kansas State University, USA)
Information Science Reference • copyright 2018 • 483pp • H/C (ISBN: 9781522526797) • US $235.00 (our price)

IGI Global
DISSEMINATOR OF KNOWLEDGE

701 East Chocolate Avenue, Hershey, PA 17033, USA
Tel: 717-533-8845 x100 • Fax: 717-533-8661
E-Mail: cust@igi-global.com • www.igi-global.com

Table of Contents

Section 2
Issues in Knowledge Sharing

Section 3
KMS Design Issues

Section 4
KM and Security

Detailed Table of Contents

The knowledge pyramid has been used for several years to illustrate the hierarchical relationships between data, information, knowledge, and wisdom. This chapter posits that the knowledge pyramid is too basic and fails to represent reality and presents a revised knowledge-KM pyramid. One key difference is that the revised knowledge-KM pyramid includes knowledge management as an extraction of reality with a focus on organizational learning. The revised pyramid includes newer initiatives such as business and/or customer intelligence, big data, analytics, internet of things. Finally, this chapter discusses how KM strategy can be generated using the final revised pyramid.

This qualitative research examined the relationship between knowledge management (KM) and systematic innovation capability in 16 Australian manufacturing and service organizations that exhibited both successful innovation and robust KM practices. A review of the literature indicated a number of areas where KM enhances and supports innovation capability. Using a multiple crosscase analysis methodology and applying a framework of systematic innovation capability, in-depth interviews were conducted with managers of the case study organizations. The analysis of the data revealed the main contributions of KM to systematic and sustained forms of innovation. Finally, a model of knowledge and innovation capability was developed to guide the development of knowledge and innovation management as a dynamic capability to support value capture, value creation, and value delivery from innovation.

A contentious merger seems to characterize the relationship between entrepreneurship, defined as the process of bringing new products and services to the market, and knowledge management. Even when discussed from the perspective of dynamic knowledge creation and innovation rather than from the codification and organization of a firm's knowledge base, the knowledge management process of new ventures often fails to keep pace with best practices. This is particularly puzzling as the competitive advantage of new ventures stems from exploiting tacit knowledge, which is at the heart of successful knowledge creation and management. This chapter investigates this contradiction and discusses challenges and opportunities to both the practitioner and the researcher.

A contingency perspective of knowledge management, as one of the popular ways of promoting decision making capabilities, recognizes the need for a fit between knowledge management solutions (KMS) and decision-making contexts which they support. In order to determine the best fit, a field survey was carried out to investigate the impact of two different types of KMS (technical and social) on decision makers' behavior and performance in different decision contexts (simple and complex). According to the results, there is a partial support for the contingency view. As expected, social KMS appears as the best fit for complex contexts, based on subjects' superior performance from comparable adoption of both KMS. In contrast, the results suggest that both KMS were an equally good fit for simple contexts, based on similar levels of subjects' performance, but social KMS was preferred in terms of adoption. These findings contribute to much necessary empirical evidence for research and provide useful guidance for practice. However, their limitations necessitate further study.

Today there is a new emphasis on a discipline related to knowledge engineering (KE): knowledge management (KM). The chapter shows that a Delphi methodology, developed and used with real success from 1986 through 2003, can be revised and up-dated to take a place in the spectrum of KE/KM. This new model combines elements of a Pareto technique at the front end of an small group Delphi paradigm/estimate-talk-estimate (SGDP/ETE) process to initially identify and sharpen the criticality of problems and their solutions. The final stage of the model uses real time Delphi (RTD) techniques. As a result, the final paradigm uses today's technology to achieve extremely accurate decision making and indicate what can be successful courses of action across many venues. Examples will be given of how this new model could be applied to current, real-life problems in three very different arenas.

Section 2
Issues in Knowledge Sharing

This chapter focuses on knowledge-sharing barriers and their implications over knowledge sharing and individual innovative behavior and explores the role of knowledge sharing on individual innovative behavior. Considering limited studies in the literature, this study provides further evidences regarding the implications of knowledge-sharing barriers and knowledge sharing in the workplace. Data was collected from 83 academic staff in a higher education institution and analyzed with Smart PLS. The results showed that organization-related knowledge-sharing barriers were negatively related to knowledge sharing. The results also indicated that individual knowledge-sharing barriers had a negative effect on individual innovative behaviors. There was no link found in this study between knowledge sharing and individual innovative behavior. These findings provided empirical evidences to the further development of knowledge management and innovation research, and insights regarding how to better foster knowledge sharing and innovative behavior in academe.

The study investigated the effects of expected reciprocity on knowledge sharing, as moderated by team and individual variables. Data (n = 84) were collected in an experimental study from undergraduate business student participants. The effects of expected reciprocity on knowledge sharing depend on the levels of individual competence, positive team attitudes, functional diversity, and demographic diversity. Implications include that the effectiveness of reciprocity in knowledge sharing depends on several factors relating to the team and individual. Encouraging reciprocity may have positive effects, but these can be overridden by poor team attitudes, low ability perceptions, and team diversity. Future research suggestions are offered.

With the global rise of the knowledge worker, there is an increasing interest in effective knowledge transfer practices within the firm. This interest is especially prevalent in regards to the transfer of tacit knowledge—knowledge that is difficult to describe and transmit to others. In this chapter, the authors consider how the emerging practice of knowledge coaching can be utilized as a means to transfer tacit knowledge between experts and protégés within the firm, and consider the individual-level factors that influence whether the practice yields success. In doing so, they report the results of a study examining the impact of individually possessed national cultural values on motivation to engage in knowledge coaching, and put forward several avenues for future research.

Section 3
KMS Design Issues

Preface

The last 10 years have brought tremendous innovation to information technology and subsequently knowledge management (KM). Key technical innovations include social media, the cloud, software as a service, mobile technologies, Internet 2.0 and collaborative technologies, unstructured data, big data, the Internet of Things, artificial intelligence, machine learning, and improved network connectivity and capacity. Additional emphasis on KM and information management issues such as governance, risk and security management, leadership, innovation, business intelligence and analytics, and strategy have gotten organizations thinking new processes and new ways in managing, transferring, and utilizing data, information, and knowledge. Researchers have published many articles in the *International Journal of Knowledge Management* exploring how these innovations impact KM. The purpose of this book is to present the best of these papers in a single location so readers can get an overview of how these technical, social, and process changes are impacting KM and changing the way KM systems (KMS) are designed, built, implemented, and used.

The book looks at four major areas where KM and KMS have seen the most change:

- Section 1, Expanding KM Boundaries and Other Trends, Chapters 1, 2, 3, 4, and 5, looks at how KM is evolving based on technical, social, decision context, and process changes and presents new views of KM and KMS, changes based on the addition of innovation and entrepreneurship to KM, and finally a survey of future trends in KM.
- Section 2, Issues in Knowledge Sharing, Chapters 6, 7, and 8, focuses on perhaps the most important issue in KM; how to get people to share and use knowledge. This section looks at key issues like barriers to knowledge sharing, reciprocity in knowledge sharing, and using knowledge coaching to improve knowledge sharing.
- Section 3, KMS Design Issues, Chapters 9, 10, 11, and 12, looks at issues in converting information technology into KMS, the impact of using knowledge ambassadors to implement KMS strategy, the impact of KM on project success, and the issue of academic research making practical impacts on KM and KMS.
- Section 4, KM and Security, Chapters 13, 14, and 15, explores the important issues of building security into KMS and using KM to improve organizational security. The section looks at security skills needed by KM practitioners and risk management.

Each chapter is an expansion of an article published in the International Journal of Knowledge Management and the chapter authors are a compendium of strong and leading KM researchers and academics.

The book is designed for all involved from KM. KM practitioners will get insight into where KM is going and how they can adapt their current KMS to future needs. KM consultants will get detailed research insight into these issues that they can use to advice clients. KM researchers will get insight into cutting edge research into KM and KMS and a basic start to current KM literature. KM academics will find a resource that can be used for advanced KM coursework.

The conclusion that can be drawn from this book is that new KM and KMS will not be the traditional technology solution, they will include many new technologies, new processes, new approaches and will be able to work with all sorts of users without relying on information technology professionals. The book will also lead readers to accept the conclusion the KM is more than technology, it is a fusion of technology, people, process, knowledge, and strategic insight.

The rest of this introduction provides short overviews of each chapter:

SECTION 1: EXPANDING KM BOUNDARIES AND OTHER TRENDS

Chapter 1, "Using A Revised Knowledge Pyramid to Redefine Knowledge Management Strategy," by Murray Jennex reviews the traditional knowledge pyramid and modifies it incorporating new technologies such as artificial intelligence, Internet of Things, big data, machine learning, and analytics. The result is a nested pyramid, one that shows the societal knowledge pyramid for the knowledge society as the overall pyramid with the KM pyramid nested inside and focused on organizational knowledge. The revised pyramid is transformed from a simple hierarchical model into a process model that shows the creation of knowledge, wisdom, and actionable intelligence through the processes of learning and organizational learning; and the application of filters for transformations between layers in the pyramid. The final revised pyramid is useful for conceptualizing new and effective KM and knowledge systems and serves as a guide for the application of new technologies to societal and organizational knowledge systems for generating wisdom and actionable intelligence for use in decision making.

Chapter 2 is "Knowledge and Innovation Management: Creating Value," by Marianne Gloet and Danny Samson. Competitive advantage in today's advanced economies is driven by innovation and the ability to manage ever-increasing forms of knowledge on a sustained basis. Knowledge intensive industries compete primarily on their capacity to innovate and thrive on cutting-edge knowledge, which drives both research and innovation. Indeed, knowledge intensive organizations (KIOs) constantly seek to reinforce sustainable links between forms of knowledge and modes of innovation. In such a dynamic environment, the proactive management of knowledge assets is essential to achieving both innovation capability and innovation performance, the ways in which organizations approach knowledge management (KM) influences innovation and becomes a source of competitive advantage (Freeze and Kulkarni, 2008). As such, KM emerges as an essential management and organizational capability in the drive to create value through knowledge. This chapter examines the ways in which KM is manifested across a range of organizations in both the manufacturing and service sectors, with a view to determining more clearly the relationship between KM and innovation. A further objective of this chapter is to investigate the extent to which KM can contribute to systematic and sustained forms of innovation within organizations. Moreover, the chapter seeks to determine the ways in which effective knowledge and innovation management can contribute to value capture and value creation in organizations.

Chapter 3 is "Knowledge Management and Entrepreneurship Research and Practice: Status, Challenges, and Opportunities," by Cesar Bandera, Katia Passerini, and Michael Bartolacci. A contentious merger seems to characterize the relationship between entrepreneurship, defined as the process of bringing new products and services to the market, and knowledge management. Even when discussed from the perspective of dynamic knowledge creation and innovation rather than from the codification and organization of a firm's knowledge base, the KM process of new ventures often fails to keep pace with best practices. This is particularly puzzling as the competitive advantage of new ventures stems from exploiting tacit knowledge, which is at the heart of successful knowledge creation and management. This chapter investigates this contradiction, and discusses challenges and opportunities to both the practitioner and the researcher.

Chapter 4 is "A Contingency Perspective for Knowledge Management Solutions in Different Decision-Making Contexts," by Kursad Ozlen and Meliha Handzic. A contingency perspective of KM, as one of the popular ways of promoting decision making capabilities, recognizes the need for a fit between the KMS and the decision making contexts which they support. In order to determine the best fit, this chapter presents a field survey carried out to investigate the impact of two different types of KMS (technical and social) on decision makers' behavior and performance in different decision contexts (simple and complex). According to the results, there is a partial support for the contingency view. As expected, social KMS appears as the best fit for complex contexts, based on subjects' superior performance from comparable adoption of both KMS. In contrast, the results suggest that both KMS were an equally good fit for simple contexts, based on similar levels of subjects' performance, but social KMS was preferred in terms of adoption. These findings contribute to much necessary empirical evidence for research and provide useful guidance for practice. However, their limitations necessitate further study.

Chapter 5 is "Knowledge Engineering, Cognitive Ergonomics, and Knowledge Management in 2017: A New Delphi Paradigm With Applications," by Ronald Lofaro. There is a new emphasis on a discipline related to knowledge engineering, (KE): KM. This Chapter shows that a Delphi methodology, developed and used with real success from 1986 through 2003, can be revised and up-dated to take a place in the spectrum of KE/KM. This new model combines elements of a Pareto technique at the front end of an small group Delphi paradigm/estimate-talk-estimate (SGDP/ETE) process, to initially identify and sharpen the criticality of problems and their solutions. The final stage of the model uses real time Delphi (RTD) techniques. As a result, the final paradigm uses today's technology to achieve extremely accurate decision making and indicate what can be successful courses of action across many venues. Examples are given of how this new model can be applied to current, real-life problems in three very different arenas.

SECTION 2: ISSUES IN KNOWLEDGE SHARING

Chapter 6 is "Exploring Knowledge-Sharing Barriers and Their Implications," by Salih Yeşil and Bengü Hırlak. Knowledge sharing is regarded as an important phase of knowledge management and determines the success of knowledge management initiatives. Knowledge sharing is defined "the exchange of knowledge between and among individuals, and within and among teams, organizational units, and organisations". Previous studies have studied the antecedents and consequences of the knowledge sharing. In reviewing the literature on knowledge sharing, and have identified a number of areas including organizational context, interpersonal and team characteristics, cultural characteristics, individual characteristics, and motivational factors that affect knowledge sharing. Many theoretical and empirical studies argued and found that knowledge sharing has various implications for individuals. Unfortunately, barriers towards knowledge sharing prevent organizations to capitalize on their knowledge assets and skills. This chapter presents an empirical study testing the effect of these barriers on knowledge sharing and individual innovative behavior.

Chapter 7 is "Team and Individual Interactions With Reciprocity in Individual Knowledge Sharing," by Megan Endres and Sanjib Chowdhury. Knowledge sharing facilitates organizational and group performance, and can be used strategically to attain competitive advantage. Motivating individual sharing, therefore, is a topic of significant interest. Reciprocity is a significant motivator that receives consistent attention in knowledge sharing research. However, despite its widespread application to organizational theories and demonstrated importance to knowledge sharing, it is surprising that the construct remains underdeveloped. In knowledge sharing research, expected reciprocity has been found to interact with a variety of variables to predict outcomes, but more research is needed to form conclusions about the wide variety of possible moderators. Additionally, most studies are cross-sectional surveys, with only a handful of experiments. Given the importance of knowledge sharing for team performance, therefore, this chapter presents an experimental study of in-person teams was used to better understand the effects of reciprocity.

Chapter 8 is "The Determinants of Interorganizational Knowledge Coaching Success: Looking Ahead to the Future of Knowledge Transfer," by Nicole Celestine and Chris Perryer. With the global rise of the knowledge worker, there is an increasing interest in effective knowledge transfer practices within the firm. This interest is especially prevalent in regards to the transfer of tacit knowledge—knowledge that is difficult to describe and transmit to others. In this chapter, the authors consider how the emerging practice of knowledge coaching can be utilized as a means to transfer tacit knowledge between experts and protégés within the firm, and consider the individual-level factors that influence whether the practice yields success. In doing so, they report the results of a study examining the impact of individually-possessed national cultural values on motivation to engage in knowledge coaching, and put forward several avenues for future research.

SECTION 3: KMS DESIGN ISSUES

Chapter 9, is "CoMIS-KMS: An Elaborated Process Model for Transitioning MIS to KMS," by Ricardo Anderson and Gunjan Mansingh. A KMS facilitates the creation, capture, refinement, storage, application, update and dissemination of knowledge. KMS may vary in sophistication and features but are generally supported by information technology. Several alternate methods and techniques have been proposed for developing knowledge based systems given observed inadequacies of traditional software development methods. Although many developing countries are still at the stage of trying to use information systems to digitize, collect and manage data, having only partial automation of tasks and limited capabilities, knowledge can be of significant value and may provide for significant improvements that may allow for better decisions and an informed way forward to prosperity. With the pervasive nature of information systems across industries and in organizations, and the lagging of developing countries in exploring some of these resources to access efficiencies, research is required to provide guidance on how to access and exploit sources of potential knowledge, and harness this knowledge into existing systems towards improvement of their capabilities. The emphasis on the value of knowledge, knowledge creation and application of knowledge across all areas of endeavor necessitates the development of research to support improvements in knowledge integration in existing information systems. This chapter presents a process model which is developed to transition existing information system environment to a KMS. The model is based on the context of a developing country environment and has been instantiated in this context. The model and its instantiation provide further contribution in design theory as well as in the descriptive domain through observation, classification and measurement.

Chapter 10 is "Knowledge Management Strategy Implementation Through Knowledge Ambassadors," by Annabeth Aagaard. KM has become a key factor in the performance of organizations and several organizations have established some sort of knowledge management strategy in an attempt to structure and support knowledge sharing across the internal and external organization. Yet, in practice many companies fail in ensuring a continuous focus on knowledge sharing, the full integration of knowledge management strategies and on continuously developing a knowledge culture. Oftentimes the 'news value' of the KM strategy and the KM initiatives wear off, when operations and daily assignments take over the employees' focus. The challenges of overcoming the obstacles to effective strategy integration and execution have been discussed theoretically and empirically by various authors. All these obstacles relate to one of four groups: the change management context, the organizational culture context, the organizational power structure context and the leadership context. Change agents have been suggested as one of the key tools in ensuring more efficient implementation of changes and in ensuring integration of strategies by a number of authors. Yet, the majority of the research performed on change agents, focuses on leaders as change agents and external change agents such as consultants and not on employees, the internal change agents. Furthermore, change agents have not been studied within the knowledge management literature or in terms of knowledge management strategy implementation. Therefore, the concept of 'knowledge ambassadors' will be introduced and explored in this chapter. Knowledge ambassadors serve as change agents in implementing changes through knowledge management strategies integration and in sustaining the organization's continuous change and development towards effective knowledge sharing in a knowledge company. In the following a framework for this examination will be presented including a mapping of the key aspects of knowledge and knowledge management, knowledge management strategies and -implementation through the use of change agents in the form of knowledge ambassadors.

Chapter 11 is "Measuring the Relationship Among Learning Enablers and IT Project Success," by Donald McKay and Timothy Ellis. IT Projects continue to fail for many of the same reasons that they did 30 years ago. These failures lead to economic consequences, for example, companies spent millions of dollars on failed ERP implementations. In the United States, the cost of failed IT projects amounts to $63 billion. It has been postulated that project knowledge issues cost 10% of the total amount of a $60 million IT project. Customers have concluded that too many of their IT projects fail. Outside of the IT industry there are also project failures due to inadequate use of existing knowledge. Fortune 500 companies lose billions of dollars every year because their employees do not share knowledge. In developing countries, a lack of project knowledge management impeded the sustainability of "reproductive health development" initiatives which, if corrected, could potentially improve the skills and knowledge of health care professionals and the quality of services to the nation. This chapter investigates how KM can be used to enhance knowledge sharing in projects and improve project success.

Chapter 12 is "Practical Relevance of Management Research: The Role of Doctoral Program Graduates," by Madora Moshonsky, Alexander Serenko, and Nick Bontis. The debate about the perceived irrelevance of academic business research dates back to the 1980s when scholars, practitioners, and public officials started criticizing scholars for placing priority on scientific rigor over relevance to industry. The disconnect between academics and practitioners has been deemed "the Great Divide" because the theoretical contributions of researchers are rarely transferred to practice. There are several factors that justify the importance of the transfer of academic knowledge to practice, also, in order to ensure the success of an academic discipline, it must have an impact on the state of both theory and practice. At the same time, managers see value in academic findings as long as these may help them improve their efficiency and effectiveness. Therefore, calls have been made for studies that examine possible transfer methods of evidence-based knowledge to practitioners. This chapter contributes to the knowledge base by exploring whether doctoral business program graduates who work in industry are knowledge ambassadors acting as an indirect channel of knowledge transfer between academics and practitioners. Particularly, the purpose of this study is to explore whether doctoral business program graduates who enter the non-academic workforce acquire, utilize, and disseminate academic knowledge in their daily decision making.

SECTION 4: KM AND SECURITY

Chapter 13 is "Integrating IS Security With Knowledge Management: What Can Knowledge Management Learn From IS Security Vice Versa?" by Murray Jennex and Alexandra Durcikova. While KM can be used to improve security performance, training and awareness, this chapter primarily focuses on the use of security in KM and touches on what IS Security can learn from KM. Knowledge has value and items of value are targets of theft and attack. This chapter posits that KM does not have close enough links with IS Security. It is posited that this is evidenced by not only a lack of research literature addressing the integration of KM and IS Security but also a lack of interest in integrating KM and IS Security by KM practitioners. To investigate the links between KM and IT Security this chapter performs a review of the KM research literature with respect to IS Security. Additionally, to assess how KM practitioners' value IS Security skills and capabilities KM job postings are analyzed to determine what skills and capabilities are desired in new KM position hires. Nest, we explore KM practitioner attitudes with respect to the role of IT Security in KM an exploratory survey is generated and presented in this chapter. Finally, before discussion, we touch upon papers that have already applied lessons learned from KM and KMS research into IS Security research. The value of this chapter is in providing insight into perceptions and attitudes with respect to integrating IS Security into KM. The concern is that there is too little integration and that KM practitioners and researchers need to put more effort into creating secure KM and KMS.

Chapter 14 is "Knowledge Management in Support of Enterprise Risk Management," by Eduardo Rodriguez and John Edwards. The separation between KM and Risk Management (RM) is part of current organizational reality. The aim of this chapter is to study how KM concepts may help improve RM, and help to turn it into true Enterprise Risk Management (ERM). It builds on previous work on knowledge-related constructs within RM. ERM implementation has evolved and studies show that organizations have expectations of improving the ERM process to perform better. This chapter begins with the identification of events that have affected the financial services industry and that indicate the need for better management of risk management knowledge. The succeeding sections introduce relevant concepts of risk management and KM, present the research model that comprises eight hypotheses, and describe the analysis of the results of two regression models that were used to test them. The final sections discuss the findings and seek to interpret their meaning.

Chapter 15 is "A Business-Driven Process Model for Knowledge Security Risk Management: Tackling Knowledge Risks While Realizing Business Benefits," by Ilona Ilvonen, Jari Jussila, and Hannu Kärkkäinen. This chapter aims to answer the research question "How can organizations manage knowledge risks and reach business benefits of changes in operation at the same time?" The chapter argues that knowledge security risks should be managed in a systematic process in line with managing business benefits, and introduce a conceptually developed process for this purpose. Several studies point out that increasing the circulation of knowledge also increases the risk of leakage. New forms of organizational operation emphasize opening up of organizational knowledge resources towards customers and other organizational stakeholders. Therefore, when making changes in practices, there is simultaneously a strong need for understanding the potential risks related to open information and knowledge flows, as well as relating these risks to the potential business benefits of the change. After introducing the theoretical background, the chapter introduces the proposed process model and discusses the relation of the process steps to previous work. After this an analytical case that illustrates the outputs of the process model from a practical perspective is presented. The paper concludes with a brief discussion on avenues for further research.

Section 1

Expanding KM Boundaries and Other Trends

Chapter 1
Using a Revised Knowledge Pyramid to Redefine Knowledge Management Strategy

Murray E. Jennex
San Diego State University, USA

ABSTRACT

The knowledge pyramid has been used for several years to illustrate the hierarchical relationships between data, information, knowledge, and wisdom. This chapter posits that the knowledge pyramid is too basic and fails to represent reality and presents a revised knowledge-KM pyramid. One key difference is that the revised knowledge-KM pyramid includes knowledge management as an extraction of reality with a focus on organizational learning. The revised pyramid includes newer initiatives such as business and/ or customer intelligence, big data, analytics, internet of things. Finally, this chapter discusses how KM strategy can be generated using the final revised pyramid.

INTRODUCTION

Much has been written on the knowledge pyramid, usually characterized as the data, information, knowledge, wisdom, DIKW, hierarchy and its use in knowledge management, KM. This chapter continues this discussion but takes a different position. It is posited that the knowledge pyramid is an artifact of KM processes and not an artifact of reality. This is a different position from authors such as Ackoff, Sharma, Bates, and Fricke that consider the pyramid as a natural expression of the relationships between DIKW and the logical progression for the generation of knowing through the creation of IKW (Ackoff, 1989; Bates, 2005; Miller, 1996; Sharma, 2004). This chapter also expands the DIKW hierarchy by considering the impact of big data, analytics, the Internet of Things, IoT, competitive intelligence, CI, and business intelligence, BI, on the DIKW hierarchy. Additionally; conversion and application processes between each layer are examined as KM processes and are incorporated into a KM process model.

DOI: 10.4018/978-1-5225-5427-1.ch001

Finally, this chapter will state definitions of terms important to KM, including definitions of data, information, knowledge, and intelligence. It is hoped that these can be accepted as working, consensus definitions, but it is recognized that these terms are philosophical in nature and can be debated as long as we want. This debate is embraced, but not encouraged, as I agree with Keen and Tan (2007) who believe that while it is important to understand KM terms, it is unproductive for researchers to get focused on trying to precisely define these terms at the expense of furthering KM research. The KM discipline needs to allow the debate but also needs to unite into a consensus set of working definitions. It is hoped this chapter will spur this consensus.

Methodology

This chapter focuses on concepts; however, the arguments made and conclusions presented are based on action research. The inspiration for this chapter comes from a project with a United States-based defense contractor. Specifics of the project and the company cannot be presented due to non-disclosure agreements. What can be said is that the company attempted to take technologies and experience developed/gained working with United States Department of Defense and other national intelligence agencies and generate a commercial knowledge management offering focused on providing KM tools and management processes. The author participated in this project as a KM academic expert responsible for providing KM focus and direction. It is participatory research per Burstein and Gregor (1999) as the author had a vested interest in the success of the project and in generating a commercial KM offering and was able to reflect on the project while participating. Specific data for this chapter comes from the company's initial presentation of what was considered to be a knowledge pyramid. This pyramid cannot be presented due to non-disclosure agreement. However, the pyramid can be described as a fusion of the traditional knowledge pyramid with KM processes and intelligence concepts. It was reflection and analysis of this pyramid and discussions with the project team that formed the basis of this chapter. The pyramid presented is the result of this reflection and is not at all similar to the project proposed pyramid.

THE TRADITIONAL KNOWLEDGE PYRAMID

References to a knowledge hierarchy can be found in the popular literature but generally Ackoff is given credit for the first academic publication. Figure 1 illustrates the traditional knowledge pyramid as originally proposed by Ackoff (1989). The inference from the figure is that data begets information begets knowledge begets wisdom. An additional inference is that there is more data than information, more information than knowledge, and more knowledge than wisdom. This model has been used in countless KM presentations and papers, and it is stated as a given truth that it is a generally accepted model showing the DIKW hierarchy (Fricke, 2007; Hey, 2008; Sharma, 2004). The model does not philosophically define data, information, knowledge, or wisdom, and it is not the purpose of this chapter to do this; there are many sources already available that make arguments supporting the various definitions. However, it is the purpose of this chapter to propose consensus, KM-focused working definitions. The traditional knowledge pyramid uses the following summarized basic definitions:

- **Data:** Basic, discrete, objective facts such as who, what, when, where, about something.

- **Information:** Data that is related to each other through a context such that it provides a useful story, as an example, the linking of who, what, when, where data to describe a specific object at a specific time.
- **Knowledge:** Information that has been culturally understood such that it explains the how and the why about something or provides insight and understanding into something
- **Wisdom:** Placing knowledge into a framework or nomological net that allows the knowledge to be applied to different and not necessarily intuitive situations

These definitions imply that there is a relationship between data, information, knowledge, and wisdom, and it is the knowledge pyramid that provides a graphical representation of these relationships as a rollup hierarchy of data leading to information leading to knowledge, and, finally, leading to wisdom (henceforth, this will be referred to as the DIKW creation flow). Houston and Harmon describe this in the use of summations (Houston & Harmon, 2002):

$$I = \sum(D),\ K = \sum(I) = \sum\sum(D),\ \text{and } W = \sum(K) = \sum\sum(I) = \sum\sum\sum(D)$$

The pyramid is typically used in instruction of database and KM concepts to reflect that "data," "information," "knowledge," and "wisdom" are different concepts and exist at increasingly higher levels of abstraction. This chapter argues that the DIKW pyramid is a misleading representation of the relationships among these concepts for information systems research purposes. Maintaining the fuzziness of these concepts in the information systems discipline leads to confusion. We need to better define these constructs to begin to eliminate some of the ambiguities in our research.

A counter point to the traditional knowledge pyramid is provided by Tuomi (2000). Tuomi (2000) states that data is not the building block for information, knowledge, and wisdom as data is not observed, collected, or recorded in a vacuum. Rather, our understanding of the world through our wisdom and knowledge drives us to collect specific information and data to support our use of our knowledge and wisdom. In this view, the hierarchy flows down the pyramid rather than up the pyramid, and data does not exist as a collection of unrelated facts as all collected facts are related to our basic knowledge

Figure 1. The knowledge pyramid
Ackoff, 1989.

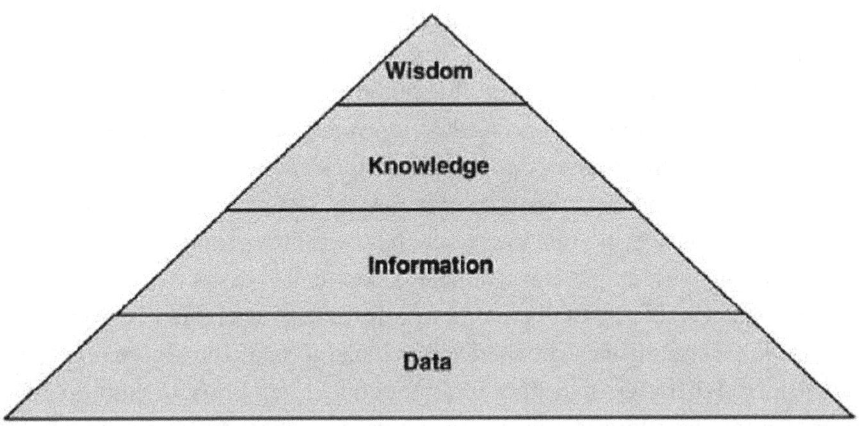

and wisdom. While this is an improvement to the knowledge pyramid, it still allows for confusion as it provides little insight and explanation for the relationships between DIKW. The proposed revision of the knowledge pyramid will show that flow is in both directions and introduces discussion on the relationships between DIKW.

THE REVISED KNOWLEDGE PYRAMID

Phase 1: A Philosophical Revision

The first phase of the revised knowledge pyramid attempts to place the knowledge hierarchy within the context of the natural or real world. Figure 2 illustrates the revised pyramid. What it shows is that data, information, knowledge, and wisdom exists in a broader context, i.e. humans are constantly gathering and processing data into information, knowledge, and wisdom. However, the data gathered and processed is not all that is available and is limited by the abilities of our sensors to detect, interpret, and capture data. Sensors can be our human senses, other's human senses, or mechanical where the sensors are anything that is not human such as a light detector, radio wave detector, pressure meter, a typed in transaction record, a query, etc. These sensors reflect that as sensor technology improve, our ability to capture more data improves. A human example of improving sensors is using LASIK to improve eyesight allowing a person to "see" much more, enriching the vision that is captured. A mechanical example is changing accuracy, range, and/or sensitivity of sensors so they can "sense" more phenomena.

The dotted arrows reflect the processing of data into information, knowledge, and wisdom using the processes of insight, analysis, and sense making. These lines are bi-directional indicating that the generation of information, knowledge, and/or wisdom may progress up the hierarchy or feedback down the hierarchy as the user understands more of what they are learning (this recognizes Tuomi's (2000) counter view). The lines between layers reflect the social networks used to transfer to different users. Social networks are being used loosely in this context and refer to any formal or informal, direct or indirect methods used to transfer data, information, knowledge and wisdom between users. Examples include classroom settings, word of mouth, published articles, conference presentations, email, etc. Ultimately, this is a representation of the knowledge hierarchy and the general learning process for people, organizations, and societies that results in multiple, large bodies of knowledge being generated and used. The endpoints (in both directions) lead to learning. Why learning? Learning has a multitude of definitions but for this model learning is defined as the acquisition of DIWK that leads to a change in behavior or expectation within the individual or group that is doing the learning.

Why is the knowledge pyramid inverted, i.e. why is there more information than data, more knowledge than information, and more wisdom than knowledge? One reason is simply mathematical. If information is the structuring of data into meaningful combinations, then the number of possible combinations for a quantity x of data is minimally x! implying there is possibly a greater amount of information than the original amount of data. Considering that users may have differing frames of reference for processing data in different disciplines of thought (for example accountants versus marketers or engineers versus biologists), it is very conceivable that the amount of information generated is greater than the original amount of data. This same argument can be used for the generation of knowledge and wisdom, especially when it is also considered that users may have different ethical, religious, or cultural belief systems that could cause them to interpret information and then knowledge differently (for example Christians may

Figure 2. Revised knowledge pyramid

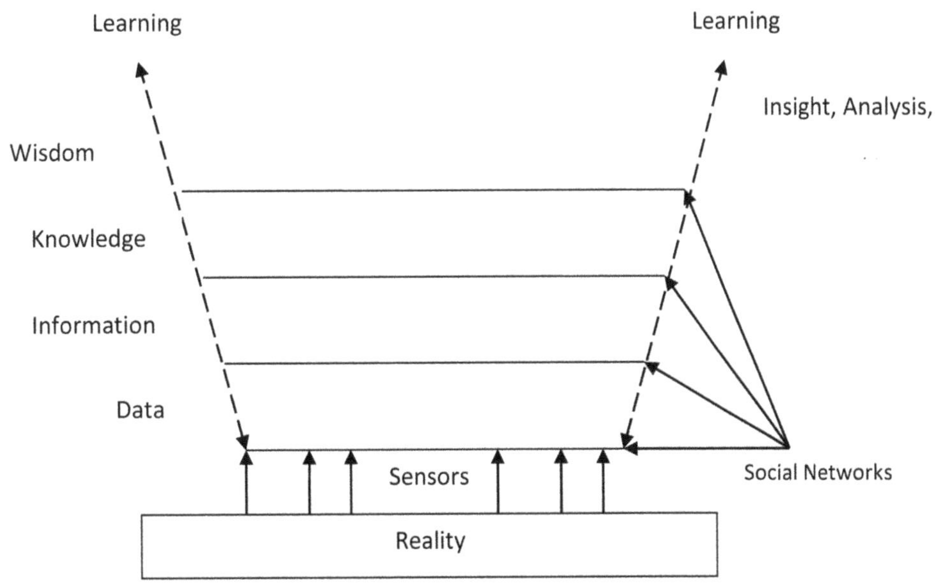

generate different wisdom from the same knowledge than Buddhists would or free societies different than totalitarian societies). This is consistent with Nonaka (1994), Jennex (2008), and others that argue that all knowledge is context- specific. An example of this idea involves technology transfer that takes DIKW from one discipline (or context) and attempts to apply it to another discipline (or context). Specifically, we can look at the case of taking the knowledge of using learning algorithms to identify meaningful sound patterns used to create sound canceling headphones and applying it to analyzing light diffraction patterns for identifying radioactive or other (such as explosive) substances.

Phase 2: Revision, Considering KM, BI, and CI

Figure 3 represents a newly proposed knowledge-KM pyramid. This pyramid is more like the traditional pyramid, albeit the "wisdom" aspect is absent and will be addressed later. The solid vertical arrows represent application of KM processes (i.e. capture, store, retrieve, apply) to the DIKW hierarchy pyramid. Jennex (2005, p. iv) defined KM as "the practice of selectively applying knowledge from previous experiences of decision making to current and future decision making activities with the express purpose of improving the organization's effectiveness." This implies that KM is not trying to address all knowledge or wisdom. Rather, KM targets specific knowledge and wisdom needed by an organization to perform specific tasks. Specific, actionable knowledge and wisdom is defined as intelligence, and it is the goal of KM to provide intelligence to the organization for use in decision making. The lines between layers reflect the filters used to focus on specific DIKW and guide the application of KM processes such as capture, store, retrieve, and apply. The purpose of KM is to make use of specific data, information, knowledge, and wisdom to generate intelligence. Additionally, KM seeks to have data, information, knowledge and intelligence shared among the right people at the right time (Jennex, et al., 2009). This implies that KM processes allow the filtering of data, information, and knowledge to generate specific, actionable intelligence that is shared with specific, limited users. Filters are placed on the social networks to limit access and to separate and capture that which is needed from that which is not. In this vein, "filters" is a fairly new term for KM. As such, it is the position of this chapter that KM filters are the implementation of KM strategy. Given this link between filters and strategy, their importance is evidenced by Jennex and Olfman (2005) who found KM strategy to be a key critical success factor. The use of intelligence leads to organizational learning. Organizational learning is defined as a quantifiable improvement in activities, increased available knowledge for decision-making, or sustainable competitive advantage (Cavaleri, 1994; Dodgson, 1993; Easterby-Smith, 1997; Miller, 1996). Huber, Davenport, and King (1998) believe an organization learns if, through its processing of DIKW, its potential behaviors are changed.

Figure 4 shows the relationship between Figures 2 and 3 by combining the two pyramids. The implication is that the knowledge-KM pyramid (Figure 3) is a subset of the revised knowledge pyramid (Figure 2). There are other differences. KM tends not to focus on wisdom but is about beginning to apply intelligence concepts and differentiates between knowledge needed to make a decision and specific actionable knowledge needed to make a specific decision in a specific context. For example, marketing knowledge is needed to create marketing campaigns but specific customer knowledge is needed to make decisions as to how to market to specific customers. For some users this is confusing as it seems to imply there are differing types of knowledge needed for KM. However, this tends to be consistent in intent to Tuomi's (2000) concept of wisdom.

This confusion between wisdom and intelligence may be a driver for those who practice Business Intelligence, BI, or Customer Intelligence, CI, etc. as they see a need for differentiating between general information and knowledge and specific decision information and knowledge within an organization.

Figure 3. The knowledge-KM pyramid

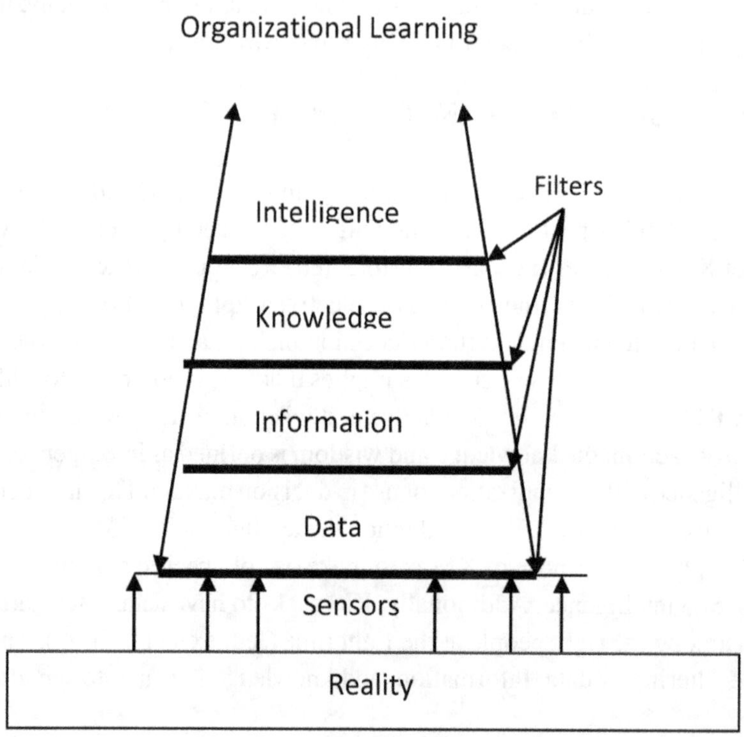

Figure 4. The revised knowledge-KM pyramid

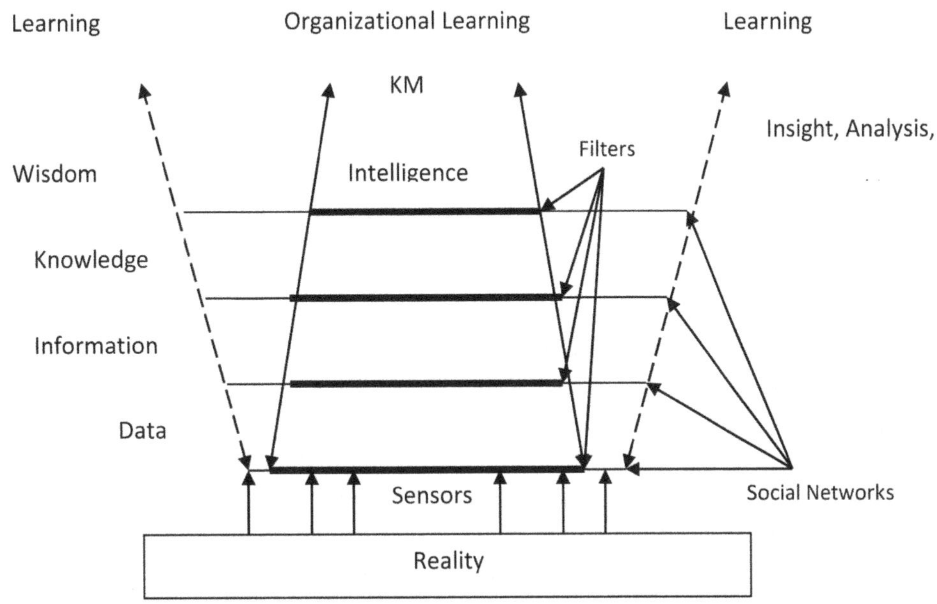

BI is generally defined as the collection, analysis, and presentation of business information for decision making (Cody et al., 2002; Negash, 2004; Watson & Wixom, 2007) while CI and Competitive Intelligence (also CI) are specialized forms of BI tailored to specific intelligence (Cody et al., 2002). Negash (2004). Figure 3 captures this idea by using the term intelligence rather than wisdom where intelligence refers to very specific actionable knowledge. The term intelligence is taken from the intelligence "activity" realm, which implies a function that seeks to generate actionable knowledge to be used in the formulation of strategies and tactics to accomplish a specific goal, such as beating an opponent during wartime or to determine what actions should be planned. This term is not chosen lightly as it does fit into the DIKW hierarchy meaning that it is the interpretation of data, information, and knowledge to create courses of action or make specific decisions. Also, the solid arrows that represent KM are shown to be double-headed as the hierarchy may travel in either direction. In fact, it is expected that while the learning process will tend to generally be a bottom-up process, meaning that it starts with the interpretation of data, organizational learning and KM will generally be top-down processes where it is first determined what actions or decisions need to take place or be made and from there it is identified what intelligence, then knowledge, then information, then data is needed to support taking the specific actions or making the specific decisions.

The other major difference is in the application of filters. While the general learning process seeks to push data, information, knowledge, and wisdom out to all who wish to use it, KM does not. As stated earlier, KM seeks to support specific decision making and thus needs specific data, information, knowledge, and intelligence (the same can be said for BI). Additionally, KM seeks to share this with the right people at the right time. This implies that KM and BI filters data, information, and knowledge to generate specific, actionable intelligence that is shared with specific, limited users. Moreover, these filters can also be described as analytical tools used to segregate DIK into that which is needed from that which is not. In this view, filters fit into the definition of BI Systems (Cody et al, 2002, p 178): (a position further supported by Negash, 2004)

BI systems combine data gathering, data storage, and knowledge management with analytical tools to present complex internal and competitive information to planners and decision makers.

A final difference between the traditional knowledge pyramid and the revised knowledge pyramid is the removal of apex's. This was done to remove confusion as an apex tends to imply that there is an ultimate point, such as the ultimate key wisdom for the traditional knowledge pyramid or a single key datum for the revised knowledge pyramid. While it is somewhat satisfying to refer to ultimate points, this chapter does not support the idea of a "big bang" theory for data or an ultimate, supreme wisdom. Rather, the chapter does support that there is some initial level of data and some ultimate amount of wisdom or intelligence. To support this position, again consider the contextual nature of knowledge and wisdom/intelligence and the multiple frames of reference and contexts that users bring to the generation of knowledge and wisdom. Also consider that most organizations do not ultimately need one key piece of intelligence as there is rarely a single decision that needs to be made. Hence, the revised knowledge pyramid is pyramidal in form as that provides a visual impact as to the relative amounts of data, information, knowledge, and wisdom/intelligence but is flat-topped, much like the Aztec and Mayan pyramids and not like the Egyptian pyramids (note that this is being used to illustrate the point and not to imply that there is more merit to one style of pyramid over another).

Phase 3: Revision, Considering Big Data, and the IoT

Wikipedia (2018) defines big data as a broad term for data sets so large and complex that traditional data analysis methods are inadequate. McAfee and Brynjolfsson (2012) expand on this by observing that big data is created at a rate that humans cannot analyze it without the aid of algorithms and analysis tools. McAfee and Brynjolfsson (2012) further state that big data is being created at a rate of 2.5 Exabytes per day and is mostly unstructured with this being the main difference between traditional data, which is structured, and Big Data which is not. Chen, et al. (2012) agree with McAfee and Brynjolfsson (2012) but suggest analytics tools such as text analytics, web analytics, network analytics, mobile analytics, and data analytics are the key to transforming big data to data, information, knowledge, and intelligence. Madden (2012) also agrees and expands the definition offered by explaining that it means data that's too big, too fast, or too hard for existing tools to process where:

- "Too big" means that organizations increasingly must deal with petabyte-scale collections of data that come from click streams, transaction histories, sensors, and elsewhere.
- "Too fast" means that not only is data big, but it must be processed quickly — for example, to perform fraud detection at a point of sale or determine which ad to show to a user on a webpage.
- "Too hard" is a catchall for data that doesn't fit neatly into an existing processing tool or that needs some kind of analysis that existing tools can't readily provide.

All agree the artifacts of knowledge are changing because of big data and look at the goals of big data being to identify intelligence for evidence based decision making, transforming intuitive based decision making to evidence based decision making, and pushing decision making to lower levels of the organization. However, big data does not remove the need for human insight and vision. Finally, Koronios, et al. (2014) found that a key success factor for big data is having a strategy for determining what big data is needed to generate business value.

Wikipedia (2018) defines IoT as the network of physical devices that connect to the web, usually through a wireless connection, and communicate with other physical devices for improving service of all devices and create value. Barnaghi, et al., (2012) expand on the IoT and discuss how the IoT generates big data. They then describe the knowledge pyramid of the IoT as being raw data leading to structured data with semantics leading to abstractions and perceptions leading to actionable intelligence (Barnaghi, et al., 2012). Gubbi, et al. (2013) and Atzori, et al. (2010) expand on the IoT and see it as a vast sensor network with devices generating tremendous amounts of data by nearly continuous recording of data reflecting the devices state and using ubiquitous analytics and cloud technology to generate value through networks of devices.

The advent of big data and the IoT lends to a revisualization of the revised knowledge KM pyramid. Two issues need to be addressed. The first is the definition of data:

Data – discrete facts such as who, what, when, and where that can be stored in a database

Big data does not fit this definition; humans cannot analyze these larger data sets. Provost and Fawcett (2013) say Data Science is about extracting information and knowledge from data. Big data uses analytic tools to process it into human understandable data chunks. Humans interpret the understandable data chunks into information and then once the information is understood as to how and why it means what it does it becomes knowledge, finally, once the knowledge is fitted into a decision context it becomes actionable intelligence. To support this assumption, consider data mining which is one of the analytics tools used to make sense of large data sets. Yu, et al. (2014) explored second-order data mining where humans determine meaning of the first order data mining outputs to produce actionable intelligence. They found that humans need to be in the process and suggest that the habitual domains of the analysts' influence data mining interpretations. Also consider Barnaghi, et al. (2012), who applied semantic operations such as ontologies to big data to produce structured data. The implication of both studies is that big data needs to be transformed to data before it can be transformed to information and so on, implying that big data is a distinct layer from data. A second implication is that since IoT is a sensor net for big data, then IoT is a distinct layer from big data. A third implication is that since IoT does more than just generate big data and there is more big data than data, then IoT may be more than big data. A fourth implication is that since IoT is a sensor net, it is a distinct layer between reality and big data.

The second issue is the changing nature of wisdom and knowledge. Chen, et al. (2012), McAfee and Brynjolfsson (2012), and Madden (2012) all agree that automated tools and analytics are changing the nature of knowledge and wisdom as they focus on producing actionable intelligence to support evidence-based decision making and automated decision making. This implies that the social networks previously used between the data, information, knowledge, and wisdom layers need to be expanded to include analytics. However, these analytics and the big data they help transform reflect the findings of Koronios, et al. (2014) that strategy is required to guide the use of big data. This implies that organizations need to focus on their big data even if they are not using KM. Additionally, Weinberger (2012) suggests that the availability of Internet based digital media sources are changing the shape, evolution and perception of knowledge resulting in the traditional pyramid of knowledge becoming a formless "network of knowledge." This is due to the ineffectiveness of filters normally used by organizations to verify sources. The use of strategy to guide acquisition of big data is an example of the application of filters, albeit weak filters, allowing big data to operate within KM function in many different ways. Thus, Weinberger's (2012) findings support the inverted pyramid for the revised model.

Figure 5 reflects the revised knowledge pyramid that includes KM, big data, analytics, and IoT. The pyramid is now an hour glass. The top inverted pyramid is as before but the transformation processes between layers now include analytics and weak filters. The bottom pyramid looks more like the traditional knowledge pyramid but now has layers of reality, IoT, and big data. The traditional pyramid shape reflects that reality still contains more data than we can record. However, the second layer, IoT, reflects that our ability to collect observations is greatly increased and results in the third layer, big data. The transformation processes between these layers reflect the use of automated analysis tools.

One major change is the renaming of the Wisdom layer to the Wisdom/Intelligence layer. This reflects the findings of Chen, et al. (2012), McAfee and Brynjolfsson (2012), and Madden (2012) of the changing nature of knowledge and wisdom to be more focused on actionable intelligence.

Another change is the redefining of the transformation processes between data, information, knowledge, and wisdom/intelligence to include Weinberger's (2012) weak filters used by humans to determine credibility while at the same time adding the more quantitative analytics processes for identifying patterns. The strong filters shown in the KM pyramid reflect the application of strategy to define what is needed from each layer and are used to prevent overload of the decision process with big data, data, information, and knowledge not needed to support the generation of specific actionable knowledge.

The final addition is the double headed arrows that link the IoT and big data layers directly to the wisdom/intelligence layer. The arrows are double headed to represent that there is automatically generated intelligence (per Chen, et al. (2012), McAfee and Brynjolfsson (2012), and Madden (2012)) that is directly input into the wisdom/ intelligence layer and that there is guidance as to what is needed from the IoT and big data layers (per Koronios, et al. (2014)). The purpose of these arrows is to reflect the trend of autonomous/automatic decision making. Examples of this include intelligent drones, other self-driving vehicles, or other automatic controllers that are programmed to react to events should connection to their human controllers be lost or they are put into an automatic action mode or should their human controllers not react to an emerging situation.

DISCUSSION

The original knowledge pyramid as defined by Ackoff (1989) and Tuomi (2000) was found wanting. Ackoff (1989) assumes data generates information which generates knowledge which generates wisdom. Tuomi (2000) believes all data and information is collected based on influence from existing knowledge and wisdom. Ackoff's (1989) pyramid flows up, Tuomi's (2000) pyramid flows down. The final revised pyramid (figure 5) incorporates both perspectives and has flow is in both directions. Tuomi (2000) is misleading in suggesting that flow is primarily down the pyramid. Basic research often times involves collecting data about something of which we do not have prior knowledge or wisdom. An example comes from physics research in the late 1800s. Scientists did not understand nor did they have knowledge about atomic structure, subatomic particles, and radiation. Data was collected and observed by scientists noting that photographic film was fogged when near certain materials. This data was gathered and then put into context as information that said specific materials caused fogging. Consideration of this information led to the formation of theories to explain why this happened, generating potential knowledge. Analysis of this knowledge led to experiments designed to prove the correctness of this potential knowledge, ultimately resulting in theories on radiation. This example shows that both Ackoff (1989) and Tuomi

Figure 5. The final revised knowledge pyramid with KM, Big Data, and IoT

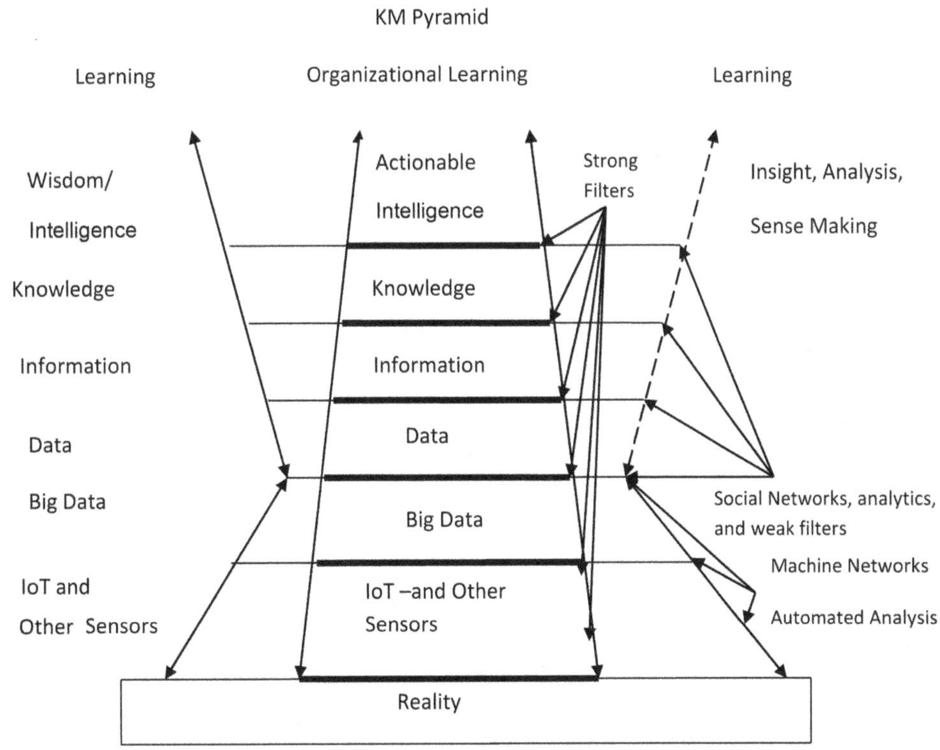

(2000) are both supported. Flow of creation is in both directions of the pyramid which is consistent with exploratory research to discover theory (flow up) and validating research to confirm theory (flow down).

Another example bolstering this argument is the use of data mining to create information by discovering patterns in data. Commonly called knowledge discovery, it is the analysis of these new found patterns, or information, and the application of context, culture, and other knowledge that creates knowledge; and the positioning of this knowledge into existing problem or competitive advantage contexts that creates actionable knowledge or intelligence. In this example, the predominant direction of creation flow is up the pyramid, but it can also be argued that previous knowledge of business needs, processes, and behavior influenced the interpretation of discovered patterns into knowledge and intelligence so that creation flow is actually in both directions. Additionally, consider analytics. Analytics seeks to gather specific data to answer specific questions. Systems are created to gather, analyze, and display specific data, information, and knowledge so that managers can monitor organizational performance and take action if warranted. Analytics is based on primarily a top-down design that starts with the intelligence needed and then determines what data, information, and knowledge to gather.

A final example comes from engineering. It is common for engineers to instrument everything they can when conducting testing, even if they are not sure of a purpose for the data they will collect. A high pressure vessel destructive test from my past illustrates this. The test called for burst testing a 2500 psi (pounds per square inch) pressure vessel using water. Standard instrumentation for measuring pressure and temperature were used plus video monitoring was added. There was no real need for collecting video data other that it was thought it would be interesting to watch. Since water, a non-compressible liquid, was being used as the pressuring medium, it wasn't expected to be a violent failure (violence in this case is represented by a very large, very fast release of energy) as it was expected that the vessel would crack and the water spill out or perhaps spray out. The test team was very surprised when the test resulted in a spectacular release of energy. The video data was what provided the story, and it was interesting to see the test setup disintegrate during the first test and then to see that the subsequent test setups were surrounded by sand bags and other retaining walls. The point of this example is that there are many situations in which we collect data for no real reason other than we can. It is only later analysis of this data that results in the generation of unexpected information, knowledge, and intelligence.

These examples bolster support for flow of creation going up and down the pyramid. Why is this important? It supports the concepts of model driven and data driven decision making. Model driven decision making where the decision is modeled to determine what data, information, knowledge, etc. is needed reflects creation flow down the pyramid and is also reflective of a strategy approach to data, information, and knowledge creation/ gathering and is reflective of Tuomi's view of flow in the pyramid. Data driven decision making comes from the identification of patterns in data resulting in the creation of information and knowledge and is representative of Ackoff's view on flow in the pyramid.

How does this affect KM? Strategy reflects planning. KM has been defined by Jennex (2005) as "the practice of selectively applying knowledge from previous experiences of decision making to current and future decision making activities with the express purpose of improving the organization's effectiveness." This means KM is purposeful planning for a strategy that will obtain actionable intelligence but KM is also about leveraging what the organization "knows." The final revised KM pyramid reflects both these perspectives. Leveraging what is known is the creation/generation of data/information/knowledge/actionable intelligence based on analysis of collected big data from IoT and the traditional collection of data.

KM is more than strategy, while strategy guides KM, KM also includes processes and technologies for storing, capturing, manipulating, and retrieving data/information/ knowledge/actionable intelligence. The final revised pyramid, through the application of weak and strong filters, machine learning and networks, social networks, analytics, and automated analysis (including capture and manipulation) tools and systems, shows that the focus for KM strategy for KM systems, KMS, now includes identifying the technologies to be used as filters as well as well identifying sources, repositories, and needed data/information/knowledge/actionable intelligence. Jennex (2010, 2012) investigated the components of a KM strategy, found having a strategy improved organizational performance, and applied it to managing technology. The basic components of a KM strategy can be generalized and made useful for managing knowledge pyramid activities including:

- Identification of users of the knowledge pyramid layers and transformation processes
- Identification of actionable intelligence needed to support organizational/societal decision making
- Identification of sources of big data, data, information, and knowledge
- Identification of big data, data, information, and knowledge to be captured
- Identification of how captured big data, data, information, and knowledge is to be stored and represented
- Identification of technologies to be used as filters to support capturing and processing big data, data, information, and knowledge
- Generation of top management support
- Establishment of metrics for big data, data, information, and knowledge use

- Establishment of feedback and adjustment process on the effectiveness of actionable intelligence use

The goal is a top down strategy approach based on the decisions to be made and identifying the technologies and decision support components needed. Additionally, the more focused the strategy the stronger the filters that are created blocking big data, data, information, and knowledge not needed to support decision making.

Strategic approaches to managing the knowledge pyramid make sense on the organizational level in support of organizational decision making, but do they work for society where society is the part of the pyramid outside the KM pyramid? The knowledge society is an approach that uses knowledge and intelligence to create processes and products that better serve citizens. Among knowledge society initiatives are KM, smart grids, electronic health, smart networks, electronic government, and crisis response and management. Many of these initiatives are being empowered by IoT, big data, and analytics as well as the rest of the knowledge pyramid layers and transformation processes to perform tasks such as more efficient energy and water management, bringing medical expertise to areas poorly served by doctors, improving efficiency of transportation and manufacturing systems.

The approach of the knowledge society leads to the addition of a direct feed of intelligence from the IoT and big data layers to the wisdom/intelligence layer. Figure 5 also shows that the bottleneck in the knowledge pyramid is human comprehensible data represented by the hour glass shape with very large amounts of observations gathered and processed automatically on one end, and very large amounts of knowledge and wisdom/intelligence generated and used by decision makers on the other end. Consequently, IoT are required to assist automatic controlled processes in utilities such as electricity, water, and communications in order to react quickly to changes in operating parameters (Gubi et al, 2013). The bottleneck in the data layer is not acceptable for such systems if the direct feed between IoT, big data, and wisdom/intelligence is to be implemented. Such a situation is undesirable for many as they need to feel that humans are in control and understand what is happening instead of worrying about various concerns. Vlacheas, et al. (2013) propose frameworks for this link to function successfully while mitigating concerns. Still, while the model could include the link, it is perhaps the most controversial addition to the revised knowledge pyramid.

CONCLUSION

What is the value of revising the knowledge pyramid? Ultimately Figure 5 showing the fully revised knowledge pyramid that includes KM, analytics, big data, IoT, and various intelligence initiatives (business intelligence, customer intelligence) reflects the way technology innovation is radically addressing society's and organizational needs for more data and more actionable intelligence. The drive for evidence based decision making rather than intuitive decision making is pushing the creation and adoption of technologies that gather and analyze observations of reality faster than and in amounts greater than what humans can process. The issues that are created are how to manage data and technologies; which technologies to use; and what data, information, knowledge, and actionable intelligence is needed for decision making. This revised knowledge pyramid provides value by helping with these three issues by: showing relationships between technologies and decision support artifacts (big data, data, information, knowledge, intelligence) and by providing a model that assists in creating a strategy for selecting and managing technologies and decision support artifacts.

This further revised knowledge-KM pyramid is more complex and, thus, less satisfying as a visual model than the original pyramid. However, we are in a period of rapid technological advancement resulting in dramatic impacts of societal and organizational decision making. We are also in a period of declining resources resulting in proponents of each technology competing with and promising more than the other technologies. This final revised knowledge pyramid attempts to mitigate the competition by showing that all these technologies can and must work together to improve the efficiency and effectiveness of decision making processes. This final revised pyramid is more satisfying conceptually as it fits our perception of reality and what technologies can do to a better degree and clarifies some of the assumed process aspects of the knowledge pyramid. Is this the final model? Probably not, but it is a further step in the right direction and a starting point for more detailed discussion on the knowledge pyramid. This final revised knowledge pyramid that includes the decision support technologies and initiatives such as KM, big data, IoT, analytics, and business/customer intelligence provides insight and guidance into how to create comprehensive decision support strategies. The model shows that decision making is a process supported by many technologies and processes. It is also hoped that this revised pyramid can be used by researchers to guide them in developing integrated decision solutions.

Finally, this chapter proposed to provide a set of what are hoped to be consensus working definitions of KM terms from the final revised knowledge-KM pyramid. These terms are summarized below:

- **Data:** Basic, discrete, objective facts such as who, what, when, where, about something. (Note that this is a shortened version of many commonly accepted definitions of data as summarized in Wikipedia)
- **Information:** Data that is related to each other through a context such that it provides a useful story, as an example, the linking of who, what, when, where data to describe a specific person at a specific time. (Note that this is a tailored version of many commonly accepted definitions of information as summarized in Wikipedia)
- **Knowledge:** Information that has been culturally understood such that it explains the how and the why about something or provides insight and understanding into something. (Note that this is a tailored version of many commonly accepted definitions of knowledge as summarized in Wikipedia)
- **Wisdom:** Placing knowledge into a framework or nomological net that allows the knowledge to be applied to different and not necessarily intuitive situations
- **Actionable Intelligence:** Specific actionable knowledge needed to make a specific decision in a specific context in a specific organization.
- **Learning:** the acquisition of DIWK that leads to a change in behavior or expectation within the individual or group that is doing the learning. (Note that this is a tailored version of many commonly accepted definitions of learning as summarized in Wikipedia)
- **Organizational Learning:** Learning that leads to a quantifiable improvement in activities, increased available knowledge for decision-making, or sustainable competitive advantage in an organization context. An organization learns if, through its processing of DIKW, its potential behaviors are changed (Cavaleri, 1994; Dodgson, 1993; Easterby-Smith, 1997; Huber, Davenport & King, 1998; Miller, 1996).

- **Social Networks:** Any formal or informal, direct or indirect methods used to transfer data, information, knowledge and wisdom between users.
- **Filters:** KM processes and technologies that limit access and separate and capture that DIKW/I which is needed from that which is not. Strong filters are usually present in KM and are designed to generate very specific data/ information/knowledge/actionable intelligence. Knowledge society filters are weak filters usually based on social networks that are used to generate consensus understanding.

The implications of this chapter are many. KMS designers can now have a better understanding of how KM fits in the general DIKW framework. They can also focus on strategy and strategy implementation through the design and implementation of KM filters. This can be further extended to the concept of generating KM policies to implement KM governance. This leads to the possibility of creating audit mechanisms that can validate that KM strategy/governance has been properly implemented via KMS which is similar in nature to auditing security policy implementations in firewalls. Another implication is a clearer understanding of the role of social networks in the creation and transfer of DIKW (where W in this case is Intelligence). KMS designers will also be focused on providing technological support for these social networks, and it clarifies that KMS are more than just knowledge storage and retrieval technologies whereas they also incorporate communication and collaboration technologies. Another implication is a new focus on identifying specific actionable knowledge for focused decision making, or, as stated in the chapter, generating intelligence. A final implication is that KM should now be recognized as an integrated part of regular DIKW (where again W refers to Intelligence) processing. KMS designers should be focusing on integrating KM processes into regular work processes.

ACKNOWLEDGMENT

This chapter is a progressive work with early versions being presented at the Hawaii International Conference on System Sciences (HICSS) (Jennex, 2009), at the SIGPHIL meeting at the Americas Conference for Information Systems (AMCIS) (Jennex, 2013), and published in the International Journal of Knowledge Management (Jennex &Bartczak, 2013) and the Data Base for Advances in Information Systems (Jennex, 2017). Comments and discussion from these meetings and with readers of the papers have been used to expand and modify the model that is presented in this paper.

REFERENCES

Ackoff, R. L. (1989). From Data to Wisdom. *Journal of Applied Systems Analysis*, *16*(1), 3–9.

Atzori, L., Iera, A., & Morabito, G. (2010). The Internet of Things: A Survey. *Computer Networks*, *54*(15), 2787–2805. doi:10.1016/j.comnet.2010.05.010

Barnaghi, P., Wang, W., Henson, C., & Taylor, K. (2012). Semantics for the Internet of Things: Early progress and back to the future. *International Journal on Semantic Web and Information Systems*, *8*(1), 1–21. doi:10.4018/jswis.2012010101

Bates, M. J. (2005). Information and Knowledge: An Evolutionary Framework for Information Science. Information Research, 10(4).

Boyd, D., & Crawford, K. (2012). Critical Questions for Big Data: Provocations for a Cultural, Technological, and Scholarly Phenomenon. *Information Communication and Society*, *15*(5), 662–679. doi: 10.1080/1369118X.2012.678878

Burstein, F. V., & Gregor, S. (1999). The Systems Development or Engineering Approach to Research in Information Systems: An Action Research Perspective. In *Proceedings of 10th Australasian Conference on Information Systems*. School of Comms & Info Mgmt Vic Uni of Wellington.

Cavaleri, S. (1994). Soft Systems Thinking: A Pre-Condition for Organizational Learning. *Human Systems Management*, *13*(4), 259–267.

Chen, H., Chiang, R. H. L., & Storey, V. C. (2012). Business Intelligence and Analytics: From Big Data to Big Impact. *Management Information Systems Quarterly*, *36*(4), 1165–1188.

Cody, W. F., Kreulen, J. T., Krishna, V., & Spangler, W. S. (2002). The Integration of Business Intelligence and Knowledge Management. *IBM Systems Journal*, *41*(4), 697–713. doi:10.1147j.414.0697

Dalal, N. (2012). Wisdom Computing: Toward a Framework for Wisdom Research in Information Systems. *AMCIS 2012 Proceedings*. Available at http://aisel.aisnet.org/amcis2012/proceedings/PerspectivesIS/14

Davenport, T. H., & Prusak, L. (1998). *Working Knowledge*. Boston, MA: Harvard Business School Press.

Dodgson, M. (1993). Organizational Learning: A Review of Some Literatures. *Organization Studies*, *14*(3), 375–394. doi:10.1177/017084069301400303

Easterby-Smith, M. (1997). Disciplines of Organizational Learning: Contributions and Critiques. *Human Relations*, *50*(9), 1085–1113. doi:10.1177/001872679705000903

Foley, E., & Guillemette, M. G. (2010). What is Business Intelligence? *International Journal of Business Intelligence Research*, *1*(4), 1–28. doi:10.4018/jbir.2010100101

Fricke, M. (2007). The Knowledge Pyramid: A Critique of the DIKW Hierarchy. *Journal of Information Science*, *35*(2), 131–142. doi:10.1177/0165551508094050

Gubbi, J., Buyya, R., Marusic, S., & Palaniswami, M. (2013). Internet of Things (IoT): A vision, architectural elements, and future directions. *Future Generation Computer Systems*, *29*(7), 1645–1660. doi:10.1016/j.future.2013.01.010

Hey, J. (2004). *The Data, Information, Knowledge, Wisdom Chain: The Metaphorical link*. Retrieved June 16, 2008 from: http://www.oceanteacher.org

Houston, R. D., & Harmon, E. G. (2002). Re-envisioning the information concept: systematic definitions. In H. Bruce, R. Fidel, P. Ingwersen, & P. Vakkari (Eds.), *Emerging frameworks and methods: proceedings of the fourth International Conference on Conceptions of Library and Information Science (CoLIS4)* (pp. 305-308). Greenwood Village, CO: Libraries Unlimited.

Huber, G. P., Davenport, T. H., & King, D. (1998). *Some Perspectives on Organizational Memory*. Presented at the 31st Annual Hawaii International Conference on System Sciences.

Jennex, M. E. (2005). What is Knowledge Management? *International Journal of Knowledge Management, 1*(4), i–iv.

Jennex, M. E. (2008). The Impact of Culture and Context on Knowledge Management. In M. E. Jennex (Ed.), *Current Issues in Knowledge Management* (pp. 6–11). IGI. doi:10.4018/978-1-59904-916-8.ch002

Jennex, M. E. (2009). Revisiting the Pyramid. In *42nd Hawaii International Conference on System Sciences, HICSS42*. IEEE Computer Society.

Jennex, M. E. (2010). Implementing Social Media in Crisis Response Using Knowledge Management. *International Journal of Information Systems for Crisis Response and Management, 2*(4), 20–32. doi:10.4018/jiscrm.2010100102

Jennex, M. E. (2012). Identifying the Components of a Knowledge Management Strategy. *18th Americas Conference on Information Systems, AMCIS18.*

Jennex, M. E. (2013). *Where is the Wisdom? In Americas Conference of AIS*. AMCIS.

Jennex, M. E. (2017). Big Data, the Internet of Things and the Revised Knowledge Pyramid. *The Data Base for Advances in Information Systems, 48*(4), 69–79. doi:10.1145/3158421.3158427

Jennex, M. E., & Bartczak, S. E. (2013). A Revised Knowledge Pyramid. *International Journal of Knowledge Management, 9*(3), 19–30. doi:10.4018/ijkm.2013070102

Jennex, M. E., & Olfman, L. (2005). Assessing Knowledge Management Success. *International Journal of Knowledge Management, 1*(2), 33–49. doi:10.4018/jkm.2005040104

Jennex, M. E., Smolnik, S., & Croasdell, D. T. (2009). Towards a Consensus Knowledge Management Success Definition. *VINE. The Journal of Information and Knowledge Management Systems, 39*(2), 174–188.

Keen, P. G. W., & Tan, M. (2007). Knowledge Fusion: A framework for extending the rigor and relevance of knowledge management. *International Journal of Knowledge Management, 3*(4), 1–17. doi:10.4018/ijkm.2007100101

Koronios, A., Gao, J., & Selle, S. (2014). Big Data Project Success – A Meta Analysis. *PACIS 2014, Proceedings*. Available at: http://aisel.aisnet.org/pacis2014/376

LaValle, S., Lesser, E., Shockley, R., Hopkins, M. S., & Kruschwitz, N. (2013). Big data, analytics and the path from insights to value. *MIT Sloan Management Review, 52*(2), 21.

Madden, S. (2012). From Databases to Big Data. *IEEE Internet Computing, 16*(3), 4–6. doi:10.1109/MIC.2012.50

Markus, B., (2005). *Learning Pyramids*. WSVA 1 – Virtual Academy and the Surveying/ Geoinformatics Community –Virtual Academy I.

McAfee, A., & Brynjolfsson, E. (2012). Big Data: The Management Revolution. *Harvard Business Review, 90*(10), 60–68. PMID:23074865

Miller, D. (1996). A Preliminary Typology of Organizational Learning: Synthesizing the Literature. *Journal of Management, 22*(3), 485–505. doi:10.1177/014920639602200305

Negash, S. (2004) Business Intelligence. *Communications of the Association for Information Systems, 13*, 177-195. Available at: http://aisel.aisnet.org/cais/vol13/iss1/15

Nonaka, I. (1994). A Dynamic Theory of Organizational Knowledge Creation. *Organization Science, 5*(1), 14–37. doi:10.1287/orsc.5.1.14

Nonaka, I., & Takeuchi, H. (1995). *The Knowledge-Creating Company – How Japanese Companies Create The Dynamics Of Innovation*. Oxford, UK: Oxford University Press.

Pomerleau, M. (2015). Cheat sheet for the Internet of Things. *GCN, Technology, Tools, and Tactics for the Public Sector*. Retrieved January 30, 2015 from http://gcn.com/articles/2015/01/30/internet-of-things.aspx?m=1

Provost, F., & Fawcett, T. (2013). Data Science and Its Relationship To Big Data And Data-Driven Decision Making. *Big Data, 1*(1), 51–59. doi:10.1089/big.2013.1508 PMID:27447038

Rowley, J. (2007). The wisdom hierarchy: Representations of the DIKW hierarchy. *Journal of Information Science, 33*(2), 163–180. doi:10.1177/0165551506070706

Sharma, N. (2004). *The Origin of the "Data Information Knowledge Wisdom" Hierarchy*. Accessed June 17, 2008: http://wwwpersonal.si.umich.edu/~nsharma/dikw_origin.htm

Tuomi, I. (2000). Data is More Than Knowledge: Implications of the Reversed Knowledge Hierarchy for Knowledge Management and Organizational Memory. *Journal of Management Information Systems, 16*(3), 103–117. doi:10.1080/07421222.1999.11518258

Vlacheas, P., Giaffreda, R., Stavroulaki, V., Kelaidonis, D., Foteinos, V., Poulios, G., & Moessner, K. (2013). Enabling smart cities through a cognitive management framework for the internet of things. *Communications Magazine, IEEE, 51*(6), 102–111. doi:10.1109/MCOM.2013.6525602

Watson, H. J., & Wixom, B. H. (2007). The Current State of Business Intelligence. *Computer, 40*(9), 96–99. doi:10.1109/MC.2007.331

Weinberger, D. (2012). *Too Big to Know: Rethinking Knowledge Now That the Facts Aren't the Facts, Experts Are Everywhere, and the Smartest Person in the Room Is the Room*. Basic Books.

Wikipedia. (2018). *Big Data*. Retrieved on March 25, 2018 from http://en.wikipedia.org/wiki/Big_data

Wikipedia. (2018). *Internet of Things*. Retrieved on March 25, 2018 from http://en.wikipedia.org/wiki/Internet_of_Things

Yu, X., Shi, Y., Zhang, L., Nie, G., & Huang, A. (2014). Intelligent Knowledge Beyond Data Mining: Influences of Habitual Domains. *Communications of the Association for Information Systems, 34*, 53. Available at http://aisel.aisnet.org/cais/vol34/iss1/53

Chapter 2
Knowledge and Innovation Management:
Creating Value

Marianne Gloet
University of Melbourne, Australia

Danny Samson
University of Melbourne, Australia

ABSTRACT

This qualitative research examined the relationship between knowledge management (KM) and systematic innovation capability in 16 Australian manufacturing and service organizations that exhibited both successful innovation and robust KM practices. A review of the literature indicated a number of areas where KM enhances and supports innovation capability. Using a multiple crosscase analysis methodology and applying a framework of systematic innovation capability, in-depth interviews were conducted with managers of the case study organizations. The analysis of the data revealed the main contributions of KM to systematic and sustained forms of innovation. Finally, a model of knowledge and innovation capability was developed to guide the development of knowledge and innovation management as a dynamic capability to support value capture, value creation, and value delivery from innovation.

INTRODUCTION

Competitive advantage in today's advanced economies is driven by innovation and the ability to manage ever-increasing forms of knowledge on a sustained basis. Knowledge intensive industries compete primarily on their capacity to innovate and thrive on cutting-edge knowledge, which drives both research and innovation. Indeed, knowledge intensive organizations (KIOs) constantly seek to reinforce sustainable links between forms of knowledge and modes of innovation. In such a dynamic environment, the proactive management of knowledge assets is essential to achieving both innovation capability and innovation performance (Kim, Lee, Chun & Benbasat, 2014; Taherparvar, Esmaeilpour & Dostar, 2014; Kuusisto & Meyer, 2003; Miles, 2007). Since KIOs play a significant role in value creation through innovation

DOI: 10.4018/978-1-5225-5427-1.ch002

(Ye, Jha & Desouza, 2015; Muller & Doloreux, 2009; Van der Aa & Elfring, 2002), the ways in which organizations approach knowledge management (KM) influences innovation and becomes a source of competitive advantage (Freeze & Kulkarni, 2008). As such, KM emerges as an essential management and organizational capability in the drive to create value through knowledge.

Arguably, as a neutral construct, knowledge achieves consequence through human action. In this context, the human values and assumptions underpinning the learning process reveal the considerable power of not only knowledge but also the processes associated with its management (Alavi, Kayworth & Leidner, 2006; Ibrahim & Reid, 2009; Nonaka and Takeuchi, 1995). Currently, many people consider knowledge as the determining factor in economic growth (Oyelaran-Oyeyinka & Sampath, 2009). For example, Storey and Barnett (2003: 146) describe knowledge as the "key competitive sustained resource" and an organization's most important asset. Knowledge is also a primary factor of production on which competitive advantage rests (Beijers, 1995; Halawi, Aronson & McCarthy, 2005). As Davenport and Prusak (1998: 161) note, the successful management of knowledge requires a particular ''combination of human, technical and economic skills", highlighting that it is neither a haphazard nor an unmanaged process. With this growing awareness of the need to manage an organization's knowledge effectively and engender a particular arrangement of people, technology and skills, KM emerged as a distinct field of study.

Since knowledge and innovation are inextricably linked, a growing body of literature focuses on the ways in which KM can enhance and support the innovation process (Jensen, Johnson, Lorenz & Lundvall, 2007; Du Plessis, 2007; Goh, 2005). However, there are challenges associated with organizational efforts to develop innovation as a core competency (Kandampully, 2002) because of the complexity of the innovation process, the diversity of knowledge assets and a broad range of approaches to KM (Malhotra & Morris, 2009; Freeze & Kulkarni, 2008). Moreover, different markets place different values on knowledge assets (Gibbons et al, 1994). This complexity combined with the interdependence that characterizes global competition compels organizations to acquire, develop and consume knowledge assets in order to achieve competitive advantage (Badaracco, 1991; Murray, 2002). Although the management of knowledge should be prioritized to the same degree as the management of an organization's human, financial and physical resources, this is not always the case. Many senior managers fail to appreciate fully the value of KM as not only a discrete management function but also a unique skill (Stewart, 1997). Indeed, KM has been found to be a significant catalyst in organizational value creation (Kianto & Ritala, 2014; Krentz, Basmer, Buxbaum-Conradi, Redlich & Wulfsberg, 2014; Weaven, Grace, Dant & Brown, 2014).

This exploratory study seeks to examine the ways in which KM is manifested across a range of organizations in both the manufacturing and service sectors, with a view to determining more clearly the relationship between KM and innovation. A further objective of this study is to investigate the extent to which KM can contribute to systematic and sustained forms of innovation within organizations. Moreover, the study seeks to determine the ways in which effective knowledge and innovation management can contribute to value capture and value creation in organizations. While literature in the field has been dominated in the past by a focus on the role of innovation in the manufacturing sector, there is a growing interest in service sector innovation (Kindstrom, Kowalkowski & Sandberg, 2013; Gallouj & Djellal, 2010; Gonzales, 2016). As such, this research investigates the relationship between KM and innovation across both the manufacturing and service sectors and explores the likelihood that intangible assets contribute to innovation in different ways depending on the setting (Hendry et al, 2008).

This study is based on the view that without knowledge, there can be no innovation (Jorna, 2006), and is significant because it highlights the role of knowledge and its management in the innovation process (Tidd & Bessant, 2014). The management of knowledge may indeed hold the key to increasing systematic innovation capability in organizational contexts. The process of innovation is influenced by people's knowledge, thinking and behavior in many different ways. Given that organizations are increasingly interested in understanding how to achieve innovation capability leading to business performance, an exploration of the ways in which knowledge is managed for innovation in business settings may yield new insights into the links between KM and innovation (Kianto & Ritala, 2014).

This study may also shed light on KM practices that contribute to sustainable forms of innovation, rather than simply aiming for haphazard forms of innovation that lack long-term benefits and which may not be easily duplicated. The goal of sustainable innovation has human, social and management dimensions (Jorna, 2006), and today's KIOs face significant challenges on all these levels in order to gain and maintain competitive advantage. For managers, this involves developing new forms of knowledge, embedding this new knowledge within organizations, as well as managing flows of information, knowledge and experience. It also involves the strategic development of knowledge relationships to ensure a broad range of knowledge sources and effective means of knowledge exchange.

LITERATURE REVIEW

Since 2000, a growing body of research and practitioner driven material on KM has penetrated the literature. In this period, diversity, specialized discourses, variations in approaches to KM and appropriate research methodologies characterize these contributions. However, as a discrete field of study, KM remains an emerging area (Heisig, 2009). As such, this growing body of literature is frequently "unconnected", due to a lack of association between ideas and more integrated approaches to the topic. Diversity in the field of KM also encouraged several debates or 'turf wars' with numerous interest groups staking their claims to various principles and practices of KM. For some people, KM is a means to make IT more efficient; for others, the focus is on developing effective ways of managing human resources (HR). Others see KM as a crucial part of records management or library and information management. Particularly in the administrative sciences, writers consider the discovery and creation of new knowledge to be at the core of effective KM.

KM is a complex and subjective endeavor that is highly contextualized and therefore highly interpretive in nature. The KM literature can be classified as belonging to various broad paradigm orientations, each representing basic theoretical assumptions, biases, beliefs and frameworks that underpin KM practices (Gloet & Berrell, 2003; Gloet & Samson, 2013). Previous research has identified three main paradigm orientations that encapsulate the basic theoretical assumptions and biases underpinning KM practices. The classification of these paradigms is based on whether KM is driven by IT, humanist or organizational culture elements. An IT orientation positions technology as the anchor for developing approaches to KM. A humanist orientation focuses on the contribution of people to KM, whereas an organizational culture approach takes a resource-based view where knowledge is regarded as a strategic resource. These three basic orientations can be identified in various organizational contexts and because each orientation involves very different approaches to KM, the tools and methodologies used to develop and sustain KM differ within and across organizations.

In the absence of any universally accepted definition of KM, it is not surprising that multiple definitions and understandings would emerge (Wallace, 2007), especially given that precise definitions of 'knowledge' and 'management' are also elusive. Indeed, linking the terms to form knowledge management', the task becomes even more challenging. The Australian KM Standard (Standards Australia, 2005:8) offers managers a functional definition of KM as a cross-disciplinary construct, which involves "the design, implementation and review of social and technological activities and processes to improve the creating, sharing, and applying or using of knowledge". In this light, KM links to "innovation and sharing behaviors" and "managing complexity and ambiguity through knowledge networks and connections". Investigating smart processes and implementing "people-centric technologies" are part of an environment where the central role of knowledge in securing and maintaining competitive advantage emerges as a constant theme (Bessant & Venables, 2008; Laperche & Uzunidis, 2008; Quinn, 1992; Quintas et al., 1997). The need for effective and strategic KM in organizations is also emphasized (Farzin et al., 2014; Yang, 2010; Wu & Lin, 2009).

The literature also abounds with definitions of innovation, although common threads emerge. Drucker (1985) for example, views innovation as a process that exploits change through specific and disciplined practices that are underpinned by learning. Generally, in an organizational context, innovation implies outcomes such as improving performance, enhancing problem-solving capabilities, adding value and creating competitive advantage. Afuah (2003) describes innovations as the use of new knowledge to offer a new product, service or process of invention along with commercialization. Therefore, the innovation process implies discovery, application and outcome. Tidd and Bessant (2014:3), provide a succinct definition of innovation as "the process of creating value from ideas". Thus, innovation *per se* is predominantly about change, and of translating knowledge into value creation.

When considering the relationship between KM and innovation, the field is characterized by multiple perspectives and a lack of rigorous empirical research. Yet the literature reveals a number of common factors in successful KM approaches to supporting innovation. In an investigation of the relationship between KM, levels of innovation and levels of competitiveness in organizations, Carneiro (2000) highlights the strategic nature of knowledge development. He suggests that a strategic approach to KM positively influences innovation and competitiveness in organizations. Keupp et al. (2010) note the value of KM in shaping and supporting innovation strategies, particularly in times of changing circumstances in both the internal and external environment. Darroch (2005) also emphasizes the strategic potential of successful KM as a coordinating mechanism to enhance and support both innovation and business performance.

Du Plessis (2007) views KM as critical because innovation is highly dependent on managing knowledge in ways that align knowledge to the innovation process. Her research identifies three significant drivers of KM in innovation. First, this involves creating and sustaining competitive advantage through the utilization of knowledge collaboration processes. A second driver involves the capacity of an organization to reduce complexity in the innovation process by managing knowledge as a strategic resource. Finally, KM can drive innovation through the process of integrating internal and external knowledge sources, thus making knowledge for innovation more accessible. For Simpson (2002), a strategic approach to KM requires organizations to hone continuously their technological, organizational and managerial processes to secure positive strategic positions that support innovation.

Other KM approaches linked to innovation success include a strong IT infrastructure within the organization (Ahmad & Schroeder, 2011) and embedding understandings about innovation deeply within an organization's culture (Donate & Guadamillas, 2010). Through collaboration and engagement with customers, KM can enhance innovation opportunities (Sawhney et al, 2005), as well as through engaging with lead users (Mahr & Lievens, 2012). Other research points to the role of KM in supporting various forms of open innovation (Chiaroni et al, 2011) and in increasing the absorptive capacity of organizations (Spithoven et al, 2011). KM can also contribute to change management and risk reduction processes in organizations and can support knowledge-based risk management frameworks that improve the success of technology-based innovations (Mu et al, 2009). The above approaches signal the importance of KM in mediating and monitoring the contextual aspects of the innovation process.

A number of studies refer to the relationship between KM and HRM in supporting innovation, including developing employee capabilities for innovation (Kong et al, 2011; Lengnick-Hall & Lengnick-Hall, 2003), enhancing organizational learning (Jimenez-Jimenez & Saint-Valle, 2011) and rewarding employees for knowledge acquisition and sharing (Jaw et al., 2010; Foss et. al., 2011). Indeed, KM has been viewed as a mediator between HRM and innovation (Lopez-Cabrales et al., 2009; Chen and Huang, 2009). Through a focus on knowledge and learning, KM can contribute to innovation and sustainability of operations and information systems (Gavronski et al., 2012). Liao and Wu's (2010) research found that by using organizational learning as a mediator, KM can improve organizational innovation. Knowledge collaboration and learning also supports innovation across a variety of platforms (Quist & Tukker, 2013).

KM supports the management of innovation processes through the systematic acquisition, sharing and dissemination of knowledge (Darroch, 2005; McAdam, 2000), as well as through managing inputs to decision making processes (Van Riel et al., 2004). In contributing to the development of a knowledge based organizational culture, KM has been found to support organizational innovation and technological performance (Donate & Guadamillas, 2010). When KM contributes to a knowledge focused culture that is underpinned by strong communication channels (Sanz-Valle et al., 2011; Foss et al., 2011), innovation performance is enhanced. Researchers also emphasize the pivotal role of KM in engendering internal work environments that support creativity and foster innovation (Zheng et al, 2010). KM plays an important role in knowledge capture to support innovation (Dougherty, 2004), including sophisticated methods of environmental scanning (Cousins et al., 2011). KM can also facilitate inter-organizational knowledge flows (Chiang & Hung, 2010), as well as increase innovation capacity by broadening knowledge sources and linking knowledge and technology more effectively (Leiponen & Helfat, 2010).

Effective knowledge and innovation management (KIM) is essential to value creation, value capture and value delivery processes within organizations, and as such, appropriate business models are needed in order to maximize the benefits of managing knowledge and innovation. Business models, business strategy and innovation are inextricably linked. As Teece (2010) points out, the core element of a business model defines and determines the way in which a business enterprise delivers value to customers and converts that value to profit. Teece argues that without a well-conceived and well-embedded business model, it is almost impossible to capture value from innovation within an organization. Although there are variations concerning the definition of business models, for Drucker (1954, 1985) it is about identifying the customer and what the customer values, while for Magretta (2002), business models consist of stories that provide an explanation of how enterprises work. Amit and Zott (2010) assert that a business model outlines the manner in which value is extracted from the exploitation of business opportunities by focusing on the content, structure and governance of transactions within an enterprise.

Essentially a business model sets out the way in which a business creates and delivers value, but a successful business model can be easily replicated by competitors, even if it does not deliver the same degree of value. So innovation in the business as well as innovation of the business model itself can be a significant competitive advantage, if it is sufficiently differentiated. A robust business model provides attractive value propositions for customers through careful consideration of cost and risk elements in order to capture value and develop sound products and services (Teece, 2010). The existence of superior human capital, technology, governance and leadership is not enough to create sustainable competitive advantage if the business model is not suited to the competitive environment. Business models and the concept of business strategy, while related, are not the same; for Casadeus-Masanell and Ricart (2010), a business model reflects the realized strategy of an enterprise. The successful management of knowledge and innovation to create value within organizations is highly dependent on having the right business model in place

There is a growing body of literature related to business models, innovation and the value created through the development of dynamic capabilities and competencies across individual elements of innovation processes and activities. For Chesbrough (2010), a business model creates value, and captures a portion of that value. In highlighting the link to innovation, he suggests that open business models capture more external ideas. However, there are significant barriers to business model innovation, including a lack of strategic agility, strong leadership, appropriate organizational processes and culture. In emphasizing the role of strategy and leadership to support innovation, Doz and Kosonen (2010) suggest that business models can be designed and renewed through increasing agility on a strategic level within an organization, and this can be accomplished through developing a strategic thrust, unified leadership and agility with respect to resources. Others argue the importance of strategic leadership in developing entrepreneurial capability, the capacity to sense, seize opportunities and ultimately transform an organization's strategies in order to gain long-term competitive advantages.

The importance of a strong customer focus and open innovation approaches that encourage knowledge sharing between various stakeholders is also highlighted. KIM focusing on customer knowledge positively impacts both innovation performance and business performance; a study by Taherparvar et. al. (2014) found that customer knowledge management has the capacity to link external and internal environments to generate new ideas and create new forms of knowledge. KIM can assist in the process of determining the balance between use of internal versus external sources of knowledge, as too much dependence on internal knowledge may hinder an organization's absorptive capacity and reduce their innovation capacity (Diaz-Diaz & De Saa-Perez, 2014). Moreover, KIM can be useful in supporting open innovation, but also in recognizing that not all stages of the innovation process are open and adjusting approaches to the management of knowledge and innovation accordingly (Hafkesbrink, 2014).

For Kianto and Ritala (2014), value creation in organizations has both static elements (such as intellectual capital assets) and dynamic components (such as KIM practices), and through careful dynamic management of static intellectual capital resources, KIM can contribute to the value creation process. Krenz et. al. (2014) also highlight the increased complexity of inter-organizational value creation networks and processes linked to innovation. Due to the potential for conflict between KIM and other general management objectives, they recommend the introduction of a knowledge intermediary support function to assist the value creation process.

In terms of other building blocks of innovation, research by Ye et. al. (2015) found that being able to effectively communicate the business value of an innovation is essential to the success of that innovation. Floren et. al. (2014) note that development of dynamic capabilities in the area of human resource management can support innovation. In a recent study on the effects of human resource and knowledge management practices on innovation performance, Donate and Guadamillas (2015) suggest that innovation capacity will be strengthened significantly through KIM, especially in terms of establishing organizational conditions to support innovation. KIM, through supporting organizational learning, has also been shown to enhance entrepreneurship and innovation in complex organizational contexts (Tobin & Pennington, 2015).

With respect to developing dynamic capabilities to support innovation, ICT services face increasing challenges in creating business value, given the complex interplay between various stakeholders, organizations and technologies involved in service delivery. While ICT can support knowledge acquisition, dissemination and exploitation within organizations, it needs to be aligned with organizational structures and is dependent on sound practices relating to the management of knowledge, technology and people (Gressgard et al., 2014). A study by Reich et. al. (2014) found that business value can be achieved through KIM in ICT-enabled projects, through increasing knowledge alignment and enhancing knowledge-intensive project management outcomes. The discussion above highlights the various areas of opportunity where KIM can assist in delivering both innovation performance and business results.

METHODOLOGY

The methodology employed in this study includes an extensive review of the literature, followed by a qualitative cross-case analysis of data derived from 16 Australian manufacturing and service organizations to investigate the relationship between KM, systematic innovation and innovation effectiveness. A conceptual model is then developed to articulate a knowledge and innovation capability (KIM) framework that can aid in value creation and value capture. The following research questions (RQs) encapsulate the scope and depth of the research:

RQ1: What Is the Nature of the Relationship Between KM and Innovation?

The first research question pertains to the ways in which KM is manifested in the organizations studied in this research. In terms of the links between KM and innovation, multiple perspectives and a lack of rigorous empirical research pose significant challenges for those seeking definitive answers. Nevertheless, the literature reveals that successful KM approaches in innovation have common factors. This leads to the next research question regarding the contribution of KM to systematic and sustained forms of innovation:

RQ2: What Is the Extent of the Contribution of KM to Systematic Innovation Capability?

The second research question seeks to apply a framework of Systematic Innovation Capability (see Figure 1) to the organizations in order to determine the extent of KM's contributions to systematic innovation capability. Based on previous research (Samson, 2010; Gloet & Samson, 2013), the Systematic Innovation Capability model identifies the processes and relationships used by successful organizations to generate a series of innovations that create value and deliver business success. The building blocks of the model are wide-ranging but include aspects of strategy, leadership, customer focus, HRM, organizational culture and approaches to sustainability. Other aspects include operations and the management of innovation, knowledge and technology. The development of systematic innovation capability requires a holistic and integrated approach to innovation across all of the above attributes. As such, systematic and sustained forms of innovation are best achieved when all of the above 'building blocks' are present. By incorporating both innovation performance and business performance as essential elements, the model views innovation from a value creation perspective, where systematic innovation capability leads to a continuous stream of innovations, rather than haphazard or unplanned innovation. A focus on innovation performance and business results also assumes that resources are committed towards achieving innovation, and that aspects of innovation performance are measured on an ongoing basis. While recognizing that significant challenges accompany innovation, the framework assumes that those organizations with a robust and systematic innovation capability have higher probabilities of success with either large scale or incremental innovation.

Since the case study organizations represented both manufacturing and service sectors, the third research question seeks to determine whether there is any significant variation across sectors in the way KM supports systematic innovation capability. Hence, we pose the following question:

RQ3: Does the Contribution of KM to Systematic Innovation Capability Vary Significantly Between Manufacturing and Service Sector Organizations?

While innovation is actively pursued across both manufacturing and service sector organizations, Lyons et al. (2007) suggest that each sector requires a different approach in order for innovation to be successful. They argue that in services, innovation efforts must permeate the organizational structure and culture in order to be successful. Emerging from the new focus on innovation in the service sector in the mid-2000s was a realization that the uniqueness of the sector required different notions about innovation than those previously applied to the manufacturing sector (Hipp & Grupp, 2005). In this light, innovation in business services presents specific challenges, which include "conceptualizing and defining innovation" and furthermore, identifying where innovation occurs within a business. Kandampully (2002) suggests that the utility of innovation *per se* in customer driven settings is its role in creating competitive advantage. Therefore, the targets for knowledge assets are customers; while technology facilitates the process, knowledge is the essential element. He therefore argues that the core competency of a service organization should be innovation. Service organizations also require less formalized approaches to innovation for successful outcomes compared to manufacturing settings (Hendry et al., 2008). In manufacturing, innovation is often a more structured, rigid and standardized process (Ettlie & Rosenthal, 2011). In

Figure 1. Systemic innovation capability model

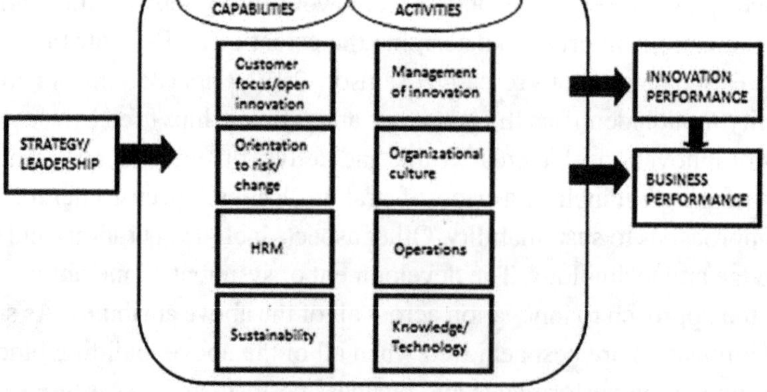

light of the discussion above, the potential of studies that investigate the relationship between KM and innovation across both manufacturing and service sector organizations is considerable.

Finally, given the close links between knowledge an innovation, we pose the following question:

RQ4: To What Extent Can Knowledge and Innovation Management Capability Contribute to Value Capture, Value Creation, and Value Delivery in Organizations?

It is important that firms wishing to capture value from innovation have a sound business model in place. The design and implementation of a viable business model is just as important to value capture as developing innovation itself; as Teece (2010, p. 190) suggests, selection and design of an appropriate business model constitutes a "key micro-foundation of dynamic capabilities". Dynamic capabilities can contribute much to an enterprise through sensing, seizing and transforming various elements within the business model, but the business model needs to be right. So, questions regarding value creation, value delivery and value capture are germane not only to extracting value from knowledge and innovation, but also to assist in developing appropriate dynamic capabilities within an organization that will ensure its sustainability and success. Given the inextricable links between knowledge and innovation, we focus our attention on the development of knowledge and innovation management (KIM) as a dynamic capability to capture value from innovation within a given business model.

The literature review was conducted primarily through Google Scholar, with the search focusing on the nexus between KM and innovation. The model of Systematic Innovation Capability (see Figure 1) guided the literature review, and KM publications relating to each of the 'building blocks' of innovation depicted in the model provided a means by which to consolidate themes relating to KM and innovation across a broad range of often unconnected literature. Moreover, literature relating to business models that support value capture, value creation and value delivery through knowledge and innovation management was also investigated.

Given that both KM and innovation represent multiple perspectives and contexts, a case study approach is applied to this research. Case study approaches are well suited to research environments characterized by complexity and overlapping approaches and scenarios (Yin, 2014), such as is the case with KM and innovation. In order to ensure methodological soundness, case studies should be based on careful selection of appropriate cases, in addition to the scientific application of case study principles and practices. Given that 16 organizations were investigated in this study, a ''cross-case analysis'' was employed to ensure a strong replication logic underpinning the cases and hence increasing construct validity, reliability and generalizability of the findings (Miles & Huberman, 1994).

A purposive sampling methodology (Recker, 2013) also enhanced reliability of the cases selected (Yin, 2014). Case study organizations were purposefully selected because they were widely recognized by business, government and professional associations to be leaders in both innovation and in the management of knowledge. These organizations demonstrated robust forms of KM where KM initiatives were planned, embedded and well resourced. Moreover, each of the selected case study organizations demonstrated their capacity for innovation across a range of indicators of innovation performance. Construct validity in multiple case studies is essential (Yin, 2014). The constructs employed here are similar to validated constructs in previous research (Samson, 2010; Gloet & Samson, 2013). Moreover, making generalizations based on statistical analysis is not the goal of this study – as Merriam (1998) suggests, "probabilistic sampling is not necessary or even justifiable'' in research of the type undertaken here. The sample also offered the researchers variance across company size, ownership type and industry sector and included foreign-owned and local organizations. The organizations consisted of seven manufacturing organizations (M1-7), seven in the services sector (S1-7) and two with a product/services mix (P1-2). It was anticipated that this range of organizations would yield insights into differences in approaches to KM and innovation across sectors, business types and organizational size. Table 1 displays the business type of each organization.

The study followed the case study protocols recommended by Yin (2014), which included using stem questions to guide participant discussion during the interviews and align the discussion to the research questions. Stem questions established the basis for analyzing systematic innovation and KM in the case organizations. Responses gave clues to organizations' approaches to both systematic innovation and KM. The stem questions are set out below:

- How is KM manifested in your organization?
- How is the relationship between KM and innovation viewed?
- What was the basis for introducing KM into your organization?
- How did your organization implement KM?
- How does KM contribute to innovation in your organization?
- What difficulties were encountered and how were these addressed?
- What type of lessons have been learned to date?
- What types of benefits have been achieved?
- What would you like to see in future iterations of KM to support innovation?

Table 1. Business type across case study organizations

Code	Business Type
M1	Bakery products
M2	Differentiated supply chain solutions/product design
M3	Technology, mining, and related products
M4	Design and production of specialty fabrics
M5	Textile inserts – product manufacturing
M6	Production of high performance textiles
M7	Automobile production
PS1	Mobile phone products and services
PS2	Travel support products and services
S1	Customized IT solutions and product development
S2	Ambulance services/community support
S3	Real estate services
S4	Environmental monitoring
S5	Revenue collection
S6	Legal services
S7	City council services and support

Depending on the size and scope of the organization, between two and six managers were interviewed in each case study organization. The multiple perspectives gleaned from the respondents essentially revealed the organization's dominant orientation to KM, as well as insights into various KM practices that support innovation. Textual analysis was applied to interpret the data gathered during the in-depth interviews, and a coding system allowed the researchers to place data in one or more predetermined categories. The data was grouped according to emerging common patterns or themes. The researchers also noted when a participant departed from the dominant pattern of responses and sought clarifications. The influence of the organizational environment helped explain divergent responses. A more comprehensive rationalization of discrepancies occurred when patterns of responses diverged significantly from previous research.

In each case, a range of documentation supported the analysis, including Annual Reports, Business Reports, performance data, project proposals and progress reports. Material available in the public domain, including the organizations' websites or via domains maintained by state and/or federal government departments also enriched the study.

MAJOR FINDINGS

The organizations in this study invested considerable resources and significant thought into supporting KM. As such, KM was a conscious and deliberate undertaking, rather than being arbitrary or haphazard. However, each of the case study organizations faced different challenges in implementing KM to support innovation. While acknowledging the importance of people to the process of KM, organizations tended to identify more strongly with IT-based forms of KM. A predominant IT orientation to KM meant that most managers were more comfortable in dealing with explicit, rather than tacit forms of knowledge. All managers in the study recognized the importance of organizational contexts in supporting KM initiatives, such as developing a strategy for KM, ensuring leadership support and shaping an organizational culture that supports KM. At the same time, the significant challenges associated with people-focused approaches to KM were clearly highlighted by the same managers.

While there are no agreed guidelines or recipes for success in creating an innovation-driven organization, the role of KM was widely regarded as being integral to the innovation process. As such, the participants, all of whom held management positions, reported practical understandings about how KM contributed to innovation across a range of activities. This includes systematically gathering, managing and disseminating information, as well as providing strong support across various business processes. KM also contributed to quality and continuous improvement, as well as shaping learning and organizational development initiatives. Managers reported that strategic KM encouraged and supported clear communication processes and assisted in growing a collaborative work environment. Through these contributions, KM emerged as a main driver of innovation that supports sustained competitive advantage.

All managers, whether in organizations with a manufacturing, services or a product/services orientation, advocated the need for a strategic thrust to drive both KM and innovation. Many advocated the need for leadership, both from the top but also from middle management in order to grow and sustain KM practices that support innovation. Regardless of their sector location, the case study organizations demonstrated a very strong customer focus, and managers regarded customers as a major source of knowledge to inform and drive innovation. The proactive management of innovation activities, along with monitoring and measurement across a range of performance indicators, were found to be essential for developing sustained forms of innovation within these organizations. Managers also identified a number of major enablers of KM to support innovation across their respective organizations. These included formal and informal KM practices, coupled with a solid IT infrastructure and centralized access to information across the organization, including repositories and intranets. A variety of channels and means by which to share knowledge were also useful, as well as the use of teams and group structures to communicate and collaborate. KM can also act as a major enabler of systematic innovation through organizational development and learning programs. A philosophy of quality and a commitment to continuous improvement across the organization assists KM to support incremental forms of innovation. Finally, employee recognition and rewards for contributions to KM and innovation assist in employee engagement and motivation, thus driving increased innovation efforts. These major enablers of KM to support innovation are listed in Table 2.

The scope and nature of KM and innovation within the case study organizations is revealed using the Systematic Innovation Capability framework that outlines the main attributes of systematic innovation capability (see Figure 1). Managers identified nine areas where their organizations rated highly on systematic innovation capability. The highest rankings came from the following areas:

- Leadership
- Customer Focus
- Strategy
- Quality processes and philosophy
- Values and culture to support innovation
- External partners/open innovation
- Change Focus
- Learning
- HR orientation to innovation

By way of contrast, areas where systematic innovation capability did not rate highly within the case study organizations include:

- Developing an appetite for risk
- Building and sustaining knowledge relationships
- Resources commitment for innovation
- Measurement of innovation

Table 2. Major enablers of KM to support innovation

• Formal and informal KM practices within the organizations
• A solid IT infrastructure
• Centralized access to information across the organization including repositories and intranets
• A variety of channels and means by which to share knowledge
• The use of teams and group structures to communicate and collaborate
• Effective learning programs to support innovation
• A philosophy of quality and a commitment to continuous improvement
• Recognition and rewards for contributions to KM and innovation

The level of embeddedness of various attributes of systematic innovation capability are displayed in Table 3. Overall, organizations in the manufacturing sector rated more strongly across the various attributes of systematic innovation capability when compared to service sector organizations. The service sector organizations failed to exhibit the same levels of organizational embeddedness for KM and innovation across the ‘building blocks’ of innovation. For instance, service sector organizations did not rate highly in the areas of strategy, open innovation, external partnerships, knowledge relationships or the management of innovation processes. Both manufacturing and service sector organizations were rated highly on customer focus and a commitment to quality processes. Organizations with a product/services mix rated poorly in terms of quality processes, learning culture and the measurement of innovation. Service sector organizations and those with a product/services mix committed less resources toward innovation when compared to manufacturing organizations. However, service sector organizations demonstrated a strong customer focus and a commitment to a learning culture. Service sector organizations in this study appeared to use less formal processes and approaches to innovation compared to the more formal innovation processes observed in the manufacturing sector. Smaller organizations also tended to use more fluid and agile KM processes as opposed to the more structured and formalized processes used in larger organizations. Interestingly, this study did not yield any significant relationship between organizational size and systematic innovation capability.

Table 3. KM support for systematic innovation capability in case study organizations

Strategic Innovation Capability Element	M1	M2	M3	M4	M5	M6	M7	PS1	PS2	S1	S2	S3	S4	S5	S6	S7
Strategy	√	√		√	√	√	√	√	√	√		√				
Leadership	√	√	√	√	√	√	√	√	√	√	√		√	√		
Customer focus	√	√		√	√	√	√	√	√		√	√	√	√	√	√
Open innovation	√	√	√	√	√	√	√	√	√	√	√		√			
Resources	√		√		√	√	√		√	√		√				
Change focus		√		√	√	√	√	√	√		√			√		√
Risk appetite			√			√			√			√	√			
Innovation culture		√		√	√	√	√	√	√	√	√	√	√			
Quality processes	√	√	√	√	√	√	√			√	√	√	√	√	√	√
HRM focus on innovation	√		√	√	√	√	√	√	√	√		√	√			√
Learning culture	√	√	√	√	√	√	√					√	√	√	√	√
Management of innovation	√	√	√	√	√	√	√		√	√		√	√			
External partnerships	√	√	√	√	√	√	√	√	√	√	√		√			
Knowledge relationships				√	√		√	√		√	√		√			
Measurement of innovation		√	√	√	√	√				√		√				

Key: √=strength of the attribute and extent of organizational embeddedness. Indicates areas where KM strongly supports systematic innovation capability.

DISCUSSION

The literature on the relationship between KM and innovation highlights the complex and multidimensional nature of both areas, characterized by multiple perspectives and differing views regarding KM and innovation. The managers in all of the case study organizations reflected these multiple views. In some organizations, KM practices were dominated by IT, while others took a more people-oriented approach to KM. In other cases, KM reflected a strategic resource based view where knowledge was regarded as a strategic resource to be managed. With respect to innovation, some organizations were more focused on research and development or developing new technologies and services, while others regarded innovation as a series of incremental or process improvements. Managers generally recognized the potential of KM to contribute to innovation, but most found it difficult to articulate the precise nature of this link, even within their own organizations. Organizations appeared to lack common definitions or understandings with respect to KM, which also extended to a lack of common articulations about innovation. The multi-dimensional and contextual nature of both innovation and KM might explain this shortcoming.

Managers generally agreed that KM approaches focused on people and organizational culture are more important when managing tacit forms of knowledge, although many pointed to the greater challenges associated with these approaches. However, the same managers remained ambiguous and somewhat uncomfortable about tacit knowledge, reporting greater feelings of confidence and success when they worked with technology-based, IT-driven KM initiatives. While managers were emphatic that the people and organizational culture aspects of KM are most important in supporting innovation, they were far more comfortable dealing with explicit rather than tacit forms of knowledge, so preferences for IT-driven approaches subtly held sway in the discussions.

Managers also identified a number of areas where KM supports innovation, including strategy, leadership, HRM, learning, IT and organizational culture. Many of these areas have also been identified in the literature. Embedding KM and innovation at the strategic level, which reinforces the pivotal role of leadership and management to support innovation, is a strong theme in the literature (Miles, 2007; Makri and Scandura, 2010). Managers spoke of the need to link KM and systematic innovation to the overall business strategy by aligning KM and innovation strategies with their organization's mission and business objectives. Such a linkage demonstrates a strong organizational commitment to both KM and innovation initiatives, and when given a strategic foothold, KM and innovation were less inclined to be treated as just another passing fad. However, managers suggested that even though a KM strategy provides direction and support for KM activities, the managers suggested that a KM strategy alone is not enough to achieve significant inroads into sustained innovation. For sustained and systematic forms of innovation to become embedded, an innovation strategy based on the overall business strategy is also required.

While managers implicitly favored IT approaches to KM and innovation, the same managers saw that developing links between KM and HRM helped address the challenges posed by managing people and organizational culture aspects related to innovation. Such linkages include the development of performance management systems that recognize and reward knowledge work leading to innovation, as well as the implementation of sophisticated HRD programs designed to increase employee engagement, develop skills and enhance capabilities that add value to an organization; these developments are also mirrored in the literature (Butler & Grace, 2010; Kong et al., 2011). Managers also reported that KM supported a strong customer focus, based on open innovation and adopting lead user strategies, thereby facilitating communication and collaboration with customers (Sawhney et al., 2005).

Despite demonstrating multiple perspectives on KM and Innovation, managers alluded to the importance of communicating clearly about KM and innovation, developing a culture underpinned by common values, goals and objectives through a shared language (Cano & Cano, 2006; Fugate et al., 2009; Befecki et al., 2009). Developing a learning culture focused on innovation was also considered to be a fundamental component in maintaining competitive advantage. By shaping a learning culture where innovation is considered a core value, KM can support the processes of knowledge generation, sharing and dissemination, all of which are fundamental to innovation.

While responses from managers in the case studies yielded some insights into the role of KM in supporting innovation generally, the salient connections between KM and systematic innovation capability are far more tenuous. The case studies clearly revealed a lack of KM support for particular innovation activities, including the commitment of resources for innovation, balancing large scale and incremental forms of innovation, developing an appetite for risk and measuring innovation. One of the main challenges faced by the organizations is the lack of resourcing for both KM and innovation. The literature indicates that investment in these areas is essential for sustained innovation performance and overall business success (Hendry et al., 2008). Perhaps most surprisingly, KM did not appear to support the development and nurturing of knowledge relationships, either with customers, suppliers or other sources, which are fundamental to systematic and sustainable forms of innovation.

In all the case study organizations, there was significant investment in technological infrastructure and significant investment in IT to support KM. However, investments in other attributes of KM and systematic innovation capability, such as HR, culture and knowledge relationships were not apparent. Another major challenge involved the development of a different attitude to risk management by recognizing that successful innovation efforts often depend on risk-taking in order to come to fruition (Befecki et al, 2009). In three of the case studies, managers reported that KM was instrumental in mitigating risk rather than encouraging an appetite for risk. The measurement of both KM and innovation was also shown to be a major shortcoming among the case study organizations, mirroring a general deficiency alluded to in the literature (Hipp & Grupp, 2005).

There were more instances where innovation focused on incremental and small-scale improvement rather than on large-scale radical transformation. This is not surprising given that all organizations were committed to pursuing quality initiatives within a strong culture of continuous improvement, which permeated each organization's unique culture. Moreover, two of the service sector case studies were in the government sector, where the main drivers of KM were continuous improvement and business excellence rather than seeking competitive advantage and creating value from innovation.

Some significant differences were noted between service and manufacturing sector organizations. On the whole, manufacturing organizations seemed to be outperforming service sector organizations or those with a product/services mix on systematic innovation capability. Two of the service sector case studies were government organizations, although this factor does not fully account for the differences. Processes and approaches supporting innovation tended to be more formal in the manufacturing sector. Moreover, manufacturing companies tended to take a more disciplined approach to innovation compared to service sector organizations, perhaps due to the tangible outcomes of manufacturing. Hence, manufacturing organizations displayed a more disciplined style of innovation. Service sector organizations favored organic approaches to the innovation process.

Even though service sector organizations are characterized by higher levels of direct interaction with customers, all organizations in the study rated highly on customer focus. What was also surprising, given that innovation in services is closely tied to customers, the service sector organizations in this study rated poorly on open innovation and development of knowledge relationships, which are a source of knowledge and hence innovation. Service sector organizations were less inclined than other organizations to measure innovation efforts. This may be due to the higher levels of direct customer interaction in services and the fact that these organizations deal more heavily in intangible forms of knowledge than manufacturing firms. The service sector organizations rated poorly in developing a strategy focus. Most did not proactively manage their innovation processes, perhaps because the nature of these processes in the services sector are more fluid.

While organizational size did not turn out to be a significant differentiator in this study, the researchers noted some differences between large and small organizations. Smaller organizations lacked maturity in systematic innovation because the process often relied on one person or a small group, thereby concentrating the risk elements attached to the process. In comparison, larger organizations were more deeply resourced and systematized, less vulnerable to a change, with the risk spread across many people and teams, thereby reducing the risks associated with innovation. However, given that all organizations generally rated poorly on developing an appetite for risk, this issue warrants further investigation.

Finally, all 16 organizations were underpinned by business models that supported innovation activities and processes. Innovation was clearly a strong focus in terms of vision, mission and strategy for most of the 16 organizations. However, knowledge and innovation management activities were not clearly focused on processes of value capture, value creation and value delivery. It may be the case that the organizations in this study had not progressed sufficiently from a focus on innovation to a more concentrated prioritization of extracting value from innovation processes and activities. It was clear that more could be done to develop knowledge and innovation management as a dynamic capability to support value capture, creation and delivery from innovation.

Capturing Value: Developing Knowledge and Innovation Management Capability in Organizations

The way in which a business competes and differentiates itself through innovation is the value proposition. In order to achieve sustained competitive advantage, firms must have a strategic understanding of their particular value proposition and constantly seek to identify new competencies to support the business model (Johnson et. al., 2008). This involves the development of a stream of dynamic capabilities, "the capacity to sense opportunities and to reconfigure knowledge assets, competencies and complementary assets and technologies to achieve sustainable competitive advantage." Teece, 1998, p. 73). Embedded within the resource-based theory of the firm (Barney, 1996), the notion of dynamic capabilities is well established in the literature as s significant source of competitive advantage, particularly within knowledge intensive organizations. Dynamic capabilities consist of various resources and management strengths that cross cut a range of business functions. Through sensing, seizing and transforming opportunities that arise in the organizational environment, firms gain competitive advantage by continually developing and reconfiguring available resources, both tangible and intangible (Teece, 2009).

For Kindstrom et. al. (2013), a dynamic capabilities approach is a useful way of improving innovation, particularly in services, across the areas of sensing, seizing and transforming the business environment. Regardless of sector or organizational type, KIM emerges as a new type of management style that recognizes the value inherent in intangible assets such as knowledge and innovation processes. We therefore propose the development of the KIM Capability Framework as a starting point for guiding the development of knowledge an innovation as a dynamic capability to capture, create and deliver value from innovation processes and activities within organizations. Derived from the literature and adapted from the Systematic Innovation Capability framework (see Figure 1), the KIM Capability Framework recognizes the various 'building blocks' of innovation. These include strategy and leadership; attitude to risk and change; customer focus; open innovation; human resource management; sustainability; innovation process management; culture and communication; operations; and knowledge and technology.

The KIM Capability Framework (see Figure 1) depicts the relationships between the various building blocks of innovation success. Strategic intent coupled with strong leadership drives capabilities such as a strong customer focus and open innovation approaches. A willingness to embrace change and to take calculated risks, coupled with HRM approaches that support innovation, can support sustainability. Effective management of innovation processes, operations, knowledge and technology can combine with a supportive culture to grow systematic forms of innovation capability within organizations. This in turn leads to innovation performance and business success. The KIM model differs from the framework of Strategic Innovation Capability in that it also highlights the need for ongoing measurement efforts, thereby insuring that the value accruing from innovation efforts is monitored in order to reinvest in the innovation cycle.

As a dynamic capability, KIM has the potential to deliver both innovation performance and business performance, thus capturing value from knowledge and innovation across the building blocks of innovation and supporting sustainable business models. However, without measurement of innovation efforts, it is not possible to determine the extent of value capture to support the business model. Hence, the KIM Capability Framework incorporates a measurement, evaluation and feedback loop in order to continually improve and refresh dynamic capabilities. The framework also advocates reinvestment of a portion of the value captured in order to further support innovation through KIM. Future research efforts should be directed to investigating the potential of the KIM Capability Model as a guide for developing knowledge and innovation as a dynamic capability to support value capture, value creation and value delivery from organizations with a strong innovation focus.

Figure 2.

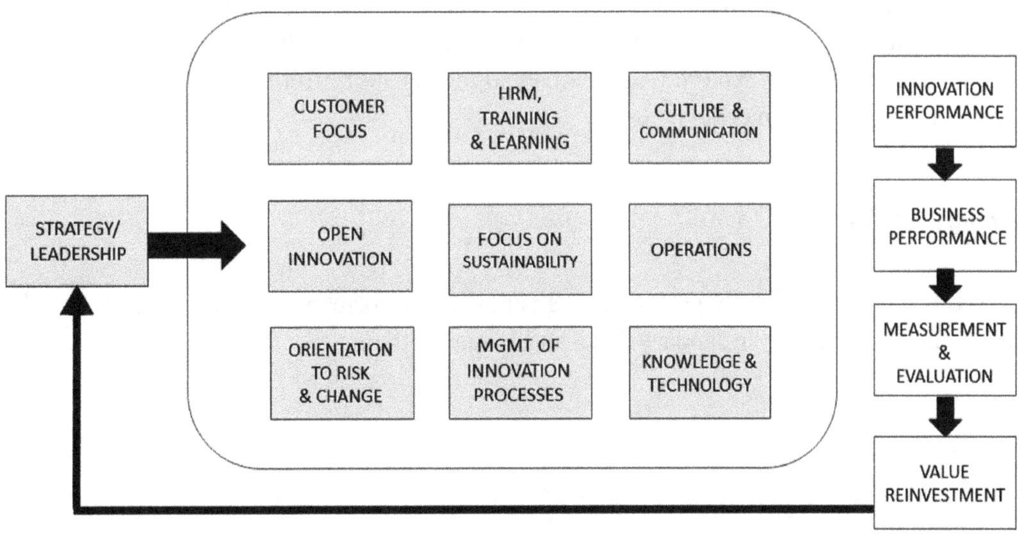

CONCLUSION

The findings suggest a strong link between KM and innovation. However, the ways in which KM contributes to systematic innovation capability is less than clear. So too are differences that might be attributed to organizational size, industry sector and business type. It may be prudent to consider whether KM and innovation have different meanings in the service sector compared to those in the manufacturing sector. In particular, further research on the nature of the relationship between KM and systematic innovation capability is required. This research was exploratory in nature and limited to the Australian context – future research should broaden the selection of organizations; thus, a larger sample would generate results that are more generalizable. As the links between KM, innovation in general and systematic innovation capability in particular represent complex relationships, the authors suggest that a combination of qualitative, quantitative and longitudinal studies will enhance the literature in the field.

The potential of KM to influence innovation and achieve competitive advantage increases significantly with effective KM approaches. Moreover, the implementation process for KM makes all the difference. The cases demonstrated sound KM practices across some but not all of the various building blocks of innovation. In order to contribute to systematic innovation capability, KM practices should address as many attributes of innovation as possible. The greater the extent to which KM supports the various attributes and activities of systematic innovation, the more likely it is that KM performs an integrated and holistic function within an organization. Organizations that embed KM practices across a range of innovation activities create a boundary spanning culture, which links various organizational domains. These links are vital to the process of increasing KM's contributions to systematic innovation capability. This exploratory research confirms the complexity of the relationship between KM and innovation process, and how people's knowledge, thinking and behavior influence the process of innovation in different ways.

This study also explored the potential of KIM as a dynamic capability to support the various building blocks of innovation and capture value from these processes and activities. The strength of the KIM Capability Model lies in the measurement and feedback loops contained within the model; without adequate measurement, evaluation and feedback processes, it is impossible to determine the value created and captured through KIM. Constant refreshing and reinvestment of value capture is also recommended in order to continually improve KIM as a dynamic capability that can contribute more fully to the value proposition and hence to a more robust business model leading to sustained competitive advantage. While the KIM Model itself was not tested in this study, future research efforts should involve the application of the KIM Model to further explore its potential and test its applicability for organizations seeking to extract value from innovation efforts.

ACKNOWLEDGMENT

Content of this paper was initially presented in an abbreviated form at the Hawaii International Conference for the System Sciences in January 2013 and January 2016.

REFERENCES

Afuah, A. (1998). *Innovation Management: Strategies, Implementation and Profits*. New York: Oxford University Press.

Ahmad, S., & Schroeder, R. (2011). Knowledge management through technology strategy: Implications for competitiveness. *Journal of Manufacturing Technology Management, 22*(1), 6–24. doi:10.1108/17410381111099789

Alavi, M., Kayworth, T., & Leidner, D. (2006). An Empirical Examination of the Influence of Organizational Culture on Knowledge Management Practices. *Journal of Management Information Systems, 22*(3), 191–224. doi:10.2753/MIS0742-1222220307

Amit, R., & Zott, C. (2010). Business Model Design: An Activity Systems Perspective. *Long Range Planning, 43*(2-3), 216–226. doi:10.1016/j.lrp.2009.07.004

Badaracco, J. L. (1991). *The Knowledge Link: How Firms Compete through Strategic Alliances*. Boston: Harvard University Press.

Barney, J. (1996). The Resource-Based Theory of the Firm. *Organization Science, 7*(5), 469–476. doi:10.1287/orsc.7.5.469

Befeki, T., Epstein, M., & Yuthas, K. (2009, February). Creating growth from effective opportunity risk management. *CWA Management*, 20-26.

Beijers, R. (1999). Questions in knowledge management: Defining and conceptualizing a phenomenon. *Journal of Knowledge Management, 3*(2), 94–109. doi:10.1108/13673279910275512

Bessant, J., & Venables, T. (2008). *Creating Wealth from Knowledge: Meeting the Innovation Challenge. Cheltenham*. Cheltenham, UK: Edward Elgar Publishing. doi:10.4337/9781848441248

Butler, T., & Grace, A. (2010). Building and Maintaining Human Capital with Learning Management Systems. In K. O'Sullivan (Ed.), *Strategic Intellectual Capital Management in Multinational Organizations: Sustainability and Successful Implications*. New York: IGI Global. doi:10.4018/978-1-60566-679-2.ch014

Cano, C., & Cano, P. (2006). Human resources management and its impact on innovation performance in companies. *International Journal of Technology Management, 35*(1-4), 11–28. doi:10.1504/IJTM.2006.009227

Carneiro, A. (2000). How does knowledge management influence innovation and competitiveness? *Journal of Knowledge Management, 4*(2), 87–98. doi:10.1108/13673270010372242

Casadeus-Masanell, R., & Ricart, J. (2010). From Strategy to Business Models and onto Tactics. *Long Range Planning, 43*(2-3), 195–215. doi:10.1016/j.lrp.2010.01.004

Chen, C., & Huang, J. (2009). Strategic human resource practices and innovation performance – the mediating role of knowledge management capacity. *Journal of Business Research, 62*(1), 104–114. doi:10.1016/j.jbusres.2007.11.016

Chesbrough, H. (2010). Business Model Innovation: Opportunities and Barriers. *Long Range Planning*, *43*(2-3), 354–363. doi:10.1016/j.lrp.2009.07.010

Chiang, Y., & Hung, K. (2010). Exploring open search strategies and perceived innovation performance from the perspective of inter-organizational knowledge flows. *R & D Management*, *40*(3), 292–306. doi:10.1111/j.1467-9310.2010.00588.x

Chiaroni, D., Chiesa, V., & Frattini, F. (2011). The Open Innovation Journey: How firms dynamically implement the emerging innovation management paradigm. *Technovation*, *31*(1), 34–43. doi:10.1016/j.technovation.2009.08.007

Cousins, P., Lawson, B., Petersen, K., & Handfield, R. (2011). Breakthrough Scanning, Supplier Knowledge Exchange, and New Product Development Performance. *Journal of Product Innovation Management*, *28*(6), 930–942. doi:10.1111/j.1540-5885.2011.00854.x

Darroch, J. (2005). Knowledge management, innovation and firm performance. *Journal of Knowledge Management*, *9*(3), 101–115. doi:10.1108/13673270510602809

Davenport, T., & Prusak, L. (1998). *Working Knowledge*. Boston: Harvard University Press.

Diaz-Diaz, N., & De Saa-Perez, P. (2014). The interaction between external and internal knowledge sources: An open innovation view. *Journal of Knowledge Management*, *18*(2), 430–446. doi:10.1108/JKM-07-2013-0257

Donate, M., & Guadamillas, F. (2010). The Effect of Organizational Culture on Knowledge Management Practices and Innovation. *Knowledge and Process Management*, *17*(2), 82–94. doi:10.1002/kpm.344

Donate, M., & Guadamillas, F. (2015). An empirical study of the relationships between knowledge management, knowledge-oriented human resource practices and innovation. *Knowledge Management Research and Practice*, *13*(2), 134–148. doi:10.1057/kmrp.2013.36

Dougherty, D. (2004). Organizing Practices in Services: Capturing Practice-Based Knowledge for Innovation. *Strategic Organization*, *2*(1), 35–64. doi:10.1177/1476127004040914

Doz, Y., & Kosonen, M. (2010). Embedding Strategic Agility: A Leadership Model for Accelerating Business Model Renewal. *Long Range Planning*, *43*(2-3), 370–382. doi:10.1016/j.lrp.2009.07.006

Drucker, P. (1954). *The Practice of Management*. New York: Harper & Row.

Drucker, P. (1985). *Innovation and Entrepreneurship*. London: Heinemann.

Du Plessis, M. (2007). The role of knowledge management in innovation. *Journal of Knowledge Management*, *11*(4), 20–29. doi:10.1108/13673270710762684

Ettlie, J., & Rosenthal, S. (2011). Services versus Manufacturing Innovation. *Journal of Product Innovation Management*, *28*(2), 285–299. doi:10.1111/j.1540-5885.2011.00797.x

Farzin, M., Kahreh, M., Hesan, M., & Khalouei, A. (2014). A Survey of Critical Success Factors for Strategic Knowledge Management Implementation: Applications for Service Sector. *Procedia: Social and Behavioral Sciences*, *109*, 595–599. doi:10.1016/j.sbspro.2013.12.512

Floren, H., Rundquist, J., Schuler, R., Bondarouk, T., & Ruel, H. (2014). HRM and innovation: Themes, contingencies and directions for future research. *European Journal of Innovation Management*, *8*(5), 570–575.

Foss, N., Laursen, K., & Pederson, T. (2011). Linking Customer Interaction and Innovation: The Mediating Role of New Organizational Practices. *Organization Science*, *22*(4), 980–999. doi:10.1287/orsc.1100.0584

Franklin, C., & Ballan, M. (2001). Reliability and Validity in Qualitative Research. In B. Thyer (Ed.), *The Handbook of Social Work Research Methods*. Thousand Oaks, CA: Sage. doi:10.4135/9781412986182.n15

Freeze, R., & Kulkarni, U. (2008). Validating Distinct Knowledge Assets: A Capability Perspective. *International Journal of Knowledge Management*, *4*(4), 40–61. doi:10.4018/jkm.2008100103

Fugate, B., Stank, T., & Mentzer, J. (2009). Linking improved knowledge management to operational and organizational performance. *Journal of Operations Management*, *27*(3), 247–264. doi:10.1016/j.jom.2008.09.003

Gallouj, F., & Djellal, F. (Eds.). (2010). *The handbook of innovation and services: a multi-disciplinary perspective*. Cheltenham, UK: Edward Elgar Publishing. doi:10.4337/9781849803304

Gavronski, I., Klassen, R., Vachon, S., & Maschado do Nascimento, L. (2012). A learning and knowledge approach to sustainable operations. *International Journal of Production Economics*, *140*(1), 183–192. doi:10.1016/j.ijpe.2012.01.037

Gibbons, M., Limoges, C., Nowotny, H., Schwartzman, S., Scott, P., & Trow, M. (1994). *The Production of Knowledge: The Dynamics of Science and Research in Contemporary Societies*. London: Sage.

Gloet, M., & Berrell, M. (2003). The Dual Paradigm Nature of Knowledge Management: Implications for achieving quality outcomes in human resources. *Journal of Knowledge Management*, *7*(1), 78–89. doi:10.1108/13673270310463635

Gloet, M., & Samson, D. (2013). Knowledge Management to Support Systematic Innovation Capability. In *Proceedings of the 46th Hawaii International Conference on the System Sciences (HICSS)*. IEEE. 10.1109/HICSS.2013.374

Goh, L. A. (2005). Fostering Innovation Through Knowledge-Centered Principles: A Case Study of Singapore Airlines. *International Journal of Knowledge Management*, *1*(4), 73–90. doi:10.4018/jkm.2005100104

Gonzales, R. V. D. (2016). Knowledge Retention in the Service Industry. *International Journal of Knowledge Management*, *12*(1), 45–59. doi:10.4018/IJKM.2016010104

Gressgard, L., Amundsen, O., Aasen, T., & Hansen, K. (2014). Use of information and communication technology to support employee-driven innovation in organizations: A knowledge management perspective. *Journal of Knowledge Management*, *18*(4), 633–650. doi:10.1108/JKM-01-2014-0013

Hafkesbrink, J. (2014). Ambidextrous Organizational and Individual Competencies in Open Innovation: The Dawn of a New Research Agenda. *Journal of Innovation Management*, *2*(1), 9–46.

Halawi, L., Aronson, J., & McCarthy, R. (2005). Resource-Based View of Knowledge Management for Competitive Advantage. *Electronic Journal of Knowledge Management, 3*(2), 75–86.

Heisig, P. (2009). Harmonisation of knowledge management - comparing 160 KM frameworks around the globe. *Journal of Knowledge Management, 13*(4), 4–31. doi:10.1108/13673270910971798

Hendry, C., Selim, G., Citron, D., Holtham, C., Brown, J., Holden, J., ... Oehlcke, F. (2008). Facilitating Innovation through the Measurement and Management of Intangibles. In H. Scarbrough (Ed.), *The Evolution of Business Knowledge*. Oxford, UK: Oxford University Press.

Hipp, C., & Grupp, H. (2005). Innovation in the service sector: The demand for service-specific innovation measurement concepts and typologies. *Research Policy, 34*(4), 517–535. doi:10.1016/j.respol.2005.03.002

Ibrahim, F., & Reid, V. (2009). What is the Value of Knowledge Management Practices? *Electronic Journal of Knowledge Management, 7*(5), 567–574.

Jaw, C., Lo, J. Y., & Lin, Y. H. (2010). The determinants of new services development: Services characteristics, market orientation, and actualizing innovation effort. *Technovation, 30*(4), 265–277. doi:10.1016/j.technovation.2009.11.003

Jensen, M., Johnson, B., Lorenz, E., & Lundvall, B. (2007). Forms of knowledge and modes of innovation. *Research Policy, 36*(5), 680–693. doi:10.1016/j.respol.2007.01.006

Jimenez-Jimenez, D., & Sanz-Valle, R. (2011). Innovation, organizational learning, and performance. *Journal of Business Research, 64*(4), 408–417. doi:10.1016/j.jbusres.2010.09.010

Johnson, M. W., Christensen, C. M., & Kagermann, H. (2008). Reinventing Your Business Model. *Harvard Business Review, 86*, 50–59. PMID:18681297

Jorna, R. (2006). *Sustainable Innovation: The Organisational, Human and Knowledge Dimension*. Sheffield, UK: Greenleaf Publishing.

Kandampully, J. (2002). Innovation as the core competency of a service organization: The role of technology, knowledge and networks. *European Journal of Innovation Management, 5*(1), 18–26. doi:10.1108/14601060210415144

Keupp, M., Palmie, M., & Gassmann, O. (2012). The Strategic Management of Innovation: A Systematic Review and Paths for Future Research. *International Journal of Management Reviews, 14*(4), 367–390. doi:10.1111/j.1468-2370.2011.00321.x

Kianto, A., Ritala, P., Spender, J.-C., & Vanhala, M. (2014). The interaction of intellectual capital assets and knowledge management practices in organizational value creation. *Journal of Intellectual Capital, 15*(3), 362–375. doi:10.1108/JIC-05-2014-0059

Kim, T., Lee, J., Chun, J., & Benbasat, I. (2014). Understanding the effect of knowledge management strategies on knowledge management performance: A contingency perspective. *Information & Management, 51*(4), 398–410. doi:10.1016/j.im.2014.03.001

Kindstrom, D., Kowalkowski, C., & Sandberg, E. (2013). Enabling service innovation: A dynamic capabilities approach. *Journal of Business Research, 66*(8), 1063–1073. doi:10.1016/j.jbusres.2012.03.003

Kong, E., Doren, C., & Ramen, R. (2011). The role of human resources practices in developing knowledge and learning capabilities for innovation: a study of service providers in India. *10th International Research Conference on Quality, Innovation and Knowledge Management.*

Krenz, P., Basmer, P., Buxbaum-Conradi, S., Redlich, T., & Wulfsberg, J.-P. (2014). Knowledge Management in Value Creation Networks: Establishing a New Business Model through the Role of a Knowledge-Intermediary. *Procedia CIRP*, *16*, 38–43. doi:10.1016/j.procir.2014.01.006

Kuusisto, J., & Meyer, M. (2003). *Insights into services and innovation in the knowledge-intensive economy. Technology Review 134/2003.* Helsinki, Finland: Tekes National Technology Agency.

Laperche, B., & Uzunidis, D. (2008). *The Genesis of Innovation.* Cheltenham, UK: Edward Elgar Publishing. doi:10.4337/9781848442856

Laursen, K., & Salter, M. (2006). Open for Innovation: The Role of Openness in Explaining Innovation Performance among U.K. Manufacturing Firms. *Strategic Management Journal*, *27*(2), 131–150. doi:10.1002mj.507

Leiponen, A., & Helfat, C. (2010). Innovation Objectives, Knowledge Sources, and the Benefits of Breadth. *Strategic Management Journal*, *31*(2), 224–236. doi:10.1002mj.807

Lengnick-Hall, M., & Lengnick-Hall, C. (2003). *Human Resource Management in the Knowledge Economy: New Challenges, New Roles, New Capabilities.* San Francisco: Berrett-Koehler.

Liao, S., & Wu, C. (2010). System perspective of knowledge management, organizational learning, and organizational innovation. *Expert Systems with Applications*, *37*(2), 1096–1103. doi:10.1016/j.eswa.2009.06.109

Lopez-Cabrales, A., Perez-Luno, A., & Cabrera, R. (2009). Knowledge as a Mediator Between HRM Practices and Innovative Activity. *Human Resource Management*, *48*(4), 485–503. doi:10.1002/hrm.20295

Lyons, R., Chatman, J., & Joyce, C. (2007). Innovation in Services: Corporate Culture and Investment Banking. *California Management Review*, *50*(1), 174–191. doi:10.2307/41166422

Magretta, J. (2002). Why business models matter. *Harvard Business Review*, *6*, 86–92. PMID:12024761

Mahr, D., & Lievens, A. (2012). Virtual lead user communities: Drivers of knowledge creation for innovation. *Research Policy*, *41*(1), 167–177. doi:10.1016/j.respol.2011.08.006

Makri, M., & Scandura, T. (2010). Exploring the effects of creative CEO leadership on innovation in high technology firms. *The Leadership Quarterly*, *21*(1), 75–88. doi:10.1016/j.leaqua.2009.10.006

Malhotra, N., & Morris, T. (2009). Heterogeneity in Professional Service Firms. *Journal of Management Studies*, *46*(6), 895–922. doi:10.1111/j.1467-6486.2009.00826.x

McAdam, R. (2000). Knowledge management as a catalyst for innovation within organizations: A qualitative study. *Knowledge and Process Management*, *7*(4), 233–241. doi:10.1002/1099-1441(200010/12)7:4<233::AID-KPM943.0.CO;2-F

Merriam, S. (1998). *Qualitative research and case study applications in education.* San Francisco: Jossey-Bass Publishers.

Miles, M., & Huberman, A. (1994). *Qualitative data analysis: an expanded sourcebook* (2nd ed.). Thousand Oaks, CA: Sage.

Miles, R. (2007). Innovation and Leadership Values. *California Management Review*, *50*(1), 192–201. doi:10.2307/41166423

Mu, J., Peng, G., & MacLachlan, D. (2009). Effect of risk management strategy on NPD performance. *Technovation*, *29*(3), 170–180. doi:10.1016/j.technovation.2008.07.006

Muller, E., & Doloreux, D. (2009). What we should know about knowledge-intensive business services. *Technology in Society*, *31*(1), 64–72. doi:10.1016/j.techsoc.2008.10.001

Myers, P. (1996). *Knowledge Management and Organizational Design*. Boston: Butterworth-Heinemann.

Nonaka, I., & Takeuchi, H. (1995). *The Knowledge Creating Company: How Japanese Companies Create the Dynamics of Innovation*. New York: Oxford University Press.

Noruzy, A., Dalfard, V., Azhdari, B., Nazari-Shirkouhi, S., & Rezazadeh, A. (2013). Relations between transformational leadership, organizational learning, knowledge management, organizational innovation, and organizational performance: An empirical investigation of manufacturing firms. *International Journal of Advanced Manufacturing Technology*, *64*(5-8), 1073–1085. doi:10.100700170-012-4038-y

Oyelaran-Oyeyinka, B., & Sampath, P. G. (2009). *Latecomer Development, States, Knowledge and Economic Growth*. London: Routledge.

Quinn, J. B. (1992). *Intelligent Enterprise: A Knowledge and Service Based Paradigm for Industry*. New York: The Free Press.

Quintas, P., Lefrere, P., & Jones, G. (1997). Knowledge management: A strategic agenda. *Long Range Planning*, *30*(3), 385–391. doi:10.1016/S0024-6301(97)90252-1

Quist, J., & Tukker, A. (2013). Knowledge collaboration and learning for sustainable innovation and consumption: Introduction. *Journal of Cleaner Production*, *48*, 167–175. doi:10.1016/j.jclepro.2013.03.051

Recker, J. (2013). *Scientific research in information systems*. Heidelberg, Germany: Springer Berlin. doi:10.1007/978-3-642-30048-6

Reich, B., Gemino, A., & Sauer, C. (2014). How knowledge management impacts performance in projects: An empirical study. *International Journal of Project Management*, *32*(4), 590–602. doi:10.1016/j.ijproman.2013.09.004

Samson, D. (2010). *Innovation for Business Success: achieving a systematic innovation capability*. Report for the Victoria Government Department of Innovation, Industry, Science and Research. Available at http://www.innovation.gov.au/Industry/IndustryInnovationCouncils

Sanz-Valle, R., Naranjo-Valencia, J., Jiménez-Jiménez, D., & Perez-Caballero, L. (2011). Linking organizational learning with technical innovation and organizational culture. *Journal of Knowledge Management*, *15*(6), 997–1015. doi:10.1108/13673271111179334

Sawhney, M., Verona, G., & Prandell, E. (2005). Collaborating to create: The Internet as a Platform for Customer Engagement in Product Innovation. *Journal of Interactive Marketing*, *19*(4), 1–15. doi:10.1002/dir.20046

Simpson, B. (2002). The Knowledge Needs of Innovating Organisations. *Singapore Management Review*, *24*(30), 51–61.

Spithoven, A., Clarysse, B., & Knockaert, M. (2011). Building absorptive capacity to organize inbound open innovation in traditional industries. *Technovation*, *31*(1), 10–21. doi:10.1016/j.technovation.2010.10.003

Standards Australia. (2005). *Australian Standard in Knowledge Management: A Guide AS 5037-2005*. Sydney: SAI Global.

Stewart, T. (1997). *Intellectual Capital: The New Wealth of Organizations*. London: Nicholas Brealey.

Storey, J., & Barnett, E. (2000). Knowledge management initiatives: Learning from failure. *Journal of Knowledge Management*, *4*(2), 145–156. doi:10.1108/13673270010372279

Taherparvar, N., Esmaeilpour, R., & Dostar, M. (2014). Customer knowledge management, innovation capability and business performance: A case study of the banking industry. *Journal of Knowledge Management*, *18*(3), 430–446. doi:10.1108/JKM-11-2013-0446

Teece, D. (1998). Capturing Value from Knowledge Assets: The new economy, markets for know-how and intangible assets. *California Management Review*, *40*(3), 55–79. doi:10.2307/41165943

Teece, D. (2009). *Dynamic Capabilities and Strategic Management: Organizing for Innovation and Growth*. Oxford, UK: Oxford University Press.

Teece, D. (2010). Business Models, Business Strategy and Innovation. *Long Range Planning*, *43*(2-3), 172–194. doi:10.1016/j.lrp.2009.07.003

Tidd, J., & Bessant, J. (2014). *Strategic Innovation Management*. Chichester, UK: John Wiley.

Tobin, T., & Pennington, W. (2015). Organizational networks and the process of corporate entrepreneurship: How the motivation, opportunity, and ability to act affect firm knowledge, learning and innovation. *Small Business Economics*.

Van der Aa, W., & Elfring, T. (2002). Realizing innovation in services. *Scandinavian Journal of Management*, *18*(2), 155–171. doi:10.1016/S0956-5221(00)00040-3

Van Riel, A., Lemmink, J., & Ouwersloot, H. (2004). High-technology service innovation success: A decision-making perspective. *Journal of Product Innovation Management*, *21*(5), 348–359. doi:10.1111/j.0737-6782.2004.00087.x

Wallace, D. P. (2007). Introduction. In D. Wallace (Ed.), *Knowledge management: historical and cross-disciplinary themes*. London: Libraries Unlimited. doi:10.4135/9781446278581.n1

Weaven, S., Grace, D., Dant, R., & Brown, J. (2014). Value creation through knowledge management in franchising: A multi-level conceptual framework. *Journal of Services Marketing*, *28*(2), 97–104. doi:10.1108/JSM-09-2013-0251

Wu, I., & Lin, H. (2009). A strategy-based process for implementing knowledge management: An integrative view and empirical study. *Journal of the American Society for Information Science and Technology*, *60*(4), 789–802. doi:10.1002/asi.20999

Yang, J. (2010). The knowledge management strategy and its effect on firm performance: A contingency analysis. *International Journal of Production Economics*, *125*(2), 215–223. doi:10.1016/j.ijpe.2010.03.012

Ye, C., Jha, S., & Desouza, K. C. (2015). Communicating the Business Value of Innovation. *International Journal of Innovation Science*, *7*(1), 1–12. doi:10.1260/1757-2223.7.1.1

Yin, R. K. (2014). *Case Study Research: Design and Methods* (5th ed.). Thousand Oaks, CA: Sage.

Zheng, W., Yang, B., & McLean, G. (2010). Linking organizational culture, structure, strategy, and organizational effectiveness: Mediating role of knowledge management. *Journal of Business Research*, *63*(7), 763–771. doi:10.1016/j.jbusres.2009.06.005

Chapter 3
Knowledge Management and Entrepreneurship Research and Practice:
Status, Challenges, and Opportunities

Cesar Bandera
New Jersey Institute of Technology, USA

Katia Passerini
St. John's University, USA

Michael Bartolacci
Penn State University – Berks, USA

ABSTRACT

A contentious merger seems to characterize the relationship between entrepreneurship, defined as the process of bringing new products and services to the market, and knowledge management. Even when discussed from the perspective of dynamic knowledge creation and innovation rather than from the codification and organization of a firm's knowledge base, the knowledge management process of new ventures often fails to keep pace with best practices. This is particularly puzzling as the competitive advantage of new ventures stems from exploiting tacit knowledge, which is at the heart of successful knowledge creation and management. This chapter investigates this contradiction and discusses challenges and opportunities to both the practitioner and the researcher.

INTRODUCTION

One could argue that that the creation of knowledge in an organization is its most crucial activity. If true, an obvious problem arises in new ventures whose flat managerial structures do not have the resources, whether they be human, financial, or technological, to devote to the capture and arrangement of knowledge in the organization. If the creation and utilization of knowledge in an organization is so

DOI: 10.4018/978-1-5225-5427-1.ch003

vital, how can a new venture survive until it has the resources and structure to effectively manage and use knowledge? Further complicating knowledge management (KM) in the new venture is the fact that some of the knowledge considered important early in the company's existence may not be relevant after pivots or changes to the business model.

In this chapter, we review the literature on KM and entrepreneurship. Most KM literature focuses on large firms – the domain in which KM was originally developed (Wiig, 1997; Prusak 2001) – and most KM literature on entrepreneurship focuses on entrepreneurial activities in large companies. An introduction to the dynamic theory of knowledge creation and its relevance and application in the field of entrepreneurship shows that knowledge socialization is the most mature knowledge stage in new ventures. We then focus on opportunities for growth and the challenges of implementing KM in new ventures that have yet to cross the "valley of death" stage of maturation (Markham, Ward et al. 2010). Finally, we propose reflections and research questions that may engage future researchers in proposing strategies to better integrate KM into the fabric of entrepreneurship. An earlier version of this work appears in Bandera, Bartolacci and Passerini (2016).

KNOWLEDGE MANAGEMENT AND NEW VENTURES IN THE LITERATURE

The notion of knowledge management in entrepreneurial companies is starting to garner an academic research focus. The number of papers published per year on knowledge management and the small and medium enterprise (SME) has been increasing linearly from almost none prior to 2000; an analysis of the publications cited in Esposito et al. (2016) indicates that the number of KM papers published in a given year on the SME is roughly that of the previous year plus one ($R^2 = 0.90$). In response to this growing interest on the topic, five literature reviews have been published in just the last four years (Durst & Runar Edvardsson, 2012; Edvardsson & Durst 2013; Garbarino-Alberti & Pastorino 2014; Esposito et al. 2016; Soares et al. 2016). These reviews agree that KM research has traditionally focused on the large company domain in which KM developed as a discipline, and only recently proceeded to address the domain of the SME.

In spite of the above activity, few rigorous works exist in the literature relating KM to entrepreneurship. This could be due in part to the fact that new ventures are characterized by severely limited resources and low probability of survival, and consequently their ability and opportunity to implement KM practices before suffering the consequences of "knowledge mismanagement" are limited. In fact, one might say that the flat managerial structure and agile business processes characteristic of entrepreneurship are in direct opposition to the somewhat time consuming and structured processes traditionally associated with KM.

One of the few research works that examines KM and entrepreneurship was by Bembenek and Piecuch (2014), who looked at the impact of KM on the formation and growth of industry clusters. They defined various types of clusters including Marshalian, hub and spoke, satellite platform, and state-anchored, and related the necessity of KM for their development and success. If one considers an industry cluster as an entrepreneurial entity in which firms either achieve synergistic effects in a short period of time or risk disbanding (Gilbert, McDougall et al. 2008; Davies, 2009; Al-Mubaraki & Busler, 2013), then this work sheds light on the importance of KM to entrepreneurship: organizational learning, which defines KM, can be accomplished within a cluster with the active transfer of information about such areas as contractors, clients, legal requirements, and potential collaboration among a cluster's membership.

Knowledge exchange within clusters is more likely in clusters with a high level of social capital; clusters offering members some type of financial support, such as state subsidized clusters, experience a greater exchange of information and cooperation. One might argue that this result goes against the notion of entrepreneurship where firms would have more autonomy within a cluster. If the cluster is viewed as a single entity, we believe that the notion of entrepreneurship may then be applied. In their examination of knowledge spillover and its relationship to entrepreneurial organizations, Ferreira, Ratten et al. (2017) find that such information gleaned from other organizations can be used to a strategic advantage. This can also work against an entrepreneurial organization that cannot properly manage and control its knowledge as it focuses on other business objectives while interacting within a cluster.

Another paper that examines the link between entrepreneurship and knowledge management is the work by Bahrami, Salehi et al. (2014), which identifies factors linking KM and corporate entrepreneurship in Iran. They state that "what underpins successful entrepreneurship is finding creative ways with the use of new technologies for better, faster, and more cost-effective marketing of new products" (pg. 7). As has been put forth previously, KM improves cost-effectiveness by helping an organization learn from its successes as well as its mistakes, which can yield significant dividends in customer discovery. The researchers also found that absorptive capacity, causal ambiguity, and an arduous relationship between a source of knowledge and its recipient are factors that hindered knowledge transfer in an organization. In a new venture with a flat organizational structure driven by entrepreneurs, one could argue that such factors will lead to its quick demise. In other words, if personalities and related human factors can be combined with a fuzzy picture of results from operations, an entrepreneurial organization might implode on itself due to the lack of learning being shared with employees and other stakeholders.

The impact of KM in new biotechnology and telecommunications ventures in Spain was examined by Palacios, Gil and Garrigos (2009). The authors, in a similar vein to Bahrami, Salehi, and Korkmaz (2014), describe obstacles to effective KM introduction in an organization. The first of these is the absence of a culture that promotes the sharing of information. A second obstacle is a lack of KM training and promotion of its benefits, possibly due to the focus on short-term survival. Using an empirically validated model, Palacios, Gil and Garrigos find a positive relationship between KM and innovation-intensive competencies. KM can thus develop skills that help the new venture survive if it has the foresight to implement KM early. For example, KM can help retain the innovation that would otherwise be lost to employee attrition common during difficult times (Schiuma, Durst et al. 2012).

An earlier case study focusing on the link between entrepreneurship and knowledge management was conducted by Guadamillas, Donate et al. (2008). They studied a successful firm in Spain that achieved growth with the use of KM and concluded that knowledge is a key strategic resource when operating in a highly competitive market. Their work also implies that knowledge evolves rapidly in such a market, and an organization will ultimately suffer without an effective structure with which to manage it.

Chaston (2012) examined the KM and entrepreneurial behavior in small UK service-sector firms. One factor relating KM and entrepreneurship brought to light in this work is that personal entrepreneurship tends to be discouraged as a firm grows. The natural outcome of this process is that idea generation and pro-activeness tend to likewise be reduced. Thus, as a firm grows, the need for KM to have already captured innovative ideas grows ever more important. Chaston concludes that in small service-sector firms in the UK, the combination of an entrepreneurial orientation and knowledge management had a positive impact on business performance, particularly during economic downturns.

THE KNOWLEDGE CREATION CYCLE AND NEW VENTURES

To review KM in the context of new ventures, we follow a foundational theory of knowledge management, the dynamic knowledge creation model, which serves as the basis for the discussion. Nonaka, Toyama and Konno's (2000) unified model for knowledge creation integrates the three core elements that are relevant to the success of any company: assets, context/environment and processes. Each of these elements plays a unique role in entrepreneurship and roughly maps the process for knowledge creation, also known as the SECI model, to entrepreneurial activities.

Leveraging Polanyi's (1967) distinction between tacit and explicit knowledge, Nonaka, Toyama and Konno define the process of knowledge creation as a progressive and recursive journey from tacit to explicit and back. With each knowledge creation cycle, the organization builds greater and more effectively organized knowledge. While the analysis is generally presented at the unit level of large organizations, such process applies to organizations of any size. Figure 1 depicts the main elements and processes of the cycle and lists the related-knowledge assets that distinguish each of the four phases in the cycle.

Knowledge creation starts with a process of socialization of individual know-how, skills, and experiences that are generally transferred through apprenticeships, mentorship and other direct encounters and face-to-face exchanges among individuals and groups. The use of direct, rather than computer-mediated, experiences is driven by the need to establish trust, understand gestures and passions, and build consensus and familiarity within the individuals exchanging knowledge. In an entrepreneurial context, establishing

Figure 1. Dynamic knowledge creation process (SECI)
Adapted from Nonaka, Toyama et al. (2000).

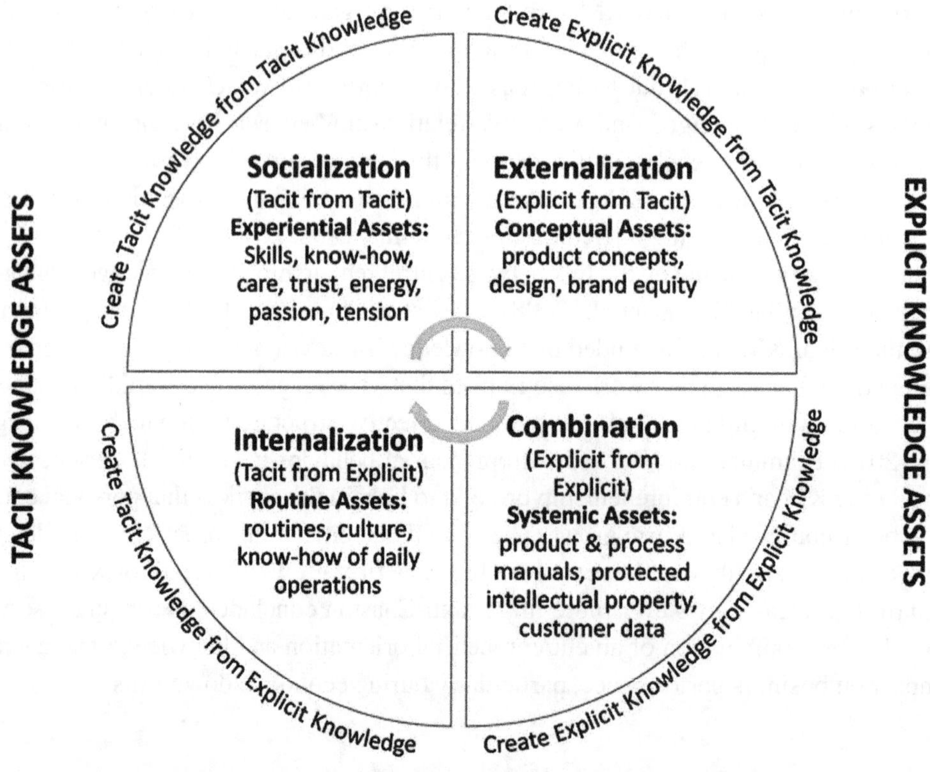

this trust is important to a small business' initial survival. Participating in networking and pitching events to find partners, promoting the company, and presenting at investors' meetings are ways entrepreneurs use to lay the sustainable foundation of the enterprise. To attract high-talented employees, small businesses use outreach events to showcase their value proposition and growth trajectory. More importantly, small businesses often join incubators, maker spaces, and innovation clusters where they can directly connect with other entrepreneurs and leverage shared resources. The emergence of entrepreneurial ecosystems such as Silicon Valley in San Jose and Silicon Alley in New York City demonstrates that the process of socialization is already significant and well-understood by small businesses (Schoonhoven & Eisenhardt, 2012).

The same understanding cannot be assumed for the other phases of the knowledge process. For example, the externalization of knowledge, which calls for documenting and codifying the tacit knowledge residing in the heads of the entrepreneur and founding team, already shows some shortcomings as the fast-paced life of the entrepreneur leaves little time for documentation efforts. While this issue could be resolved by shadowing programs that pair the head of the organization with apprentices tasked with documenting processes and procedures, this approach is rarely adopted. Concerns about maintaining secrecy motivate some new ventures to "fly under the radar" and intentionally avoid the externalization of knowledge while they prepare applications for intellectual property protection or for being first-to-market (Ries, 2011).

The combination stage of knowledge creation, which requires organization, optimization, cataloguing and sharing of already codified knowledge assets, is also low priority in small businesses as this process presupposes an investment in content management technology and, much more costly the dedication of someone's time to conduct this procedure. New ventures are generally focused on processes directly tied to their product or service; rules and procedures, when they exist, are more flexible and adaptable. This creates advantages (e.g., enables rapid pivots) and disadvantages (e.g., a bottleneck with the knowledge owner at its center) characteristic of small enterprises. Combination and codification may occur when obligated to do so, such as when competing for funding from government agencies, foundations, or equity investors, or when applying for patents.

Finally, the internalization phase calls for the dissemination of company practices and procedures through application and direct experiences. Entrepreneurs can foster the acquisition of company knowledge via "learning-by-doing," experiential activities, and "reflection-in-action." As with the other stages of the cycle, the resource-constrained and fast-paced environment in which the new venture operates may leave little room for such experimentation and reflection.

While unique knowledge and competencies inspire the formation of a new venture, the looming exhaustion of resources compels it to focus on implementation rather than engaging in the planning and reflection important to sustainable growth. This shortcoming may also be one of the contributing reasons for the high mortality rate of new ventures.

The above suggests that, while knowledge management is particularly arduous for new ventures, it may represent important leverage for their success. This contradiction – the need for knowledge management and its weak adoption among practitioners – is even more striking when noting that small businesses often rate highly in the key success factors of knowledge creation: autonomy, creative chaos, redundancy, love, care, trust and commitment (Nonaka, Toyama et al. 2000). Autonomy of decision-making may be higher in a small organization than in a large hierarchical business. Creative chaos and tension to deliver constitute the essence of entrepreneurial activities and are subject to a sense of urgency. Redundancy and duplication occur due to the lack of functional specialization when dealing with small entities as every employee needs to perform multiple roles. Finally, love, care, trust, and commitment are also key attitudinal motivators to launch a new business. Individuals quit lower risk and higher paying jobs to pursue their passions and devote themselves to the venture. This generally leads to higher satisfaction with the job despite longer work hours (Rowden, 2002) and greater but not necessarily perceived risk (Das & Teng, 1997).

Why, then, is knowledge management underutilized, overlooked, and is far from the lexicon of entrepreneurs? Why is it not embedded in entrepreneurship education and training programs? In the next sections, we provide evidence of the limited applications of KM in entrepreneurship by reviewing extant literature, propose solutions, and suggest future research opportunities.

Knowledge Management, Entrepreneurship, and the Valley of Death

In this review of issues connecting KM and entrepreneurship, we focus on the domain of the venture that precedes the infamous valley of death, whose crossing few ventures survive (Figure 2). This valley represents a line of demarcation in many disciplines. In finance, it is the transition from R&D funding to sales revenue (Murphy & Edwards, 2003; Wessner, 2008; Saxton, Saxton et al. 2010). In business organization and in the mindset of the founder, it is the transition from innovation to efficiency (Flamholtz & Randle, 2012; Frese & Gielnik, 2014). In business strategy, it is the transition from verifying and refining a plausible business model (e.g., pivoting) to executing one that has been empirically verified (Boisot & MacMillan, 2004; Ries, 2011). In KM, it is the transition from tacit knowledge to explicit knowledge (Gaimon & Bailey, 2013).

The aforementioned transitions are also not limited to nascent firms; any large firm may experience one or some of them at some point in time. However, the newly formed venture, i.e., the startup, will likely have to endure all of them and at the same time. Our unit of study is thus the KM requirements of the startup, and its entrepreneurial founder(s), from inception up to and including the valley of death.

Entrepreneurship research uses several taxonomies with which to classify entrepreneurs. Schumpeter (1947) describes the innovation-driven entrepreneur who creates an opportunity and disrupts markets through "creative destruction" (Mazzucato, 2011), whereas Kirzner (1997) describes the entrepreneur who discovers an existing opportunity. Either type of entrepreneur can chose a strategy of continuous growth or of sustaining a particular lifestyle. Schumpeterian growth ventures, such as technology startups, tend to bear greater risk because their business models are typically less certain than those of Kirznerian ventures, and require more investment per capitalization than lifestyle ventures (Leach & Melicher, 2014). Nevertheless, whether Schumpeterian, Kirznerian, growth or lifestyle, all entrepreneurs face the valley of death make risky investments based on tacit knowledge. The information asymmetry that gives rise to the entrepreneur's competitive advantage stems from a scarcity of explicit relevant knowledge (Smith, Matthews et al. 2009). However, this advantageous condition – which is an asset in the formation of a new venture – is also a source of uncertainty and risk, and becomes a liability when trying to cross the valley of death.

KM practices can help the entrepreneur make the transition from tacit knowledge to explicit knowledge, and help the venture make the crossing. Entrepreneurs and startup administrators have ample literature with insights from prominent entrepreneurs or results from empirical studies to help cross the valley of death. Because this crossing involves so many information-centric disciplines, the literature is diverse and usually includes aspects of knowledge creation. For example, both the lean startup methodology (Ries, 2011) and the SECI model present iterative processes of actions (underlined italics in Figure 3)

Figure 2. Lifecycle of a venture depicting the valley of death stage

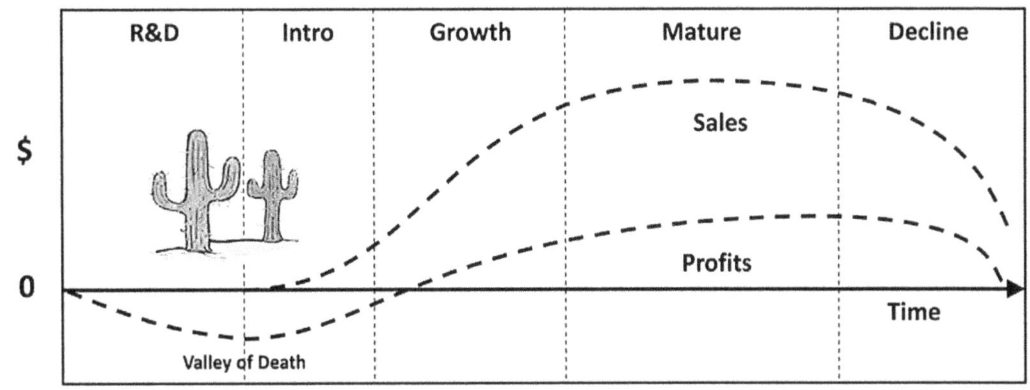

and work products (bold) for the creation and retention of tacit and explicit knowledge. The lean startup methodology deploys iterations of a minimum viable product (MVP) to collect information with which to verify the new venture's business model; as with the SECI model, it grows tacit knowledge (an idea) into explicit actionable analytics (data) that feeds back into greater tacit knowledge.

The lean startup methodology and the SECI model were conceived in different environments that represent opposite ends of the valley of death: the former in a software startup and the latter in large industry. Unfortunately, the literature typically presents them in their corresponding traditional contexts, leaving to the entrepreneur the non-trivial task of making the connection between them. For example, the SECI model is applicable to all stages of the company but is more abstract, whereas the lean startup methodology describes practical procedures for business model validation but not for subsequent stages of firm maturity. Entrepreneurship educators and trainers, many of whom already subscribe to the lean startup methodology, should explicitly identify to their audience the KM in the processes they teach, and that the value of KM goes beyond the startup stage of a venture.

Figure 3. Lean startup methodology and SECI model process cycles

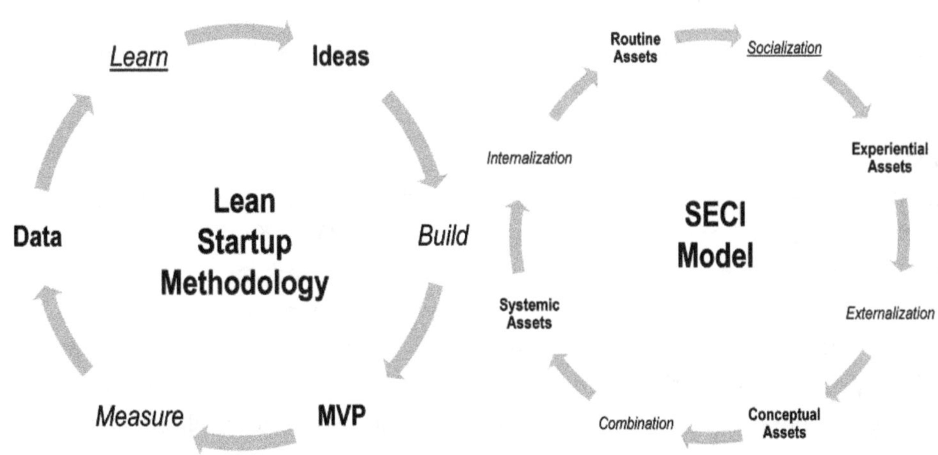

THE NEED FOR KM IN ENTREPRENEURSHIP AND ITS TRANSLATIONAL CHALLENGES

KM in entrepreneurship may be facing translational challenges, whereby valuable implementation lags behind rapidly advancing research. KM is not a common topic in entrepreneurship education or training, and few entrepreneurs recognize the term or its potential benefits (uit Beijerse 2000, Yew Wong & Aspinwall, 2005). Boisot and MacMillan (2004) argue that KM "has been underselling itself as well as the scope that it has for broadening its agenda. Knowledge management can and should restore epistemic legitimacy to an entrepreneurial mindset… This is partly a matter of cognitive style but partly also a matter of education."

Fortunately, current topics in entrepreneurship education, training, and research fit nicely in the path of knowledge evolution modelled by the SECI model and staged models of entrepreneurship (Gaimon & Bailey, 2013; Bygrave & Zacharakis, 2014). Moreover, entrepreneurs value knowledge, particularly the tacit knowledge they derived from their social capital that enables their initial competitive advantage (Cope, Jack et al. 2007; Kreiser, Patel et al. 2013), and deploy tools of knowledge management such as repositories and intranets, albeit without KM procedures in place (Yew Wong & Aspinwall, 2005).

So why does the startup entrepreneur need KM? This question can also be answered in the context of the valley of death. We previously discussed various impediments in startups to the four stages of knowledge creation (socialization, externalization, combination, and internalization). While focusing on the challenges of launching a new venture, entrepreneurs often underappreciate how different these challenges are from those of growing the venture (Isenberg, 2012; Douglas, 2013), and are ill-prepared to cross the valley; Isenberg reminds us "anyone who has been a parent knows that the long and complicated job of growing a healthy, educated and moral child is vastly more challenging than giving birth."

To scale the competitive advantage that inspires venture formation, KM theory dictates that tacit knowledge be codified and shared within the growing firm's ecosystem, thereby converting tacit knowledge into a growing set of explicit knowledge (Nonaka, Toyama et al. 2000; Schiuma, Durst et al. 2012; Gaimon & Bailey, 2013). The ecosystem includes internal assets such as employees that can access and refine online standard operating procedures, and external assets such as the reviews of customers - particularly early adopters. Even the social capital that contributed to the tacit knowledge has to adapt to growth (Pirolo & Presutti, 2010). The importance of explicit knowledge thus increases in the growth stage of the venture.

As cited above, researchers from different disciplines including KM describe the valley of death in their terms and recommend procedures from their disciplines to help the entrepreneur survive its crossing. Unfortunately, had entrepreneurs heeded these disciplines, the valley would not be so notorious. Consequently, we rephrase the last question as: can KM be any better than other disciplines in helping the entrepreneur?

KM is more abstract than most business disciplines; instead of dealing with tangible assets like money or inventory, it deals with intangible assets like tacit knowledge. One could speculate this would dampen enthusiasm for its implementation among entrepreneurs. In their work on the differences in managerial and entrepreneurial KM, Boisot and MacMillan (2004, p. 521) concede that "Questions of epistemology may strike practicing managers as somewhat remote from their practical day-to-day concerns" and they "hope that these principles will bring out the practical value to managers of taking a theoretical subject like epistemology seriously." However, Boisot and MacMillan also quote psychologist Kurt Lewin (1951,

p. 169) "There is nothing so practical as a good theory," who also prophetically wrote long before the rise of KM "Experience alone does not create knowledge."

Lewin implies that the abstractness and large business origins of KM models may facilitate their application to entrepreneurship. We address abstractness and origin separately. Their abstractness means that KM models encompass more startups than models of other disciplines. Financial models, for example, are very specific to the type of startup, with innovation-driven, lifestyle, growth-based, and harvest-focused startups requiring different financial models, whereas the modern telecommunications industry, regardless of specific market segment (e.g., mobile, quantum, or data security to name a few) owes its existence to Claude Shannon's 1948 abstract model of information (Verdu 1998, Stonier 2012). Any of the cited KM models may be relevant to all entrepreneurs, thereby avoiding confusion about which to use. Entrepreneurs might even agree on the same critical success factors (Yew Wong and Aspinwall 2005). The challenge to the KM researcher is that of translating KM models into the lexicon of the entrepreneur.

The origin of KM may also be fortuitous in that addressing the startup domain with a model initially created in the large business domain can be easier than the opposite: applying to the large business domain a model created in the startup domain. Entrepreneurs may agree that the distant side of the valley of death, where KM originated, represents their ultimate goal. That agreement is exploitable when deploying KM to entrepreneurship. The opposite would be more difficult: what appeal would a company that has crossed the valley of death find in a discipline designed for companies who have yet to cross it?

Coincidentally, not only does KM research benefit the startup domain, it can benefit *from* the startup domain as well. In their critical analysis of KM, Fahey and Prusak (1998) list eleven errors endemic to the field. Some of these errors stem from conditions found in large companies but not in new ventures. For example, the errors "viewing knowledge as existing predominantly outside the heads of individuals" and "paying little heed to the role and importance of tacit knowledge" are rarely made by entrepreneurs, who instead value and are driven by tacit knowledge. Moreover, in their efforts to grow their tacit knowledge and competitive advantage through face-to-face networking, entrepreneurs rarely make the mistake of "substituting technological contact for human interface."

The entrepreneur's goal to commercially exploit her/his tacit knowledge precludes her/him from committing two other errors on the list: "emphasizing knowledge stock to the detriment of knowledge flow" and "disentangling knowledge from its uses." Similarly, by focusing on their future goal, entrepreneurs tend to avoid the error of "focusing on the past and the present and not the future." The lean startup methodology focuses on the prevention of the error "failing to recognize the importance of experimentation" which, in the context of entrepreneurship, refers to empirical validation of the business model. However, not all entrepreneurs have the discipline to empirically validate their business model, and these run the risk of realizing too late that there is no market for their offering (i.e., no resources remain with which to conduct pivots).

We see that in the startup domain, six of the errors endemic to KM in large company settings are intrinsically mitigated by entrepreneurship; a seventh is mitigated as well, but only among disciplined entrepreneurs. What constitutes an intrinsic behavior among entrepreneurs may in fact be an aspirational behavior among large business managers.

OPPORTUNITIES FOR RESEARCH INTEGRATION

The above discussions lead us to consider four opportunities for KM researchers interested in the startup domain of the entrepreneur. We discuss these below, and for each opportunity suggest related research questions.

Opportunity #1: Refine the Unit of Study

First and foremost, the growing activity of research into the role of KM in the SME domain must define the unit of study more precisely so that it distinguishes startups from the remaining SMEs. In spite of the aforementioned activity in the field, KM literature rarely makes the distinction between SMEs and entrepreneurs. Entrepreneurship literature also covers diverse domains including the nascent venture and the established firm, but commonly considers innovation-based startups as the domain of the entrepreneur and contrasts this domain with that of the lifestyle company and SME (Carland, Hoy et al. 1984; Aulet & Murray, 2013).

The need to refine the unit of study in entrepreneurship research is pervasive. There is even little agreement between governments on the legal distinction between a micro, small, and medium company (Potter, Lunati et al. 2010). A 2008 survey of US firms that received Small Business Innovation Research funding reveals that over 40% of these recipients had five or fewer employees – far less than the US Small Business Administration's definition of an SME, which is 500 or fewer employees (Wessner, 2008). Surely, while important to both, KM in a five-person company is different from that in a 500 person company. If we consider a multidimensional feature-space with traditional corporate measures as dimensions (e.g., revenue × number of employees × valuation × age), it is not hard to visualize that companies will form separable clusters that are market-specific and present KM challenges that differ between clusters but are consistent within.

The valley of death, a major concern of entrepreneurs (if not the primary concern), is a line of demarcation defining used by schemas in many business disciplines. This in turn may facilitate cross-disciplinary research. Because they are inextricably linked, the unit of study should include both the startup and the entrepreneur.

Few entrepreneurs recognize KM as a discipline, and thus do not appreciate its value once they pass the valley, nor their current vulnerability as measured by KM. To address this, suggested topics of research include mapping entrepreneurs' critical success factors to KM model elements, techniques whereby the entrepreneur can determine in which KM model element s/he resides, and KM policies for resource-constrained domains.

Key research questions related to the unit of study may include:

- Do KM best practices differ between startup categories?
- How can KM help the entrepreneur decide if s/he should persevere or pivot?
- How much time do KM practices require before providing accurate and actionable information to the entrepreneur?

Opportunity #2: Express KM Models Using the Lexicon the Entrepreneur

Entrepreneurship research has grown rapidly and in doing so has developed an lexicon that is widely accepted by both researchers and practitioners (Ács & Audretsch, 2006; Busenitz, Plummer et al. 2014). To facilitate KM's contribution to this domain, KM researchers will have to adopt this domain's lexicon and not impose that of the large company. This may also increase the appeal of KM research to entrepreneurship practitioners, who can contribute to KM research activity as they do in the field of entrepreneurship research (Steffens, Weeks et al. 2014).

Key research questions related to vocabulary may include:

- How do KM models integrate multiple stakeholders in the startup?
- Are KM models for growth, lifestyle, and social entrepreneurship the same or do they differ?

Opportunity #3: Maintain the Breadth of KM

While focusing on KM for the entrepreneur, the researcher must not lose sight of the fact that KM models apply to small ventures and large companies. This may seem to contradict the challenge to refine the unit of study, but in fact it builds on the strength of KM. As with any powerful theory, the relevance of KM to diverse domains encourages us advance this line of research. Isenberg (2012) points out that "In focusing entrepreneurship policy almost exclusively on startups we are favoring quantity of startup at the expense of quality of scale-up." KM is inherently focused on quality.

Key research questions related to the breadth of the domain may include:

- How does the knowledge creation spiral of the startup evolve from opportunity recognition to harvest event?
- Does the knowledge creation spiral of the serial entrepreneur differ from that of the novice?

Opportunity #4: Extend the Depth of KM

The opportunity of cross-disciplinary research, enabled by a common unity of study and lexicon, permits using research methods and results from other disciplines to deepen the field of KM. For example, the relationship between startup success, ecosystems such as startup accelerators, business incubators, innovation clusters, and technology parks, and the policies that govern such ecosystems is an active field of entrepreneurship research (Hackett & Dilts, 2004; Lambooy, 2005; Al-Mubaraki & Busler, 2013); the study of knowledge transfer, generation, and codification in these environments would likewise deepen the understanding of KM's role in this relationship.

Social capital is often a new venture's only asset, and can be more indicative of startup success than intellectual property (Hayton, 2005). Another multidisciplinary research topic is thus the role of KM in the formation and evaluation of the relationships that form social capital. For example, whereas entrepreneurship research makes the distinction between strong and weak ties in models of social capital, KM has yet to make this distinction or recommend strategies for relationship management (Kreiser, Patel et al. 2013; Stam, Arzlanian et al. 2014).

The cultures of startups and the disposition of their founders tend to be more strongly correlated than the cultures of large companies and the disposition of their CEOs (Casson, 2010). Consequently, a founder's attitudes will influence the use and perceived need of KMS. The entrepreneurial mindset is one of the most jointly researched topics in the fields of entrepreneurship and education, yielding increasingly sophisticated models of behavior and intent (Chell, 2008). Startup performance and opportunity recognition have been found to be heavily influenced by the entrepreneur's cognitive biases (Li, Huang et al. 2009; Dang & McKelvey, 2016) and networking (Andersson, Formica et al. 2009; Bembenek & Piecuch, 2014; Jiao, Cui et al. 2014). However, these behavioral models have not yet incorporated KM model elements, making this topic another candidate for undoubtedly fruitful and impactful multidisciplinary research.

Key research questions related to the depth of the domain may include:

- How can KM model the tacit knowledge and social capital of the entrepreneur?
- In what ways do entrepreneurial traits influence the adoption of KM in their startups?
- Can cluster, incubator, and accelerator administrators use KM to improve economic impact?
- In what ways can a startup use KM in opportunity recognition? Are they effective?

CONCLUSION

When arriving at the lecture on growth, entrepreneurship professors have been known to tell their students that "entrepreneurs obsess so much over survival, that they are often ill-prepared for success." Another relevant anecdote may be Stephen Covey's seventh habit of effective people - "sharpen the saw" (Coveym 1989). By transcending the survival and growth stages of a new venture, KM is well suited to contribute to the field of entrepreneurship research, and in the process enjoy an active research agenda over the foreseeable future. Moreover, the unique attributes of the startup domain might direct KM researchers and practitioners away from some reportedly common errors. In this paper, we posed the question: can KM be any better than other disciplines in helping the entrepreneur? We conclude that the abstractness and large business origins of KM potentiate its contribution to the field of entrepreneurship research and suggest integration opportunities and areas for further exploration.

REFERENCES

Abereijo, I. O. (2015). Transversing the 'Valley of Death': Understanding the determinants to commercialization of research outputs in Nigeria. *African Journal of Economic and Management Studies*, *6*(1), 90–106. doi:10.1108/AJEMS-10-2012-0066

Ács, Z. J., & Audretsch, D. B. (2006). *Handbook of entrepreneurship research: An interdisciplinary survey and introduction*. Springer Science & Business Media.

Al-Mubaraki, H. M., & Busler, M. (2013). Business Incubation as an Economic Development Strategy: A Literature Review. *International Journal of Management, 30*(1 Part 2), 362–372.

Andersson, T., Formica, P., & Curley, M. G. (2009). *Knowledge-driven entrepreneurship: the key to social and economic transformation*. Springer Science & Business Media.

Aulet & Murray. (2013). *A Tale of Two Entrepreneurs: Understanding differences in the Types of Entrepreneurship in the Economy*. Available at SSRN 2259740.

Bahrami, M., Salehi, M., & Korkmaz, S. (2014). A Study of the Driving Factors of Knowledge Management and Corporate Entrepreneurship in Iran. *IUP Journal of Knowledge Management, 12*(1), 13.

Bandera, C., Bartolacci, M. R., & Passerini, K. (2016). Knowledge Management and Entrepreneurship: A Contradictory Recipe. *International Journal of Knowledge Management, 12*(3), 1–14. doi:10.4018/IJKM.2016070101

Beijerse, R. P. (2000). Knowledge management in small and medium-sized companies: Knowledge management for entrepreneurs. *Journal of Knowledge Management, 4*(2), 162–179.

Bembenek, B., & Piecuch, T. (2014). Knowledge Management in Industry Clusters As an Indication of Entrepreneurship. *CBU International Conference Proceedings*. 10.12955/cbup.v2.441

Boisot, M., & MacMillan, I. C. (2004). Crossing epistemological boundaries: Managerial and entrepreneurial approaches to knowledge management. *Long Range Planning, 37*(6), 505–524. doi:10.1016/j.lrp.2004.09.002

Busenitz, L. W., Plummer, L. A., Klotz, A. C., Shahzad, A., & Rhoads, K. (2014). Entrepreneurship research (1985–2009) and the emergence of opportunities. *Entrepreneurship Theory and Practice, 38*(5), 981–1000. doi:10.1111/etap.12120

Bygrave, W., & Zacharakis, A. (2014). *Entrepreneurship*. Hoboken, NJ: John Wiley & Sons, Inc.

Carland, J. W., Hoy, F., Boulton, W. R., & Carland, J. A. C. (1984). Differentiating entrepreneurs from small business owners: A conceptualization. *Academy of Management Review, 9*(2), 354–359.

Casson, M. (2010). Entrepreneurship, business culture and the theory of the firm. In Handbook of entrepreneurship research. Springer.

Cerchione, R., Esposito, E., & Spadaro, R. M. (2016). A literature review on knowledge management in SMEs. *Knowledge Management Research and Practice, 14*(2), 169–177. doi:10.1057/kmrp.2015.12

Chaston, I. (2012). Entrepreneurship and knowledge management in small service-sector firms. *Service Industries Journal, 32*(6), 845–860. doi:10.1080/02642069.2011.559224

Chell, E. (2008). *The entrepreneurial personality: A social construction*. Routledge. doi:10.4324/9780203938638

Cope, J., Jack, S., & Rose, M. B. (2007). Social Capital and Entrepreneurship An Introduction. *International Small Business Journal, 25*(3), 213–219. doi:10.1177/0266242607076523

Costa, E., Soares, A. L., & de Sousa, J. P. (2016). Information, knowledge and collaboration management in the internationalisation of SMEs: A systematic literature review. *International Journal of Information Management, 36*(4), 557–569. doi:10.1016/j.ijinfomgt.2016.03.007

Covey, S. R. (1989). *The 7 habits of highly effective people: Powerful lessons in personal change.* New York: Fireside.

Dang & McKelvey. (2016). Knowledge management processes and the formation of entrepreneurial opportunities. *Journal of Innovation Economics & Management*, (1), 31-59.

Das, T. K., & Teng, B.-S. (1997). Time and entrepreneurial risk behavior. *Entrepreneurship Theory and Practice, 22*(2), 69–88. doi:10.1177/104225879802200206

Davies, M. (2009). *Mixed-use Incubator Handbook: A Start-up Guide for Incubator Developers.* Washington, DC: The World Bank Group.

Douglas, E. J. (2013). Reconstructing entrepreneurial intentions to identify predisposition for growth. *Journal of Business Venturing, 28*(5), 633–651. doi:10.1016/j.jbusvent.2012.07.005

Durst, S., & Runar Edvardsson, I. (2012). Knowledge management in SMEs: A literature review. *Journal of Knowledge Management, 16*(6), 879–903. doi:10.1108/13673271211276173

Edvardsson, I. R., & Durst, S. (2013). Does knowledge management deliver the goods in SMEs? *Business and Management Research, 2*(2), 52. doi:10.5430/bmr.v2n2p52

Fahey, L., & Prusak, L. (1998). The eleven deadliest sins of knowledge management. *California Management Review, 40*(3), 265–276. doi:10.2307/41165954

Ferreira, J. J., Ratten, V., & Dana, L.-P. (2017). Knowledge spillover-based strategic entrepreneurship. *The International Entrepreneurship and Management Journal, 13*(1), 161–167. doi:10.100711365-016-0415-6

Flamholtz, E. G., & Randle, Y. (2012). *Growing pains: Transitioning from an entrepreneurship to a professionally managed firm.* John Wiley & Sons.

Frese, M., & Gielnik, M. M. (2014). The psychology of entrepreneurship. *Annual Review of Organizational Psychology and Organizational Behavior, 1*(1), 413–438. doi:10.1146/annurev-orgpsych-031413-091326

Gaimon, C., & Bailey, J. (2013). Knowledge Management for the Entrepreneurial Venture. *Production and Operations Management, 22*(6), 1429–1438. doi:10.1111/j.1937-5956.2012.01337.x

Garbarino-Alberti, H., & Pastorino, S. A. (2014). Knowledge management steps, tools, techniques and influencing factors in SMEs: Systematic literature review. *Computing Conference (CLEI), 2014 XL Latin American.* 10.1109/CLEI.2014.6965108

Gilbert, B. A., McDougall, P. P., & Audretsch, D. B. (2008). Clusters, knowledge spillovers and new venture performance: An empirical examination. *Journal of Business Venturing*, *23*(4), 405–422. doi:10.1016/j.jbusvent.2007.04.003

Guadamillas, F., Donate, M. J., & Pablo, J. D. (2008). Knowledge management for corporate entrepreneurship and growth: A case study. *Knowledge and Process Management*, *15*(1), 32–44. doi:10.1002/kpm.297

Hackett, S., & Dilts, D. (2004). A Systematic Review of Business Incubation Research. *The Journal of Technology Transfer*, *29*(1), 55–82. doi:10.1023/B:JOTT.0000011181.11952.0f

Hayton, J. C. (2005). Competing in the new economy: The effect of intellectual capital on corporate entrepreneurship in high-technology new ventures. *R & D Management*, *35*(2), 137–155. doi:10.1111/j.1467-9310.2005.00379.x

Isenberg, D. (2012). Focus Entrepreneurship Policy on Scale-Up, Not Start-Up. *Harvard Business Review*, 21.

Jiao, H., Cui, Y., Zhu, Y., & Chen, J. (2014). Building entrepreneurs' innovativeness through knowledge management: The mediating effect of entrepreneurial alertness. *Technology Analysis and Strategic Management*, *26*(5), 501–516. doi:10.1080/09537325.2013.872774

Kirzner, I. M. (1997). Entrepreneurial discovery and the competitive market process: An Austrian approach. *Journal of Economic Literature*, *35*(1), 60–85.

Kreiser, P. M., Patel, P. C., & Fiet, J. O. (2013). The Influence of Changes in Social Capital on Firm-Founding Activities. *Entrepreneurship Theory and Practice*, *37*(3), 539–568. doi:10.1111/etap.12039

Lambooy, J. (2005). Innovation and knowledge: Theory and regional policy. *European Planning Studies*, *13*(8), 1137–1152. doi:10.1080/09654310500336444

Leach, J. C., & Melicher, R. W. (2014). *Entrepreneurial finance*. Cengage Learning.

Lewin, K. (1951). *Problems of research in social psychology. In Field theory in social science: Selected theoretical papers* (pp. 155–169). New York: Harper & Row.

Li, Y.-H., Huang, J.-W., & Tsai, M.-T. (2009). Entrepreneurial orientation and firm performance: The role of knowledge creation process. *Industrial Marketing Management*, *38*(4), 440–449. doi:10.1016/j.indmarman.2008.02.004

Markham, S. K., Ward, S. J., Aiman-Smith, L., & Kingon, A. I. (2010). The Valley of Death as Context for Role Theory in Product Innovation. *Journal of Product Innovation Management*, *27*(3), 402–417. doi:10.1111/j.1540-5885.2010.00724.x

Mazzucato, M. (2011). *The entrepreneurial state*. London: Demos.

Murphy, L. M., & Edwards, P. L. (2003). *Bridging the valley of death: Transitioning from public to private sector financing*. Golden, CO: National Renewable Energy Laboratory.

Nonaka, I., & Takeuchi, H. (1995). *The knowledge-creating company: How Japanese companies create the dynamics of innovation*. Oxford University Press.

Nonaka, I., Toyama, R., & Konno, N. (2000). SECI, Ba and leadership: A unified model of dynamic knowledge creation. *Long Range Planning*, *33*(1), 5–34. doi:10.1016/S0024-6301(99)00115-6

Palacios, D., Gil, I., & Garrigos, F. (2009). The impact of knowledge management on innovation and entrepreneurship in the biotechnology and telecommunications industries. *Small Business Economics*, *32*(3), 291–301. doi:10.100711187-008-9146-6

Pirolo, L., & Presutti, M. (2010). The Impact of Social Capital on the Start-ups' Performance Growth. *Journal of Small Business Management*, *48*(2), 197–227. doi:10.1111/j.1540-627X.2010.00292.x

Polanyi, M. (1967). The Tacit Dimension. London: Routledge & K. Paul.

Potter, J., Lunati, M., Marchese, M., Martinez-Fernandez, C., & Noya, A. (2010). *SMEs, Entrepreneurship and Innovation*. OECD Publishing.

Prusak, L. (2001). Where did knowledge management come from? *IBM Systems Journal*, *40*(4), 1002–1007. doi:10.1147j.404.01002

Ries, E. (2011). *The lean startup: How today's entrepreneurs use continuous innovation to create radically successful businesses*. Random House LLC.

Rowden, R. W. (2002). The relationship between workplace learning and job satisfaction in US small to midsize businesses. *Human Resource Development Quarterly*, *13*(4), 407–425. doi:10.1002/hrdq.1041

Saxton, M. K., Saxton, T., Steen, J., & Verreynne, M. (2010). Failure to advance: Resource logic for early venture failure. *Prometheus*, *28*(1), 15–27. doi:10.1080/08109021003668620

Schiuma, G., Durst, S., & Wilhelm, S. (2012). Knowledge management and succession planning in SMEs. *Journal of Knowledge Management*, *16*(4), 637–649. doi:10.1108/13673271211246194

Schoonhoven, C. B., & Eisenhardt, K. M. (2012). Regions as Industrial Incubators of Technology-based Ventures. In Sources of Metropolitan Growth. Transaction Publishers.

Schumpeter, J. A. (1947). The creative response in economic history. *The Journal of Economic History*, *7*(2), 149–159. doi:10.1017/S0022050700054279

Smith, B. R., Matthews, C. H., & Schenkel, M. T. (2009). Differences in Entrepreneurial Opportunities: The Role of Tacitness and Codification in Opportunity Identification*. *Journal of Small Business Management*, *47*(1), 38–57. doi:10.1111/j.1540-627X.2008.00261.x

Stam, W., Arzlanian, S., & Elfring, T. (2014). Social capital of entrepreneurs and small firm performance: A meta-analysis of contextual and methodological moderators. *Journal of Business Venturing*, *29*(1), 152–173. doi:10.1016/j.jbusvent.2013.01.002

Steffens, P. R., Weeks, C. S., Davidsson, P., & Isaak, L. (2014). Shouting from the ivory tower: A marketing approach to improve communication of academic research to entrepreneurs. *Entrepreneurship Theory and Practice*, *38*(2), 399–426. doi:10.1111/etap.12079

Stonier, T. (2012). *Beyond information: The natural history of intelligence*. Springer Science & Business Media.

Verdu, S. (1998). Fifty years of Shannon theory. *IEEE Transactions on Information Theory*, *44*(6), 2057–2078. doi:10.1109/18.720531

Wessner, C. W. (2008). *An Assessment of the SBIR Program*. National Academies Press (US) National Research Council Committee on Capitalizing on Science, Technology, and Innovation.

Wiig, K. M. (1997). Knowledge management: Where did it come from and where will it go? *Expert Systems with Applications*, *13*(1), 1–14. doi:10.1016/S0957-4174(97)00018-3

Yew Wong, K., & Aspinwall, E. (2005). An empirical study of the important factors for knowledge-management adoption in the SME sector. *Journal of Knowledge Management*, *9*(3), 64–82. doi:10.1108/13673270510602773

Chapter 4
A Contingency Perspective for Knowledge Management Solutions in Different Decision-Making Contexts

Kursad Ozlen
CENTRUM Católica, Peru

Meliha Handzic
International Burch University, Bosnia and Herzegovina

ABSTRACT

A contingency perspective of knowledge management, as one of the popular ways of promoting decision making capabilities, recognizes the need for a fit between knowledge management solutions (KMS) and decision-making contexts which they support. In order to determine the best fit, a field survey was carried out to investigate the impact of two different types of KMS (technical and social) on decision makers' behavior and performance in different decision contexts (simple and complex). According to the results, there is a partial support for the contingency view. As expected, social KMS appears as the best fit for complex contexts, based on subjects' superior performance from comparable adoption of both KMS. In contrast, the results suggest that both KMS were an equally good fit for simple contexts, based on similar levels of subjects' performance, but social KMS was preferred in terms of adoption. These findings contribute to much necessary empirical evidence for research and provide useful guidance for practice. However, their limitations necessitate further study.

INTRODUCTION

Fostering decision making capabilities are always one of the main targets of today's organizations. Scholars and practitioners suggest and improve Knowledge management solutions (KMS) as the best useful tools in order to reach this target. Contingency perspective of knowledge management is one of the emerged popular ways (Hansen et al., 1999; Snowden, 2002; Becerra et al., 2004) of identifying the

DOI: 10.4018/978-1-5225-5427-1.ch004

alignment of KMS with decision contexts and personal preferences of decision makers. This view is also advantageous to clarify the seamless integration between knowledge management and the business environment (El Sawy, 2003).

The contingency viewpoint presents new challenges for researchers and practitioners regarding its theoretical and practical consequences for decision making support. This work aims to clarify these challenges by addressing the fit between organizational knowledge management solutions (KMS) and their decision-making contexts. In order to reach the best solution-context-user fit, an improved understanding is necessary for different KMS and decision contexts and their impact on pre-decisional behavior and performance of decision makers.

KMS are an emerging class of decision aids, especially in target managerial work, by focusing on enabling and facilitating creation, sharing, retention and discovery of required knowledge for decision making. KMS are helpful in reducing/eliminating decision biases (Arnott, 2002) to improve users' decision-making capabilities. There are different recommended social and technical solutions in terms of this support (Handzic, 2004; 2007).

Technical or systems-orientated solutions generally describe KMS in terms of various ICT- based systems designed to facilitate the management of knowledge by enhancing the accumulation and transfer of knowledge through the organization (Sambamurthy & Subramani, 2005). In contrast, social or human-orientated approaches to KMS emphasize on organizational leadership, culture, structure and measurement as key enablers of processes im enhancing knowledge development (Holsapple, 2003). The literature strongly suggests that the potential return from KMS implementations can be enormous if they are properly designed and implemented (Alavi & Leidner, 2001). Therefore, understanding the decision-making context is helpful in determining the best KM solution(s) before the selection of proper KMS.

The literature distinguishes between two main positions on "proper" KMS design and implementation. While the universalistic view is offering one single best approach which should be adopted in all circumstances, the contingency view suggests that no one approach is best under all circumstances. Collectively, the proponents of the contingency view (Hansen et al., 1999; Snowden, 2002; Becerra et al., 2004) suggest a series of knowledge, task, organization and environment characteristics as the potential influencers of alternative KMS implementations. However, there needs to be more empirical evidence to support this proposition.

Therefore, the purpose of this study is to remove this weakness and contribute to the improved understanding of the issue of KMS fit to context. In particular, the study aims to empirically examine the impact of different KMS approaches on decision-making behavior and performance in different decision contexts. The improved understanding of the issue will serve as a foundation for better KMS design and implementation.

LITERATURE REVIEW

This study focuses on two key factors: type of KMS approach and complexity of decision context and their role in decision support. The following sections provide a literature review regarding different KMS approaches, decision contexts and the use of KM in various decision contexts to set up the research model and the questions.

KMS Approaches

Knowledge is contextual information. The implementation of knowledge management targets enhancing decision making beside many other advantages (Khasawneh & Abu-Shanab, 2013). On the other hand, the knowledge-based perspective of the firm suggests knowledge as an important asset to reach the strategic goals of the firm in a more effective way in order to gain sustainability, competitive advantage and other long-term benefits (Raj, 2012; Sajeva & Jucevicius, 2010). A survey result by Reuters suggests that 90% of the companies implementing knowledge management solutions reach better decision making (Raj, 2012).

Knowledge management solutions are set of socio-technical enablers and facilitators of knowledge processes that modify and move knowledge stocks and thus foster knowledge development. Literature proposes a wide variety of social and technical mechanisms and initiatives that have the potential to help knowledge flow and grow, and consequently lead to enhanced or innovative performance.

With respect to socially orientated solutions, Handzic (2007) recommends four major classes: culture, leadership, structure and measurement. Organizational culture is amongst the most important factors in enabling a knowledge-conducive environment. Therefore, many social initiatives try to nurture a knowledge culture by promoting espoused values, systems, structures, and artefacts in order to achieve a desired mindset in staff members (Handzic & Agahari, 2003). Also, rewards and incentive systems are helpful in motivating knowledge sharing and knowledge contribution (Evangelou & Karacapilidis, 2005). According to Hauschild et al. (2001), successful companies encourage employees for seeking, sharing, and creating knowledge instead of pushing knowledge down from the top by employing different incentive systems.

Furthermore, literature suggests that successful knowledge management implementation requires strong leadership to guide an organization towards managing its knowledge resources for leveraging performance (Holsapple, 2003). Strong interpersonal, communication and change management skills, an understanding of the business, technological expertise, and the ability to build relationships are important leadership characteristics. Leaders act as catalysts by inspiring, mentoring, setting examples, listening, and engendering trust and respect. They also influence organizational culture, structure and measurement (Handzic, 2011).

With respect to structure, a variety of organizational forms are useful to create a collaborative and knowledge sharing environment. In general, networked structures (or communities of practices-CoPs), are believed to offer the ability for individuals to work together and encourage open communication and learning based on their common interests (Wenger, 1998). On the other hand, bureaucratic structures that emphasize hierarchies and command and control over individuals are believed to discourage creativity and innovation (Lesser & Storck, 2001).

There is a widespread consensus in literature that organizations have difficulty in managing knowledge without addressing the measurement issue, and vice versa. The intellectual capital (IC) perspective embraces these two together by linking knowledge with strategic business objectives and learning processes (Zhou & Fink, 2003). In terms of competitiveness, organizations need to measure their current knowledge and target knowledge by knowing what they know and what they must know. Measurement initiatives provide metrics and feedback to management. The outcome of such a measurement exercise is expected to be a more effective knowledge management approach.

Technical KMS occupy a variety of information and communication technologies and software applications. There are four main classes of ICT-based KMS depending on their knowledge-enabling function (Handzic, 2004). These classes are knowledge retention, sharing, discovery and generation systems. Knowledge retention systems enable capturing and storage, as well as subsequent access to recorded knowledge. Knowledge sharing systems facilitate transfer of knowledge through interaction and collaboration among people. Knowledge discovery systems allow the identification of hidden patterns in data, their interpretation and prediction. Finally, knowledge generation systems support creativity and research that enable new knowledge development. These four classes of systems jointly support processes of exploitation and exploration of explicit and tacit knowledge.

Wide range of specific technologies and commercial KM software available on the market are helpful while implementing the above four classes of KMS (Alavi & Leidner, 2001; Tsui, 2003). Thus, knowledge retention systems may involve various storage and retrieval technologies and business intelligence systems including databases, data warehouses, data marts etc. Knowledge sharing systems may use a variety of communication and collaboration technologies including emails, forums, audio and video conferencing, social networks etc. Knowledge discovery systems may rely on different data and text mining technologies, business analytics tools, visualization etc. Finally, knowledge generation systems may incorporate a selection of simulation games, online tutorials, virtual experiments etc.

One review regarding KM frameworks (Heisig, 2009) reports these technical initiatives as critical factors of KM success in more than 50% of the frameworks. A number of other investigations report KMS system and/or service quality as the determinants of KMS adoption and success (Liu et al., 2005; Xu & Quaddus, 2005). Empirical studies link KMS sophistication to improved performance (Salleh et al., 2010) and reveal that the effectiveness of decision making was influenced by the availability of required technologies (Mohsen et al., 2011).

In summary, the reviewed knowledge management concepts and models, as well as prior empirical research suggest that the positive impact of both social and technical initiatives on decision makers' pre-decisional knowledge processes and thus on improved individual decision making. However, from the contingency perspective, the effect of different approaches may have different outcomes in different decision-making situations. According to the review, the benefits expected from adopting social KMS are greater if the situation is complex rather than simple. On the other hand, the expected benefits as a result of adopting technical KMS may not be as targeted. This study has a target to examine the issue empirically.

DECISION CONTEXTS

The term "context" describes "the set of circumstances that form the setting for an event" (Dictionary.com, 2017). The literature recommends a variety of informational, operational, organizational, environmental, historical, attentional, behavioral and causal aspects that comprise the decision makers' context (Salleh et al., 2010). Scholars have also suggested that the context model should include only contextual elements in terms of satisfying the decision maker's requirements. For the purpose of modeling a decision context, some investigators (Wood, 1986; Campbell, 1988) have taken into account three elements: decision task, decision environment and decision maker. The characteristics of these three elements together determine the level of decision context complexity.

With respect to decision task, literature evaluates complexity in terms of the objective properties of the task and the subjective reaction of the individual. Among objective complexity properties, Wood (1986) has introduced component, coordinative and dynamic dimensions of the task. These refer to a number of cues or acts that form and strength their relationships and change over time. On the other hand, Campbell (1988) has proposed task complexity as a primary psychological experience that may be evoked for reasons other than task, such as anxiety and fear.

Investigators have also recommended a number of environmental factors regarding their contribution to context complexity. Some of these include time and money constraints, significance, irreversibility and accountability. In general, constraints are recognised as stressors, while others may be related to important status or financial consequences for the decision maker or client. In addition, literature identifies personal knowledge, ability and motivation as those individual characteristics that make the decision situation more or less complex. With knowledge and ability comes an opinion about appropriateness of a strategy and likelihood of application.

From the perspective of complexity theory, the difficulty of the decision situation is expected to increase with the objective and/or perceived complexity due to decision task, environment and/or decision maker. This, in turn, is expected to affect individual decision behavior and subsequent decision performance. The behavioral decision theory proposes a positive relationship between the level of decision context complexity and the level of analytical complexity of decision making strategy.

The definition of the context is essential to identify the critical success factors, drivers and constraints, relevant to the implementation of Knowledge Management (Jafari et al, 2008; Choo & de Alvarenga Neto, 2010). Regarding the requirements and challenges to knowledge management for decision support, Smirnov and Sandkuhl (2015) suggest context-orientation as a key feature of modern approaches. Moreover, Snowden (2004) uses complexity theory as a basis to differentiate between known, complicated, complex and chaotic knowledge domains that determine KM strategies. In addition to knowledge characteristics (explicit/tacit, declarative/procedural), Becerra et al. (2004) suggest a number of other contingency factors that contribute to context complexity and influence knowledge processes. These include task uncertainty and interdependence, organization size, business strategy and environmental uncertainty.

In summary, the above review suggests that context complexity may be an important factor in determining the suitability of KMS support for decision making. However, different studies make different claims regarding the role of context in KMS impacts. The contingency view suggests that resolving complex situations requires greater reliance on people, while simple situations can be handled easier with technology. The current study proposes to examine the issue empirically.

The Use of KMS in Decision-Making Contexts

Decision making is the selection of the best feasible solution among the alternatives. Knowledge management solutions (KMS) are necessary to cultivate the selection process by simplifying the problem through the use of various technical and social elements. Technical KMS is better for simple contexts by means of fostering the automation of the processes. For complex contexts, it is better using more social KMS.

Scholars suggest KMS among the fundamental enablers to effective decision making (McCall et al., 2008). In support of this view, a recent survey conducted by Reuters (Raj, 2012) reports that 90% of the companies that utilised KMS exhibits better decision-making capabilities, and 81% has increased productivity after such utilisation. Therefore, using KMS capabilities are necessary to provide timely and relevant knowledge to support organizational decision-making activities and improve overall organisational performance.

Prior studies indicate that introducing KMS into decision making resulted in the proper usage of the right knowledge to make the right decisions. This, in turn, improved overall organizational performance in terms of quality, productivity and satisfaction. Specifically, the integration of KMS enabled finding, sharing and developing needed knowledge through active networks, reflective practice, helpful relationships, robust infrastructure, accessible technology, supportive leadership, effective communication, and trusting climate (Khasawneh & Abu-Shanab, 2013).

KMS in relation to the decision-making process begins with the acquisition of necessary knowledge (e.g. know-what, know-how, know-why and care-why) for the initial situation diagnosis, it is followed by searching and setting the solution alternatives, choosing the most suitable alternative and introducing it into action. For decision making based on knowledge, a company needs to take into account organizational issues related to the company: management, i.e. employees, processes, equipment; regular identification and update of critical knowledge of a company; continuous conduction of critical knowledge areas analyses; and monitoring and development of knowledge assets and their content (Litvaj & Stancekova, 2015).

Past empirical research provides that integrating knowledge management systems with decision support systems during the decision-making process contribute to the improved quality of decisions in both private businesses and public sector organizations. Thus, a recent case study reports that the knowledge management system was successfully implemented in a prosperous multinational company - AIRBUS (Litvaj & Stancekova, 2015). The company's KMS portfolio covers where, what, who and how types of knowledge provided by business search across databases, lessons learnt capture and reuse, yellow pages and professional networks. With respect to the public sector, a descriptive study of KMS success in Turkish healthcare organizations also reveals positive effects of KMS on decision performance in terms of better, more efficient, effective and innovative decisions (Handzic & Ozlen, 2012). Finally, a much wider investigation into the effectiveness of KMS in managerial decision making across sectors further confirms their significant benefits to decision quality, satisfaction and confidence (Ozlen et al., 2014).

The majority of prior research investigating KMS in decision making addresses the issue from the generic perspective, without taking into account potential differences among KMS and/or decision contexts. Yet, KMS differ widely in their level of sophistication, and decision-making contexts differ in their level of complexity. The contingency approach suggests the consideration of these differences in order to find the best fit between a knowledge management solution and the decision-making activity which it supports. The aim of this study is to accomplish this via empirical examination.

Research Model and Objectives

Drawing insights from knowledge management and decision-making literature, the following research model is proposed and presented in Figure 1. The model depicts four interrelated variables: (i) KMS approach, (ii) decision context, (iii) knowledge processes and (iv) decision performance.

KMS approach is defined in terms of an organization's dominant orientation towards technical or social mechanisms and practices in solutions design. Decision context is conceptualized as a set of task, environment and personal attributes that characterize the overall complexity of the decision situation faced by decision makers. Knowledge processes reflect the degree of decision makers' KMS adoption for decision support. Decision performance is regarded as the ultimate business outcome of knowledge management.

Figure 1. Proposed research model

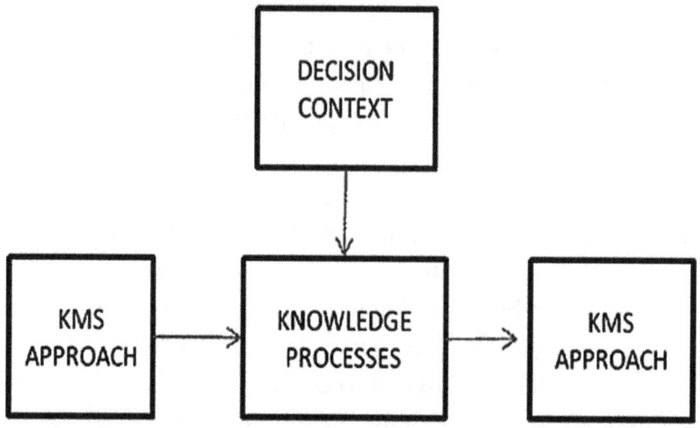

The model further proposes several plausible relationships between model variables. More specifically, it posits that KMS approach and decision context mutually influence knowledge processes and through knowledge processes impact decision performance. On the other hand, from the perspective of contingency theory, KMS approach has a differential impact on knowledge processes and thus decision performance dependent upon the decision context. The current study tests these assumptions empirically.

RESEARCH METHODOLOGY

A field study was conducted among decision makers employed in 22 Turkey-based organizations that were purposely selected by the researchers due to their knowledge-intensive character and implementation of KMS. Out of 372 responses received, 73 were excluded from further consideration due to missing data in the answers. Therefore, the analyses were performed with the remaining 299 useful responses. The study focuses on individual levels of analysis, taking as sample units decision makers (respondents) who needed to assess whether or not to adopt a KMS to support their decision making.

A survey form, designed to capture the participants' perceptions, consists of their decision context complexity, their opinions about the organization's KMS, as well as to self-report the extent of pre-decisional knowledge processes and perceived decision performance. All relevant items, used to measure the variables included in the research model, are provided in the list available in the Appendix. Measurement validity was established by factor analysis and reliability by Cronbach's alpha values over 0.7 threshold (Nunnally, 1978). In replying to the questionnaire, the respondents rated their agreement with each given statement relative to negative and positive end-points of a 7-point Likert scale.

Two independent variables, KMS approach and decision context, were evaluated in the following way. KMS approach was classified as dominant technical KMS or social KMS, based on a ratio of the mean score on items of organizational leadership, structure, culture & measurement and the mean score on items for organizational business intelligence, business analytics, communication & collaboration, creativity & e-learning initiatives. Values below 0.5 denoted dominant technical KMS and values above 0.5 denoted prevalent social KMS. There were no undecided cases. Decision context was measured by an average rating score for decision task, decision environment and decision maker items. Values below

3 (out of 7) denoted simple context, while values between 3 and 5 denoted complex contexts (although only moderately). There were no scores higher than 5, therefore no cases of extreme complexity.

Two dependent variables were knowledge processes and decision performance. Knowledge processes were evaluated in terms of the extent of acquiring and sharing existing knowledge and generating and discovering new knowledge. The variable was measured by an average rating score of these four process items. Decision performance was evaluated in terms of perceived decision quality, confidence and satisfaction operationalized by respective items' rating scores. Decision performance was considered the ultimate indicator of KMS value for decision support.

All the collected and calculated responses were encoded, entered into a computer and combined into one file. The participants were grouped by KMS approach (technical or social) and decision context (simple or complex) based on their scores. Then, their behavioral and performance responses were analyzed using a SPSS statistical package. The results are reported in the next section.

RESULTS

Descriptive results (means & standard deviations) for knowledge processes and decision performance by KMS and context are presented in Table 1. Results of further analysis of the collected data by ANOVA are reported in Table 2. The analysis found some significant results.

Table 1. Descriptive statistics for dependent variables by KMS and context

Dependent Variable	KMS Approach	Decision Context	Mean	Std. Dev.	N
Knowledge processes	Technical KMS	Simple context	4.5705	1.44180	39
		Complex context	4.9966	1.12692	73
		Total	4.8482	1.25578	112
	Social KMS	Simple context	5.1288	1.31430	66
		Complex context	4.9332	1.27711	121
		Total	5.0022	1.29024	187
	Total	Simple context	4.9214	1.38298	105
		Complex context	4.9571	1.22017	194
		Total	4.9445	1.27751	299
Decision performance	Technical KMS	Simple context	5.6072	1.21118	39
		Complex context	5.2611	1.25997	73
		Total	5.3816	1.24875	112
	Social KMS	Simple context	5.9595	.95098	66
		Complex context	5.5816	1.00334	121
		Total	5.7150	.99916	187
	Total	Simple context	5.8287	1.06325	105
		Complex context	5.4610	1.11462	194
		Total	5.5901	1.10909	299

Table 2. Summary results of ANOVA tests of between-subjects effects

Source	Dependent Variable	df	Mean Square	F	Sig.
KMS Approach (KA)	Knowledge processes	1	3.903	2.407	.122
	Decision performance	1	7.214	6.087	.014**
Decision Context (DC)	Knowledge processes	1	.847	.522	.471
	Decision performance	1	8.354	7.049	.008***
KA * DC	Knowledge processes	1	6.158	3.797	.052*
	Decision performance	1	.016	.014	.907
Error	Knowledge processes	295	1.622		
	Decision performance	295	1.185		
Total	Knowledge processes	299			
	Decision performance	299			

* p<0.10, ** p<0.05, *** p<0.01

Knowledge Processes

The study tested whether there would be any difference in decision makers' knowledge processes due to different KMS approach, regardless of their decision contexts. The results in Table 2 indicate that KMS had no overall impact on knowledge processes. The main effect of KMS on process scores was not significant (F=2.407, p=0.122). However, there was a significant interaction effect between KMS and context (F=3.797, p=0.052). This interaction effect is presented graphically in Figure 2.

The follow-up analyses by context showed that KMS significantly influenced knowledge processes in simple, but not in complex decision contexts. In simple contexts, the subjects self-reported less ex-

Figure 2. KMS effects on knowledge processes by contexts

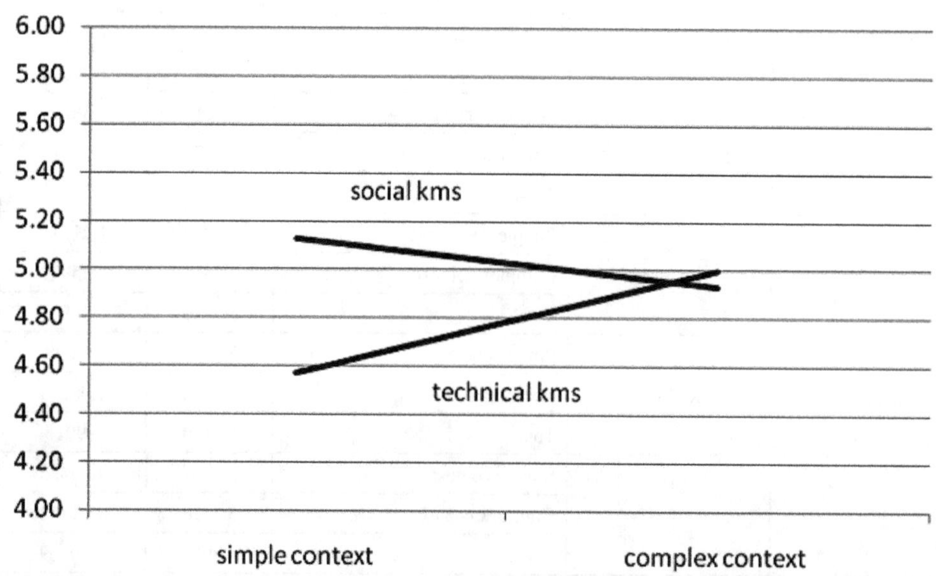

tensive knowledge processing when enabled by technical than by social KMS (4.57 vs 5.13, t= -1.980, p=0.051). In complex contexts, they self-reported similarly extensive knowledge processes irrespective of KMS (5.00 vs. 4.93, t=0.361, p=0.719).

These results are contrary to the expected greater reliance on technical KMS in simple and on social KMS in complex decision contexts. However, the fact that all mean scores were greater than 4 (out of 7) indicates that both approaches to KMS were fairly useful in supporting subjects' pre-decisional knowledge processes.

Decision Performance

The study also examined whether there would be any difference in subjects' decision performance due to different KMS approach irrespective of their decision contexts. The analysis found that KMS had a significant overall impact on decision performance. Table 2 shows a highly significant main effect of KMS on performance scores (F=6.087, p=0.014). However, further analysis by contexts revealed that the effect was significant only in complex contexts, but not in simple contexts.

In simple contexts, the subjects self-reported similar levels of decision performance irrespective of KMS approach (5.61 vs. 5.96, t= -1.555 p=0.125). In complex contexts, they reported significantly lower levels of performance with technical KMS than with social KMS (5.26 vs. 5.58, t= -1.848, p=0.067).

Mean scores for two KMS approaches by contexts are presented graphically in Figure 3. The figure shows that all scores were above 5 (out of 7). This meant that subjects performed quite well irrespective of KMS and context. Less than perfect scores indicate room for further improvement.

Figure 3. KMS effects on decision performance by contexts

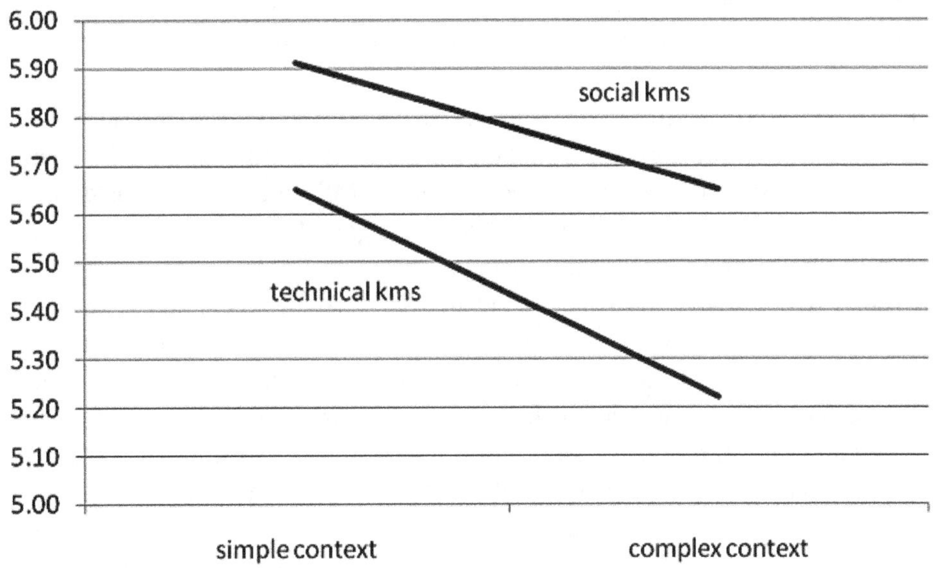

DISCUSSION

Main Findings

The main findings of this study provide partial support for the contingency view of KM. The contingency perspective proposes that the impact of KMS approach (social or technical) on decision making is dependent upon decision context (simple or complex). In summary, the proposition is supported when decision making was evaluated in terms of pre-decisional knowledge processes in simple contexts (but not complex), and decision performance in complex contexts (but not simple).

More specifically, the study has demonstrated that social KMS inspired greater learning effort than technical KMS in simple contexts. This was evidenced in higher mean scores of knowledge processes found among social than technical KMS subjects. However, when faced with complex situations, subjects tended to rely heavily on whatever dominant KMS approach was implemented in their organisation for supporting pre-decisional knowledge processing. This is visible in comparable mean scores for complex context subjects' processes irrespective of KMS.

The nature of subjects' behaviour revealed in the study is contrary to expectations. In general, literature advocates the use of technology-based codification strategies as more suitable than people-orientated personalisation strategies in ordered task domains where patterns are known or knowable and can be easily reused or analysed with technology (Hansen et al. 1999, Snowden 2002). However, this study found evidence of the opposite behaviour.

The fact that subjects tended to rely significantly more on humans than systems to support personal learning activities in simple contexts may be potentially explained by collectivistic rather than individualistic Turkish culture (Hofstede, 2013). In such cultures, people prefer to work in groups and initially invest time and effort to establish peer relationships. The other potential explanation may be geographical proximity of the subjects. Each participating organisation had one single office space located in Turkey. Therefore, people working in such close proximity could have found it easier to talk to colleagues than to use the systems for assistance.

With respect to decision performance, the study revealed that the type of adopted KMS approach made no difference to performance in simple contexts. This was demonstrated by comparable mean performance scores between subjects supported by social and technical KMS. In contrast, decision performance in complex contexts was highly dependent upon the type of adopted KMS approach, with social KMS being more effective than technical ones in enhancing performance. This was evidenced by significantly higher mean scores of decision performance found among social than technical KMS subjects.

The explanation for the discovery made for complex contexts can be found in the contingency perspective (Khalifa et al., 2008). Such a finding is in full accordance with the suggested increasing importance of the human factor and diminishing value of technology in unordered decision domains with more uncertain or novel problems. The fact that the adoption of social KMS did not translate to significant improvement in performance in simple contexts means that savings can be made by implementing technical KMS as a less costly solution for reuse of known or analysis of knowable domains and without any loss of performance (Hansen et al., 1999; Snowden, 2002). However, the final choice of the most suitable KMS approach for simple contexts may require a cost/benefit analysis of economic factors and decision makers' preferences.

IMPLICATIONS, LIMITATIONS, AND FUTURE RESEARCH

Knowledge of the factors that influence the adoption and effectiveness of KMS in decision making have important implications for research. Because KMS support and extend decision- making capability, a thorough understanding of the underlying processes is required to provide constructive guidance for the design and implementation of such initiatives. To this end, the current study contributes empirical evidence of partial support for the contingency view of KM, by identifying the moderating effect of context in the adoption and effectiveness of KMS in simple and complex situations, respectively.

The findings also have some important implications for practical implementations of KMS in organizations. Performance-wise, this study revealed a clear advantage of social over technical KMS initiatives for supporting decision making in difficult, novel situations with significant consequences. This should be sufficient justification for managers to implement strategies that emphasize knowledge leadership, network structures, learning cultures and continued measurement and feedback.

When it comes to decision making in simple, routine and familiar situations, managers have a choice between two equally effective strategies: social as described above, or technical, involving various information and communication technologies that facilitate knowledge accumulation and transfer. When individual pre-decisional behaviors are taken into account, social initiatives were clearly preferred over technical ones, as they inspired more learning activities. This implies that managers should consider giving higher priority to social than technical initiatives in order to satisfy decision makers' preferences. Nevertheless, cost- and geography-related considerations may change management priorities in favor of technical KMS.

The above findings and recommendations need to be interpreted and considered with care due to a number of study limitations. Firstly, the study examined only cases with a dominant social or technical approach to KMS. The results could be different in the case of integrated socio-technical KMS. Next, the lack of true experimental design, unequal groups, subjective measures and parametric statistical tests may limit the validity of the study findings. In addition, exhibited preferences and behaviors of the Turkey-based participants may not generalize to other geographical and cultural settings.

Therefore, further studies need to be carried out to address the above limitations and establish validity and generalizability of the current findings. Furthermore, the instrument to assess context is not well documented in literature. In future, criteria other than complexity applied in this study may be used to classify different contexts. Finally, the current study focused on the individual behavior and performance. Future research needs to extend the present investigation to collective issues at group, organization, industry, region or global level.

CONCLUSION

The findings of this study make a significant contribution to KM (and DSS) research by providing partial empirical support for the contingent impact of KMS on individual decision making. More specifically, the study establishes that social KMS are more effective than technical KMS for aiding complex decision-making situations. The opposite is true for simple decision-making situations. Although less favored by subjects, technical KMS appear more appropriate for aiding simpler decision making, given the same performance impact as social KMS (but at a lower cost and global reach).

These findings imply that organizations that implement technical rather than social KMS will generate less employee engagement in learning activities, but may expect equal performance in simpler, known or knowable decision circumstances. In other, more complex, unknown or chaotic decision circumstances, organizations do require socially-orientated approaches to achieve superior performance.

Given the limited nature of current findings there is a need for further investigation with an aim of enriching the collective understanding of KM phenomena. Several possible directions are suggested for future research that would replicate and extend current investigation to other contexts, subjects, initiatives and measures in order to verify and generalize these findings. Future research is also recommended to address the open questions regarding the concept of context, as well as to broaden the investigation from individuals to collectives.

REFERENCES

Alavi, M., & Leidner, D. E. (2001). Knowledge management and knowledge management systems: Conceptual foundations and research issues. *Management Information Systems Quarterly*, *25*(1), 107–136. doi:10.2307/3250961

Arnott, D. (2002). *Decision biases and decision support systems development* (Working Paper. No. 2002/04). Monash University, Decision Support Systems Laboratory.

Becerra-Fernandez, I., Gonzales, A., & Sabherwal, R. (2004). *Knowledge management: Challenges, solutions, and technologies*. Upper Saddle River, NJ: Pearson Education, Inc.

Campbell, D. J. (1988). Task Complexity: A Review and Analysis. *Academy of Management Review*, *13*(1), 40–52.

Choo, C.-W., & de Alvarenga Neto, R.-C.-D. (2010). Beyond the ba: Managing enabling contexts in knowledge organiza5ons. *Journal of Knowledge Management*, *14*(4), 592–610. doi:10.1108/13673271011059545

Dictionary.com. (2017, August 30). *Dictionary.com*. Retrieved from http://www.dictionary.com/browse/context?s=t

El Sawy, O. A. (2003). The IS Core IX: The 3 Faces of IS Identity: Connection, Immersion, and Fusion. *Communications of the Association for Information Systems*, *12*(1), 39.

Evangelou, C., & Karacapilidis, N. (2005). On the interaction between humans and knowledge management systems: A framework of knowledge sharing catalysts. *Knowledge Management Research and Practice*, *3*(4), 253–261. doi:10.1057/palgrave.kmrp.8500076

Handzic, M. (2004). *Knowledge Management: Through the Technology Glass (Series on Innovation and Knowledge Management)*. Singapore: World Scientific Publishing.

Handzic, M. (2007). *Socio-technical knowledge management: Studies and initiatives*. Hershey, PA: IGI Global Publishing. doi:10.4018/978-1-59904-549-8

Handzic, M. (2011). Integrated Socio-Technical Knowledge Management Model: An Empirical Evaluation. *Journal of Knowledge Management*, *15*(2), 198–211. doi:10.1108/13673271111119655

Handzic, M., & Agahari, D. (2004). Knowledge sharing culture: A case study. *Journal of Information &.Knowledge. Management*, *3*(2), 135–142.

Handzic, M., & Ozlen, K. (2012). A Descriptive Analysis of KMS Success in Turkish Healthcare Organisations. In J. G. Cegarra (Ed), *Proceedings of the 13th European Conference on Knowledge Management:* Vol 1 (pp. 441-449). Polytechnic University of Cartagena.

Hansen, M. T., Nohria, N., & Tierney, T. (1999, March). What's your strategy for managing knowledge? *Harvard Business Review*, 106–116. PMID:10387767

Hauschild, S., Licht, T., & Stein, W. (2001). Creating a knowledge culture. *The McKinsey Quarterly*, *1*, 74–81.

Heisig, P. (2009). Harmonisation of knowledge management – comparing 160 KM frameworks around the globe. *Journal of Knowledge Management*, *4*(13), 4–31. doi:10.1108/13673270910971798

Hofstede, G. (2013). *National Culture - Countries: Turkey*. The Hofstede Centre. Retrieved July 2, 2013, from http://geert-hofstede.com/turkey.html

Holsapple, C. W. (2003). *Knowledge Management Handbook*. Berlin: Springer. doi:10.1007/978-3-540-24748-7

Jafari, M., Fathian, M., Jahani, A., & Peyman, A. (2008). Exploring the contextual dimensions of organization from knowledge management perspective. *Vine: The Journal of Information and Knowledge Management Systems, 38*(1), 53 – 71.

Khalifa, M., Yan, A., & Shen, K. N. (2008). Knowledge management system success: A contingency perspective. *Journal of Knowledge Management*, *1*(12), 119–132. doi:10.1108/13673270810852430

Khasawneh, R., & Abu-Shanab, E. (2013). Using KMS as a tool to improve decision making process. In S. Saed & I. Alsmadi (Eds.), Knowledge-Based Processes in Software Development (pp. 1-11). Hershey, PA: Information Science Reference, IGI Global. doi:10.4018/978-1-4666-4229-4.ch001

Lesser, E. L., & Storck, J. (2001). Communities of practice and organisational performance. *IBM Systems Journal*, *40*(4), 831–841. doi:10.1147j.404.0831

Litvaj, I., & Stancekova, D. (2015). Decision - Making, and Their Relation to The Knowledge Management, Use of Knowledge Management in Decision – Making. *Procedia Economics and Finance*, *23*, 467–472. doi:10.1016/S2212-5671(15)00547-X

Liu, S. C., Olfman, L., & Ryan, T. (2005). Knowledge Management System Success: Empirical Assessment of a Theoretical Model. *International Journal of Knowledge Management*, *1*(2), 68–87. doi:10.4018/jkm.2005040106

McCall, H., Arnold, V., & Sutton, S. (2008). Use of knowledge management systems and the impact on the acquisition of explicit knowledge. *Journal of Information Systems*, *22*(2), 77–101. doi:10.2308/jis.2008.22.2.77

Mohsen, Z. A., Ali, M., & Jalal, A. (2011). The Significance of Knowledge Management Systems at Financial Decision Making Process. *International Journal of Business and Management*, *6*(8), 130–142. doi:10.5539/ijbm.v6n8p130

Nunnally, J. C. (1978). *Psychometric Theory*. New York: McGraw-Hill.

Ozlen, K., Handzic, M., & Durmic, N. (2014). Empirical Study of KMS Impact on Decision Support. *Eurasian Journal of Business and Economics*, *7*(13), 133–144.

Raj, P. (2012). Knowledge management – A road map for winning organization. *International Journal of Research in Economics & Social Sciences*, *2*(2), 363–373.

Sajeva, S., & Jucevicius, R. (2010). Determination of essential knowledge management system components and their parameters. *Social Sciences*, *1*(67), 80–90.

Salleh, N. A., Jusoh, R., & Isa, C. R. (2010). Relationship between information systems sophistication and performance measurement. *Industrial Management & Data Systems*, *110*(7), 993–1017. doi:10.1108/02635571011069077

Sambamurthy, V., & Subramani, M. (2005). Editorial: Special issue on information technologies and knowledge management. *Management Information Systems Quarterly*, *29*(1), 1–7. doi:10.2307/25148665

Smirnov, A. V., & Sandkuhl, K. (2015). Context-Oriented Knowledge Management for Decision Support in Business Networks: Modern Requirements and Challenges. *Joint Proceedings of the BIR 2015 Workshops and Doctoral Consortium co-located with 14th International Conference on Perspectives in Business Informatics Research (BIR 2015)*, *1420*, 9-23.

Snowden, D. (2002). Complex acts of knowing: Paradox and descriptive self-awareness. *Journal of Knowledge Management*, *6*(2), 110–111. doi:10.1108/13673270210424639

Tsui, E. (2003). Tracking the Role and Evolution of Commercial Knowledge Management Software. In C. W. Holsapple (Ed.), *Handbook on Knowledge Management 2: Knowledge Directions* (pp. 5–27). New York: Springer. doi:10.1007/978-3-540-24748-7_1

Wenger, E. (1998). *Communities of practice*. Cambridge, UK: Cambridge University Press. doi:10.1017/CBO9780511803932

Wood, R. E. (1986). Task Complexity: Definition of the Construct. *Organizational Behavior and Human Decision Processes*, *37*(1), 60–82. doi:10.1016/0749-5978(86)90044-0

Xu, J., & Quaddus, M. (2005). Adoption and diffusion of knowledge management systems: An Australian survey. *Journal of Management Development*, *24*(4), 335–361. doi:10.1108/02621710510591343

Zhou, A., & Fink, D. (2003). The intellectual capital Web: A systematic linking of intellectual capital and knowledge management. *Journal of Intellectual Capital*, *4*(1), 34–48. doi:10.1108/14691930310455379

APPENDIX: LIST OF RESEARCH VARIABLES AND MEASURES

Table 3.

Decision Context	**Decision Task**
	Most decision problems that I solve are complicated/complex
	In my organisation, I encounter a lot of problems with uncertain/changing causal links
	In my organisation, many of my decision tasks are rather ambiguous/unclear
	My decision problems are often novel/unfamiliar/unknown to me
	Decision Environment
	I have limited time & money to spend on making my decisions
	My decisions have significant personal & organisational consequences
	I am solely accountable for all my decisions
	Most of my decisions are irreversible and can not be easily corrected
	Decision Maker
	I have the necessary knowledge and skills to perform my decision tasks
	I am able to solve decision problems that I encounter
	My motivation to do well is high
	I learn quickly from experience
KMS Approach	**Technical KMS**
	In my organisation, KMS has sophisticated business intelligence components
	My organisation's KMS incorporates intelligent business analytics tools
	In my organisation, KMS comprises excellent systems for communication & collaboration
	In my organisation, KMS includes advanced e-learning and creativity support features
	Social KMS
	Leadership of my organisation is visionary
	My organisation is organised as a network structure/form
	In my organisation, there is a knowledge-friendly culture
	My organisation has developed a knowledge measurement system
	Knowledge Processes
	I use/rely on KMS to:
	access captured internal/external knowledge and gather intelligence
	uncover and interpret hidden patterns in data and extract new knowledge
	exchange ideas and share knowledge with my colleagues and experts
	close gaps in my own knowledge and look for new innovative ideas
	Decision Performance
	I am more confident in the quality of my decisions
	I am more satisfied with the process/outcome of my decision making
	My efficiency/effectiveness of decision making has improved

Chapter 5
Knowledge Engineering, Cognitive Ergonomics, and Knowledge Management in 2017:
A New Delphi Paradigm With Applications

Ronald John Lofaro
Embry-Riddle Aeronautical University, USA

ABSTRACT

Today there is a new emphasis on a discipline related to knowledge engineering (KE): knowledge management (KM). The chapter shows that a Delphi methodology, developed and used with real success from 1986 through 2003, can be revised and up-dated to take a place in the spectrum of KE/KM. This new model combines elements of a Pareto technique at the front end of an small group Delphi paradigm/estimate-talk-estimate (SGDP/ETE) process to initially identify and sharpen the criticality of problems and their solutions. The final stage of the model uses real time Delphi (RTD) techniques. As a result, the final paradigm uses today's technology to achieve extremely accurate decision making and indicate what can be successful courses of action across many venues. Examples will be given of how this new model could be applied to current, real-life problems in three very different arenas.

INTRODUCTION

KE was defined in 1983 by Edward Feigenbaum and Pamela McCorduck as follows: "KE is an engineering discipline that involves integrating knowledge into computer systems in order to solve complex problems normally requiring a high level of human expertise." In 2017 one finds that an expansion in the field called knowledge engineering (KE) has occurred and is still in progress. It seems that KE now stands alone and has some five to seven Journals dealing with KE research, models and applications. In fact, KE is now such an enormous field and endeavor; there are even two Journals with the same title:

DOI: 10.4018/978-1-5225-5427-1.ch005

International Journal of Knowledge Engineering; one from IGI Global and the other put out by Bioinfo Publications. Knowledge engineering is also linked to cognitive science and socio-cognitive engineering where the knowledge is produced by socio-cognitive aggregates (mainly humans); A newer term, cognitive engineering (CE), includes mental workload, decision-making, skilled performance, human-computer interaction, human reliability, work stress and training as these may relate to human-system design. There is not only great overlap with KE, but almost an isomorphism.

Cognitive Ergonomics deals with decision-making and focuses on the fit between human cognitive abilities and limitations and the task. As such, it can be viewed as a portion of a Venn diagram Universe that includes Cognitive Engineering, again with definite overlapping.

Today, there is a new emphasis on a discipline related to KE: knowledge management (KM). Knowledge management (KM) has been defined as "...the practice of selectively applying knowledge from previous experiences of decision making activities with the express purpose of improving the organization's effectiveness." (Jannex, 2014). The main difference between KM and KE seems to be that the (knowledge) manager establishes the direction the process should take, where the (knowledge) engineer develops the means to accomplish that direction. Therefore, accurate decision-making plays large in both the knowledge manager's and the knowlede engineer's road to success/failure. Aside: The author has many articles (e.g., Lofaro & Smith 1999; 2000; 2001a; 2001b; 2003; 2009) and a book chapter, (Lofaro & Smith, 2010) in print on decision-making in the unforgiving aeronautical environment.

Currently, the KM aspect of KE has become a multidisciplinary field, bringing in concepts and methods from several computer science domains such as artificial intelligence (AI), databases, expert systems, decision support systems and geographic information systems.

There is also a somewhat new emphasis in the KE/KM fields on ontology, a term that comes from philosophy. In philosophy, it is the study of the nature of being, becoming, existence, or reality, as well as the basic categories of being and their relations; it often deals with how entities may be grouped, related within a hierarchy, and subdivided according to similarities and differences. In KE/KM, an ontology is a formal naming and definition of the types, properties, and interrelationships of the entities that really or fundamentally exist for a particular domain of discourse. It is thus a practical application of philosophical ontology, with a taxonomy (see later in Chapter). A KE/KM ontology compartmentalizes the variables needed for some set of computations and establishes the relationships between them; thus, an ontology is used to limit complexity and to organize and structure information. It has now found practical applications in AI, information science and technology; in short: in much of KM and KE.

For a more complete overview and discussion on KE, differing views and uses, the reader is referred to Studer, Benjamins and Fensel (1998). For works on KM, Amazon's web site has 50 or more, including the ubiquitous *The Complete Idiot's Guide to Knowledge Management* (Rumizen, 2001). At this point, the author believes that readers of this Chapter, as they are and have been in these fields for years, need no further citations.

HISTORY AND RATIONALE

As to why this Chapter came to be written: It really began in 1986 with a modified Delphi as will be shown; this modification was termed a small-group Delphi paradigm (SGDP) (Lofaro, 1992a) using intact, face-to-face groups of subject matter experts (SME's) to work on specified tasks. Circa 2010, the author realized that this 1986 SGDP could nowadays be seen as a KE effort. Additionally, the author found out that, some 20 plus years after the original SGDP effort, that mini-Delphi/ETE was now "acceptable" and in the literature. At this point, the author believed that some ideas could be sketched out on using the SDGP as a part of an ETE (Lofaro, 2015). The author is here attempting a similar effort to show how such an SGDP/ETE amalgam with a modified Real Time Delphi (RTD) could be used on some specific problems, as well as adding the use of a mini-Pareto as a start point. This Chapter will briefly explicate what began, 30 years ago as a cognitive task analysis (CTA) for identifying critical skills needed for skilled performance, with decision-making involved in every step. Over the years, it evolved into a Delphi paradigm that extended far beyond a CTA.

This Chapter is a natural progression/extension of the author's Chapter entitled "Knowledge Engineering: A Methodology and Examples" (Lofaro, R.J., 2013) in the 3rd edition of *The Encyclopedia of Information Science and Technology* (2014). In that Chapter the author presented both the Small Group Delphi as developed and expanded over the years and the newer Estimate-Talk-Estimate variant of the Delphi process. At that point, elements of both Delphi processes were blended to produce an initial look, with some methodological avenues, for use in a paradigm in tune with the tremendous speed, instantaneity and scope of today's technological advances in communication. What now? The author believes that another process for analysis needed to be front-loaded into any SGDP/ETE; elements of a Pareto analysis. That is to say, the author believes the next step is to combine elements of a Pareto technique at the front end of an SGDP/ETE process, using RTD in the final step and resulting in a new paradigm.

All this is done in the hopes that such efforts might be seen and be of some interest, even be of some help by the practioners and professionals, for lack of a better word, in KE and/or KM. It is hoped to bring some or all of the above together as the Chapter moves on. To that end, one more term followed by look back in history.

THE TRADITIONAL DELPHI

The Delphi method was developed by Project RAND during the 1950-1960s. Project RAND began as an organization formed immediately after World War II to connect military planning with research and development decisions and later became an independent, nonprofit organization, the RAND Corporation... which derived its name from a contraction of the terms research and development. Delphi techniques, a subset of CE/KE, have become common methodologies for eliciting analyses, expert opinions and evaluations on a variety of topics. In general, standard Delphi techniques include anonymity of response, multiple iterations, convergence of the distribution of answers and a statistical group response, preserving intact a distribution that still may be a wide one (Judd, 1972). However, as Meister (1985) had noted, "The (Delphi) methodology is by no means fixed...[it] is still evolving and being researched." This is as true now as it was when Meister stated it. In point of fact, with the leaps in communication methods and related computer technology, this is even more true today. The rationale for this statement is actually two-fold. The first being that in the last 10 to 15 years, there has been quantum leaps in computer memory/power as well as communication technology that uses desk-top computer, even iPhone technology. These leaps and advances seem now to occur almost daily. Concomitantly, Delphi techniques have recently begun to look at and attempt to take full advantage of these advances.

It is posited, and hopefully shown in this Chapter that a new Dephi methodology, developed and used ...with real success...from 1986 through 2003 can be revised and up-dated to take a place in the spectrum of KE/KM uses.

Delphi Characteristics

The following key characteristics of the Delphi method help the participants focus on the issues at hand and, are what separates Delphi from other methodologies:

1. **Anonymity of the Participants:** Usually all participants remain anonymous. Their identity is not revealed, even after the completion of the final report. This prevents the authority, personality, or reputation of some participants from dominating others in the process. Arguably, it also frees participants (to some extent) from their personal biases, supposedly minimizes the "bandwagon effect" or "halo effect," (Bandwagon Effect: A psychological phenomenon whereby people do something primarily because other people are doing it, regardless of their own beliefs, which they may ignore or override; Halo Effect: This effect was given its name by psychologist Edward Thorndike. It is a cognitive bias in which one's judgments of a person's character can be influenced by one's overall impression of him or her. It can be found in a range of situations from the courtroom to the classroom and in everyday interactions).Thus, the Delphi allows free expression of opinions, encourages open critique, and facilitates admission of errors when revising earlier judgments.
2. **Structuring of Information Flow:** The initial contributions from the experts are collected in the form of answers to questionnaires and their comments to these answers. The panel director/facilitator controls the interactions among the participants by processing the information and filtering out irrelevant content. This avoids the possible negative effects of face-to-face group discussions and thereby, solves any problems resulting from group dynamics.
3. **Regular Feedback:** Participants comment on their own responses, the responses of others and on the progress of the group/panel as a whole. At any moment they can revise their earlier statements. Again, the claim is made that, while in face-to-face group meetings, participants tend to stick to previously stated opinions and often conform too much to the group leader.
4. **Role of the Facilitator:** The person coordinating the Delphi method can be known as a facilitator or leader, and facilitates the responses of their panel of experts, who are selected for a reason, usually that they hold knowledge on an opinion or view. The facilitator sends out questionnaires, surveys etc. and, if the panel of experts accept, they follow instructions and present their views. Responses are collected and analyzed, then common and conflicting viewpoints are identified. If consensus is not reached, the iterative process continues to gradually work towards synthesis and building consensus.

What has been presented above is the standard description and rationale for the Delphi process.

THE SMALL GROUP DELPHI PARADIGM: OVERVIEW

A modification to Delphi processes is the small group Delphi paradigm (SGDP). The SGDP took the Delphi process in another direction by modifying it via merger with elements of group dynamics in order to have interactive (face-to-face) Delphi workshops. The development of this modified Delphi, the SGDP, involved the merger of a specific knowledge engineering technique (Delphi), with Fleischmann's theories of underlying abilities (Fleishmann and Quaintance, 1984; revised 2000) and some principles of group dynamics. This modification resulted in a paradigm for using small groups of subject matter experts

(SMEs: An individual who exhibits the highest level of expertise in performing a specialized job, task, or skill within the organization; Related definition: Domain Expert: A person with special knowledge or skills in a particular area of endeavor. The term *domain expert* is frequently used in expert systems software development.) for any project that requires that a set of SMEs be used to identify, evaluate, and criticality rank tasks (an extended and enhanced task analysis), identify core needs/skills, recommend modifications to equipment, procedures and training. Finally, the SGDP can be used to sharpen, modify and revise existing methodologies. Thus, traditional Delphi processes were modified into a new paradigm. This effort and subsequent use (Please see The Small Group Delphi Paradigm: Past Uses Section for these citations thus escaping redundancies) of the resultant paradigm, SGDP, in other, varied venues have produced both new modifications to the SGDP and highly accurate data that were operationally implemented. The extensions over the years of the paradigm indicated that it had an applicability over many domains. These many SGDP efforts also resulted in sharpening, modifying and revising the original methodology. That the core SGDP has been used in many environments demonstrated a robust flexibility and generalizability of the paradigm. One last point: it may be be usual to try a locate, as it were, an article(s) within the literature; in this case, the Delphi. The only existing articles on the SGDP, from 1986 through 1999, were written by the author with, on several occasions, co-authors.

The Small Group Delphi Paradigm: Origin

This Chapter was over 30 years in the making. It begins with a specific problem and task: to develop for the U.S. Army Aviation Command an aviator candidate selection test that also showed which of the current rotorcraft (the U.S. Army uses this term, not helicopter as do other U.S. military services.) would be the optimum operational aircraft placement for the candidate upon completion of initial pilot training. The U.S. Army Aviation Center, located at Ft. Rucker, AL, conducts all initial rotorcraft and transition training for pilot candidates and was revising almost completely its rotorcraft pilot training system. This revision was a unparalled effort because the sequence up to then had been, after completion of initial training, that the pilot was assigned to a field (operational) unit to gain some "seasoning," i.e., experience in actual operations. Only after that, to come back for transition training. In short, Army Aviation wanted to initially train pilots and immediately afterward put them in transition training for one of the then operational rotorcraft, based solely on a test. In KMS terminology, the U.S. Army Aviation Command was the organization. Much of KMS today has one focus on the use in/to an organization. The use for the testing and scoring protocols developed from the SGDP was to quickly and accurately select the optimum aviator candidates for initial training and then place them in the optimum rotorcraft for transition training; optimum being a performance-based word.

To return to the Delphi; at that time, 1985/6, there was some literature on the Delphi. The RAND Corporation was a leader in its use and what was seen as a seminal paper on the Delphi process was written by a then-Rand employee (Brown, 1968). However, there were several articles that voiced strong objections to a Delphi not being anonymous. The major one being that having participants see and interact with each other would at best, dilute results, and at worst, negate them. As an example, Pill (1970) had said that this may dilute the opinions of the real expert. Therefore, such objections had to be addressed.

To begin with, the objections seemed to the author to be somewhat bizarre. First, the subject matter experts (SMEs) selected ARE the real experts. However, if what is meant is that one or more persons in a group may have more expertise in a specific area that is being worked on and the group would defer to them. The reality (based on the author's having conducting several workshops using SME's to focus on and solve problems) was this: The other SMEs recognize, welcome and use another SME's expertise–as their goal is the best result/product possible.

Another objection was that the group dynamics may force ratings and analyses towards a mean or middle ground that does not fully reflect all the SMEs' views. In response, true SMEs will not allow that to happen for personal reasons and pride. They will not "go along to get along."

In the author's opinion, the objections to face-to-face interaction could not only can be overcome but the use of group dynamics coupled with traditional Delphi processes could add robustness and a completeness, as it were, to results obtained from this coupling. The instruction in group dynamics and group processes, the trained facilitator and the iterative methodology used in accomplishing the sub-objectives/objectives were all structures in the SGDP process designed to ensure that the prior objections could be met.

At this point in 1985/6, the author, who was experienced in team training and group function/dynamics, decided to develop a modification to Delphi processes wherein the abilities approach of Fleishmann was used with face-to-face, small groups (five and ten) of Army aviators, who were recognized by their peers as exceptional pilots in their particular rotorcraft from all Army posts world-wide, functioning as SME's. These SME's would be given a 30 page read-ahead package. After some weeks, these SME's would be brought to Ft. Rucker where they would have group function/dynamics training, and begin facilitated group work on the objectives specified in the read-ahead. This endeavor came to be called the small group Delphi paradigm/SGDP (Lofaro,1992a). Thus, the SGDP took the Delphi process in another direction by modifying it via merger with elements of group dynamics in order to have interactive (face-to-face) Delphi workshops. The modification resulted in a paradigm for using small groups of subject matter experts (SMEs) in a workshop setting. In short; there are alternatives, other pathways and modifications to the traditional Delphi process....as will be again seen.

The use of the underlying abilities approach as delineated in their book, *Taxonomies of Human Performance,* Fleischmann and Quaintance (1984, revised 2000) state that, "specific tasks are said to require certain ability profiles if performance is to be maximized. Abilities provide a natural basis for describing and classifying tasks in terms of human performance requirements." Fleischmann and Quaintance defined ability as "an innate enduring attribute." As such, abilities are indicators and predictors of later skill development. Individuals can be taught skills and, with practice, these skills can be enhanced. However, when an individual possesses an underlying ability that directly relates to a skill/set of skills, they will be enabled to perform tasks based on that skill/set of skills at a higher level. Therefore, the underlying abilities of the individual have a significant impact on the individual's level of performance on any given task. Put another way, the ability requirements theory/approach is away of describing and classifying human tasks. In this approach, tasks are described, contrasted, and compared to the abilities required of the individual performing a specific task. Once a set of tasks is identified, a human performance taxonomy can be developed from it. Taxonomy, as used here, denotes a system that classifies and describes human tasks according to a particular focus, such as the abilities seen as essential to specific task performance. This approach became a cornerstone in the SGDP as will be seen.

In the U.S. Army Aviation Center modified Delphi, the SGDP, there were four workshops, one for each of the then-operational U.S. Army rotorcraft. These workshops had small groups of aviator subject matter experts (SMEs). The workshops were highly structured, face-to-face sessions. The goal of the workshops were to change the tacit information possessed by the SME's into explicit information focused on identifying the specific skills needed to perform well as a pilot in the different U.S. Army rotorcraft. The structure of the workshops, the clear objectives, the clear stipulation of the process to be used to achieve the objectives and the facilitation allowed the SME's to develop expicit data to meet the objectives.

The basic categories the author and associates looked at in the original 1986 SGDP were psychomotor (hand/eye coordination as an example); cognitive; personality characteristics. Thus, we were establishing what would be called today an ontological scheme to reduce complexity and structure the data. After the SME's identified the skill sets ...following the Fleishmann/Quaintance concept of taxonomies...this completed the ontological taxonomy for each helicopter. Sub-tests were then chosen (or modified from existing ones) to provide measures on skill sets found in the workshops. Finally, a unified test and scoring algorithms were developed. The use of multivariate discriminant analyses resulted in a "profile" of the (weighted) factors that made for success for pilots, but in only three of the four rotorcraft. It was found that what had happened was that each centroid contained the same set of what we called then basic rotorcraft piloting skills, common across all centroids and second sets of skills, particular to only one rotorcraft. It was determined that this common set of skills were those needed to pilot the UH-I Iriquois (called the "Huey"), at that time the basic Army operational rotorcraft. This finding was both vital and important. It showed what we had thought and what, informally, the SME's had brought up: there was a set a basic pilot skills needed for all rotorcraft. This boded well for the selection aspect of the test. Secondly, each rotorcraft has a secondary set of skills specific to that rotorcraft; this boded well for the placement aspect of the test. And a real serendipity: when the Army was going to go to the new training system, they were going to stop using the then initial training rotorcraft, the TH-55, a piston-engined relic. The UH-1, which was, as were all Army operational rotorcraft, jet engined, was to now be the initial training rotorcraft. The skills common across all three centroids were extracted and formed a fourth centriod; the UH-I skill set centroid.

Each candidate's algorithmic score(s)/profile is compared with the composite profile of those pilots already successfully flying that specific rotorcraft. The result was the development and fielding of a computerized test battery and scoring algorithms for selecting U.S. Army aviator candidates for initial training and one that also indicated which of the types of operational rotorcraft was the optimum for the pilot to be transitioned trained. The placement portion of the tests also indicated a second choice rotorcraft for transition training. There followed validations of sorts: in the first, the unified test was given to hundreds of current Army rotorcraft pilots to see if the test could place them in the rotorcraft they were already flying; it did 95% of the time. Next, the test was given to an extremely large sample of Army personnel to see they would be grouped by rotorcraft; the resultant four centroids showed they were differentiated (minimal or no centroid overlap) by rotorcraft type. Thereupon, The U.S. Army Aviation Center made the test battery and set of algorithms (now called Multitrack, as was the new training system) operational. It was successfully used in selecting rotorcraft aviator candidates for initial rotorcraft training and later, placement in their operational rotorcraft for more than five years.

At this point we must mention validity. Note: complete statistics on the extremely high predictive validity are available (Lofaro, Intano & House, 1990). Of particular import are both the high "pass" (success) rate when the pilots were placed for transition training in the operational rotorcraft Multitrack indicated versus the high "fail" rates when, for some (bureaucratic?) reason, the pilots were not placed in the optimum first choice or even the second choice rotorcraft. When not placed in either of these two Multitrack-indicated rotorcraft, the failure rate was over 90%. When some of the articles cited below were presented/published, there were some who claimed we must have "altered" the data to achieve those results. Then, as now: not true! The author left Ft. Rucker in very late 1989 to return to the FAA.

However the author worked closely with the listed co-authors in preparing all articles dealing with the SGDP and/or Multitrack.

To return: During initial work on SGDP and Multitrack, the concern was with more of an operational/use/use-value type validity. Following Cook and Campbell (1979), it was seen as the degree to which a study accurately reflects or assesses the specific concept that the researcher is attempting to measure. There are many types of validity, most subsumed under internal and external. The original concern was neither with internal nor with external validity. i.e., if the results of a study are generalizable or transferable...most discussions of external validity focus solely on generalizability (Campbell and Stanley, 1966.). However, as will be shown below, the work was both generalizable and transferable. The researchers main concern in 1986 and afterwards, was with predictive validity: the degree to which the operationalization can predict (or correlate with) with success of the construct in operational use. There is one point to be made here: over eighty percent of the various items in the skill sets as identified by the SMEs turned out to be valid indicators of what was needed for success in the piloting each particular aircraft. That is to say, scores on tests to evaluate such skills/set "shook out" as weighted variables in the Multitrack scoring algorithms. The remaining less than twenty percent did not add enough scoring power to the algorithms and were not used. But, all items in all skill sets were agreed on by the SME's as needed for some level of success in each helicopter.

This initial SGDP R&D effort has been extensively presented (Intano, Howse & Lofaro, 1991; Intano & Lofaro, 1990; Lofaro, Intano & Howse, 1990; Intano & Lofaro, 1989; Lofaro & Intano, 1989). In these articles and reports, the success rate in predicting candidates' completion of initial flight training was almost 99%; more importantly, the success rate in predicting the candidates' then successful completion of transition training in the operational helicopter type indicated by the Multitrack algorithm was well over 95%. This level of predictive validity was why The US Army Aviation command used the tests/algorithms for well over 5 years. In that time, the US Army developed and fielded new rotorcraft types, necessitating a new set of workshops to modify Multitrack; this was not done. Aside: The author was later told, when teaching for Embry-Riddle Aeronautical University (circa 2008) at the U.S. Coast Guard Training Center that some? all? of "Multitrack" is again in use by the Army. He was told this by United States Coast Guard (USCG) Instructor Pilots who had come over from Army Aviation.

In summary: The SGDP accomplished what had never been done before. It delivered an operational, computerized set of tests with scoring algorithms that not only selected the best aviator candidates for initial pilot trading but also showed which of the four then-existing rotorcraft types these candidates should be placed for transition training in upon completion of initial training. This effort was successful in that the U.S. Army Aviation Command not only used Multitrack but, there was a very high predictive validity as to both selection and placement

The Small Group Delphi Paradigm: Past Uses

In 1987-1988, the SGDP was used in a series of six workshops designed to identify the tasks that are essential and/or unique to each U.S. Army rotorcraft in the mission environment, with criticality ratings. The result was a training syllabus which contained the identified tasks; the number of iterations for each task needed to achieve an optimum performance level; the number times per year such training was to occur.

In 1991, the paradigm was used in a United States Federal Aviation Administration-sponsored project that developed a computerized air traffic control specialist (ATCS) selection test battery (Gibb and Lofaro, 1993). This battery was used by New Zealand Air Traffic Service.

The SGDP was used in an Air Transport Association/Federal Aviation Administration workshop in 1992 (Lofaro,1992b) to work on a mission performance model for flight crew resource management integration and evaluation.

In 1993/4, Lofaro and Gibb (Lofaro, Gibb and Garland, 1994; Gibb and Lofaro, 1994) used the SGDP techniques to develop airline baggage x-ray screener abilities and a prototype selection test.

[Note: In this endeavor, Lofaro gave training to Gibb and associates in doing the SGDP as he was not able to facilitate the Delphi's himself. Lofaro also provided Gibb with a written precis of the SGDP technique. Gibb and associates did the actual SGDP workshops themselves at various locations. However, the use of SGDP techniques by others indicate that the techniques are not dependent on who administers them; rather, they not only are effective across venues but also with different personnel directing/facilitating the SGDP workshops.]

In 1998 and 1999, Lofaro used the SGDP in a series of workshops to develop sets of managerial and employee core competencies. For managers, each set was specific to only one position; for employees, each set was specific to only one general classification. (Lofaro, 1998, 1999).

In the early 2000's Lofaro, working with Helene Maliko-Abraham, did four more applications of the SGDP, ranging from selection through training to useability testing (Lofaro & Maliko-Abraham, 2002; Maliko-Abraham & Lofaro, 2001a, 2001b, 2003)

These successful adaptations, modifications, implementations, across many venues, showed that the SGDP is both transferable and generalizable and possesses external validity. The SGDP can be used for any project that requires that a set of SMEs be used to identify, evaluate and criticality rate tasks (an enhanced task analysis); to identify core needs/skills; to recommend modifications to equipment, procedures and training. Finally, the SGDP can be used to sharpen, modify and revise existing methodologies.

THE SGDP AS A KM ENDEAVOR

KMS today places emphasis on the success of using a KMS process. The SGDP process had levels of success; the internal success, so to speak, was the high level of agreement of the skills/sets identified via the SGDP process and the skill/sets used and tested for in the Multitrack algorithms. The external / organizational success was the acceptance and use of Multitrack by the U.S. Army Aviation Command.

While the definition of measures of KM success did not exist in 1986 (and not even throughout the ensuing use of the SGDP into 2003), nor did the Jennex and Olfman KM Success Model (Jennex, 2013), it would seem, that in part, these measures were somewhat met by the Mutitrack test/scoring algorithms resulting from the SGDP.

Since KMS success is "...a multidimensional concept defined by capturing the right knowledge, getting the right knowledge to the right user and using this knowledge to improve organizational and/or individual performace..." (Jennex, 2012 citing himself, Smolinik and Croasdell, 2009), the author posits the SGDP was a KMS success.

Examples: A KM Success Model measure of KM success is system quality. In the SGDP the SME's created/produced knowledge that was stored, able to be retrieved and was applied. That knowledge probably could not have been otherwise captured as the SME's may not have been available later, the need for their specific expertise had not, nor would be asked for again... but Army Aviation needed it exactly then. Knowledge quality was achieved as the knowledge was shown to be useful as to correctness and inclusion. The service quality was seen in the performance impact of the using (making operational) Multitrack and subsequent U.S. Army Aviation Command satisfaction.

The Small Group Delphi Paradigm: 2016 Technological Modifications

Some twenty-five years after the SGDP was devised, used and appeared in multiple publications, it has been re-discovered, as it were, and is now recognized as an acceptable CE/KE method. The use of face-to-face groups in a Delphi is now a fact and is termed Estimate-Talk-Estimate (ETE). New communication capabilities and technologies seem to have driven the development of what are generically referred to as a mini-Delphi or ETE, with many variants. There is also the Real Tine Delphi (RTD) that maintains anonymity but uses computer linking for a level of instantaneity. (Gordon, & Pease, 2006). Other innovations come from the use of computer-based (and later web-based) Delphi conferences. One example of a difference in a type of ETE (a computer-based Delphi) versus either a traditional or SGDP Delphi is the iteration structure used in the traditional or SGDP Delphis. Such those iterations which are divided into three or more discrete rounds, can be replaced by a process of continuous (roundless) interaction, enabling SMEs to change their evaluations at any time. As Bolognini (2001) has said, the involvement of a large number of participants, the use of two or more panels representing different groups (such as policy-makers, experts, citizens), which the administrator can give tasks reflecting their diverse roles and expertise, and have them interact within ad hoc communication structures: these are strengths of computer and web-based ETEs. The reader is referred to the work of Turoff and Hiltz (1996) on computer-based Delphis.

In Sum: While the SGDP structure and processes are still relevant, they can be and are in real need of some level of integration with ETE and RTD techniques and current advances in computer and communication abilities and techniques. The author's belief is that the result of an integration would be a revised SGDP that will produce the same level, if not a higher level, of accurate information and products in the aviation arena, or many other venues. In modifying the SGDP for use with today's technology and advances in KE/KM, any problems to be investigated would require, The Chapter article will now present a "how to" for using this modification on a specific problem. As before, It must be strongly noted that the Chapter is not a report on the results of prior experimentation. Rather it is an attempt to indicate the structure of a new, modified Delpi and meld that modified Delphi with today's KE/KM efforts and today's technology. More importantly, it is also a call for research efforts to try the new procedures and validate (or not) them. Would that the author were still in prior positions as government agency researcher/manager as the author could have done these efforts; this is not the case today. The stage has been set for the remainder of the Chapter.

A Quick Look at Core SGDP Components

What follows is a very short synopsis because there are space constraints on this paper. For a more complete SGDP write-up, email the author; *lofaro@msn.com.*

1. **SME Selection:** Identify the best SMEs for the SGDP Workshop(s).
2. **The SGDP Objectives:** Develop the objectives and sub-objectives so that all the components needed to achieve the whole objective will be in place.

3. **The Read-Ahead Package:** It must be prepared and sent to the Workshop participants at least three weeks in advance.
4. **The SGDP Process:** Upon participant arrival, proceed with instructions and exercises in group dynamics.

DEVELOPING A TEMPLATE FOR A SGDP/ETE/RTD... AND BEYOND

The core SGDP structures to be maintained are: the careful selection of a limited number of SMEs (however, there can be many small(er) SME groups functioning at one time), the use of an extensive read-ahead package for the SMEs, the use of some facilitation and group dynamics instruction, combined with some type face-to-face sessions. One example of a difference in an ETE or RTD (a computer-based Delphi with anonymity) versus either a traditional or SGDP Delphi: the iteration structure used in the traditional or SGDP Delphi, which is divided into three or more discrete rounds, can be replaced by a process of continuous (roundless) interaction, enabling SMEs to change their evaluations at any time. A new SGDP/ETE/RTD model would be computer-based. The reader is referred to the work of Turoff and Hiltz (1996) on computer-based Delphis.

This Chapter will show how to integrate the elements of the paradigm cited above with elements of Pareto Analysis and real time Delphi (RTD.) The resultant model uses today's technology to achieve extremely accurate decision making and indicate courses of action across many venues. This Chapter will take that template and integrate it with elements of Pareto techniques, to initially identify and sharpen criticality of possible problems/solutions of issues. The selection and the rationale for the selection of the elements of both the Pareto analysis and for the elements of the SGDP will be explicated. At that point, the new Pareto/SGDP/ETE process could be implemented to achieve guidelines for achievable resolution of a variety of problems requiring accurate decision-making. The decision-making will be of a distributed nature, enabled by computers and the Internet.

Finally, this Chapter will examine three current, problematic areas and outline how this now tri-partite paradigm can be used on these three discrete, unrelated area showing generalizability and transferability of the paradigm

The Integration Process

We already have ETE-type Delphis done using web sites where the process is conducted in real-time. The ETE type would not involve selecting elements thereof but would be chosen on the basis of what data came out of the Pareto/SGDP. Re-stated: As to which type of ETE would become part of this new paradigm, that is dependent on what elements of the Pareto and SGDP are to be part of the new paradigm. The questions to be answered are which elements of Pareto analysis are to be used (with rationale) as well as which elements of a SGDP are to be incorporated, again with rationale Integration of the Pareto/SGDP with a ETE/RTD (*sans* anonymity) approach can be achieved thusly: all participants can be logged on simultaneously, each participant can briefly state their name and credentials, the group dynamics instruction can be done by the facilitator to all simultaneously (Aside: it would seem that a linked network of all SMEs is possible and even de rigueur*)*. The SGDP face-to-face group meetings would now be done on-line. This will allow for instantons feedback by any SME during a session, as well as discussions). The iteration structure used in SGDP, which is divided into as many discrete rounds as needed for consensus, can be replaced by a process of continuous (round less) on-line interaction, enabling SMEs to change their evaluations at any time and give a rationale with ensuing discussion in real-time. Finally, the statistical group(s) response(s) can be updated in real-time and shown whenever a SME, or a group, provides a new evaluation. It is clear that "face-to-face" discussion will be virtual. This is, to the author, a real and significant loss. But, the speed, multiple iterations, real-time, access to a large number of SMEs and other aspects to be gained cannot be ignored. Another possible modification is a multi-tiered SGDP/ETE/RTD where the use of two or more groups working different problems can be convened and given objectives based on their expertise. As these groups come to consensus on their objectives, these new data can be integrated, built into a new re-ahead package and made available to a new set or participants with new or prior SMEs.

A second modification would be use of only specific elements of a Pareto Analyis as the first step after the group dynamics instruction. This is because it would identify problems, then, sharpen (focus in on the ones that are amenable to resolution) while, at the same time, providing an ordering of which should be worked on via criticality ratings. The elements of the Pareto Analysis would be the first three steps in such an analytic technique. Step 1: Identify and list problems to be examined. Or, if and only if one problem/issue exists, a break-out of the sub-problems could be done. Step 2: identify causal factors inherent in each problem. Step 3: Score (in this case, criticality rate) the problems, resulting is a somewhat rough ontology. The Pareto, done in the context of a SGDP/ETE/RTD could also allow for many possible solutions to be evaluated by many types of SMEs with differing areas of expertise but areas germane to the problem. This would result in a winnowing down of courses of action to those that were realiostic and possible of success. The advantages of using a Pareto, core SGDP structure and ETE/RTD (all computer/Internet driven) are that they are all content-area neutral and, in a real sense, generic in application.

General Methodology for Pareto/SGDP/ETE/RTD

All participants can be logged on simultaneously, each participant can briefly state their name and credentials, the group dynamics instruction can be done by the facilitator to all simultaneously (aside: it would seem that a linked network of all SMEs is possible and even *de rigueur.* This will allow for instantons feedback by any SME during a session, as well as discussions). The iteration structure used in SGDP, which is divided into as many discrete rounds as needed for consensus, can be replaced by a process of continuous (roundless) interaction enabling SMEs to change their evaluations at any time and give a rationale with ensuing discussion in real-time. Finally, the statistical group(s) response(s) can be updated in real-time and shown whenever a SME, or a group, provides a new evaluation. It is clear that "face-to-face" discussion will be virtual. This is, to the author, a real and significant loss. But, the speed, multiple iterations, real-time, access to a large number of SMEs and other aspects to be gained cannot be ignored. Another possible modification is a multi-tiered Pareto and then, SGDP where the use of two or more groups working different problems can be convened and given objectives based on their expertise. As these groups come to consensus on their objectives, these new data can be integrated, built into a new re-ahead package and made available to a new SGDP set with new or prior SMEs.

THREE EXAMPLES OF USING AN SGDP/ETE WITH PARETO TECHNIQUES: 2017 AND BEYOND

An initial thought, with the author's background in education (mathematics degree; taught mathematics and math ed) was the current and it seems now controversial United States federal Common Core Standards. While these Standards were federally developed (actually done under contracts), the States were to develop (or not) curricula based on these standards. One set of issues seem to revolve around whether the Common Core Standards are an attempt by the federal government, through federally-funded interstate consortia assessment entities, to make the States hew to the Standards. This would make the Standards, *de facto*, mandatory at the State level. Another issue is the critiques of the Standards themselves, especially it seems the Mathematics and English ones, by educators, scholars and parent interest groups. At the federal level, it seems the Standards are a *fait accompli,* thus precluding any Pareto/SGDP/ETE efforts. However, at the State level, such Delphis could help in analyses of the Federal Standards; development of State-specific curricula and pedagogy for teaching the new State curricula.

Such an effort would involve multiple subject area-specific groups, each composed of educators (classroom teachers as well as University teacher education professors), parents and some legal experts. The goal would be a State-wide curricula, with assessment techniques designed to both accurately evaluate student progress and needs as well as pass muster with the Smarter Balanced Assessment Consortia (SBAC) and the Partnership for Assessment for Colleges and Careers (PARCC); these are the federally funded assessment consortia mentioned above. While this seems to possibly be a moot effort as there are federal Standards, some States (Oklahoma, Texas, Virginia, Alaska, Nebraska and Indiana) have already indicated they will not use the federal Standards and Minnesota will not use the federal Mathematics Standards. So, it seems that such States might support such Pareto/SGDP/ETE efforts.

The author must begin with a *caveat*: What has been shown above is a template for decisions and actions that have not yet and may never be accomplished. Thus, it is in the realm of projection and in a word, conjecture. That having been said, here are thoughts on how the described effort can be done, then seen in the light of KM. It is suggested that the effort be limited to Mathematics and English, with the knowledge manager for each discipline being arrived at by consensus from the participants. The knowledge engineers, again a two discipline division as above, would use the SGDP/ETE/RTD/Pareto model as guidelines, making such modifications as were/were not deemed necessary for task completion. As a first step, the building a model of a knowledge domain; then defining the terms inside that domain and the relationships among them...developing a taxonomic ontology. The knowledge manager establishes the direction the process should take; this direction would differ by State but generally would be the development of State-specific K-12 curricula...by the subject areas above with assessment tools. Therefore, accurate decision-making plays large in both the knowledge manager's and the knowledge engineer's road to success/failure.

One rough measure of success would be if modifications/changes in existing Common Core standards were developed while the development of State (and other interest groups) K-12 English and Mathematics curricula might be another. As per the Jennex and Olfman KM Success Model (Jennex, 2013), internal success would be shown if, when the multiple working groups met to finalize their products, the elements of these curricula were basically extremely similar. would be one, if not the only, measure of success...in this case, modifications/changes in existing Common Core standards could indicate one measure of success while the development of State (and other interest groups) K-12 English and Mathematics curricula might be another Organizational external success would be shown if that mathematics and English curricula developed would be adapted by the State for which they were developed A KM Success Model measure of KM success is system quality. This could be measured if the State-specific Cullicula developed, when used, resulted in the test scores achieved by students were at least equal, if not superior to the scores achieved by States using the federal standards/curricula.

Additionally, this effort, if done would create/produce knowledge that could be stored and able to be retrieved . Such knowledge probably could not have been otherwise captured as the federal Core Curriculum effort is in a finalization stage, and the possibility for such an Pareto/SGDP/ETE/RTD would not occur again.

Example Two

The author selected, as another exemplar (as the author has over 26 years of experience in aviation (United States Air Force, United States Army Aviation Command and FAA), that choice of issues was simple. What are some current 2017 aviation issues (as well as prior but unresolved ones) that are both vital to aviation safety and amenable to some form of the SGDP/ETE/RTD/Pareto ?

Initally, the author considered that a look at air carrier upset training might be fruitful. The Congress passed the Airline Safety and FAA Extension Act of 2010, which mandated that the Federal Aviation Administration (The FAA regulates aviation in the USA, from air traffic to civil aviation security, the operation of air carriers, pilot training and more) require pilots to complete 1,500 flight hours before they're allowed to fly commercially, up from just 250 hours before the act. However, the FAA, while not yet issuing an Advisory Circular (AC) or a Federal Aviation Regulation (FAR), has issued a document called Airline Upset Recovery Training Aid, version 2. It is believed that a modified SGDP/ETE/RTD, using senior pilots as SME's could point a way to types/procedures and applicability for upset training. Upon a re-look, the area outlined may seem fruitful but it is fraught with regulatory and legal problems. That having been said, since the upset training is in the FAA rule-making arena, any Pareto/SGDP/ETE/RTD efforts would probably be moot… albeit both germane and instructive and there seemingly is a definitive need for such efforts. However, we go to another aviation problem area which may be more amenable and less encumbered.

An aviation problem area that is of long standing and, has been often researched: flight crew fatigue. Fatigue from flying has been an issue since World War I, yet no resolution has been found. A fairly recent example is the United Parcel Service (UPS) flight 1354 crash in Birmingham, Alabama, in late 2013. At this time, it seems that the UPS pilots, NTSB and Airbus (the type aircraft involved was an Airbus A300-600), are in, to put it mildly, some disagreement as to causation. Remember that UPS has an FAA-managed aircrew fatigue risk management program (Croft, John, 2014c). If memory serves, (the author was with the FAA at the time), the FAA, the National Aviation and Aerospace Administration (NASA) and United Air Lines were involved in a study on the fatigue effects of flying "long haul," Trans Pacific flights; these were sometimes 15 or more hours long. As one result, the FAA proposed a draft AC called Controlled Rest on the Flight Deck; the air carrier industry opposed it and it was shelved. Here we are, in 2017 and aviation still is working on this issue while lives are being lost. Thus let us look at aircrew fatigue. What would be the design and process for a SGDP/ETE to work this safety issue?

The SME's must include, in the main, current air carrier pilots; both U.S. flag-carriers and regional air carriers pilots must be in any/every working group. The goal of such a effort would be to develop a set of recommendations for an FAA FAR on dealing with aircrew fatigue as a recognized safety problem, with rationale. The FAA rule-making process is long and complex. It does have methods for aviation industry input as well as aviation industry objections/agreement. With that in mind, each group in this effort must have an FAA attorney as a member. The material for every group should include accident investigation data (for the prior 30 plus years) where aircrew fatigue was identified as a causal or contributing factor

The author must begin with a *caveat*: What has been shown above is a template for decisions and actions that have not yet and may never be accomplished. Thus, it is, in a word, conjecture. That having been said, here are thoughts on how the described effort can be done, then seen in the light of KM.

Assuming that such an effort would be both approved and aided by both the FAA and the U,S. air carrier community, what could be expected as results and possible products? The *sine qua non* would be the development and approval for the long-delayed and controversial FAA AC on flight crew (controlled) rest to combat he effects of fatigue. Possibly the flight crew rest findings/recommendations of such a effort could become an FAR with the AC as the concomitant "how to." The knowledge manager would indicate that the above are the goals/direction for the workgroup(s)

Since the FAA now has a monitored fatigue risk management program, this program should go over to the air carriers, as part of either an FAA FAR or/and an FAA AC. The knowledge engineer would specify the Pareto/SGDP/ETE/RTD as the process for the work groups…with such modifications seen as necessary. Remember that the SGDP underwent modifications and alterations as it was used over the years. In all the efforts proposed in this Chapter, one could expect no less in the process.

The process here would face many obstacles as is always the case when the FAA and air carrier industry look at what seems to be the same problem and try to find an agreed-on resolution. In KM terms, the best (only?) measure of success would be an approved FAR with an accompanying AC. Possibly an AC alone with air carrier buy-in could be an acceptable success story; both results would entail a internal/external success. System quality, albeit the last 40 year US air carrier accident rate is already an extremely low plateau, would be a reduction in fatigue-related accidents

Example Three

A final candidate for this Chapter's paradigm is Operational Testing (OT); specifically OT&E (Operational Testing & Evaluation). The author worked for some years with a former graduate student of his in this area. The goal of any OT is to determine the system's capability to perform its mission in the operational setting and to both determine and evaluate core operational effectiveness and suitability problems (Maliko-Abraham, Helene and Lofaro, R.J., 2003). Operational tests and evaluations are conducted on production, or production-representative articles, to determine whether systems are operationally effective and suitable. Such testing entails operational testing of production representative test articles and uses typical operational scenarios that are as realistic as possible. It is here that the Pareto/SGDP/ETE/RTD can be modified so that SMEs can use it in OT.

Dumas and Redish (1999) indicate that to be a good OT testing scenario, it needs to be short, developed in the end-user's terminology and, so clear that all participants using the scenario will understand it. A small set of carefully selected SMEs would be used to face-validate the existing scenarios. This SME set also would ensure that all the necessary operational issues were embedded in the scenarios. A second group of SMEs would also assist in the test sequencing as well as the techniques/scalings to be used in workload analysis. While the process of multiple SGDP workshops may seem lengthy, it is not the case. The Pareto/SGDP/ETE/RTD sessions would run consecutively, each new one beginning with the data from the prior one---and each session should need approximately 4 days to complete. The experience and insight as to the realities of the operational arena of the SMEs are both needed and invaluable. Upon completion of the actual OT, all SME members of the two previous sessions would be convened to interpret the results and to make a set of recommendations. These recommendations may include retrofit, modification, and training. The two groups would work independently at first. Then, as they finished, they would convene into one group to finalize their results.

Space/word count limitation preclude all but a bare bones look at the KM aspects of this example. In this effort, the KM processes would be this: the (knowledge) manager would work with the company to define the measures of system(s) operationally effectiveness and suitability. It is here that the Pareto/SGDP/ETE/RTD can be modified so that SMEs can use it in OT's. The knowledge engineer (again, using SME's and the model) would face validate existing (proposed?) scenarios or, much to be preferred, would develop the operational scenarios for the testing that are as realistic as possible. The created/produced knowledge can stored, will be able to be retrieved and will be applied to the OT&E. The testing would have a good chance of minimal iterations as the scenarios would have a high degree of verisimilitude to the actual conditions of operational use.

CONCLUSION

The model explicated above, with exemplar uses, offers significant advances in diversified, accurate decision-making needed over a broad spectrum of problem areas; defining and explicating standards as well as the training needed to achieve resolution. Again, it must be strongly noted that the Chapter is not a report on the results of prior experimentation, rather it is an attempt to meld a modified Delpi procedure with today's KE/KM efforts and today's technology. It provides a start-point template and is also a call for research efforts using this template to validate, or not, it.

FUTURE RESEARCH DIRECTIONS

The Future Research Directions flow from the Conclusions. The model shown needs to be implemented; perhaps as a University or think-tank project...even as a government (at some level) one. The results, pro or con, need to be published and, as always, suggested modifications need to be listed, with rationale, for any future work. It seems that the newest Delphi technique, a web-based communication structure (Hyperdelphi) involving a large number of participants; the use of two or more panels representing different groups (such as policy-makers, experts, citizens), which the administrator can give tasks reflecting their diverse roles and expertise, and make them to interact within *ad hoc* communication structures. These web-based variable communication structures, are designed to make Delphi conferences more fluid and adapted to the hypertextual and interactive nature of digital communication. Future research can revolve around comparison of the accuracy of results using a traditional Delphi, the SGDP, various ETE Delphis (computer and/or web-based) and the model presented in this Chapter. Comparisons can also be made among Hyperdelphi results and those of other Delphi techniques. Finally, new Delphi techniques may emerge from such research and comparisons.

REFERENCES

Bolognini, M. (2001). *Electronic Democracy. Delphi Method and Public Policy-Making*. Rome: Carocci Editore.

Brown, B. B. (1968). Delphi Process: A Methodology for the Elicitation of Opinions of Experts. Santa Monica, CA: The Rand Corporation. (Possible ASMTE Vectors, February, 1968)

Campbell, D. T., & Stanley, J. C. (1966). *Experimental and Quasi-experimental Designs for Research*. Chicago: Rand McNally.

Cook, T. D., & Campbell, D. T. (1979). *Quasi-Experimentation: Design and Analysis for Field Settings*. Boston, MA: Houghton Mifflin Company.

Croft, J. (2014c). Rest Assured. *Aviation Week & Space Technology, 176*(18), 38.

Dumas, J. S., & Redish, J. C. (1999). *A Practical Guide to Usability Testing* (revised ed.). Portland, OR: Intellect Books.

Feigenbaum, E. A., & McCorduck, P. (1983). *The Fifth Generation* (1st ed.). Reading, MA: Addison-Wesley.

Fleishman, E. A., & Quaintance, M. K. (2000). *Taxonomies of human performance: The description of human tasks*. Potomac, MD: Management Research Institute, Inc.

Gibb, G. D., & Lofaro, R. J. (1993). Computer-based Assessment of Air Traffic Controller Abilities: Experimental Selection Tasks. *Proceedings of 1993 Aerospace Medical Association Annual Symposium*.

Gibb, G. M., & Lofaro, R. J. (1994). *Airport Security Screener and Checkpoint Security Supervisor Training Workshop. DOT/FAA/CT-94/*109. Springfield, VA: National Technical Information Service.

Gordon, T. J., & Pease, A. (2006). RT Delphi: An Efficient, "Round-less," Almost Real Time Delphi Method. *Journal of Technological Forecasting and Social Change, 73*(4).

Intano, G. P., Howse, W. R., & Lofaro, R. J. (1991). The Selection of an Experimental Test Battery for Aviator Cognitive, Psychomotor Abilities and Personal Traits. U.S. Army Institute for the Behavioral and Social Sciences RN 91-21.

Intano, G. P., & Lofaro, R. J. (1989). Developmental Research and Initial Validation of the U.S. *Army Aviator Classification Process in Military Testing Association 1989 National Conference Proceedings*.

Intano, G. P., & Lofaro, R. J. (1990). Initial Validation of the Army Aviator Classification Process in Human Factors. *Society. Bulletin, 33*(4).

Jennex, M. (2013). Knowledge Management Success in an Engineering Firm. *Management Review, 2*(2), •••.

Jennex, M. (2014). Knowledge Management. In Wiley Encyclopedia of Management. Somerset, NJ: Wiley and Sons.

Jennex, M., Smolinik, S., & Croasdell, D. (2009). Towards a Consensus Knowledge Management Success Definition. *Vine*, *39*(2), 174–188. doi:10.1108/03055720910988878

Judd, R. L. (1972). *Use of Delphi Methods in Higher Education in Technological Forecasting and Social Change, 4*. Reed Elsevier.

Lofaro, R. J. (1992a). A Small Group Delphi Paradigm. *The Human Factors Society Bulletin, 35*(2), 1-4.

Lofaro, R. J. (1992b). *Workshop on Integrated Crew Resource Management (CRM). DOT/FAA/RD-95/5*. Springfield, VA: National Technical Information Service.

Lofaro, R. J. (1998). Identifying and Developing Managerial and Technical Competencies: Differing Methods and Differing Results. Proceedings of 42nd annual Human Factors and Ergonomics Society.

Lofaro, R. J. (1999). Core Competencies and Individual Differences: Chasm or Linkage? Human Factors and Ergonomics Society/HF&ES Individual Differences in Performance News, 12(2).

Lofaro, R. J. (2013). Knowledge Engineering: A Methodology and Examples. In Encyclopedia of Information Science and Technology, 3rd Edition. IGI-Global.

Lofaro, R. J. (2015). Cognitive Engineering: What's Old is New Again. *Proceedings of the 18th International Symposium on Aviation Psychology (ISAP)*.

Lofaro, R. J., Gibb, G. M., & Garland, D. (1994). *A Protocol for Selecting Airline Passenger Baggage Screeners. DOT/FAA/CT-94/110*. Springfield, VA: National Technical Information Service.

Lofaro, R. J., & Intano, G. P. (1989). Exploratory Research and Development: Army Aviator Candidate Classification By Specific Helicopter *Proceedings of 5th International Symposium of Aviation Psychologists.*

Lofaro, R. J., Intano, G. P., & Howse, W. R. (1990). Final Validation of the Army Aviator Multitrack Algorithm. *Proceedings of the 1990 Annual Military Testing Association Conference.*

Lofaro, R. J., & Maliko-Abraham, H. (2002). Techniques for Developing High Predictive Validity Selection and Screening Test Batteries. *Proceedings of the 2002 World Aviation Congress.*

Lofaro, R. J., & Smith, K. M. (1999). Operational Decision-Making (ODM) and Risk Management (RM): Rising Risk, the Critical Mission Factors and Training. *Proceedings of Tenth International Symposium on Aviation Psychology.*

Lofaro, R. J., & Smith, K. M. (2001a). A Paradigm for Developing Operational Decision-making (ODM). *Proceedings of 2001 SAE World Aviation Congress (WAC) conference.*

Lofaro, R. J., & Smith, K. M. (2001b). Operational Decision-Making: Integrating New Concepts into the Paradigm. *Proceedings of Eleventh International Symposium on Aviation Psychology.*

Lofaro, R. J., & Smith, K. M. (2003). The Finalized Paradigm for Operational Decision-Making (ODM) Paradigm: Components and Placement. *Proceedings of the 12th International Symposium on Aviation Psychology.*

Lofaro, R. J., & Smith, K. M. (2009). Professionalism in Airline Operations…and Accident Investigations? *Proceedings of the 15th International Symposium on Aviation Psychology*.

Lofaro, R. J., & Smith, K. M. (2011). The Aviation Operational Environment: Integrating a Decision-Making Paradigm, Flight Simulator Training and an Automated Cockpit Display for Aviation Safety. In Technology Engineering and Management in Aviation: Advancements and Discoveries. IGI Global.

Maliko-Abraham, H., & Lofaro, R. J. (2001a). A Screening Test for Military Air Traffic Controllers. Proceedings of the 2001 International Military Testing Association (IMTA) Conference.

Maliko-Abraham, H., & Lofaro, R. J. (2001b). Air Traffic Delays: New ATC Equipment Coupled With New ATC Candidate Selection Methods May Be One Answer. *Proceedings of the 2001 National Air Traffic Controllers Association (NATCA)*.

Maliko-Abraham, H., & Lofaro, R. J. (2003). Usability Testing: Lessons Learned and Methodology. *Proceedings of the 2003 International Military Testing Association (IMTA) Conference*.

Meister, D. (1985). *Behavioral Analysis and Measuremen*. New York, NY: Wiley Press.

Pill, J. (1970). *The Delphi Method: Substance, Context: Critique and an Annotated Bibliography*. Technical Memorandum 183. Cleveland, OH: Case Western Reserve University.

Rumizen. (2002). *The Complete Idiot's Guide to Knowledge Management*. New York: Penguin Group/ Alpha Books.

Studer, R., Benjamins, V. R., & Fensel, D. (1998). Knowledge Engineering. *Principles and Methods in Data & Knowledge Engineering*, *25*(1-2), 161–197. doi:10.1016/S0169-023X(97)00056-6

Turoff, M., & Hiltz, R. S. (1996). Computer-based Delphi processes. In Gazing Into the Oracle: The Delphi Method and Its Application to Social Policy and Public Health (pp. 56-88). London: Jessica Kingsley Publishers.

KEY TERMS AND DEFINITIONS

Common Core State Standards (United States Initiative): The Common Core State Standards Initiative is an educational assembly in the United States that details what K-12 students should know in subject areas focused on English language arts and mathematics at the end of each grade. What has been produced is a set of standards by subject and federal government will fund states to incorporate these standards into teaching curricula. This has drawn criticism and generated controversy as a method of federal control of education.

Estimate-Talk-Estimate (ETE) Delphi Process: This Delphi process/technique has lately been adapted for use in face-to-face meetings, and is then called mini-Delphi or Estimate-Talk-Estimate (ETE) Delphi.

Pareto Analysis: A technique used for decision making based on the Pareto Principle, known as the 80/20 rule. It is a decision-making technique that separates a limited number of input factors as having the greatest impact on an outcome, either desirable or undesirable. In its simplest terms, Pareto analysis will typically show that a disproportionate improvement can be achieved by ranking various causes of a problem and by concentrating on those solutions or items with the largest impact. This type of decision-making can be used in many fields of endeavor, from government policy to decision-making.

Real-Time Delphi (RTD): The real-time Delphi is a relatively new and efficient method for collecting and synthesizing expert opinions. RTD is a "roundless" Delphi. The respondents participate by filling out an online questionnaire, and the results—both numerical and qualitative—are updated as responses are recorded in "real time." Respondents can—and are encouraged to—revisit the questionnaire as many times as they want. Each time, they are shown their own responses as well as the updated answers of the others, and they can revise and change their own inputs based on this feedback.

United States Federal Aviation Administration: (FAA): Is the national aviation authority of the United States. As an agency of the United States Department of Transportation, it has authority to regulate and oversee all aspects of American civil aviation.

United States Federal Aviation Administration (FAA) Advisory Circulars (AC): The FAA normally publishes advisory circulars (ACs) designed to provide assistance and guidelines in complying with FARs; they serve as a "how to" template. The AC's are titled and numbered and are not legally binding.

United States Federal Aviation Administration Regulations (FAR): All are in the Combined Federal Regulations (CFR), Title 14, Aeronautics and Space; each a descriptive title and number; they are legally binding.

Section 2
Issues in Knowledge Sharing

Chapter 6
Exploring Knowledge-Sharing Barriers and Their Implications

Salih Yeşil
Kahramanmaraş Sütçü İmam University (KSU), Turkey

Bengü Hırlak
Kilis 7 Aralık University, Turkey

ABSTRACT

This chapter focuses on knowledge-sharing barriers and their implications over knowledge sharing and individual innovative behavior and explores the role of knowledge sharing on individual innovative behavior. Considering limited studies in the literature, this study provides further evidences regarding the implications of knowledge-sharing barriers and knowledge sharing in the workplace. Data was collected from 83 academic staff in a higher education institution and analyzed with Smart PLS. The results showed that organization-related knowledge-sharing barriers were negatively related to knowledge sharing. The results also indicated that individual knowledge-sharing barriers had a negative effect on individual innovative behaviors. There was no link found in this study between knowledge sharing and individual innovative behavior. These findings provided empirical evidences to the further development of knowledge management and innovation research, and insights regarding how to better foster knowledge sharing and innovative behavior in academe.

INTRODUCTION

Organizations have become interested and invested in knowledge management because of increasing competitive pressures for efficiency, growth and cycle time reduction, and flexibility (Ryan *et al.*, 2012; Wang & Noe, 2010). Inkinen (2016) regarded the knowledge management as one of the most discussed new management methods. Knowledge management involves a set of processes aimed at maximizing the outcomes of the knowledge produced within a business unit, a firm, a network of firms (Filieri, 2010). The development and practice of knowledge management related activities is continuously and dramatically increasing in various types of organizations (Halawi *et al.*, 2005; Ramjeawon & Rowley,

DOI: 10.4018/978-1-5225-5427-1.ch006

2017; Sözbilir & Yeşil, 2016) due to its link with innovation and performance (Filieri, 2010; Inkinen, 2016; Ramjeawon & Rowley, 2017) .

Knowledge sharing is regarded as an important phase of knowledge management (Chong *et al.,* 2014; Ramjeawon & Rowley, 2017; Wang & Noe, 2010) and determines the success of knowledge management initiatives (Wang & Noe, 2010). Knowledge sharing is defined "the exchange of knowledge between and among individuals, and within and among teams, organizational units, and organisations" (Khalil & Shea, 2012). Previous studies have studied the antecedents and consequences of the knowledge sharing. In reviewing the literature on knowledge sharing, Wang and Noe (2010) have identified a number areas including organizational context, interpersonal and team characteristics, cultural characteristics, individual characteristics, and motivational factors that affect knowledge sharing. Many theoretical and empirical studies argued and found that knowledge sharing has various implications for individuals (Aulawi *et al.,* 2009; Mura *et al.,* 2013; Radaelli *et al.,* 2014; Wang & Noe, 2010), teams (Maccurtain *et al.,* 2008; Weber *et al.,* 2011) and organizations (Chong *et al.,* 2014; Kamasak & Bulutlar, 2010; Lin, 2007; Mehrabani & Mohamad, 2011; Sözbilir &Yeşil, 2016). Unfortunately, barriers towards knowledge sharing prevent organizations to capitalize on their knowledge assets and skills (Athar & Evans, 2015; Dee & Leisyte, 2017; Wang & Noe, 2010).

The literature on knowledge management identifies individual (lack of time, trust, social network as well as difference in age, education and expertise), organization (lack of organizational vision, strategy and structure along with hierarchy and competition between departments) and technology (inadequate training and technical support regarding IT and various system related issues) related barriers that inhibit knowledge sharing and management initiatives and implementations (Athar & Evans, 2015; Chong *et al.,* 2014; Dee & Leisyte, 2017; Khalil & Shea, 2012; Ramjeawon & Rowley, 2017; Riege, 2005). Among the others, knowledge sharing barriers can also diminish innovative capability of the individuals in the workplace (Athar & Evans, 2015; Patterson *et al.,* 2009).

Current study concentrates on the knowledge sharing barriers and their impact on knowledge sharing and individual innovative behavior, and also explores the role of knowledge sharing on individual innovative behavior. Data for the research was collected from academics working in a public higher education institution in Turkey. The data was analyzed by using Smart PLS and the results were presented and discussed with regard to theoretical and managerial implications. The findings of the study are expected to contribute to the related literatures in many ways as explained below.

Despite the presence of the knowledge sharing barriers discussed and identified in the literature (Akgün *et al.,* 2017; Athar & Evans, 2015; Barson *et al.,* 2009; Lilleoere & Hansen, 2011; McLaughlin *et al.,* 2008; Riege, 2005; Santos *et al.,* 2012), there seem to be very few empirical studies testing the effect of these barriers on knowledge sharing (e.g., Khalil & Shea, 2012), thus showing the importance of undertaking this type of research. Besides, to our knowledge, there seems to be no empirical study investigating the effect of knowledge sharing barriers on individual innovative behavior. Hence, this provides a great impetus to undertake the current study.

Moreover, studying the effect of knowledge sharing on individual innovative behavior can contribute to the literature because only few studies (Hu *et al.,* 2009; Kang & Lee, 2017; Radaelli *et al.,* 2014; Mura *et al.,* 2013) have attempted to look at the proposed link through empirical studies despite the theoretical association between these two constructs (Aulawi *et al.,* 2009; Kang & Lee, 2017). Supporting this, Radaelli *et al.,* (2014:400) recently argued that "in the literature, the potential link between employee's knowledge sharing and their own innovative behaviors remains both under-theorized and empirically untested", thus encouraging future studies in this research domain.

In addition, having considered relatively few studies of knowledge management coming from public organizations and from developing countries (Chiu & Chen, 2016; Monavvarian & Kasaei, 2007; Ramjeawon & Rowley, 2017; Syed-Ikhsan & Rowland, 2004), this study may provide important insights regarding the knowledge management in different cultural and organizational settings. To our knowledge, current study appears to be one of the few studies (e.g., Akgün *et al.,* 2017; Işık & Aydın, 2016; Sezgin *et al.,* 2015) related to knowledge sharing and knowledge sharing barriers and individual innovative behavior conducted in Turkey. Thus, findings from this study conducted in a public university in a developing country is expected to bring new insights into the understanding knowledge sharing barriers and its implications.

The fact that this study was conducted in a higher education institution is also important because higher education institutions are also knowledge-driven organizations (Fullwood *et al.,* 2013; Ramjeawon & Rowley, 2017; Duderstadt, 2000) through knowledge creating, sharing and dissemination (Chong *et al.,* 2014; Dee & Leisyte, 2017; Fullwood *et al.,* 2013; Ramjeawon & Rowley, 2017). Knowledge sharing is also integral for the success of knowledge management in higher education institutions (Chong *et al.,* 2014; Dee &Leisyte, 2017; Khalil & Shea, 2012) yet, presence of different barriers may undermine knowledge sharing (Chong *et al.,* 2014; Khalil & Shea, 2012; Ramjeawon & Rowley, 2017; Wang *et al.,* 2010). Empirical studies in academic context seem to be limited compared to other organizational settings (Chong *et al.,* 2014; Fullwood *et al.,* 2013), thus increasing the values of current study. Supporting our study, Ramjeawon and Rowley (2017) encouraged further research on knowledge management processes and policies in universities in both developed and developing countries to better understand and enhance our knowledge in this area.

Like the other sectors, higher education also competes on a number of issues such as obtaining funds from different sources and attracting best qualified students. Societies expect universities to play a leading role in helping the organizations to be more innovative (Powell, 2012). These issues along with globalization and increasing need, expectation and demand of various stakeholders force universities and their members to redesign their functions and to be more flexible and more sensitive to innovation, innovation related activities and behavior (Powell, 2012; Ramjeawon & Rowley, 2017; Saginova & Belyansky, 2008). Moreover, being a knowledge-driven organizations, higher education institutions need to pay attention to the enablers and barriers of knowledge sharing and their implications in order to capitalize on their knowledge assets and skills (Chong *et al.,* 2004; Ramjeawon &Rowley, 2017). The argument presented above are also true for the Turkish higher education institutions. In order to contribute to the literature, this study was designed to better identify and understand the knowledge sharing barriers and their implications in a higher education institution in Turkey.

In short, conducting an empirical study with objectives and reasoning outlined above is expected to provide further empirical evidences to knowledge management and innovation research in general, and insights regarding how to better foster knowledge sharing and innovative behavior in higher education institutions. Findings from this study regarding the identifying and removing barriers toward knowledge sharing and individual innovative behaviors will eventually improve the institutional capability for the better knowledge management and innovation performance.

The following sections explain the theoretical underpinnings of the study, hypotheses development, methodology, results, discussion and conclusions respectively

BACKGROUND

Knowledge Sharing

Both academics and practitioners have shown great interest in knowledge management research (Inkinen, 2016; Jennex *et al.,* 2014) because of the fact that knowledge management plays a crucial role in determining the success of the organizations (Inkinen, 2016; Jennex *et al.,* 2014; Syed-Ikhsan and Rowland, 2004). Rapid transformation of customer requirements, shorter product life cycles, and the growing costs of technologies have also prompted the organizations to invest in knowledge management (Ryan *et al.,* 2012; Wang & Noe, 2010). Different definitions of knowledge management exist in the literature depending on the researcher and academic discipline (Filieri, 2010). Filieri (2010) defined knowledge management (KM) as "a set of processes aimed at maximizing the outcomes of the knowledge produced within a business unit, a firm, a network of firms". Knowledge management in organizations aims to make knowledge available to the right person with right timing and form (Filieri, 2010). Knowledge sharing plays crucial role in determining the success of knowledge management initiatives (Radaelli *et al.,* 2014; Wang & Noe, 2010).

Knowledge sharing constitutes one of the most important parts of knowledge management (Chong *et al.,* 2014; Gupta & Govindrajan, 2000; Riege, 2005). As an intangible asset of the organisations, the management of knowledge is more complex than any other physical assets (Filieri, 2010). Knowledge sharing is defined as "the exchange of different types of knowledge between individuals, groups, units, and organisations" (Khalil & Shea, 2012). Knowledge sharing is mainly regarded as the process of knowledge-exchange in a group or an organization (Bircham- Connoly *et al.,* 2005: Cheng et al., 2009; Lilleoere & Hansen, 2011). Knowledge sharing also refers to the "process of capturing knowledge or moving knowledge from a source unit to a recipient unit" (Bircham- Connoly *et al.,* 2005:1).

Previous studies have studied the antecedents and consequences of the knowledge sharing. In reviewing the literature on knowledge sharing, Wang &Noe (2010) have identified a number antecedents including organizational context (e.g., organizational culture and climate, management support, rewards and incentives, organizational structure), interpersonal and team characteristics (e.g., team characteristics and processes, diversity), cultural characteristics, individual characteristics, and motivational factors (e.g., interpersonal trust and justice, individual attitudes) that affect knowledge sharing For example, Cheng *et al.* (2009) looked at the individual, organizational and technology factors affecting the willingness to share knowledge. They found that incentive systems and personal expectation are two leading factors in driving academics to engage in knowledge sharing activity. "Forced" participation is not an effective way in terms of knowledge sharing among academics. In another study, Jain *et al.* (2007) reported that rewards and performance appraisal were positively related to knowledge sharing among academic staff. Please see the work of Wang & Noe (2010) for the review of related empirical studies in different organizational settings.

In terms of consequences of knowledge sharing, many theoretical and empirical studies argued and found that knowledge sharing has various implications for individuals (Aulawi *et al.,* 2009; Chong *et al.,* 2014; Mura *et al.,* 2013; Radaelli *et al.,* 2014; Wang & Noe, 2010), teams (Curado *et al.,* 2017; Maccurtain *et al.,* 2008; Weber *et al.,* 2011) and organizations (Dee & Leisyte, 2017; Eidizadeh *et al.,* 2017; Kamasak & Bulutlar, 2010; Lin, 2007; Mehrabani & Mohamad, 2011; Sözbilir& Yeşil, 2016).

In relation to how knowledge is shared in organization, McDermott and O'Dell (2001) noted that there is not one right way to get people to share, but many different ways depending on the values and style of the organization. Van den Hooff and Van Weenen (2004) made a distinction between two different forms of knowledge sharing: knowledge donating (communication to others what one's personal intellectual capital is) and knowledge collecting (consulting colleagues in order to get them to share their intellectual capital) and argued that knowledge needed is obtained using these two methods. This study adopts the two dimensional knowledge sharing (knowledge collecting and knowledge donating) identified by Van den Hooff and Van Weenen (2004) and were used in several studies with high validity and reliability (e.g., Chong *et al.,* 2014; Lin *et al.,* 2007; Sözbilir & Yeşil, 2016). With respect to what type of knowledge is shared among the academics, Fullwood *et al.* (2013) noted that knowledge shared among the academics tend to be research, teaching and learning based.

Knowledge Sharing Barriers

In response to the growing importance of knowledge management, researchers have begun to study knowledge sharing barriers at different organizational and industrial settings (Akgün *et al.,* 2017; Athar &Evans, 2015; Barson *et al.,* 2009; Huang & Davison, 2008; Kaps, 2011; Lilleoere & Hansen, 2011; McLaughlin *et al.*, 2008; Shahid & Alamgir, 2011; Szulanski, 1996; Kim & Joh 2005; Riege, 2005; Santos *et al.* 2012). They have argued and identified many barriers that prevent knowledge sharing activities in the organizations.

Contrary to conventional wisdom that blames primarily motivational factors, findings of Szulanski (1996) showed that the major barriers to internal knowledge transfer were knowledge-related factors such as the recipient's lack of absorptive capacity, causal ambiguity, and an arduous relationship between the source and the recipient. Kim and Joh (2005) conducted a study in construction organizations and found seven key barriers to knowledge sharing; lack of time, poor communication skills, lack of up-to-date knowledge, intolerance of failure, lack of transparent rewards and recognition systems, shortage of infrastructure supporting knowledge sharing, lack of integration of IT systems and processes. Huang and Davison (2008) reported that loss of knowledge power is the only potential barrier in the organization. Results from the study of McLaughlin *et al.* (2008) showed that several knowledge sharing barriers (distance, trust, risk, and available technology etc.) affect knowledge creation and sharing. Barson *et al.* (2009) found three types of knowledge sharing barriers related to technology, organization and people. They reported that people related barriers tend to be the main barrier toward knowledge sharing in their study. Shahid and Alamgir (2011) distinguished three types of barriers to knowledge sharing; organizational, individual and technological. They specifically looked at the role of ICT in creating knowledge sharing barriers and reducing the other barriers. Kaps (2011) argued that cultural differences in terms of the communication style (explicit or implicit communication in low or high context cultures) or perception of risk or power often hinder people in sharing their knowledge. Lilleoere and Hansen (2011) reported that knowledge sharing barriers has negative influence on knowledge sharing practices. Santos *et al.* (2012) studied knowledge sharing barriers in the context of project management. The barriers identified through their research are; codification process, inadequate information technology, lack of initiative and strategy by the workers, and lack of time and resources. In a recent study, Athar and Evans (2015) conducted a empirical study in the pharmaceutical industry and found nine categories of knowledge sharing barriers (include high cost of sharing knowledge, information technology limitations, knowledge-hiding, lack of socialization, lack of trust culture, non-educational mindset, organizational politics, poor leadership and time pressure). In a recent study Akgün *et al.* (2017) found knowledge, individual, and organization related barriers in software development teams.

Despite various studies in different organizational and industrial settings discussed above, only few studies were conducted in higher education institutions (e.g., Chong *et al.,* 2014; Khalil & Shea, 2012; Ramjeawon & Rowley, 2017; Wang *et al.,* 2010). Wang *et al.* (2010) conducted an empirical study on research and innovation teams in a university and found psychological, managerial and communication related knowledge sharing barriers. Based on the empirical study in a higher education institution, Khalil and Shea (2012) identified four important barriers to knowledge sharing; bounded individual capacity, inadequate organizational capability, fear of knowledge revelation, and knowledge nature. Chong *et al.* (2014) found several individual, organizational and technological barriers (especially lack of rewards and recognition, inadequacy of IT systems and trust in and relationships with colleagues) in higher education institutions. In another study, Ramjeawon and Rowley (2017) found several barriers (a lack of policies and appropriate reward mechanisms, resources, data, funding and time for research, frequent leadership changes, a lack of a knowledge-sharing culture and research repositories and weak industry-academia linkages) in a university.

The barriers identified in the previous theoretical and empirical studies tend to concentrate around individual, organizational and technological related factors. Following this, our study adopts three different types of knowledge sharing barriers namely, individual, organizational and technological, identified by Riege (2005). According to Riege (2005), individual barriers are related to lack of time, trust, social network as well as difference in age, education and expertise. Organizational barriers concern lack of organizational vision, strategy and structure along with hierarchy and competition between departments. Technology related barriers involve inadequate training and technical support regarding IT and various system related issues.

Individual Innovation Behavior

Innovation is defined "as the development and implementation of new ideas by people who over time engage in transactions with others within an institutional order" (Van De Ven, 1986:590). Amabile (1996:1) defined innovation as "the successful implementation of creative ideas within an organization". Almost everybody from academic and practitioner side tends to agree the importance of innovation for the competitiveness of organizations as well as social and economic development of societies (Eidizadeh *et al.,* 2017; Kim, 1997; Salaman & Storey, 2002; Scholl, 2005; Terziovski, 2007). Despite the risk and uncertainty, successful innovation can have a significant effect on firms' financial and economic performance (Unsworth & Parker, 2003; Marques & Ferreira, 2009).

Individual's ability to innovate in the workplace is one of the most important ways to produce innovation in the organizations (De Jong & Hartog, 2007; Radaelli *et al.,* 2014; Xerri & Brunetto, 2011; Wang & Zhu, 2012). The willingness and ability of individuals to innovate ensure the flow of innovation in organizations. Many researchers in the literature believe in the importance of innovative work behavior for the performance and survival of the organizations (Carmeli *et al.,* 2006; De Jong & Hartog, 2007; Janssen, 2000; Radaelli *et al.,* 2014; Scott & Bruce, 1994; Unsworth & Parker, 2003). Organizations cope with the changes in the business environment through emphasizing human resources and capitalizing their innovation ideas and behavior (Unsworth & Parker, 2003).

Due to its rich and elusive nature, many names (e., individual innovation, individual innovation behavior, innovative behavior, innovative work behavior etc.) and definitions of individual innovation behavior (e.g., De Jong & Hartog 2008; Janssen, 2000; Scott & Bruce, 1994; Wu *et al.,* 2011) exist in the literature. For instance, Janssen (2000:288) defined innovative work behavior, "as the intentional creation, introduction and application of new ideas within a work role, group or organization, in order to benefit role performance, the group, or the organization" based on West and Farr (1989) and West (1989). Innovative behavior is defined in the health sector as "the introduction and implementation of new ideas, processes, products, or procedures designed to significantly benefit the patient" (Weber *et al.,* 2011). Scott and Bruce (1994) views individual innovation as a multistage process with different activities and different individual behaviors necessary at each stage. They outlined three stages relevant to individual innovations, namely idea generation, coalition building and implementation. Wu *et al.* (2011) argued that in contrast to innovation at the team or organization level, individual innovation behavior is based on an individual's engagement in generating and applying new ideas and approaches in the workplace. De Jong and Hartog (2008:5) argued that "innovative work behavior typically includes exploration of opportunities and the generation of new ideas (creativity related behavior), but could also include behaviors directed towards implementing change, applying new knowledge or improving processes to enhance personal and/or business performance (implementation oriented behavior)". They distinguished four related dimensions of innovative work behavior: opportunity exploration, idea generation, championing and application. In this study, we use the individual innovation behavior concept developed by De Jong and Hartog (2008) as one dimension, measured through combination of opportunity exploration, idea generation, championing and application. Individual innovative behavior has also been measured and investigated as one dimension in several studies (e., De Jong & Hartog, 2008; Janssen, 2000; Scott & Bruce, 1994; Kleysen & Street, 2001). Introducing innovative ideas, contribution to implementing new ideas, searching out new working methods, techniques, and generating 201 solutions to the problems are some of the behaviors considered and named as individual innovative behavior from academic staff in this study.

Researchers have interested in individual innovative behaviors in terms of antecedent, construct itself and outcomes (De Jong & Hartog, 2008). Studies looking at the antecedent of individual innovative behavior looked at the various factors. In reviewing the literature, Parzefall *et al.* (2008) looked at the main organizational, team, job and individual level factors as antecedents of employee innovativeness. Leader-Member-Exchange (LMX), satisfaction with HR practices (employee influence, flow, rewards and work content) (Sanders *et al.,* 2010), leadership, individual problem-solving style, and work group relations (Scott & Bruce, 1994), knowledge sharing (Hu *et al.,* 2009), creative self-efficacy (Hsu *et al.,* 2011) need for cognition (Wu *et al.,* 2011), self-leadership (Carmeli *et al.,* 2006), participative leadership, external work contacts (De Jong & Hartog, 2008), individual and organizational learning (Xiaojun & Peng, 2010), personality (Yeşil & Sözbilir, 2013) and job autonomy and learning goal orientation (Sazandrishvili, 2009) has been found to affect individual innovative behavior.

While some researchers have interested in explaining and validating the individual innovative behavior (e.g., De Jong & Hartog, 2008; Scott & Bruce, 1994; Wu *et al.,* 2011), others have looked at the implications of individual innovative behaviors (e.g. De Jong & Hartog, 2008). Individual innovative behaviors were positively related to innovation output (suggestions and implemented innovations) in a study conducted by De Jong & Hartog (2008).

HYPOTHESIS DEVELOPMENT

Knowledge Sharing Barriers and Knowledge Sharing

It is argued that if the organizations want to be successful in knowledge sharing, they need to pay attention to the knowledge sharing barriers (Athar &Evans, 2015; Chong *et al.,* 2014; Dee & Leisyte, 2017; Riege, 2005). To extend the knowledge on the implications of knowledge sharing barriers, Riege (2005) suggested to further research the role of knowledge sharing barriers in organizational process and outcomes. Therefore, this study focuses on knowledge sharing barriers and how they affect individuals from collecting and donating knowledge in the workplace.

Previous studies identified a number of barriers for effective knowledge sharing in different organizations (Athar & Evans, 2015; Barson *et al.,* 2009; Huang & Davison, 2008; Kaps, 2011; Lilleoere & Hansen, 2011; McLaughlin *et al.,* 2008; Riege, 2005; Santos *et al.,* 2012; Shahid & Alamgir, 2011). The barriers found in previous studies have been discussed in detail in the previous sections. Just to give a few example from higher education institutions, Chong *et al.* (2014) found several individual, organizational and technological knowledge sharing barriers (especially lack of rewards and recognition, inadequacy of IT systems and trust in and relationships with colleagues) in higher education institutions. In another study, Ramjeawon and Rowley (2017) found several knowledge sharing barriers (a lack of policies and appropriate reward mechanisms, resources, data, funding and time for research, frequent leadership changes, a lack of a knowledge-sharing culture and research repositories and weak industry-academia linkages) in a university.

Despite the presence of the barriers identified in the literature, there seem to be very few empirical studies testing the effect of these barriers on knowledge sharing (e.g., Khalil & Shea, 2012). Khalil and Shea (2012) found that knowledge sharing barriers are negatively related knowledge sharing effectiveness measures. Therefore, it is expected that individual, organization and technology-based knowledge sharing barriers have the potential to negatively affect knowledge sharing (knowledge collecting and donating) among the academic staff. Based on these arguments and findings, following hypothesis is suggested:

H1: Knowledge sharing barriers (individual, organizational and technological related) are negatively related to knowledge sharing (knowledge collecting and knowledge donating).

Knowledge Sharing Barriers and Individual Innovative Behavior

Individuals in the workplace is one of the most important sources to produce innovation (Carmeli *et al.,* 2006; De Jong & Hartog, 2008; Radaelli *et al.,* 2014; Ramamoorthy *et al.,* 2005; Xerri & Brunetto, 2011). The willingness and ability of staff to innovate ensure the flow of innovation in the organizations (Åmo, 2006; Kessel *et al.,* 2012; Radaelli *et al.,* 2014). Effective knowledge sharing has the capacity to foster individual and organizational learning, performance, competitiveness and innovation (Alam *et al.,* 2009; Angel, 2006; Caloghirou *et al.,* 2004; Kamasak & Bulutlar, 2010; Lin, 2007; MacCurtain *et al.,* 2008; Mehrabani & Mohamad, 2011; Riege, 2005; Wang & Noe, 2010). It also helps individuals to develop and facilitate individual innovative capability and behavior (Aulawi *et al.,* 2009; Radaelli *et al.,* 2014; Xerri & Brunetto, 2011). From these argument, one can suggest that any types of barriers toward knowledge sharing can also undermine the innovation related activities (introducing innovative ideas, searching out new method and techniques, and generating original solutions to the problems etc.), namely individual innovative behavior, from academics. In this study, individual related knowledge sharing barriers (measured as lack of trust, time, social network etc.), organizational barriers (measured as hierarchy, top-down communication etc.) and technology related barriers (measured as lack of training and technical support for IT, mismatch between individual needs and IT applications) are also argued to have a negative influence on individual innovative behavior in academe.

To support the proposed link between knowledge sharing barriers and individual innovative behavior, a numbers of arguments from the related literature are presented. For instance, lack of trust, time and social network among many other factors may reduce the social relations among the academic staff, which in turn may diminish individual innovative capability. To back up this argument, Patterson *et al.* (2009) noted that "specifically, the exposure to a variety of approaches and perspectives within or outside a group is likely to facilitate cognitive processes related to creativity and innovation, such as divergent, autonomous, and flexible thinking". Therefore, barriers toward social interactions such as lack of time and trust can reduce creativity and innovative behavior of the academics. On the other hand, hierarchy and top-down communication are the characteristics, which may lead to take away the motivation and courage of the academics to initiate change and implement new ideas. Supporting this view, Patterson *et al.* (2009) argued that encouragement of risk-taking, an open style of communication, participative and collaborative style, giving autonomy and freedom, support for innovation are likely to enhance innovative behavior. Moreover, mismatch between individual needs and IT applications, lack of training and technical support for IT as well as lack of integration and compatibility among the IT systems can have detrimental effect on innovative initiatives from academic staff. It is known that IT is significantly associated with knowledge sharing (Riege, 2005) and innovation (Austin, 2007; Chen & Tsou, 2007). Through routinizing, digitizing, and automating, IT creates value and help organizations to implement new ideas (Austin, 2007). Chen and Tsou (2007) raised the importance of IT for innovation related initiatives, reported that adoption of information technology is positively related to service innovation and reflected upon the importance of IT for innovation related initiatives. Riege (2005:30) noted that "technology now is a main driver in most companies and industry sectors that most day-to-day activities highly depend on". IT is also important in academic environment for teaching and researching. Day-to-day activities of academics depends highly on IT and its effectiveness. Duderstadt (2000) also mentioned how IT can increase capacity of the staff in academe in term of knowing and doing things, communicating and collaborating with others. Hence, any problems associated with it prevent academic to better perform their job and undermine their knowledge creation, sharing, learning and problem solving.

In short, individual, organizational and technology related barriers toward knowledge sharing have also the capacity to block any innovation related initiatives of academics in higher education institutions. Based on these arguments, the following hypothesis is proposed:

H2: Knowledge sharing barriers (individual, organizational and technological) is negatively associated with individual innovative behavior.

Knowledge Sharing and Individual Innovative Behavior

Many researchers contend that knowledge sharing has an important role in determining innovation success of the firms (Bhirud *et al.,* 2005; Kamaşak & Bulutlar, 2009; Keyes, 2008; Mehrabani & Mohamad, 2011; Wangpipatwong, 2009), top management teams (Maccurtain *et al.,* 2008), diverse teams (Weber *et al.,* 2011) and individuals (Aulawi *et al.,* 2009; Mura *et al.,* 2013; Radaelli *et al.,* 2014; Rebernik &

Širec, 2007). In this study, knowledge sharing is proposed to have a positive influence on individual innovative behavior of academic staff.

Knowledge and knowledge sharing are particularly important for successful innovation in organizations. Supporting this view, Nonaka (1994) argued that obtaining and exploiting of new source of information is necessary for innovative behavior of the firms. Similarly, MacCurtain *et al.* (2008) noted that motivation regarding knowledge sharing has positive impact on innovation within top management teams. Likewise, Keyes (2008) contended that knowledge sharing in organizations can be translated into innovation, productivity and competitive advantage.

Knowledge sharing has also the potential to positively affect individual innovative behavior. According to Aulawi *et al.* (2009), knowledge sharing can stimulate individuals to think more critically and more creatively. Similarly, Rebernik and Širec (2007) argued that the exchange of knowledge among different agents is of crucial importance for innovative behavior in organizations. In the same way, both Radaelli *et al.* (2014) and Mura *et al.* (2013) argued that knowledge sharing increases individual innovativeness with regard to propensity and capacity to generate, promote and implement new ideas.

Despite the theoretical association between these two constructs, only few empirical studies have attempted to look at the link between knowledge sharing and individual innovative behavior (Radaelli *et al.,* 2014). For instance, Hu *et al.* (2009) reported that knowledge sharing (consist of two parts; symbiosis and reputation and altruism) is positively related to employee service innovation behaviors. In a recent study, Mura *et al.* (2013) found that knowledge sharing increases individual innovativeness in terms of promoting and implementing new ideas. While Radaelli *et al.* (2014) also found the positive link between knowledge sharing behavior and individual innovative behavior, Kang and Lee (2015) reported indirect effect of knowledge sharing on individual innovative behavior through absorptive capacity.

Based on these arguments, we propose that knowledge sharing affects individual innovative behaviour in academe and suggest the following hypothesis:

H3: Knowledge sharing (knowledge donating and knowledge collecting) is positively related to individual innovative behavior.

METHODOLOGY

Sample and Data Collection Instrument

The participant consisted of academic staff from four faculties in a higher education institution. The institution is located in Kilis in Turkey. The institution is a research and teaching based university and has four faculties. It has both undergraduate and postgraduate programs with approximately 230 academic staff. All available academic staff was surveyed during one week of data collection process. A questionnaire was used to collect the data (Hair *et al.,* 2006) because of its characteristics (e.g., relatively less time-consuming, inexpensive and easy to administer). One of the researchers distributed questionnaire and collected them later on. The participation in the study was voluntary. Before administering the questionnaire, a pilot study was conducted to make sure that all the questionnaire items are valid, understandable and acceptable to the intended participants. Eighty-three usable questionnaires were returned with a 36% response rate.

Measures

The questionnaire items were derived mainly from previous studies and modified to fit to the nature of this study. This study employs individual barriers, organizational barriers and technology barriers taken from Riege (2005) to measure knowledge sharing barriers among the academic staffs in a higher education institution. Knowledge sharing barriers have thirty-eight statements (17 items for individual barriers; 13 items for organizational barriers; 8 items for technological barriers). All knowledge collecting and donating items were taken from Van Den Hooff and De Ridder (2004) and Lin (2007). Six items for knowledge donating were taken from Van Den Hooff and De Ridder (2004) and one item from Lin (2007). Researchers also added two items for the knowledge donating scale. Four items for knowledge collecting were taken from Van Den Hooff and De Ridder (2004) and one item was added by the researchers. Some items taken from the previous studies were modified to the research context, for instance, department and university were used instead of business unit and organization respectively when necessary. Innovation behavior scale of academics was measured by taking nine items from the study of De Jong and Hartog (2008). A Likert type scale with five response options ranging from strongly disagree to strongly agree was used for measuring all the items. Some of the items were dropped during the pilot study and later in the data analyses to meet significance level and reliability measures. All questionnaire items were originally written in English. Original items were translated into Turkish through translation-back-translation procedure with the help of two Turkish-English bilingual lecturers in management field with experience in this type of research. First, one bilingual speaker translated the English version of the items into the Turkish version. Then, the other bilingual lecturer back translated the Turkish version of items into the English version to make sure that the English and Turkish versions of the items reflect a high degree of comparable accuracy.

Data Analysis

All analyses were performed based on the data collected through a survey by using PLS-Graph (build 1126), a Partial Least Squares (PLS) Structural Equation Modelling (SEM) tool (Ringle *et al.,* 2005).

RESULTS

Table 1 summarizes respondents' demographic profile. According to the table, majority of the respondents are male (71.1%), mostly married (51.8%), 25 to 29 years old (41%), and has 1-3 years of work tenure (48.2%). Majority of the participants work for faculty of arts and science (33.7%) and faculty of education (32.5%). Most of the respondents work as a research assistant (49. 4%) and assistant professors (34. 9%).

The research model reflecting the research hypotheses H1 through H3 was analyzed using Smart PLS 2.0. Smart PLS simultaneously assesses the psychometric properties of the measurement model and estimates the parameters of the structural model. Reliability results of testing measurement model are shown in Table 2.

The results reveal that the measures are robust in terms of their internal consistency reliabilities as indexed by their composite reliabilities. The composite reliabilities of different measures in the model range from 0.85 to 0.91, which exceeds the recommended threshold value of 0.70 (Nunnally, 1978). The average variance extracted (AVE) for each measure exceeds 0.50, consistent with recommendation

Table 1. Respondents' demographic profile

Demographic Characteristic	Frequency	Percentage (%)	Demographic Characteristic	Frequency	Percentage (%)
Gender			**Faculty**		
			Faculty of Arts and Science	28	33,7
Male	59	71,1	Faculty of Engineering and Architecture	14	16,9
Female	24	28,9	Faculty of Economy and Administrative Sciences	14	16,9
Total	83	100,0	Faculty of Education	27	32,5
			Total	83	100,0
Marital Status			**Staff Title**		
			Professor	2	2,4
Married	43	51,8	Associate Professor	5	6,1
Single	40	48,2	Assistant Professor	29	34,9
Total	83	100,0	Research assistant	41	49,4
			Instructor	6	7,2
			Total	83	100,0
Age			**Work Tenure**		
20-24	7	8,4			
25-29	34	41,0	Less than one year	18	21,7
30-34	12	14,5	1-3 year	40	48,2
35-39	11	13,3	4-6 year	10	12,0
40-44	12	14,5	7-9 year	7	8,5
45+	7	8,3	10 year and older	8	9,6
Total	83	100,0	**Total**	83	100,0

Table 2. Reliability assessment of the measurement model

	AVE	Composite Reliability	R Square	Cronbach's Alpha	Ind Bar	Org Bar	Tech Bar	Know Co	Know Do	Inno Behav
IndBar	0,5263	0,8535	0,0000	0,8119	**0,7254**					
OrgBar	0,5198	0,8942	0,0000	0,8701	0,6412	**0,7206**				
TechBar	0,5650	0,9003	0,0000	0,8709	0,3531	0,4030	**0,7516**			
KnowCo	0,6630	0,9075	0,1943	0,8728	-0,2069	-0,4288	-0,2077	**0,8242**		
KnowDo	0,5343	0,9110	0,1215	0,8906	-0,1835	-0,3443	-0,1562	0,6797	**0,7309**	
InnoBehav	0,5055	0,9004	0,2239	0,8763	-0,3005	-0,1934	0,1782	0,1449	0,2190	**0,7109**

Note: IndBar: Individual Barriers, OrgBar: Organisational Barriers, TechBar: Technological Barriers, KnowCo: Knowledge Collecting, KnowDo: Knowledge Donating, InnoBehav: Innovative Behavior

of Fornell and Larcker (1981). Table 2 also shows the test results regarding discriminant validity of the measure scales. The bolded elements in the matrix diagonals, representing the square roots of the AVEs, are greater in all cases than the off-diagonal elements in their corresponding row and column. This result supports the discriminant validity of the scales.

Convergent validity is tested with Smart PLS by extracting the factor loadings and cross loadings of all indicator items to their respective latent construct. According to the results (Results are available from the researcher if asked), all the items loaded on their respective construct from lower bound of 0.53 to an upper bound of 0.86 and more highly on their respective construct than on any other construct. All items load more highly on their respective construct than the other construct showing convergent validity. Most of the items with few exceptions loaded above or close to the threshold level of 0.50 (Hair *et al.,* 2006). Furthermore, each item's factor loading on its respective construct was highly significant (P<0.01). The loadings found through the analysis confirm convergent validity of measures for the latent constructs. Please note that some of the items of research constructs were dropped from the model due to their low factor loading or reflect high loading on more than one factor.

The results of the structural model were depicted in Figure 1. The beta values of path coefficient shows the direct influences of predictor upon the predicted latent constructs. Although individual, organizational and technological barriers are hypothesized to negatively affect knowledge collecting and knowledge donating; only organizational barriers have a negative effect on both knowledge collecting (β= -.486, $p < 0.05$) and knowledge donating (β= -.376, $p < 0.05$). This result gives partial support for H1. The results also show that only individual knowledge sharing barriers but not the others (organizational and technical) is negatively related to individual innovative behavior (β= -.364, $p < 0.05$), partially confirming H2. There was an interesting result regarding the link between knowledge sharing barriers and individual innovation behavior. Despite the expectation that all the knowledge sharing barriers have negative effect on individual innovative behavior, technological barriers were positively associated with individual innovative behavior. The result also revealed that knowledge sharing process variables (knowledge collecting and donating) did not influence individual innovative behavior, thus H3 is rejected.

CONCLUSION

This study aimed to explore the role of knowledge sharing barriers on knowledge sharing (knowledge collecting and donating) and individual innovative behavior, and investigate the influence of knowledge

Figure 1. The structural model with path coefficients
*Note: * Significant at p<0.05*

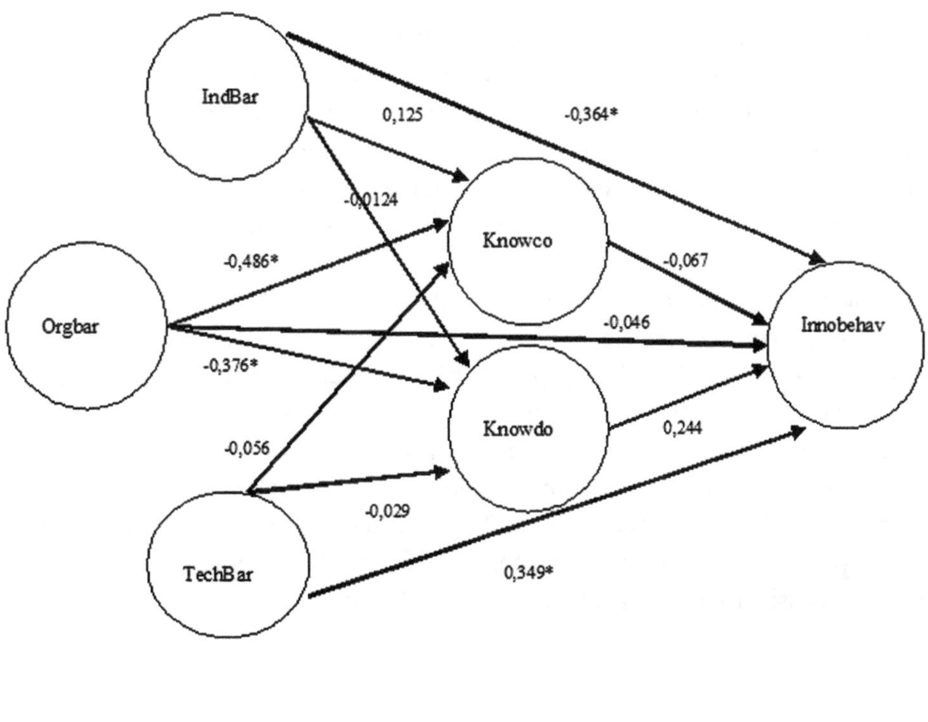

sharing on individual innovative behavior. The data, collected from the eighty-three academics from a university located in Kilis city of Turkey, was analyzed by using Smart PLS.

The findings from the current study indicated that only organizational barriers had a negative effect on both knowledge collecting and knowledge donating. This result gives partial support for H1, supports the theoretical arguments underlying the relationship between knowledge sharing barriers and knowledge sharing (e.g. Barson *et al.*, 2009; Dee & Leisyte, 2017; Huang & Davison, 2008; Kaps, 2011; Lilleoere & Hansen 2011; Ramjeawon & Rowley, 2017; Riege, 2005; Santos *et al.*, 2012; Shahid & Alamgir, 2011) and is consistent with previous empirical research. The previous studies found several knowledge sharing barriers affecting knowledge sharing (Dee & Leisyte, 2017; Khalil & Shea, 2012; Lilleoere & Hansen, 2011; McLaughlin *et al.*, 2008). Yet, there is no study empirically testing the effect of knowledge sharing barriers on knowledge sharing apart from one study in which knowledge sharing barriers were negatively related knowledge sharing effectiveness measures (Khalil & Shea, 2012). Contrary to other studies in the literature that found the several barriers toward knowledge sharing, this study found empirical evidence that organizational knowledge sharing barriers were negatively related to knowledge sharing. Our finding reinforces the importance of knowledge sharing barriers for effective execution of knowledge sharing in academic organizations. Knowledge sharing barriers inhibit academic staff to collect and donate knowledge within organizations. Particularly, lack of leadership, strategy and vision, environment in term of knowledge sharing as well as strong hierarchy and its related outcomes all lead to reduce knowledge sharing potential among the academic staff. Higher education institutions need to pay attention to different organizational aspects that help to create an environment where knowledge sharing can easily take place. This type of organizations needs to adjust their organizational structure, systems and policies for the effective knowledge sharing (Dee & Leisyte, 2017; Khalil & Shea, 2012; Wang *et al.*, 2010). Although it is not confirmed in our study, individual and technological barriers play an important role in knowledge sharing (Riege, 2005; Cheng *et al.*, 2009; Chong *et al.*, 2014; Khalil & Shea, 2012), therefore, any individual and technological knowledge sharing barriers need to be approached with caution bearing it mind that they can be detrimental to effective knowledge sharing. The finding from our study also provides evidence to the types of knowledge sharing barriers that exist in higher education institutions and their effect on knowledge sharing. It seems that organization related barriers rather than individual and technological related barriers toward knowledge sharing is the most active player in term of creating negative effect on knowledge sharing in our study. Previous studies conducted in academic environment found various individual, organizational and technological barriers affecting knowledge sharing (Dee & Leisyte, 2017; Ramjeawon &Rowley, 2017). In a study conducted Khalil and Shea (2012) found that fear of knowledge revelation was found to be the most influential barrier on the effectiveness of knowledge sharing. Thus, universities need to find out the barriers and their relative importance, concentrate their capabilities and resources to eliminate these barriers, and create an environment for knowledge sharing to take place and flourish among academics (Dee & Leisyte, 2017; Ramjeawon & Rowley, 2017; Wang *et al.*, 2010). The fact that different barriers require different solutions (McLaughlin *et al.*, 2008) needs attention from the policy and decision makers. Organizational structures and culture, people and their relationship need to be considered when knowledge sharing barriers are being identified and removed, and solutions are executed (McLaughlin *et al.*, 2008). Paying attention to knowledge sharing barriers is not sufficient as they are also affected by knowledge sharing enablers discussed in detail by Lilleoere and Hansen (2011). Both barriers and enablers need to be taken into account when considering knowledge sharing activities in academe.

The findings from this study show that only individual knowledge sharing barriers but not the others (organizational and technological barriers) were negatively related to individual innovation behavior, partially confirming H2. Although this finding is supported by the theoretical argument in the literature (e.g. Angel, 2006; Patterson *et al.,* 2009), it is really hard to assess the relative importance and significance of this finding due to limited empirical work in this area. On the other hand, it is theoretically and empirically known that knowledge sharing is viewed as an important mechanism to improve individual and organizational learning and innovation (Dee & Leisyte, 2017; Kamasak & Bulutlar, 2010; Radaelli *et al.,* 2014; Lin, 2007; Mura *et al.,* 2013; Riege, 2005; Wang & Noe, 2010), thus barriers toward knowledge sharing can be suggested to have negative effect on innovation capability of both individual and organizations. Our study provides indirect support for this kind of arguments. The findings from our study also showed the importance of knowledge sharing barriers for individual innovative behavior. Individual related factors rather than organizational and technological factors operate as barriers toward individual innovative behavior in academe. It is interesting to note that even though individual related knowledge sharing barriers had no effect on knowledge sharing as discussed above, they reflected a negative effect on individual innovative behavior. Several individual related barriers such as lack of time, trust, social network, and flexibility are among the many factors that inhibit ability of academic staff to produce innovative behavior. Therefore, these barriers in academe need to be carefully identified, evaluated and addressed so that innovative behavior can be expected from the academic personnel. In a general sense, elimination of all types of barriers including knowledge sharing barriers at least creates a barrier free environment for knowledge and innovation management to flourish in academe. Another finding from the current study was that technological knowledge sharing barriers were positively associated with individual innovative behavior. This finding was contrary to our expectations and theoretical arguments. In order to better evaluate such finding, we looked at the zero-order correlation between these two variables. The result revealed that there was no significant relationship exists between two variables. It appears that a suppression effect was operating and that there was no relationship between technological knowledge sharing barriers and individual innovative behavior. Another interesting finding was that organization related knowledge sharing barriers did not have any effect on individual innovation behavior, yet they negatively affected knowledge sharing. Organization related knowledge sharing barriers may have negative effect over the individual innovative behavior through its effect on knowledge sharing in the long run. Therefore, organizational knowledge sharing barriers as well others and their effect on individual innovative behavior should be carefully monitored and managed. Considering the limited empirical studies, the findings from our study encourages further empirical studies in academic environment to identify different knowledge sharing barriers and their effect on individual innovative behavior.

The last findings from the study was related to the link between knowledge sharing and individual innovative behavior. There are different antecedents of individual innovative behavior (De Jong & Hartog, 2007), one of which is knowledge sharing as investigated by our study. In our study, knowledge sharing measured through knowledge collecting and donating had no effect on individual innovative behavior, thus rejecting H3. Although there is a certain level of individual innovative behavior from academic staff (observed through descriptive statistics), these behaviors are not affected by knowledge sharing dimensions in our study. This result contradicts with the theoretical arguments (Aulawi *et al.*, 2009; Mura *et al.*, 2013; Radaelli *et al.,* 2014; Rebernik & Širec, 2007) and empirical studies (Hu *et al.,* 2009; Mura *et al.*, 2013; Radaelli *et al.,* 2014). In an academic environment, it is also argued that knowledge sharing enables academic staff to be more effective (Jain *et al.,* 2007). Although knowledge collecting and donating exist between academic staff (as evidenced by the descriptive statistics) in our study, this collective asset is not being capitalized for the innovative behavior of academic staff. Although some studies in different organizational setting found the positive effect of knowledge sharing on individual innovative behavior (Hu *et al.*, 2009; Mura *et al.*, 2013; Radaelli *et al.*, 2014), organizational innovation (Kamaşak & Bulutlar, 2009), organizational innovation capability (Lin, 2007) and competitive advantage (Eidizadeh *et al.*, 2017; Sözbilir & Yeşil, 2016), to our knowledge, there seems to be no study exploring the effect of knowledge sharing on individual innovation behavior in academic environment. Therefore, it is really difficult to evaluate the relative importance and significance of this finding in academic environment. Perhaps, our findings may provide indirect support to only one study in which Zhou and Uhlaner (2009) found that internal knowledge sharing does not play a key role in fostering SMEs' innovativeness. The fact that there was no link between knowledge sharing and individual innovative behavior deserve some explanation. This finding of the current study might be the result of other antecedents playing more prominent role on individual innovative behavior. For example, other individual (e.g., personality, motivation, commitment, job satisfaction etc.) and organizational factors (e.g., culture, structure, leadership, IT etc.) may be more important than knowledge sharing to stimulate innovative behaviors in academe. In a recent study done by Radaelli *et al.* (2014) emphasized two preconditions for individual knowledge sharing (namely, employees' ability and opportunity) that affect innovative behaviors. On the other hand some individual and organizational-level interventions (e.g., incentives and rewards, knowledge education and training, and a supportive climate and workload) need to be initiated to foster knowledge sharing and innovative behavior (Radaelli *et al.*, 2014; Ramjeawon & Rowley, 2017). It is also possible that there might be mediators, playing an important role between knowledge sharing and individual innovative behavior. One study reported that absorptive capacity is an intervening factor between knowledge sharing capability and innovation capability (Andrawina & Govindaraju, 2009). However, this indirect effect of knowledge sharing on individual innovative behaviour was not found in another study conducted by Radaelli *et al.* (2014). These conflicting results might be due to the different conceptualization, measurement and research context (please see the cited researches and Lilleoere & Hansen, 2011). The findings from our research emphasizes the importance of research context, conceptualization and measurement and shows their role in the relationship among the constructs studied. In any case, our finding raise some doubts regarding the implications of knowledge sharing on innovative behavior in academe that will perhaps encourage further investigation. Supporting this, Radaelli *et al.* (2014) encourages this types of researches in different professional settings so that results can be compared in terms of theoretical and practical implications.

To conclude, this paper contributes to the existing body of knowledge in terms of identifying the types of knowledge sharing barriers, their relative importance, their effect on knowledge sharing and individual innovative behavior in an academic environment. The current research also explored the link between knowledge sharing and individual innovative behavior in academe. To make the last concluding remarks, we also agree on the statement made by Chong *et al.* (2014:218) that "Universities need to be consciously and explicitly managing the knowledge-sharing activities associated with the creation of their knowledge assets and to recognize the value of their intellectual capital to their continuing role in society and in a wider global marketplace for international competitiveness". In doing so, they need to pay attention the individual, organizational, and technological knowledge sharing barriers and make sure that these barrier are identified and removed so that effective knowledge sharing and innovation related activities and behaviors can flourish in organizations.

Like other studies, there are some limitations in the current study that need to be taken into account when evaluating the results. One limitation is that the study concerns only academic staff from one institution with relatively small sample size. Therefore, this creates barriers to generalize the findings to the other working staff (e.g., administrative staff) and other educational institutions. This study adopts the cross-sectional research design that investigates the opinions of academic staff regarding the research variables at one point in time, therefore, the directional relationships obtained through this study need to be viewed with caution. Another limitation is that the data was collected via a self-report questionnaire. The single-method variance might have also affected the results of the study. To account for the detriment of single-method bias, researchers took some preliminary cautions during the data collection process. Following Podsakoff *et al.* (2003), the researchers ensured the respondents with information on the front page of the questionnaire regarding the confidentiality of their individual responses. To reduce respondents' concern about being evaluated, we also assured the participants that the questionnaire had no right or wrong answers. Spector (1987, 2006) argued that that well-developed measures with sound psychometric properties seemed free of this problem. All the measures in this study came from previous studies with sound psychometric properties.

Some recommendation can be given based on the limitations and findings of this study. Future studies may involve other staff and different universities (both private and public) across the country and other organizational settings with big sample size. This would also enable the researchers to compare the findings and to resolve the conflict encountered in this study. Potential future studies may also include other individual (e.g., personality, motivation, and job satisfaction) and organizational factors (e.g., culture and leadership), and mediators that affect knowledge sharing and individual innovation behaviors in higher education institutions. Other parts of knowledge management (e.g., knowledge creation and application) can also be included in this type of research to explore the role of knowledge sharing barriers over these parts. Additionally, knowledge sharing and individual innovativeness are dynamic rather than static process, it is thus essential to conduct research in longitudinal and qualitative nature to capture the significance of the proposed relationship. Although individual innovation behavior is treated as one dimension in this study, future studies may include multi-dimensions measurement of individual innovative behavior.

REFERENCES

Akgün, A. E., Keskin, H., Ayar, H., & Okunakol, Z. (2017). Knowledge sharing barriers in software development teams: A multiple case study in Turkey. *Kybernetes*, *46*(4), 603–620. doi:10.1108/K-04-2016-0081

Alam, S. S., Abdullah, Z., Ishak, N. A., & Zain, Z. M. (2009). Assessing knowledge sharing behavior among employees in smes: An empirical study. *International Business Research*, *2*(2), 115–122. doi:10.5539/ibr.v2n2p115

Amabile, T. A. (1996). *Creativity and innovation in organizations*. Harvard Business School Press.

Åmo, B. (2006). Employee innovation behaviour in health care: The influence from management and colleagues. *International Nursing Review*, *53*(3), 231–237. doi:10.1111/j.1466-7657.2006.00455.x PMID:16879187

Andrawina, L., & Govindaraju, R. (2009). Knowledge sharing capability, absorptive capacity, and innovation capability: An empirical study of Indonesia's information and communication technology industry. *Journal of ICT*, *8*, 85–102.

Angel, R. (2006). Putting an innovation culture into practice. *Ivey Business Journal.* Retrieved from http://www.iveybusinessjournal. com/topics/innovation/putting-an-innovation-culture-into-practice

Athar, M. A. Q., & Evans, N. (2015). Deterrents to knowledge-sharing in the pharmaceutical industry: A case study. *Journal of Knowledge Management*, *19*(2), 296–314. doi:10.1108/JKM-09-2014-0391

Aulawi, H., Sudirman, I., Sudirman, I., Suryadi, K., & Govindarajn, R. (2009). Literature review towards knowledge enablers which is assumed significantly influences ks behavior. *Journal of Applied Sciences Research*, *5*(12), 2262–2270.

Austin, R. (2007, January). *The role of IT in ınnovation-based value creation. CIO Digers.*

Barson, R. J., Foster, G., Struck, T., Ratchev, S., Pawar, K., Weber, F., & Wunram, M. (2009). Inter-and intra-organisational barriers to sharing knowledge in the extended supply-chain. *Proceedings of the International Conference on e-Business and e-Work*, 367-373.

Bhirud, S., Rodrigues, L., & Desai, P. (2005). Knowledge sharing practices in km: A case study in Indian software subsidiary. *Journal of Knowledge Management Practice.* Retrieved from http://www. tlainc. com/articl103.htm

Bircham-Connolly, H., Corner, J., & Bowden, S. (2005). An empirical study of the impact of question structure on receipient attitude during knowledge sharing. *Electronic Journal of Knowledge Management*, *32*(1), 1–10.

Caloghirou, Y., Kastelli, I., & Tsakanikas, A. (2004). Internal capabilities and external knowledge sources: Complements or substitutes for innovative performance? *Technovation*, *24*(1), 29–39. doi:10.1016/S0166-4972(02)00051-2

Carmeli, A., Meitar, R., & Weisberg, J. (2006). Self-leadership skills and innovative behavior at work. *International Journal of Manpower*, *27*(1), 75–90. doi:10.1108/01437720610652853

Chen, J.S. & Tsou, H.T. (2007). Information technology adoption for service innovation practices and competitive advantage: The case of financial firms. *Information Research, 12*(3).

Cheng, M. Y., Ho, J. S. Y., & Lau, P. M. (2009). Knowledge sharing in academic institutions: A study of multimedia university Malaysia. *Electronic Journal of Knowledge Management*, *7*(3), 313–324.

Chiu, C.-N., & Chen, H.-H. (2016). The study of knowledge management capability and organizational effectiveness in Taiwanese public utility: The mediator role of organizational commitment. *SpringerPlus*, *5*(1), 1520. doi:10.118640064-016-3173-6 PMID:27652093

Chong, C. W., Yuen, Y. Y., & Gan, G. C. (2014). Knowledge sharing of academic staff: Acomparison between private and public universities in Malaysia. *Library Review*, *63*(3), 203–223. doi:10.1108/LR-08-2013-0109

De Jong, J. P. J., & Den Hartog, D. N. (2007). How leaders influence employees' innovative behavior. *European Journal of Innovation Management, 10*(1), 41-64. 10.1108/14601060710720546

De Jong, J. P. J., & Hartog, D. N. D. (2008). *Innovative work behavior: Measurement and validation.* Scientific Analysis of Entrepreneurship and SMEs. Retrieved from http://www.entrepreneurship-sme.eu/pdf-ez/ h200820.pdf

Dee, J., & Leisyte, L. (2017). Knowledge sharing and organizational change in higher education. *The Learning Organization, 24*(5), 355–365. doi:10.1108/TLO-04-2017-0034

Duderstadt, J. J. (2000). *The future of higher education in the knowledge-driven, global economy of the 21st century.* Retrieved from http://milproj.ummu. umich.edu/publications/toronto/

Eidizadeh, R., Salehzadeh, R., & Esfahani, A. C. (2017). Analysing the role of business intelligence, knowledge sharing and organisational innovation on gaining competitive advantage. *Journal of Workplace Learning, 29*(4), 50–267. doi:10.1108/JWL-07-2016-0070

European Commission. (2001). *European innobarometer survey*. Luxembourg: European Commission.

Filieri, R. (2010). *Overcoming knowledge sharing barriers through communities of practice*. Cambridge Scholars Publishing.

Fornell, C., & Larcker, D. F. (1981). Evaluating structural equation models with unobservable variables and measurement error. *JMR, Journal of Marketing Research, 18*(1), 39–50. doi:10.2307/3151312

Fullwood, R., Rowley, J., & Delbridge, R. (2013). Knowledge sharing amongst academics in UK universities. *Journal of Knowledge Management, 17*(1), 123–136. doi:10.1108/13673271311300831

Gupta, A. K., & Govindarajan, V. (2000). Knowledge flows within multinational corporations. *Strategic Management Journal, 21*(4), 473–496. doi:10.1002/(SICI)1097-0266(200004)21:4<473::AID-SMJ843.0.CO;2-I

Hair, J. F. Jr, Black, W. C., Babin, B. J., Anderson, R. E., & Tatham, R. L. (2006). *Multivariate data analysis* (6th ed.). Prentice-Hall PTR.

Halawi, L., Aronson, J., & McCarthy, R. (2005). Resource-Based View of Knowledge Management for Competitive Advantage. *Electronic Journal of Knowledge Management, 3*(2), 75–86.

Hsu, M. L. A., Hou, S., & Fan, H.-L. (2011). Creative self-efficacy and innovative behavior in a service setting: Optimism as a moderator. *The Journal of Creative Behavior, 45*(4), 258–272. doi:10.1002/j.2162-6057.2011.tb01430.x

Hu, M. M., Horng, J., & Sun, Y. C. (2009). Hospitality teams: Knowledge sharing and service innovation performance. *Tourism Management, 30*(1), 41–50. doi:10.1016/j.tourman.2008.04.009

Huang, Q., & Davison, R. (2008). Knowledge sharing barriers at the individual level in a Chinese bank. *PACIS 2008 Proceedings*.

Inkinen, H. (2016). Review of empirical research on knowledge management practices and firm performance. *Journal of Knowledge Management, 20*(2), 230–257. doi:10.1108/JKM-09-2015-0336

Işık, C., & Aydın, E. (2016). Bilgi paylaşımının yenilikçi iş davranışına etkisi: Ayder yaylası konaklama işletmeleri üzerine bir uygulama. *Girişimcilik ve İnovasyon Yönetimi Dergisi, 5*(2), 75–103.

Jain, K. K., Sandhu, M. S., & Sidhu, G. K. (2007). Knowledge sharing among academic staff: A case study of business schools in klang valley, Malaysia. *Research Papers* (Vol. 2). Retrieved from http:// www. ucsi.edu.my/cervie/ijasa/volume2/pdf/08A.pdf

Janssen, O. (2000). Job demands, perceptions of effort-reward fairness and innovative work behaviour. *Journal of Occupational and Organizational Psychology, 73*(3), 287–302. doi:10.1348/096317900167038

Jennex, M. E., Smolnik, S., & Croasdell, D. (2014). Knowledge Management Success in Practice. In *47th Hawaii International Conference on System Sciences*. IEEE Computer Society.

Kamasak, R., & Bulutlar, F. (2010). The influence of knowledge sharing on innovation. *European Business Review, 22*(3), 306–317. doi:10.1108/09555341011040994

Kang, M., & Lee, M.-J. (2017). Absorptive capacity, knowledge sharing, and innovative behaviour of R&D employees. *Technology Analysis & Strategic Management Journal, 29*(2), 219–232. doi:10.1080 /09537325.2016.1211265

Kaps, I. V. (2011). Barriers in intercultural knowledge sharing. learning's from an international plant engineering project. *Journal of Knowledge Management*, 3.

Kessel, M., Hannemann-Weber, H., & Kratzer, J. (2012). Innovative work behavior in healthcare: The benefit of operational guidelines in the treatment of rare diseases. *Health Policy (Amsterdam), 105*(2-3), 146–153. doi:10.1016/j.healthpol.2012.02.010

Keyes, J. (2008). *Identifying the barriers to knowledge sharing in knowledge intensive organizations.* Retrieved from http://www.newarttech. com/KnowledgeSharing.pdf

Khalil, O. E. M., & Shea, K. T. (2012). Knowledge sharing barriers and effectiveness at a higher education institution. *International Journal of Knowledge Management, 8*(2), 43–64. doi:10.4018/jkm.2012040103

Kim, L. (1997). *Imitation to innovation: The dynamics of Korea's technological learning (management of innovation and change).* Harvard Business School Press.

Kim, T. W., & Joh, H. W. (2005). *Key barriers and their strategic responses to activate knowledge sharing in construction organizations.* 1st Annual ICCEM (International Conference on Construction Engineering and Management), Seoul, South Korea.

Kleysen, R. F., & Street, C. T. (2001). Toward a multi-dimensional measure of individual innovative behavior. *Journal of Intellectual Capital, 2*(3), 284-296.

Lilleoere, A., & Hansen, E. (2011). Knowledge-sharing enablers and barriers in pharmaceutical research and development. *Journal of Knowledge Management, 15*(1), 53–70. doi:10.1108/13673271111108693

Lin, H. (2007). Knowledge sharing and firm innovation capability: An empirical study. *International Journal of Manpower, 28*(3/4), 315–332. doi:10.1108/01437720710755272

Maccurtain, S., Flood, P. C., Ramamoorthy, N., West, M. A., & Dawson, J. F. (2008). *Top team trust, knowledge sharing and innovation.* Link Working Paper Series WP 05-08.

Marques, C. S., & Ferreira, J. (2009). SME innovative capacity, competitive advantage and performance in a 'traditional' industrial region of Portugal. *Journal of Technology Management & Innovation*, *4*(4). doi:10.4067/S0718-27242009000400005

Martensson, M. (2000). A critical review of knowledge management as a management tool. *Journal of Knowledge Management*, *4*(3), 204–216. doi:10.1108/13673270010350002

McDermott, R., & O'Dell, C. (2001). Overcoming cultural barriers to sharing knowledge. *Journal of Knowledge Management*, *5*(1), 76–85. doi:10.1108/13673270110384428

McLaughlin, S., Paton, R. A., & Macbeth, D. K. (2008). Barrier impact on organizational learning within complex organizations. *Journal of Knowledge Management*, *12*(2), 107–123. doi:10.1108/13673270810859550

Mehrabani, E. S., & Mohamad, A. N. (2011). Singapore the role of 'training activities' and 'knowledge sharing' in the relationship between leadership development and organizational effectiveness. *International Conference on Sociality and Economics Development IPEDR*, *10*, 172-177.

Monavvarian, A., & Kasaei, M. (2007). A KM model for public administration: The case of Labour Ministry. *The Journal of Information and Knowledge Management Systems*, *37*(3), 348–367.

Mura, M., Lettieri, E., Radaelli, G., & Spiller, N. (2013). Promoting professionals' innovative behaviour through knowledge sharing: The moderating role of social capital. *Journal of Knowledge Management*, *17*(4), 527–544. doi:10.1108/JKM-03-2013-0105

Nonaka, I. (1994). A dynamic theory of organisational knowledge creation. *Organization Science*, *5*(1), 14–37. doi:10.1287/orsc.5.1.14

Nunnally, J. C. (1978). *Psychometric theory*. New York, NY: McGraw-Hill.

Pangil, F., & Mohd, N. A. (2009). Assessing the relationship between organisational commitment and knowledge sharing behaviour. *Malaysian Management Journal*, *13*(1 & 2), 35–50.

Parzefall, M. R., Seeck, H., & Leppänen, A. (2008). Employee innovativeness in organizations. A review on the antecedents. *Finnish Journal of Business Economics*, *2*, 165–182.

Patterson, F., Kerrin, M., Gatto-Roissard, G., & Crowley, T. (2009). Characteristics & behaviours of innovative people in organisations: from theory to policy and practice. *Proceedings of the European Work and Organizational Psychology Conference*.

Podsakoff, P. M., MacKenzie, S. B., Lee, J. Y., & Podsakoff, N. P. (2003). Common method biases in behavioral research: A critical review of the literature and recommended remedies. *The Journal of Applied Psychology*, *88*(5), 879–903. doi:10.1037/0021-9010.88.5.879 PMID:14516251

Radaelli, G., Lettieri, E., Mura, M., & Spiller, N. (2014). Knowledge sharing and innovative work behaviour in healthcare: A micro-level investigation of direct and indirect effects. *Creativity and Innovation Management*, *23*(4), 400–414. doi:10.1111/caim.12084

Ramamoorthy, N., Patrick, C., Flood, S. T., & Sardessai, R. (2005). Determinants of innovative work behavior: Development and test of an integrated model. *Creativity and Innovation Management*, *14*(2), 142–150. doi:10.1111/j.1467-8691.2005.00334.x

Ramjeawon, P. V., & Rowley, J. (2017). Knowledge management in higher education institutions: Enablers and barriers in Mauritius. *The Learning Organization*, *24*(5). doi:10.1108/TLO-03-2017-0030

Rebernik, M., & Širec, K. (2007). Fostering innovation by unlearning tacit knowledge. *Kybernetes*, *36*(3/4), 406–419. doi:10.1108/03684920710747039

Riege, A. (2005). Three dozen knowledge sharing barriers managers must consider. *Journal of Knowledge Management*, *9*(3), 18–35. doi:10.1108/13673270510602746

Ringle, C. M., Wende, S., & Will, A. (2005). *SmartPLS 2.0 (beta)*. University of Hamburg. Retrieved from http://www.smartpls.de/

Ryan, S. D., Zhang, X., Prybutok, V. R., & Sharp, J. H. (2012). Leadership and knowledge management in an e-government environment. *American Scientist*, *2*(1), 63–81.

Saginova, O., & Belyansky, V. (2008). Facilitating innovations in higher education in transition economies. *International Journal of Educational Management*, *22*(4), 341–351. doi:10.1108/09513540810875671

Salaman, J. G., & Storey, J. (2002). Managers' theories about the process of innovation. *Journal of Management Studies*, *39*(2), 147–165. doi:10.1111/1467-6486.00286

Santos, V. R., Soares, A. L., & Carvalho, J. Á. (2012). Knowledge sharing barriers in complex research and development projects: An exploratory study on the perceptions of project managers. *Knowledge and Process Management*, *19*(1), 27–38. doi:10.1002/kpm.1379

Sazandrishvili, N. (2009). *Contextual and personal antecedents of innovative behavior* (Master's thesis). University of Twente.

Scholl, H. (2005). Innovationas drivers of growth. In *Proceedings of the CBI Conference Synopsis Sharing Knowledge with the Business and Academic Communities*. Carnegie Bosch Institute for Applied Studies in International Management.

Scott, S. G., & Bruce, R. A. (1994). Determinants of innovative behavior: A path model of individual innovation in the workplace. *Academy of Management Journal*, *37*(3), 580–607. doi:10.2307/256701

Sezgin, O. B., Uçar, Z., & Duygulu, E. (2015). Güven, yenilikçi iş davranişi ilişkisinde bilgi paylaşiminin aracilik rolü. *İşletme Fakültesi Dergisi, 16*(2), 1-20.

Shahid, A., & Alamgir, R. (2011). *ICT enabled knowledge sharing–impact of ict on knowledge sharing barriers: The case of avanade* (Master Thesis).

Sözbilir, F., &Yeşil, S. (2016).The Impact of Knowledge Management on Competitive Advantage: An Empirical Study in Turkey. *Bilgi,* (32), 92-116.

Spector, P. E. (1987). Method variance as an artifact in self-reported affect and perceptions at work: Myth or significant problem? *The Journal of Applied Psychology*, *72*(3), 438–443. doi:10.1037/0021-9010.72.3.438

Spector, P. E. (2006). Method variance in organizational research: Truth or urban legend? *Organizational Research Methods*, *9*(2), 221–232. doi:10.1177/1094428105284955

Syed-Ikhsan, S. O. S., & Rowland, F. (2004). Knowledge management in a public organization: A study on the relationship between organizational elements and the performance of knowledge transfer. *Journal of Knowledge Management, 8*(2), 95–111. doi:10.1108/13673270410529145

Szulanski, G. (1996). Exploring internal stickiness: Impediments to the transfer of best practice within the firm strategic. *Management Journal, 17*, 27–43.

Terziovski, M. (2007). *Building innovation capability in organizations - an international cross-case perspective*. Imperial College Press. doi:10.1142/p492

Unsworth, K. L., & Parker, S. K. (2003). Proactivity and innovation: Promoting a new workforce for the new workplace. In D. Holman, T. D. Wall, C. W. Clegg, P. Sparrow, & A. Howard (Eds.), *The new workplace: A guide to the human impact of modern working practices* (pp. 175–196). Chichester, UK: Wiley.

Van de Ven, A. H. (1986). Central problems in the management of innovation. *Management Science, 32*(5), 590–607. doi:10.1287/mnsc.32.5.590

Van den Hooff, B., & De Ridder, J. A. (2004). Knowledge sharing in context: The influence of organizational commitment, communication climate and cmc use on knowledge sharing. *Journal of Knowledge Management, 8*(6), 117–130. doi:10.1108/13673270410567675

Van Den Hooff, B., & Van Weenen, F. (2004). Committed to share: Commitment and cmc use as antecedents of knowledge sharing. *Knowledge and Process Management, 11*(1), 13–24. doi:10.1002/kpm.187

Wang, L., Wang, H., & Wang, K. (2010). The investigation analysis on knowledge-sharing barriers and incentives of university scientific research and innovation teams management and service science (MASS). *Proceedings of the International Conference on*, 1-4.

Wang, S., & Noe, R. A. (2010). Knowledge sharing: A review and directions for future research. *Human Resource Management Review, 20*(2), 115–131. doi:10.1016/j.hrmr.2009.10.001

Wang, Y., & Zhu, Y. (2012). Organizational socialization, trust, knowledge sharing and innovation behavior: Research on mechanism and path. *R & D Management*.

Wangpipatwong, S. (2009). Factors influencing knowledge sharing among university students. *Proceedings of the 17th International Conference on Computers in Education*, 800-807.

Weber, H. H., Kessel, M., Budych, K., & Schultz, C. (2011). Shared communication processes within healthcare teams for rare diseases and their influence on healthcare professionals' innovative behavior and patient satisfaction. *Implementation Science; IS, 6*(40). Retrieved from http://www.implementation-science.com/content/6/1/40 PMID:21510848

West, M. A. (1989). Innovation among health care professionals. *Social Behaviour, 4*, 173–184.

West, M. A., & Farr, J. L. (1989). Innovation at work: Psychological perspectives. *Social Behaviour, 4*, 15–30.

Wu, C., Parker, S. K., & De Jong, J. P. J. (2011). Need for cognition as an antecedent of individual innovation behavior. *Journal of Management, 21*, 1–24.

Xerri, M., & Brunetto, Y. (2011). Fostering the innovative behaviour of sme employees: A social capital perspective. *Research and Practice in Human Resource Management*, *19*(2), 43–59.

Xiaojun, L., & Peng, L. (2010). *The impact of learning culture on individual innovative behaviour*. Institute of Electrical and Electronics Engineers.

Yeşil, S., & Sözbilir, F. (2013). An Empirical Investigation into the Impact of Personality on Individual Innovation Behaviour in the Workplace. *Procedia: Social and Behavioral Sciences*, *81*, 540–551. doi:10.1016/j.sbspro.2013.06.474

Zhou, H., & Uhlaner, L. (2009). *Knowledge management as a strategic tool to foster innovativeness of SMEs*. ERIM Report Series Reference No. ERS- 2009-025-ORG. Retrieved from http://ssrn.com/abstract=1410468

ADDITIONAL READING

Becerra-Fernandez, I., & Sabherwal, R. (2010). *Knowledge management systems and process*. New York: M.E. Sharpe, Inc.

Collison, C., & Parcell, G., (2004). Learning to Fly: Practical Knowledge Management from Leading and Learning Organizations.

Davenport, T. H., & Prusak, L. (2000). *Working knowledge: How organizations manage what they know*. Boston: Harvard Business School Press.

Gurteen, D. (1999), 'Creating a knowledge sharing culture', Knowledge Management Magazine, http://www.gurteen.com/gurteen/gurteen.nsf/id/ksculture

Jennex, M. E., (2008). Knowledge Management: Concepts, Methodologies, Tools and Applications (6-volume set).

Nonaka, I., & Takeuchi, H. (1995). *The Knowledge-Creating Company*. Oxford University Press.

KEY TERMS AND DEFINITIONS

Individual Innovative Behavior: Creating and applying ideas, methods, solutions, and approaches in the workplace to create value.

Knowledge Collecting: Collecting knowledge from others.

Knowledge Donating: Donating knowledge to others.

Knowledge Management: Management of the processes and procedures to capitalize on knowledge in organizations.

Knowledge Sharing: Exchange of knowledge between individuals, groups, and organizations.

Knowledge-Sharing Barriers: Individual, organization, and technology-related factors that hinder flow of knowledge between individual, groups, and organizations.

Chapter 7
Team and Individual Interactions With Reciprocity in Individual Knowledge Sharing

Megan Lee Endres
Eastern Michigan University, USA

Sanjib Chowdhury
Eastern Michigan University, USA

ABSTRACT

The study investigated the effects of expected reciprocity on knowledge sharing, as moderated by team and individual variables. Data (n = 84) were collected in an experimental study from undergraduate business student participants. The effects of expected reciprocity on knowledge sharing depend on the levels of individual competence, positive team attitudes, functional diversity, and demographic diversity. Implications include that the effectiveness of reciprocity in knowledge sharing depends on several factors relating to the team and individual. Encouraging reciprocity may have positive effects, but these can be overridden by poor team attitudes, low ability perceptions, and team diversity. Future research suggestions are offered.

INTRODUCTION

Knowledge sharing facilitates organizational and group performance and can be used strategically to attain competitive advantage (Erickson & Rothberg, 2011; Nahapiet & Ghoshal, 1998; Wan, Haggerty, & Wang, 2015; Yang, 2007). Motivating individual sharing, therefore, is a topic of significant interest.

Reciprocity is a significant motivator that receives consistent attention in knowledge sharing research (Chen & Hung, 2010; Cho, Li, & Su, 2007; Di Gangi, Wasko, & Tang, 2012; Wu, Hom, Tetrick, Shore et al., 2006). However, despite its widespread application to organizational theories and demonstrated importance to knowledge sharing, it is surprising that the construct remains underdeveloped (Wu et al., 2006).

DOI: 10.4018/978-1-5225-5427-1.ch007

An important part of construct development is identification of interactions that act as contingencies (Johns, 2017). In knowledge sharing research, expected reciprocity has been found to interact with a variety of variables to predict outcomes, but more research is needed to form conclusions about the wide variety of possible moderators (e.g., Haeussler, Jiant, Thursby, & Thursby, 2014; Zhao, Detlor, & Connelly, 2016). In addition, most studies are cross-sectional surveys, with only a handful of experiments (e.g., Hung, Durcikova, Lai, & Lin, 2011). Given the importance of knowledge sharing for team performance (e.g., Killingsworth, Xue, & Liu, 2016), therefore, an experimental study of in-person teams was used to better understand the effects of reciprocity.

First, the reciprocity construct in knowledge sharing research is presented, and variables are proposed to interact with reciprocity in affecting knowledge sharing. Experimental methods and measures are presented, and hypotheses are tested. Results and applications to organizations, management education, and researchers are given. Finally, strengths and limitations of the study are provided.

REVIEW OF RELEVANT LITERATURE

Knowledge sharing is defined as occurring "when an individual is willing to assist as well as to learn from others in the development of new competencies" (Yang, 2007, p. 83). Reciprocity is implied in the definition as necessary – a give-and-take of knowledge. Researchers typically rely on social exchange theory (Coyle-Shapiro & Conway, 2005), social capital theory (Nahapiet & Ghoshal, 1998), and theory of planned behavior (Ajzen, 1985) in framing the effects of reciprocity on knowledge sharing. From these theoretical bases, the expectation of reciprocity increases motivation to share knowledge. In recent years, reciprocity has received increasing attention in the literature, with the use of sophisticated modeling techniques and attention to moderators.

The relationship between reciprocity and knowledge sharing is most often studied solely as a direct effect on outcomes, typically including attitudes toward sharing, satisfaction with sharing, intention to share, and actual sharing behaviors. Most studies found a positive relationship between reciprocity and knowledge sharing outcomes (e.g., Lin, Lee, & Wang, 2009; Tamjidyamcholo, Baba, Tamjid, & Gholipour, 2013), while a few found nonsignificant (e.g., Chen & Hung, 2010; Hung et al., 2011) or negative results (e.g., Chen & Hung, 2010; Wasko & Faraj, 2005). Knowledge sharing reciprocity studies are primarily cross-sectional surveys of either within-organizational groups (e.g., Akhavan & Hosseini, 2016) or online communities (e.g., Tamjidyamcholo et al., 2013). Few are experimental studies (e.g., DiGangi et al., 2012; Hung et al., 2011) or simulations (e.g., Caimo & Lomi, 2015).

Few researchers investigated reciprocity as part of a moderating relationship, although the number of studies have steadily increased in recent years. However, there is still not a strong body of evidence to draw general conclusions about moderating influences because of the wide variety of potential interactions that are yet to be tested. Findings of these studies suggest that many relationships between reciprocity and knowledge sharing may have contingencies that are not identified, including the person, environment, or the type of knowledge shared.

The variables studied as interacting with reciprocity can be classified as characteristics of the individual (Chang, Hsu, Shaiu, & Tsai, 2015; Haeussler, 2011; Haeussler et al., 2014; Hung, Lai, & Chao, 2015; Lai & Chen, 2014; Zhang, Chen, & Vogel, 2009; Zhao, Detlor, & Connelly, 2016), of the sharing environment (Caimo & Lomi, 2015; DiGangi et al., 2012; Jeon, Kim, & Koh, 2011; Kang, Kim, &

Bock, 2010; Kankanhalli et al., 2005), and of the knowledge itself (Cho et al., 2007; Kordzadeh, Liu, Au, & Clark, 2014).

The moderator most often studied is the characteristics of the individual. These vary widely in type but can be grouped as personal attributes and individual roles. Researchers have considered a variety of personal attributes. Chang et al. (2015) studied nationality of the sharer (U.S. vs. China) as a potential moderator in effects on intentions to share knowledge. Reciprocity did not interact with nationality to affect intentions to share.

Exchange ideology, an individual belief about sharing norms, is central to social exchange theory and was investigated as a moderator by Zhang et al. (2009). Reciprocity positively influenced knowledge sharing for individuals with low emotional intelligence (EI), who shared regardless of their team members' treatment of them. Reciprocity did not influence knowledge sharing for those with high EI who shared only when they felt team members' treatment was fair. Last, Zhao et al. (2016) studied an individual difference variable, knowledge self-efficacy, or self-belief in competency in the task, which in this case was IT knowledge. When knowledge self-efficacy increased, knowledge sharing attitudes increased only when users expected low reciprocity from other members in the online community. When users expected high reciprocity, knowledge self-efficacy had no effect on attitudes. Although the authors hypothesized that reciprocity would interact with task enjoyment to predict attitudes, this was not supported.

Studies have considered roles of individual sharers to be important variables interacting with reciprocity to predict outcomes. Two studies investigated academic scientists versus industrial scientists as potential moderating influences on knowledge sharing (Haeussler, 2011; Haeussler et al., 2014). Job type interacted with reciprocity so that expected reciprocity increased sharing only for industrial, not academic, scientists. Two studies classified participants as "lurkers" (less active) or "posters" (more active) in an online community (Hung et al., 2015; Lai & Chen, 2014). Reciprocity interacted with the sharer role so that reciprocity positively influenced attitudes toward knowledge sharing only for lurkers (Hung et al., 2015), and positively influenced intentions to share knowledge (Lai & Chen, 2014). Posters relied on the enjoyment of helping and knowledge self-efficacy to influence attitudes (Hung et al., 2015) and intentions (Lai & Chen, 2014), but not on reciprocity. Zhao et al. (2016) similarly found that attitudes decreased for active members as reciprocity increased, but attitudes increased for less active members as reciprocity increased.

Characteristics of the sharing environment form another category of variables considered as interacting with reciprocity. Caimo and Lomi (2015) found that reciprocity increased knowledge sharing between subunits in an organization, but not within subunits. The size of the audience interacted with reciprocity to affect knowledge sharing (DiGangi et al., 2012). Reciprocity positively affected sharing with a smaller audience (reach of information), but did not affect sharing with a larger audience. Jeon et al. (2011) discovered that reciprocity had a stronger positive relationship with knowledge sharing attitudes for formal communities of practice versus informal ones. The type of sharing (closed versus open) also interacted with reciprocity to predict knowledge sharing (Kang et al., 2010). Reciprocity positively predicted knowledge sharing for one-on-one relationships but not one-to-many relationships. Kankanhalli et al. (2005) considered the sharing norms of the environment to interact with reciprocity to predict use of electronic knowledge repositories. If the norms for sharing were strong, reciprocity did not predict sharing. If the norms were weak, however, reciprocity was important in predicting sharing.

The knowledge that is shared may have variations that interact with reciprocity in affecting sharing. Cho et al. (2007) found that reciprocity positively affected knowledge sharing intentions for externally shared and explicit knowledge only (versus internally-shared and tacit knowledge). Researchers have paid little attention to time effects, and Kordzadeh et al. (2014) discovered that in the short-term, reciprocity positively affected user participation in a virtual health community.

A number of variables interacted with reciprocity to predict knowledge sharing outcomes. Few of the variables have repeated findings, however, so the studies basically support that there are contingencies on the relationship between reciprocity and knowledge sharing.

Research Model

Researchers overwhelmingly have found that expected reciprocity positively affects knowledge sharing outcomes (e.g., Tamjidyamcholo et al., 2013; Lin, Lee, & Wang, 2009), strongly supporting the first hypothesis:

Hypothesis One: Expected reciprocity will be positively related to knowledge sharing intent.

Moderator studies suggest a variety of boundary conditions on knowledge sharing including the individual, the environment and the knowledge itself. Moderators that were not addressed in prior studies include self-efficacy, team attitudes, and team diversity.

Competence

Individual motivation to share is often due to high perceived competence in the task (e.g., Cabrera, Collins, & Salgado, 2006; Cho et al., 2007; Hung et al., 2015; Kankanhalli et al., 2005; Lai & Chen, 2014; Quigley, Tesluk, Locke, & Bartol, 2007). Therefore, with high ability levels, reciprocity may not be necessary to motivate a person to share knowledge (Wasko & Faraj, 2005).

Researchers typically study competence in terms of knowledge self-efficacy, or the "confidence to provide knowledge that is valuable" (Kankanhalli et al., 2005, p. 123). This measure relies on a person's expertise and ability to share the knowledge as one construct. In one study, knowledge self-efficacy interacted with reciprocity to predict attitudes toward knowledge sharing (Zhao et al., 2016). Knowledge self-efficacy was a better predictor of attitudes when expected reciprocity was weak. The authors suggested that reciprocity (as a reward) undermined the effects of self-efficacy.

A task-specific self-efficacy measure to directly address a person's perceived capability to perform the task would be most consistent with Bandura (1997). This measure would more specifically target the perceived expertise only. Based on past research, it is expected that reciprocity will have less influence on knowledge sharing in the case of high self-perceived individual competence in the specific task. Therefore, the following hypothesis is offered:

Hypothesis Two: Individual competence will moderate the positive relationship between expected reciprocity and knowledge sharing intent so that the relationship is stronger for individuals with low competence than for those with high competence.

Positive Team Attitudes

Higher levels of knowledge sharing result from positive team attitudes variables such as commitment (e.g., Cabrera et al, 2006; C. Lin, 2007; Ranasinghe & Dharmadasa, 2013; Wasko & Faraj, 2005), trust (e.g., Akhavan & Hosseini, 2016; Cleveland & Ellis, 2015; Killingsworth et al., 2016; Shanshan, 2014; Xue, Liang, Hauser, & O'Hara, 2012) and affiliation, or sense of togetherness (e.g., Bock, Zmud, & Kim, 2005; Jeon et al., 2011; Khalil et al., 2014; Killingsworth et al., 2016).

Knowledge sharing is critical for structured, in-person teams (e.g., Killingsworth et al., 2016). Positive team attitudes are important in facilitating sharing, especially due to the ambiguity in developing exchange relationships (Kankanhalli et al., 2005; Lin & Huang, 2010). Relying on the theory of reasoned action (Ajzen, 1985), researchers find that positive team attitudes and high reciprocity in knowledge sharing go hand-in-hand (e.g., Ranasinghe & Dharmadasa, 2013; Tamjidyamcholo et al., 2013; Xue et al., 2012).

Reciprocity should become stronger in ongoing relationships, especially those in which the task is interdependent (Frenkel & Sanders, 2007). "When individuals are in a high-quality relationship, they will behave in ways that will benefit their exchange partners, such as performing better and exerting extra effort, even if they are not immediately rewarded for such behavior" (Lin & Huang, 2010, p. 189). In this type of team situation, positive team attitudes should enhance work and benefit the group task even further in the presence of expected reciprocity (e.g., Frenkel & Sanders, 2007; Taber & Deosthali, 2014). When an individual expects reciprocal knowledge, a positive team environment adds fuel to the intention to share knowledge.

The relationship between expected reciprocity and knowledge sharing can be maintained based on the value of the exchanged knowledge, even in a situation of low team attitudes (Coyle-Shapiro & Conway, 2005; Wu et al., 2006). For example, perceptions of team member competence are a strong predictor of knowledge sharing, even in situations without the influence of positive team attitudes (Moreland & Myaskovsky, 2000). In a give-and-take exchange, sharing knowledge is based on receiving knowledge and is not enhanced by positive attitudes toward each other.

In summary, the relationship between reciprocity and knowledge sharing should be stronger when team attitudes are positive. However, the relationship should remain even in the case of poor team attitudes simply based on expectations in give-and-take relationships. Therefore, the following hypothesis is offered:

Hypothesis Three: Team attitudes will moderate the relationship between expected reciprocity and knowledge sharing intent so that the relationship is stronger in teams with high team attitudes than in those with low team attitudes.

Team Diversity

Team diversity can be defined "differences among team members on any characteristics people might use to describe how they and another person are different" (De Wit & Greer, 2008, pp. 1-2). Hoffman and Maier (1961) originally argued that diversity enhances team perspectives, cognitive resources, and problem-solving abilities. Members perceive the diversity of their fellow group members, particularly others' competence and fit to the group task requirements. Many researchers support these propositions, showing that diversity improves team outcomes through broader perspectives and more alternative solutions (Knouse & Chretien, 1996; Milliken & Martins, 1996).

Knowledge sharing is a process variable that may lead to higher team performance, but deals with the functional interactions of team members. Although past researchers found both positive and negative relationships between team demographic diversity and team outcomes (Chowdhury, 2005; Wei, Liu, &

Chen, 2015), there is little research investigating the relationships between team demographic diversity and knowledge sharing (e.g., Aktharsha & Anisa, 2011), especially when considering reciprocity.

When team members must interact, such as during knowledge sharing activities, diversity may negatively affect the intention to share knowledge, whether the diversity is task-related or non-task related (e.g., De Wit & Greer, 2008). Since diversity promotes conflict in team performance situations, also, and whether the conflict promotes task accomplishment (Pelled, Eisenhardt, & Xin, 1999), it is reasonable to expect that a disruption in interaction would negatively affect knowledge sharing. Meta-analyses regarding team performance reveal that the effects of gender diversity were negative, age diversity was nonsignificant (Bell, Villado, Lukasik, Briggs, & Belau, 2011; Schneid, Isidor, Steinmetz, & Kabst, 2016), and cultural diversity was mixed (Stahl, Maznevski, Voight, & Jonsen, 2010). Functional, or job-related, diversity was not related to team outcomes (Sivasubramaniam, Liebowitz, & Lackman, 2012). The nonsignificant and mixed findings suggest that team diversity can suppress positive effects of reciprocity on knowledge sharing. Based on this research evidence, the following hypotheses are offered:

Hypothesis Four: Team functional diversity will moderate the relationship between expected reciprocity and knowledge sharing intent so that the relationship will be stronger in teams with low functional diversity than in those with high functional diversity.

Hypothesis Five: Team age diversity will moderate the relationship between expected reciprocity and knowledge sharing intent so that the positive relationship will be stronger in teams with low age diversity than in those with high age diversity.

Hypothesis Six: Team gender diversity will moderate the relationship between expected reciprocity and knowledge sharing intent so that the positive relationship will be stronger in teams with low gender diversity than in those with high gender diversity.

Hypothesis Seven: Team ethnic diversity will moderate the positive relationship between expected reciprocity and knowledge sharing intent so that the relationship will be stronger in teams with low ethnic diversity than in those with high ethnic diversity.

METHODOLOGY

An experimental study was conducted to test the hypotheses. An instruction-based manipulation was used and, therefore, was determined to be well-suited for data collection in an undergraduate business class. Although many knowledge sharing studies collect data in the corporate environment, an experiment allows for control of extraneous variance (e.g., Di Gangi et al., 2012). Student subjects were told that they would have a guest speaker regarding an important course-related topic. Students were told they would be awarded with extra credit for their participation based on their performance on a quiz at the end of class. Procedures included:

1. One of the experimenters delivered a 20-minute lecture regarding how to read statistical output from surveys. She showed sample output from an employee satisfaction survey and indicated what parts of the output were important. She first worked through a practice quiz that asked questions about the sample employee satisfaction survey, such as: "What is the range of answers to the question of pay satisfaction?" and "With what aspect of work are employees least satisfied?"

2. Next, participants individually completed a quiz. After 15 minutes, student participants graded their own quizzes as the experimenter read answers aloud.
3. Next, the experimenter told students they would complete another quiz with their teams, with whom they had worked regularly during the semester.
4. Before the group quiz, students individually filled out a written survey measuring: (a) self-efficacy to successfully perform on the quiz, (b) intent to share knowledge with group members, (c) perceived impact of one's knowledge sharing on the group, and (d) perception that others in my group will share knowledge with me.
5. Students were dismissed and told that there would be no group quiz, and that they would each receive the maximum number of extra credit points.

Researchers have suggested that student subjects are appropriate in research focused on psychological and behavioral processes versus on the larger organizational context (Greenberg, 1987; Kruglanski, 1975). Subjects in the experiment were students in an undergraduate management class at a large Midwestern university. A total of 84 students participated in the study, 43 of whom were male (51.2%). A total of 21 student teams had been randomly formed by the professor for in-class work with four students per team. Participants were 10 weeks into a 15-week semester. In these groups, students completed in-class assignments over the course of the semester on a regular basis. The average student age was 23.14 (s.d. = 3.97), with average work experience of 6.46 years (s.d. = 3.87).

Knowledge sharing measures were taken from Bock et al. (2005). *Intent to share knowledge* displayed good reliability ($\alpha = 0.87$) and was measured with three questions on a scale of 1 (strongly disagree) to 7 (strongly agree), including questions such as: "On the task today, I intend to share my notes with other members." *Impact of my knowledge sharing* was deemed reliable ($\alpha = 0.85$) and was measured with three items regarding perceived impact of one's knowledge sharing on the group, including questions such as: "I believe the knowledge I share with my group today will improve our work processes on this task."

A confirmatory factory analysis (CFA) on the three knowledge sharing variables was performed using LISREL 8.8 to establish their separate meaning. According to Kenny, Kashy, & Cook (2006), the CFA model indicated acceptable goodness of fit statistics ($\chi 2 = 95.89$; $df = 51$; $p < 0.001$, CFI = 0.96; RMSEA = 0.10; SRMR = 0.06).

Intent and impact are normally treated as separate outcome variables in knowledge sharing research, but the measure here emphasizes intent to share useful knowledge and this distinguishes the variable from other team interaction processes. Examples of this type of measure can be found in organizational research. These include normative expectations by the degree of motivation to comply (Krueger, Reilly, & Carsrud, 2000), instrumentality by valence in expectancy theory (Mathieu, Tannenbaum, & Salas, 1992), and goal achievement by perceived importance (Cording, Christmann, & King, 2008). Therefore, intent to share impactful knowledge = intent to share knowledge x impact of knowledge sharing. Hereafter, the intent to share impactful knowledge construct is referred to only as knowledge sharing *intent*.

Expectation of reciprocal knowledge sharing was measured with three questions on a scale of 1 (strongly disagree) to 7 (strongly agree), including: "On the task today, I believe my fellow team members intend to share their notes with other members." The measure displayed good reliability ($\alpha = 0.95$).

The independent variables included competence, team attitudes, and team diversity. Competence was measured in two ways: task self-efficacy and the student's overall GPA. Task self-efficacy is one's perceived ability to perform the actual task for which knowledge is being shared (Quigley et al., 2007). Five questions (such as, "I can identify the range in survey results,") used, a scale from 0 (no confidence at all) to 100 (complete confidence), in 10-point intervals to measure task self-efficacy. The measure was deemed reliable (α = 0.86). Each self-efficacy scale was individualized to a specific task (Bandura, 1997). A validity check on the task self-efficacy construct is its ability to predict actual performance. Although performance is not a variable in hypothesized relationships, self-efficacy did predict student grades on the study quiz ($F = 2.27, p < 0.03$).

Performance was measured from a quiz relating to in-class learning, so the student's *GPA* was used as an objective measure of individual competence. GPA was chosen as a measure of competence because researchers have shown it to be positively and significantly related to performance self-efficacy (task-related), academic self-efficacy (general), and effort regulation (Richardson, Abraham, & Bond, 2012).

Due to their frequent use in knowledge sharing research, measured team attitude variables were chosen. These included trust (Chen & Hung, 2010; Kankanhalli et al., 2005; Lin & Huang, 2010; Lin et al., 2009; Quigley et al., 2007), commitment (Cabrera et al., 2006; C. Lin, 2007; Wasko & Faraj, 2005), and affiliation (Bock et al., 2005). The variables were averaged to create one composite called team attitudes. In various areas of study, researchers combine measures to create a composite that represents similar attributes (e.g., Gold, Malhotra, & Segars, 2001; Pineda & Lerner, 2006).

Team attitudes were measured using previously validated measures, each using a scale from 1 (strongly disagree) to 7 (strongly agree):

- **Trust (Quigley et al., 2007):** Three questions, including: "I have confidence that my team members will get this task done right" (α = 0.93).
- **Commitment (Mowday, Steers, & Porter, 1979):** Three questions, including: "I am extremely glad that I am in my team, versus another" (α = 0.60).
- **Affiliation (Bock et al., 2005):** Three questions, including: "Members of my team consider other members' standpoints highly" (α = 0.84).

As expected, team attitude variables were significantly intercorrelated ($p < .01$). The Cronbach's alpha for all team attitude items together is high (α = 0.88).

Four aspects of team diversity were measured – functional, age, gender and ethnicity. Categorical heterogeneity variables were calculated as in past research (Chowdhury, 2005). Blau's index has been cited as reliable and consistent with other acceptable indices of heterogeneity (Bantel & Jackson, 1989). It is calculated as: $1\text{-}\sum p_i^2$, where p is the proportion or percent of team members in a category and i is the number of different categories represented in the team.

Team functional diversity was measured as heterogeneity in the student's field of study. Students were aware of each other's majors in their teams. One's major field as a senior in college reflects the definition of functional diversity as an area of expertise and as reflecting a person's skills, knowledge, and ability (Bunderson & Sutcliffe, 2002). Functional background is a categorical variable and, therefore, its heterogeneity within teams is calculated using Blau's index (1977). Limited range in team diversity can limit observable variation in the dependent measures. Functional diversity in the teams ranged from 0.00 to 0.63 (mean = 0.42, s.d. = 0.19).

Team ethnic diversity and team gender diversity were also measured using Blau's Index (1977). Teams exhibited a wide range of both ethnic diversity, ranging from 0.00 to 0.67 (mean = 0.34, *s.d.* = 0.24), and gender diversity, ranging from 0.00 to 1.00 (mean = 0.42, *s.d.* = 0.27). Team age diversity was calculated by the standard deviation of ages in one's team and ranged from 0.50 to 9.77 (mean = 3.10, *s.d.* = 2.55).

Finally, teams were numbered and coded, but were not related to the dependent measure. Gender, ethnicity, age, degree type and work experience information were gathered but none was related to the dependent variable measures.

DATA ANALYSIS AND RESULTS

Table 1 provides descriptive variables and correlations. Expected reciprocity is positively related to knowledge sharing intent ($r = 0.45, p < 0.01$), suggesting support for Hypothesis 1. For analysis of interactions in Hypotheses 3-7, variables were centered to lower the correlation between the independent measures and the interaction terms (Kenny & Judd, 1984).

Hypothesis 1 stated that expected reciprocity will be positively related to knowledge sharing outcomes. A regression test confirmed that expected reciprocity predicted intent to share impactful knowledge ($R^2 = 0.21, F(1, 83) = 21.04, p < 0.01$). Hypothesis 1 is supported.

Hypothesis 2 stated that individual competence will moderate the positive relationship between expected reciprocity and knowledge sharing intent so the relationship is stronger for individuals with low competence versus high competence. Table 2 shows support for both the interaction of reciprocity with task self-efficacy ($p < 0.01$) and with GPA ($p < 0.01$). Figures 1 and 2 illustrate the interactions.

Hypothesis 3 proposed that team attitudes will moderate the relationship between expected reciprocity and knowledge sharing intent so that the relationship is stronger in teams with high team attitudes than in those with low team attitudes. The hypothesis is supported, as shown in Table 3 ($p < 0.01$), and illustrated in Figure 3.

Hypothesis 4 stated that team functional diversity will moderate the relationship between expected reciprocity and knowledge sharing intent so that the relationship will be stronger in teams with low functional diversity than in those with high functional diversity. In support, Table 4 shows the significant interaction ($p < 0.05$). Figure 4 illustrates the relationship.

Hypothesis 5 proposed that team age diversity will moderate the relationship between expected reciprocity and knowledge sharing intent so that the positive relationship will be stronger in teams with low

Table 1. Study variables: Descriptives and correlations

	Mean	S.D.	1	2	3	4	5	6	7	8
1. KS[a] intent	33.60	9.21	-							
2. Reciprocal KS	5.52	1.20	0.45**	-						
3. Task SE[b]	81.86	14.47	0.36**	0.11	-					
4. GPA	3.15	0.45	-0.19†	-0.27*	0.01	-				
5. Team attitudes	5.00	0.97	0.53**	0.65**	0.12	-0.29**	-			
6. Functional diversity	0.42	0.19	0.18†	0.05	-0.17	-0.03	0.13	-		
7. Age diversity	3.10	2.55	-0.07	-0.21†	0.13	-0.05	-0.24*	-0.15	-	
8. Gender diversity	0.42	0.27	-0.26*	-0.13	-0.04	0.19†	-0.19†	0.03	0.42**	-
9. Ethnic diversity	0.34	0.24	0.03	0.20*	0.01	-0.09	0.11	0.05	-0.12	0.11

[a]KS = Knowledge Sharing;[b]SE = Self-efficacy
**$p < .01$; *$p < .05$; †$p < .10$

Table 2. Multiple regression of expected reciprocity and individual competence on KS[a] intent

IV	DV: KS Intent
Reciprocal KS	0.95**
Task Self-efficacy	0.03**
Reciprocal KS X Task Self-efficacy	-0.11**
R^2	0.30**
F	11.23**
Reciprocal KS	4.17
GPA	-0.60**
Reciprocal KS X GPA	-5.47**
R^2	0.29**
F	10.28**

[a]KS = Knowledge Sharing
**$p < 0.01$; *$p < .05$; †$p < 0.10$

Figure 1. Interaction of task self-efficacy and reciprocal knowledge sharing on knowledge sharing intent

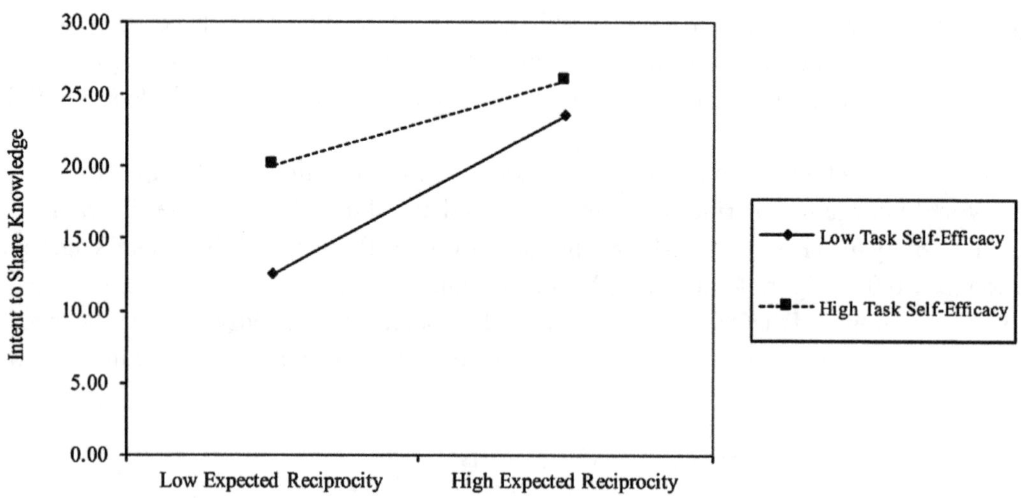

age diversity than in those with high age diversity. The interaction between reciprocity and age diversity is significant ($p < 0.01$), as shown in Table 4 and illustrated in Figure 5.

Hypothesis 6 stated that team gender diversity will moderate the relationship between expected reciprocity and knowledge sharing intent so that the positive relationship will be stronger in teams with low gender diversity than in teams with high gender diversity.

The interaction between gender diversity and reciprocity is not significant ($p < 0.26$); therefore Hypothesis 6 is not supported.

Hypothesis 7 proposed that team ethnic diversity will moderate the positive relationship between expected reciprocity and knowledge sharing intent so that the relationship will be stronger in teams with low ethnic diversity than in those with high ethnic diversity. As shown in Table 4, the interaction

Figure 2. Interaction of GPA and reciprocal knowledge sharing on knowledge sharing intent

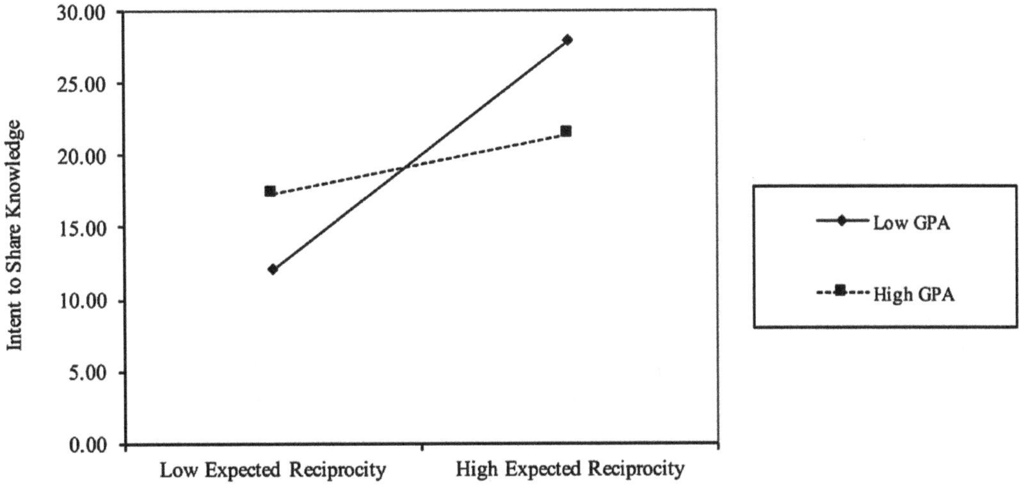

Table 3. Multiple regression of positive team attitudes on KS[a] intent

IV	DV: KS Intent
Reciprocal KS	3.15**
Team Attitudes	3.44**
Reciprocal KS X Positive Team Attitudes	2.26**
R^2	0.42**
F	18.95**

[a]KS = Knowledge Sharing
**$p < 0.01$; *$p < .05$; †$p < 0.10$

Figure 3. Interaction of team attitudes and reciprocal knowledge sharing on knowledge sharing intent

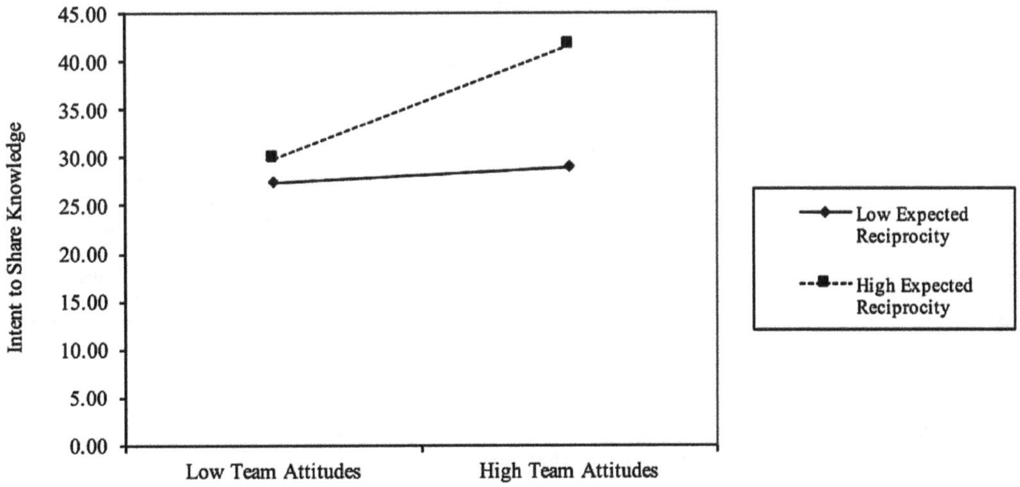

Table 4. Multiple regression of team diversity variables on KS[a] intent

IV	DV: KS Intent
Reciprocal KS	3.63**
Functional Diversity	6.09**
Reciprocal KS X Functional Diversity	-8.60*
R^2	0.28**
F	10.22**
Reciprocal KS	3.91**
Age Diversity	-0.28
Reciprocal KS X Age Diversity	-1.09**
R^2	0.29**
F	12.04**
Reciprocal KS	5.06
Gender Diversity	**-6.43†
Reciprocal KS X Gender Diversity	-4.75
R^2	0.25**
F	8.40**
Reciprocal KS	4.15**
Ethnic Diversity	-4.08
Reciprocal KS X Ethnic Diversity	9.51**
R^2	0.25**
F	9.71**

[a]KS = Knowledge Sharing
**$p < 0.01$; *$p < .05$; †$p < 0.10$

Figure 4. Interaction of functional diversity and reciprocal knowledge sharing on knowledge sharing intent

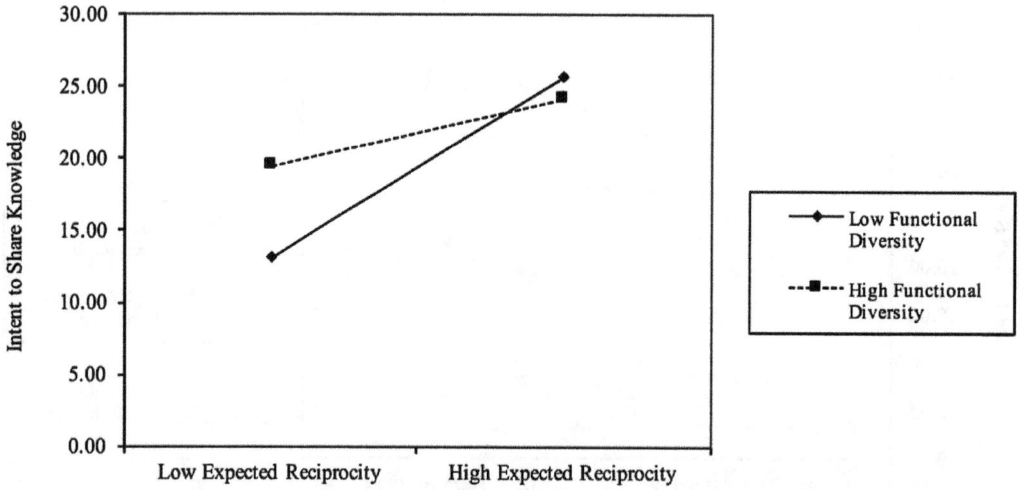

Figure 5. Interaction of age diversity and reciprocal knowledge sharing on knowledge sharing intent

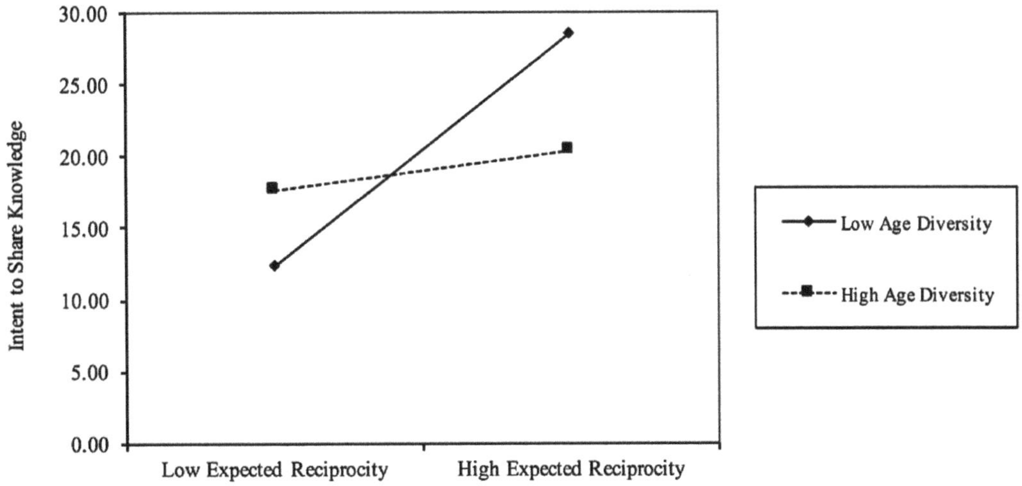

Figure 6. Interaction of ethnic diversity and reciprocal knowledge sharing on knowledge sharing intent

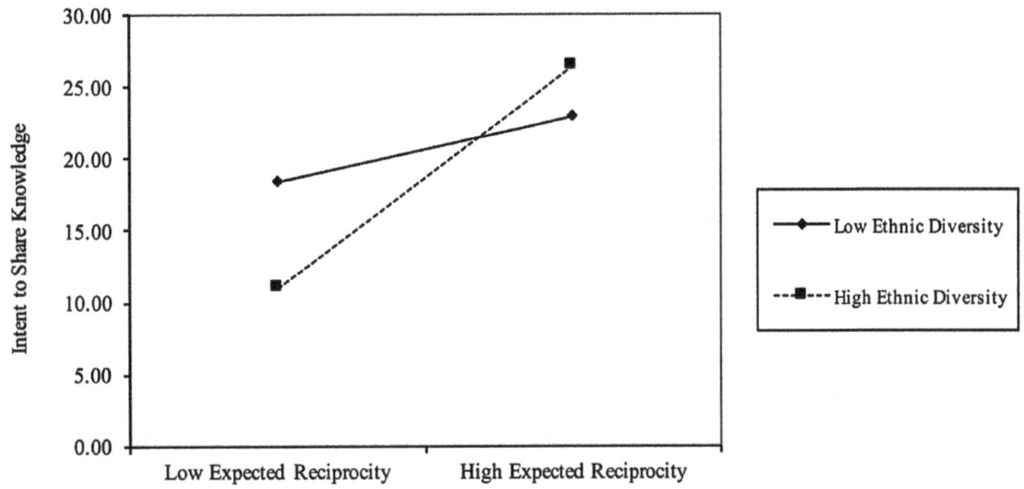

between reciprocity and team ethnic diversity is significant ($p < 0.01$). However, the interaction term is positive, indicating that the finding is opposite to the hypothesis. Team ethnic diversity did act as a moderator, but the relationship between reciprocity and intent was stronger for teams with high ethnic diversity versus those with low ethnic diversity, as shown in Figure 6.

DISCUSSION AND IMPLICATIONS

The purpose of this study was to investigate the expected reciprocity in knowledge sharing. The specific goal was to study the role of commonly studied variables in knowledge sharing research and how reciprocity may interact with these variables to affect sharing outcomes. It was proposed that these variables would moderate the positive relationship between expected reciprocity and knowledge sharing intent. The hope was that the findings here would help explain past contradictions in knowledge sharing research.

First, the dependent measure was defined as intent to share impactful knowledge. Most researchers investigate knowledge sharing intent as simply whether individuals will share knowledge with others (e.g., Chen & Hung, 2010; Cho et al., 2007; Lai & Chen, 2014). While this is a valuable measure, it was important to distinguish the simple intent to share from other team processes.

As predicted, expected reciprocity was positively related to knowledge sharing intent. Individual competence also moderated the positive relationship between expected reciprocity and knowledge sharing intent. Competence was measured as one's perceived competence in the task (task self-efficacy) and in an objective performance measure (GPA). For both measures, the relationship between reciprocity and intent was stronger for individuals with low competence than for those with high competence. When people feel that they are competent to perform a task, they are inherently motivated to share knowledge (e.g., Hung et al., 2015; Lai & Chen, 2014; Wasko & Faraj, 2005). Therefore, reciprocity is more helpful in encouraging knowledge sharing for those with low competence, who are not already motivated to share.

Despite the higher benefit to sharing for those with low self-perceived ability, the highest intent to share was reported by those with high self-perceived ability for both low and high expected reciprocity situations. Similarly, highest intention to share was reported by those with a high GPA at low levels of reciprocity. However, highest intention to share was reported by those with a low GPA at levels of high expected reciprocity. One explanation is that GPA is not as task-specific as self-efficacy. Therefore, knowledge of one's precise ability on the given task may have exerted a stronger effect on continuing intention to share, while precise knowledge of one's deficiencies in low self-efficacy may have a stronger muting effect on one's intention, regardless of reciprocity. Self-efficacy was also salient to the individual because performance was so recent. Positive feelings from the measure may have inflated or deflated intentions. As an objective measure, GPA may not have suffered from this effect.

The positive effects of reciprocity on knowledge sharing are dependent on the fact that team attitudes are positive. As the findings show, expected reciprocity exerts an impressively stronger, more positive effect on knowledge sharing intent when team attitudes are high. In contrast, expected reciprocity exerts little effect on knowledge sharing when team attitudes are low. Negative team attitudes suppress the positive effects of reciprocity on sharing. The lack of team trust, commitment and cooperation taint the motivating value of reciprocal knowledge. Encouraging reciprocity, therefore, is futile without positive team attitudes.

Four types of team diversity were investigated: functional background, age, gender and ethnicity. We proposed that each type of team diversity would moderate the relationship between expected reciprocity and knowledge sharing intent. There is disagreement on the positive versus negative effects of team diversity on team performance (Chowdhury, 2005; Stahl et al., 2010), but overall negative effects were proposed of both task-related and non-task-related diversity based on the disruption of interpersonal conflict (e.g., Pelled et al., 1999) and, therefore, disruption of the knowledge sharing process.

Functional and age team diversity each moderated the relationship between expected reciprocity and knowledge sharing intent. In situations of low team diversity, increased expected reciprocity resulted in the largest gains in sharing intent. Also notable is the nature of the interaction at low levels of expected reciprocity. High team functional and team age diversity resulted in intent to share that was higher than the low diversity situations. In high diversity teams, participants were willing to share knowledge, but increased expected reciprocity did not improve that willingness. In low diversity teams, participants were less willing to share knowledge at low levels of expected reciprocity versus high diversity teams.

The reason may be related to expectations of those perceived to be like oneself. The participant may react negatively to a similar other who does not share. A participant may also see the inequity of others who contribute less and feel demotivated. There may be specific stereotypes related to age or functional differences (e.g., "accounting students are better in math" or "older students study harder"). In this study, the mere existence of variation was important and not the direction of bias was not important in this study. Whether there are specific biases relating to age and functional differences in team knowledge sharing situations would provide an interesting opportunity for future researchers, however.

Hypotheses for team gender diversity and team ethnic diversity were not confirmed. First, team gender diversity did not significantly moderate the relationship between expected reciprocity and sharing intent. One reason that gender diversity may have failed to have significant effects is simply due to the lack of statistical range, or the dichotomous nature of the gender variable (Bell et al., 2011). In addition, the interaction was in the expected direction of the hypothesis.

In contrast, ethnic team diversity did significantly moderate the relationship between expected reciprocity and knowledge sharing, but in the opposite direction than expected. In the case of low expected reciprocity, high diversity negatively affected intent to share. In the case of high expected reciprocity, high team ethnic diversity positively affected intent to share.

Notably, the two types of diversity with unexpected results could be termed non-task-related types of team diversity, and the two types of diversity with expected results could be termed task-related diversity. Van Knippenberg, De Dreu, and Homan (2004) emphasized that the type of diversity is meaningful in disrupting group processes to affect the group. These explanations may assist in understanding how ethnicity and gender have little or no effect on knowledge sharing. However, the non-task-related team ethnic diversity was seen as positive. With high levels of team ethnic diversity, expected reciprocity flourished in affecting intent to share. With low levels of team ethnic diversity, expected reciprocity lost its motivational value. Participants seemed to use ethnic diversity as an indicator of valuable input to the task, or as motivation to share knowledge.

The mixed findings relating to team ethnic diversity could also be related to the nature of the sample. Although the two largest groups were Caucasian (n = 51, 64.6%) and African American (n = 18, 22.8%), the sample included members of five other ethnic groups (n = 10, 12.6%). Students are accustomed to working in teams with students of various nationalities and backgrounds, and this may make their expectations more positive. The nature of the task may have affected results, as well. Higher diversity may have accentuated the expectation that those of other ethnicities may have stronger backgrounds or knowledge of statistical concepts.

Implications for Research and Practice

Applications of the study findings are wide-ranging for both managers and higher education. Clearly, the direct relationships of commonly studied variables on knowledge sharing intent are altered when expected reciprocity is considered. Failure to consider reciprocity may result in overestimation or underestimation of those direct effects.

Facilitating reciprocity should be an active process, and team members should be encouraged to openly discuss their intentions to share with the rest of the group. Facilitating the perception that reciprocity is an expected part of group work may also be a facet of organizational climate, encouraged through management's rhetoric and in basic discussions of team planning and team documentation. Moreland and Myaskovsky (2000) suggested that awareness of others' competence can be as important as increasing interactions between members in order to foster knowledge sharing. Individuals can be made aware of others' competences through the use of leader introductions that formally present team members' backgrounds as applicable to the team task and by publicizing accomplishments. In a management classroom, an activity such as building a team resume can lead to awareness of others' capabilities and willingness to share. An important part of this resume activity may be to indicate what part of the team task one is most experienced in or most enthusiastic about, tying that person to a special position in the team from which he/she can provide information.

Managers and professors with teamwork in their classes should be aware that increasing positive team attitudes such as trust, commitment and affiliation does not result in higher knowledge sharing without the expectation of reciprocal knowledge. Fostering positive team attitudes results in positive team outcomes, but to increase knowledge sharing among members, reciprocation is critical. Increasing reciprocation in a team can boost knowledge sharing considerably, even in the absence of positive team attitudes. Teams that are short-term, have rotating membership, or some adversarial relationships may still share knowledge effectively with the expectation of valuable knowledge from others. Teams in university classrooms are often short-term and composed of strangers. Although researchers often propose building trust or commitment in these teams (Hansen, 2006), a more effective route may be to build expectation of reciprocal knowledge sharing.

Diverse teams must deal with the ambiguity of others' differences and their applicability to the task situation. Managers and educators must analyze which differences are perceived as strengths as well as those which may be perceived as challenges. As noted by van Knippenberg et al. (2004), it is the team members' perceptions of the situation that matters, not the actual effect on team outcomes. For teams with a high level of task-related diversity that positively affects team performance, encouraging reciprocity is not as beneficial as in a team with low levels of task-related diversity. Consider that the task-related diversity, such as varied functional backgrounds, may create the perception of having extra resources. Teams without these 'resources' may need extra encouragement that they will, in fact, be receiving reciprocal knowledge from members who are perceived to have fewer resources to bring to the task.

Managers and educators should identify forms of diversity that are not related to the task, as well. These types of diversity may be associated with bias toward the members' worth in adding to the team's performance. Instead of working on diversity sensitivity, the results suggest that a positive step toward team knowledge sharing is to encourage the expectation that members will receive reciprocal knowledge from each other. With high levels of non-task-related diversity, in fact, the expectation of reciprocal knowledge appears to protect teams from all negative effects and to share as much as teams with low non-task-related diversity.

Limitations and Implications

As in any research study, there are strengths and limitations. This study added to the knowledge sharing literature with an in-depth investigation of reciprocal knowledge sharing. A wide range of variables were investigated, including both objective and subjective measures. The study examined each type of diversity separately, as suggested by current research (Bell et al., 2011). A new dependent measure was introduced as intent to share impactful knowledge. This new measure gave depth to the traditional intent measure.

The experimental design allowed more control over the task and interactions, as in other experimental knowledge sharing studies (Di Gangi et al., 2012). Most knowledge sharing studies focus on sharing information amid other competing issues in organizations (e.g., Lin & Huang, 2010) and/or in virtual communities (e.g., Chen & Hung, 2010; Xue et al., 2012). The face-to-face, controlled environment used here should give a valuable perspective to the literature.

This research study is not without limitations. The teams had been formed based on class needs and their prior experiences may have tainted experimental performance and perceptions. These may be labeled short-term project teams, and the outcomes of an experiment using longer-termed teams may differ. This provides an opportunity for future researchers to compare team type and tenure. In addition, the study did not measure variables that may have helped understanding of results in some cases, such as climate.

Many researchers have discussed the limitations of using student subjects. In the current study, an instruction-based task made a learning environment appropriate for studying knowledge sharing behavior. As suggested by Greenberg (1987), the goal of research is not always to generalize directly to the organizational context, but often to understand basic human behaviors and psychological processes. In fact, researchers showed that students are appropriate subjects when studying human behavior because they often exhibit various attitudes of the society in general (Gordon, Slade, and Schmitt, 1986; Krughanski, 1975).

The task itself may have limitations in its generalizability directly to organizations. Again, however, the goal of the research was not to generalize the task to a corporate environment but to better understand the human response to a team knowledge sharing situation. The goals of this study required the ability to control external variance and observe knowledge sharing behavior in a team environment. Results should be generalizable to adults who learn new, complex material in a team environment. In fact, the control should not be a limitation in this study, but a strength in removing unrelated factors from the environment that may obscure the manipulation's effects.

Because relationships were observed one at a time, in separate hierarchical models, some variables that overlap may have been missed and this may have compounded or minimized effects. In future research, a structural model with more data points should be used. Team-level contextual factors may add significantly to the understanding of team-level climates for knowledge sharing and regarding team diversity (e.g., Joshi & Roh, 2009).

A certain subset of measures was chosen to represent competence, team attitudes and diversity. Other measures are potential causes of knowledge sharing, such as knowledge sharing self-efficacy (Chen & Hung, 2010). The reciprocity measure is a subjective expectation. Comparison to actual reciprocity would be a valuable method for determining bias. Perceived impact of the expected reciprocity would also give depth to the measure.

Future researchers should consider different types of tasks and stages of task performance. The novelty and complexity of tasks affects performance perceptions and can distort how one views the task (Endres, Chowdhury, & Alam, 2008; Endres, Chowdhury, & Milner, 2009). The types of rewards, meaningfulness of the task, and degree of interdependence can all affect how individuals share knowledge with others and expect reciprocity.

REFERENCES

Ajzen, I. (1985). From intentions to actions: A theory of planned behavior. In J. Kuhl & J. Beckmann (Eds.), *Action control: From cognition to behavior* (pp. 11–39). Berlin: Springer-Verlag. doi:10.1007/978-3-642-69746-3_2

Akhavan, P., & Hosseini, S. M. (2016). Social capital, knowledge sharing, and innovation capability: An empirical study of R&D teams in Iran. *Technology Analysis and Strategic Management*, *28*(1), 96–113. doi:10.1080/09537325.2015.1072622

Aktharsha, U., & Anisa, H. (2011). Knowledge management system and learning organization: An empirical study in an engineering organization. *IUP Journal of Knowledge Management*, *9*(2), 26–43.

Bandura, A. (1997). *Self-efficacy: The exercise of control*. New York: Freeman.

Bantel, K. A., & Jackson, S. E. (1989). Top management and innovations in banking: Does the composition of the top team make a difference? *Strategic Management Journal*, *10*(S1), 107–124. doi:10.1002mj.4250100709

Bell, S. T., Villado, A. J., Lukasik, M. A., Briggs, A. L., & Belau, L. (2011). Getting specific about demographic diversity variable and team performance relationships: A meta analysis. *Journal of Management*, *37*(3), 709–743. doi:10.1177/0149206310365001

Blau, P. M. (1977). *Inequality and heterogeneity*. New York: The Free Press.

Bock, G. W., Zmud, R. W., Kim, Y. G., & Lee. (2005). Behavioral intention formation in knowledge sharing: Examining the roles of extrinsic motivators, social-psychological forces, and organizational climate. *Management Information Systems Quarterly*, *29*(1), 87–111. doi:10.2307/25148669

Bunderson, J. S., & Sutcliffe, K. M. (2002). Comparing Alternative Conceptualizations of functional diversity in management teams: Process and performance effects. *Academy of Management Journal*, *45*(5), 875–893. doi:10.2307/3069319

Cabrera, A., Collins, W. C., & Salgado, J. F. (2006). Determinants of individual engagement in knowledge sharing. *International Journal of Human Resource Management*, *17*(2), 245–264. doi:10.1080/09585190500404614

Caimo, A., & Lomi, A. (2015). Knowledge sharing in organizations: A Bayesian analysis of the role of reciprocity and formal structure. *Journal of Management*, *41*(2), 665–691. doi:10.1177/0149206314552192

Chang, Y., Hsu, P., Shiau, W., & Tsai, C. (2015). Knowledge sharing intention in the United States and China: A cross-cultural study. *European Journal of Information Systems*, *24*(3), 262–277. doi:10.1057/ejis.2014.28

Chen, C. J., & Hung, S. W. (2010). To give or to receive? Factors influencing members' knowledge sharing and community promotion in professional virtual communities. *Information & Management*, *47*(4), 226–236. doi:10.1016/j.im.2010.03.001

Cho, N., Li, G., & Su, C. (2007). An empirical study on the effect of individual factors on knowledge sharing by knowledge type. *Journal of Global Business and Technology*, *3*(2), 1–15.

Chowdhury, S. (2005). Demographic diversity for building an effective entrepreneurial team: Is it important? *Journal of Business Venturing*, *3*(6), 727–746. doi:10.1016/j.jbusvent.2004.07.001

Cleveland, S., & Ellis, T. J. (2015). Rethinking knowledge sharing barriers: A content analysis of 103 studies. *International Journal of Knowledge Management*, *11*(1), 28–51. doi:10.4018/IJKM.2015010102

Cording, M., Christmann, P., & King, D. R. (2008). Reducing causal ambiguity in acquisition integration: Intermediate goals as mediators of integration decisions and acquisition performance. *Academy of Management Journal*, *51*(4), 744–767.

Coyle-Shapiro, J., & Conway, N. (2005). Exchange relationships: Examining psychological contracts and perceived organizational support. *The Journal of Applied Psychology*, *90*(4), 774–781. doi:10.1037/0021-9010.90.4.774 PMID:16060794

DiGangi, P., Wasko, M. M., & Tang, X. (2012). Would you share?: Examining knowledge type and communication channel for knowledge sharing within and across the organizational boundary. *International Journal of Knowledge Management*, *8*(1), 1–21. doi:10.4018/jkm.2012010101

Endres, M. L., Chowdhury, S. K., & Alam, I. (2008). Gender effects on bias in complex financial decisions. *Journal of Managerial Issues*, *20*(2), 238–254.

Endres, M. L., Chowdhury, S. K., & Milner, M. (2009). Ambiguity tolerance and accurate assessment of self-efficacy in a complex decision task. *Journal of Management & Organization*, *15*(1), 31–46. doi:10.1017/S1833367200002868

Erickson, G. S., & Rothberg, H. N. (2011). Assessing knowledge management needs: A strategic approach to developing knowledge. *International Journal of Knowledge Management*, *7*(3), 1–10. doi:10.4018/jkm.2011070101

Frenkel, S. J., & Sanders, K. (2007). Explaining variations in co-worker assistance in organizations. *Organization Studies*, *28*(6), 797–823. doi:10.1177/0170840607073079

Gold, A. H., Malhotra, A., & Segars, A. H. (2001). Knowledge management: An organizational capabilities perspective. *Journal of Management Information Systems*, *18*(1), 185–214. doi:10.1080/07421222.2001.11045669

Gordon, M. E., Slade, L. A., & Schmitt, N. (1986). The “science of the sophomore” revisited: From conjecture to empiricism. *Academy of Management Review*, *11*(1), 191–208.

Greenberg, J. (1987). The college sophomore as guinea pig: Setting the record straight. *Academy of Management Review*, *12*, 157–159.

Haeussler, C. (2011). Information-sharing in academia and the industry: A comparative study. *Research Policy*, *40*(1), 105–122. doi:10.1016/j.respol.2010.08.007

Haeussler, C., Jiang, L., Thursby, J., & Thursby, M. (2014). Specific and general information sharing among competing academic researchers. *Research Policy*, *43*(3), 465–475. doi:10.1016/j.respol.2013.08.017

Hansen, R. S. (2006). Benefits and problems with student teams: Suggestions for improving team projects. *Journal of Education for Business*, *82*(1), 11–19. doi:10.3200/JOEB.82.1.11-19

Hoffman, L. R., & Maier, N. R. F. (1961). Quality and acceptance of problem solutions by members of homogeneous and heterogeneous groups. *Journal of Abnormal Psychology*, *62*(2), 401–407. doi:10.1037/h0044025 PMID:13715029

Hung, S., Durcikova, A., Lai, H., & Lin, W. (2011). The influence of intrinsic and extrinsic motivation on individuals' knowledge sharing behavior. *International Journal of Human-Computer Studies*, *69*(6), 415–427. doi:10.1016/j.ijhcs.2011.02.004

Hung, S., Lai, H., & Chou, Y. (2015). Knowledge-sharing intention in professional virtual communities: A comparison between posters and lurkers. *Journal of the Association for Information Science and Technology*, *66*(12), 2494–2510. doi:10.1002/asi.23339

Jeon, S., Kim, Y., & Koh, J. (2011). An integrative model for knowledge sharing in communities-of-practice. *Journal of Knowledge Management*, *15*(2), 251–269. doi:10.1108/13673271111119682

Johns, G. (2017). Incorporating context in organizational research. *Reflections: The SoL Journal.*

Joshi, A., & Roh, H. (2009). The role of context in work team diversity research: A meta analytic review. *Academy of Management Journal*, *52*(3), 599–627. doi:10.5465/AMJ.2009.41331491

Kang, M., Kim, Y., & Bock, G. (2010). Identifying different antecedents for closed vs open knowledge transfer. *Journal of Information Science*, *36*(5), 585–602. doi:10.1177/0165551510375667

Kankanhalli, A., Tan, B. C. Y., & Wei, K. K. (2005). Contributing knowledge to electronic knowledge repositories: An empirical investigation. *Management Information Systems Quarterly*, *29*(1), 113–143. doi:10.2307/25148670

Kenny, D. A., & Judd, C. M. (1984). Estimating the nonlinear and interactive effects of latent variables. *Psychological Bulletin*, *96*(1), 201–210. doi:10.1037/0033-2909.96.1.201

Kenny, D. A., Kashy, D. A., & Cook, W. L. (2006). *Dyadic data analysis*. New York: Guilford Press.

Khalil, T., & Atieh, K., Mohammed, A. U., & Bagdadlian, F. (2014). Examining the social and technical factors influencing school teachers knowledge sharing intentions in a teachers online professional community. *Electronic Journal of Knowledge Management*, *12*(3), 157–165.

Killingsworth, B., Xue, Y., & Liu, Y. (2016). Factors influencing knowledge sharing among global virtual teams. *Team Performance Management*, *22*(5), 284–300. doi:10.1108/TPM-10-2015-0042

Knouse, S. B., & Chretien, D. (1996). Workforce diversity and TQM. In S. B. Knouse (Ed.), *Human resources management perspectives on TQM: Concepts and practices* (pp. 261–274). Milwaukee, WI: American Society for Quality Control Press.

Kordzadeh, N., Liu, C. Z., Au, Y. A., & Clark, J. G. (2014). A multilevel investigation of participation within virtual health communities. *Communications of the Association for Information Systems*, *34*, 494–512.

Krueger, N. F. Jr, Reilly, M. D., & Carsrud, A. L. (2000). Competing models of entrepreneurial intentions. *Journal of Business Venturing*, *15*(5/6), 411–432. doi:10.1016/S0883-9026(98)00033-0

Kruglanski, A. W. (1975). The human subject in the psychology experiment: Fact and artifact. In L. Berkowitz (Ed.), Advances in Experimental Social Psychology (Vol. 8, pp. 101–147). New York: Academic Press. doi:10.1016/S0065-2601(08)60249-X

Lai, H., & Chen, T. T. (2014). Knowledge sharing in interest online communities: A comparison of posters and lurkers. *Computers in Human Behavior*, *35*, 295–306. doi:10.1016/j.chb.2014.02.004

Lin, C. (2007). To share or not to share: Modeling knowledge sharing using exchange ideology as a moderator. *Personnel Review*, *36*(3), 457–475. doi:10.1108/00483480710731374

Lin, H., Lee, H., & Wang, D. (2009). Factors influencing knowledge sharing based on a fuzzy AHP approach. *Journal of Information Science*, *35*(1), 25–44. doi:10.1177/0165551508091310

Lin, T. C., & Huang, C. C. (2010). Withholding effort in knowledge contribution: The role of social exchange and social cognitive on project teams. *Journal of International Management*, *47*, 188–196.

Mathieu, J. E., Tannenbaum, S. I., & Salas, E. (1992). Influences of individual and situational characteristics on measures of training effectiveness. *Academy of Management Journal*, *35*(4), 828–847. doi:10.2307/256317

Milliken, F. J., & Martins, L. L. (1996). Searching for common threads: Understanding the multiple effects of diversity in organizational groups. *Academy of Management Review*, *21*(2), 402–433.

Moreland, R. L., & Myaskovsky, L. (2000). Exploring the performance benefits of group training: Transactive memory or improved communication? *Organizational Behavior and Human Decision Processes*, *82*(1), 117–133. doi:10.1006/obhd.2000.2891

Mowday, R. T., Steers, R. M., & Porter, L. (1979). The measurement of organizational commitment. *Journal of Vocational Behavior*, *14*(2), 224–247. doi:10.1016/0001-8791(79)90072-1

Nahapiet, J., & Ghoshal, S. (1998). Social capital, intellectual capital, and the organizational advantage. *Academy of Management Review*, *23*(2), 242–266.

Pelled, L. H., Eisenhardt, K. M., & Xin, K. R. (1999). Exploring the black box: An analysis of work group diversity, conflict, and performance. *Administrative Science Quarterly*, *44*(1), 1–28. doi:10.2307/2667029

Pineda, R. C., & Lerner, L. D. (2006). Goal attainment, satisfaction and learning from teamwork. *Team Performance Management*, *12*(5/6), 182–191. doi:10.1108/13527590610687938

Quigley, N., Tesluk, P., Locke, E., & Bartol, K. (2007). A multilevel investigation of the motivational mechanisms underlying knowledge sharing and performance. *Organization Science*, *18*(1), 71–92. doi:10.1287/orsc.1060.0223

Ranasinghe, S. B., & Dharmadasa, P. (2013). Intention to knowledge sharing: From planned behavior and psychological needs perspectives. *International Journal of Knowledge Management, 9*(4), 33–50. doi:10.4018/ijkm.2013100103

Richardson, M., Abraham, C., & Bond, R. (2012). Psychological correlates of university students' academic performance: A systematic review and meta-analysis. *Psychological Bulletin, 138*(2), 353–387. doi:10.1037/a0026838 PMID:22352812

Schneid, M., Isidor, R., Steinmetz, H., & Kabst, R. (2016). Age diversity and team outcomes: A quantitative review. *Journal of Managerial Psychology, 31*(1), 2–17. doi:10.1108/JMP-07-2012-0228

Shanshan, S. (2014). A comprehensive relational model of factors influencing knowledge sharing: An empirical study. *International Journal of Knowledge Management, 10*(1), 1–25. doi:10.4018/ijkm.2014010101

Sivasubramaniam, N., Liebowitz, S. J., & Lackman, C. L. (2012). Determinants of new product development team performance: A meta-analytic review. *Journal of Product Innovation Management, 29*(5), 803–820. doi:10.1111/j.1540-5885.2012.00940.x

Stahl, G. K., Maznevski, M. L., Voight, A., & Jonsen, K. (2010). Unraveling the effects of cultural diversity in teams: A meta-analysis of research on multicultural work groups. *Journal of International Business Studies, 41*(4), 690–709. doi:10.1057/jibs.2009.85

Taber, T. D., & Deosthali, K. (2014). Analysis of self-reported motives for task-related helping: Implications for an integrated theory of helping. *Journal of Business and Psychology, 29*(3), 343–366. doi:10.100710869-013-9327-4

Tamjidyamcholo, A., Baba, M. S. B., Tamjid, H., & Gholipour, R. (2013). Information security—Professional perceptions of knowledge-sharing intention under self efficacy, trust, reciprocity, and shared-language. *Computers & Education, 68*, 223 232.

van Knippenberg, D. A., De Dreu, C. K. W., & Homan, A. C. (2004). Work group diversity and group performance: An integrative model and research agenda. *The Journal of Applied Psychology, 89*(6), 1008–1022. doi:10.1037/0021-9010.89.6.1008 PMID:15584838

Wan, Z., Haggerty, N., & Wang, Y. (2015). Individual level knowledge transfer in virtual settings: A review and synthesis. *International Journal of Knowledge Management, 11*(2), 29–61. doi:10.4018/IJKM.2015040103

Wasko, M. M., & Faraj, S. (2005). Why should I share? Examining knowledge contribution in networks of practice. *Management Information Systems Quarterly, 21*(9), 35–57. doi:10.2307/25148667

Wei, X., Liu, Y., & Chen, S. (2015). A meta-analysis of the relationship between team demographic diversity and team performance. *Acta Psychologica Sinica, 47*(9), 1172–1187. doi:10.3724/SP.J.1041.2015.01172

Wu, J. B., Hom, P. W., Tetrick, L. E., Shore, L. M., Jia, L., Chaoping, L., & Song, L. J. (2006). The norm of reciprocity: Scale development and validation in the Chinese context. *Management and Organization Review*, *2*(3), 377–402. doi:10.1111/j.1740-8784.2006.00047.x

Xue, Y., Liang, H., Hauser, R., & O'Hara, M. (2012). An empirical study of knowledge sharing intention within virtual teams. *International Journal of Knowledge Management*, *8*(3), 47–61. doi:10.4018/jkm.2012070103

Yang, J. (2007). The impact of knowledge sharing on organizational learning and effectiveness. *Journal of Knowledge Management*, *11*(2), 83–90. doi:10.1108/13673270710738933

Zhang, X., Chen, Z., Vogel, D., & Guo, C. (2009). Exchange ideology as a moderator of knowledge sharing in virtual teams: A social exchange theory perspective. *International Journal of Internet and Enterprise Management*, *6*(2), 143–163. doi:10.1504/IJIEM.2009.023927

Zhao, L., Detlor, B., & Connelly, C. E. (2016). Sharing knowledge in social Q&A sites: The unintended consequences of extrinsic motivation. *Journal of Management Information Systems*, *33*(1), 70–100. doi:10.1080/07421222.2016.1172459

Chapter 8
The Determinants of Interorganizational Knowledge Coaching Success:
Looking Ahead to the Future of Knowledge Transfer

Nicole A. Celestine
University of Western Australia, Australia

Chris Perryer
University of Western Australia, Australia

ABSTRACT

With the global rise of the knowledge worker, there is an increasing interest in effective knowledge transfer practices within the firm. This interest is especially prevalent in regards to the transfer of tacit knowledge—knowledge that is difficult to describe and transmit to others. In this chapter, the authors consider how the emerging practice of knowledge coaching can be utilized as a means to transfer tacit knowledge between experts and protégés within the firm, and consider the individual-level factors that influence whether the practice yields success. In doing so, they report the results of a study examining the impact of individually possessed national cultural values on motivation to engage in knowledge coaching, and put forward several avenues for future research.

INTRODUCTION

With the rise of the new century, there is a growing recognition that the sustainability and success of the firm relies not on the physical availability of resources—whether they be physical labour or materials—but the knowledge embedded within these resources. We see evidence of this paradigm shift in the proliferation of new terminology such as 'knowledge worker' and 'knowledge capital', indicating a movement away from thinking of value in quantitative terms, and towards thinking of it in qualitative

DOI: 10.4018/978-1-5225-5427-1.ch008

terms. Further, with the growing rollout of automation, artificial intelligence, and machine learning across industries worldwide, some have predicted a future world of work far removed from our current employment landscape. Indeed, it has been suggested that by 2025, knowledge work automation tools and systems could take on tasks equal to the output of 110 million to 140 million full-time roles (Manyika, Chui, Bughin, Dobbs, Bisson, & Marrs, 2013).

Given that even traditional knowledge industries are now being influenced by automation, we are likely to see rapid and continuous change in these industries, whereby the greatest contributions of the individual will be qualitative in nature. I.e. possessing creative and innovative competencies will be vital. As such, in order to create and sustain value, the knowledge industries of the future will need to develop competencies around the development and dissemination of these creative skills amongst workers. In recognition of this emerging, knowledge-based view of the firm, scholars have therefore attempted to uncover the factors that influence the willingness of knowledge workers to disseminate their knowledge—a crucial consideration for human capital organisations that rely on scarce expert talent as a means of offering value to their customers (Foss, 2007).

In light of this emphasis on individually-possessed knowledge capital, one approach has been to examine the effects of an array of stable-individual traits, to better understand the influence they bear upon knowledge transfer (KT) behaviours. To illustrate, Cabrera, Collins, and Salgado (2006) found that self-efficacy and openness to experience influenced self-reported participation in knowledge exchange. Similarly, Cho, Zheng, and Su (2007) found that self-efficacy and agreeableness were positively related to preferences to use knowledge sharing mechanisms. These are just two of many studies that have examined the impact of trait variables on willingness to pass on knowledge (see also Lu, Leung, & Koch, 2006; Matzler, Renzl, Mooradian, von Krogh, & Mueller, 2011; Matzler, Renzl, Müller, Herting, Mooradian, 2008; Wang, Noe, & Wang, 2011). However, to the best of these authors' knowledge, there has yet to be an examination of the impact of values on individual-level KT.

As it stands, there exists a plethora of research around the factors that influence inter-organisational KT effectiveness (see Van Wijk, Jansen, & Lyles, 2008). Among these, we commonly see consideration of national cultural values as influencers of KT occurring across borders. For instance, in a study of a Japanese pharmaceutical company's foreign subsidiaries, Magnier-Watanabe and Senoo (2010) found that national cultural values exhibited effects on the local subsidiaries' knowledge management practices, including knowledge diffusion within the firms. While these and other findings shed light on the substantial predictive power of national culture in organisation-level models of KT effectiveness (e.g. Bhagat, Kedia, Harveston, & Triandis, 2002), some studies have suggested that the real dissension exists at the individual level. For instance, in their meta-analytic review, Van Wijk et al. (2008) found that cultural distance acts as the greatest barrier to KT across units *within*, not outside, the firm. It would therefore seem that firms acquiring knowledge from external partners have developed expertise in dealing with different cultures or limit their KT practices to partners with whom they are most familiar, suggesting that an examination of national culture's impact on intraorganisational KT is timely and most necessary.

By framing our discussion according to the principles of Rational Choice theory (Scott, 2000), we put forward a theoretical model illustrating how individually-possessed national cultural values will impact KT processes through several avenues. First, we argue that individually-possessed values will affect the perceived valence attached to KT incentives, influencing motivation to invest effort in these processes. Second, we argue that national cultural values will impact upon KT processes themselves, as knowledge sources will exhibit differences in the types of behaviours they view as appropriate, and subsequently employ in the process of transferring their knowledge. We go on to test these propositions via an empirical study that examines KT through the lens of knowledge coaching relationships, and knowledge-coaching behaviour—a broad bundle of activities that are highly effective for transferring tacit knowledge. In other words, the empirical portion of this chapter will combine research in the areas of national culture, motivation, and knowledge management.

We concede that there is a certain ecological clumsiness involved in combining these three areas. Studies on national culture are usually focused on the country level (Hofstede, 2001); the overwhelming majority of the knowledge management literature is on the organisational level (Minbaeva, Mäkelä, & Rabbiosi, 2012); while studies on motivation are usually on the individual level (Ryan & Deci, 2000). This chapter contributes to the knowledge management literature by examining all three of these factors at the person level, in order to better understand the forces that facilitate or inhibit the effective transfer of individually-possessed knowledge. Such an approach is advantageous, as it allows us to avoid fallacious treatments of culture as unvarying and monolithic—a growing criticism of extant research, arguing that globalisation is in fact dissolving country-level cultural differences (Friedman, 2005). The overarching objective of this chapter is therefore to shed light on how the areas of motivation, national culture, and KT can interact to result in optimal knowledge dissemination within the firm.

BACKGROUND

Since the turn of the century, there has been a sudden reemergence of interest in knowledge management, sparking efforts to enhance construct clarity around the topic (Fahey & Prusak, 1998). As such, we assume Nonaka, Umemoto, and Senoo's (1996, p. 205) definition of knowledge as "a meaningful set of information that constitutes a justified true belief and/or an embodied technical skill". Under the umbrella of 'knowledge', we see reference to two key types of knowledge: tacit and explicit. Tacit knowledge, is that which is "difficult to describe and transmit to others" (Chilton & Bloodgood, 2008, p. 77), characterised by its symbiotic rooting in experience and practice (Venkitachalam & Busch, 2012), and its inability to be entirely divorced from the knower. Conversely, explicit knowledge is codifiable and can be more easily passed on to others (Kogut & Zander, 1992).

When considering the transfer of knowledge between individuals, we have empirical reason to believe that a consideration of the type of knowledge being transferred is crucial for two reasons, and therefore emphasise knowledge *type* as a key variable in our theoretical model (see Figure 1). Firstly, it has been shown that due to the inherent 'stickiness' associated with tacit knowledge, it can be exceptionally difficult to transfer. The term 'stickiness' refers to a potentially wide array of challenges associated with KT, including but not limited to a lack of motivation within the knowledge source and/or recipient, a weak relationship between the parties, or an inability to retain the new knowledge through routine use. Due to its inherent causal ambiguity, these problems become amplified when dealing with tacit knowledge (Szulanski, 2000). It is therefore not uncommon for knowledge sources to adjust their willingness to share tacit knowledge according to its stickiness. Secondly, the mechanisms for transferring tacit and explicit knowledge differ markedly—an issue which we will discuss next.

Figure 1. Theoretical model

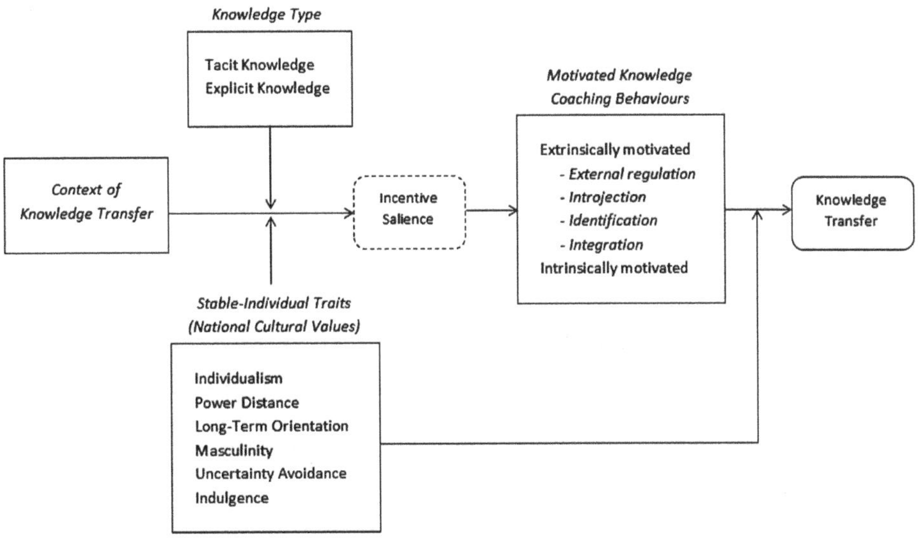

KNOWLEDGE TRANSFER

As one of several processes within the domain of knowledge management, KT has become a widely debated topic of discussion. We define KT in accordance with Ko, Kirsch, and King (2005, p. 62), conceptualising it as "the communication of knowledge from a source so that it is learned and applied by a recipient". To be sure, we also make a distinction between KT and knowledge sharing. Whereas the former is implies the unidirectional movement of knowledge, the latter is typically viewed as bidirectional and reciprocal (Liyange, Elhag, Ballal, & Li, 2009). Given that in a knowledge coaching relationship (discussed next), the majority of the knowledge will be flowing from the knowledge coach to the protégé in a unidirectional fashion, a discussion of KT as opposed to knowledge sharing is better aligned with the terminology usage of other scholars (e.g. Ko, et al., 2005; Sarker, Sarker, Nicholson, & Joshi, 2005).

Managers are often motivated to endorse the transfer of tacit knowledge, as workers possessing this deeply-held form of expertise can perform complex tasks in an effective manner (Lin & Lee, 2004). For instance, organisational members with job-relevant tacit knowledge can rapidly diagnose problems or opportunities in a work environment, and extrapolate from past experience to arrive at a suitable response (Leonard & Swap, 2004). While developing this highly effective form of tacit knowledge can apparently take up ten years or more, the learning process can be accelerated with the help of a knowledge coach (Leonard & Swap, 2005), which we define as *an individual who attempts to recreate their tacit knowledge in the mind of a less-experienced individual using active learning techniques.* Knowledge coaching is hands-on, closely resembles on-the-job training, and emphasises the improvement of employee performance and capabilities. It is also posited to be one of the most efficient means of facilitating tacit KT (Leonard & Swap, 2005), and a means that many organisations unwittingly benefit from as a result of formal and informal mentoring relationships.

Knowledge coaching assumes that the most effective way of re-creating knowledge is not to teach in the usual sense, but to systematically guide the experience of a protégé. Pairing an organisation's knowledgeable experts with novices therefore presents unique opportunities for KT that extend beyond the traditional student-teacher paradigm. To illustrate, Leonard and Swap (2005) put forward a continuum of potential modes for transferring knowledge from a knowledge coach to a protégé, which vary according to how suitable they are for transferring tacit knowledge. Among the least suitable methods, knowledge coaches can provide lectures or explicit directives to a protégé—strategies that are more appropriate for transferring explicit knowledge, as they involve the coding of knowledge into a written or oral form. Somewhat more effective strategies involve the recommendation of rules of thumb, the presentation of illustrative stories, or Socratic questioning. These methods require a protégé to engage in an active process of thinking and problem-solving, as they determine how morals and rules apply to present-moment scenarios (Leonard & Swap, 2005). The most fruitful means of transferring tacit knowledge involve an active process of learning-by-doing, whereby a knowledge coach and protégé take part in guided practice, guided observation, guided problem solving, and/or guided experimentation (Leonard & Swap, 2005).

The key message, therefore, is that the more active and involved a protégé is in the learning process, the richer the knowledge they will retain following the experience. However, when we consider the highly-involved, complex process associated with tacit KT, it becomes apparent that a large investment may be required from the parties involved—particularly the knowledge coach, who may have multiple protégés under their wing, or other time-consuming responsibilities on their plate. We therefore argue that individual motivation is a vital ingredient for knowledge coaching to yield success.

Individual Motivation

It is a reality that without strong individual motivation, KT will seldom occur (Vera-Munoz, Ho, & Chow, 2006). By framing the individual as a self-interested agent, Rational Choice Theory (Scott, 2000) posits that people will balance the anticipated costs and gains associated with a particular behaviour, to arrive at action that maximises personal advantage (Friedman, 1953). It is from this balancing of gains and losses that motivation (or a lack thereof) will be generated. To think about this in the context of knowledge coaching, there are a broad array of factors that may be viewed as costs or gains associated with the process, which may be salient to either the knowledge coach or the protégé. For instance, the parties may view the knowledge coaching relationship as an opportunity—perhaps as an avenue to build social capital and strengthen ties with one another. Alternatively, the parties may view the process as overly time-consuming, and hence too costly to invest in. In some cases, there may even be perceived risk associated with transferring one's knowledge (Becerra, Lunnan, & Huemer, 2008). For instance, a knowledge coach may be dissuaded from transferring their knowledge due to concerns about compromising their own personal source of value within the firm (i.e., possessing unique competencies that no one else has; Anaza & Nowlin, 2017). In regards to tacit KT, it is therefore likely that knowledge sources will want to ensure they are adequately compensated for their investment in the process. This compensation, however, need not always be monetary in nature.

In order to better understand how those involved in KT may perceive the presence of potential gains deriving from knowledge coaching, we draw upon the principles of self-determination theory (SDT; Ryan & Deci, 2000), and suggest that motivation to engage in KT may stem from one or more sources, categorised under the umbrellas of extrinsic and intrinsic motivation (see Figure 1). Beginning with the former, the SDT taxonomy of human motivation proposes that extrinsic motivation can vary greatly in the degree to which it is autonomous, forming four key categories. At the lowest level of autonomy, externally regulated behaviours are performed to satisfy external demands, such as if one's supervisor were to reprimand them for not transferring knowledge to a colleague. Introjection stems from a motivation to maintain one's self-esteem, or feelings of self-worth, such as if one sought to avoid feelings of guilt associated with neglecting to engage in KT. Identification represents a more autonomous form of motivation, whereby an individual personally values a behaviour, or recognises it as important. Lastly, integrated regulation, denoting the most autonomous form of extrinsic motivation, reflects a total assimilation of behaviour with the needs of the self. Whereas these all constitute forms of extrinsic motivation, intrinsically motivated behaviours are undertaken because they are experienced as fun, enjoyable, or interesting (Ryan & Deci, 2000); and behaviour that facilitates KT can be motivated by a combination of these extrinsic and intrinsic drivers. For instance, a knowledge coach may not only be conscious of an externally-imposed deadline that is externally regulating their coaching behaviour, but may simultaneously derive enjoyment from the process itself, suggesting the presence of intrinsic motivation. Given that those considering whether to engage in KT will develop expectations around these differential gains and losses associated with the process, a key question arises: Which types of individuals will perceive which types of gains?

Key to a consideration of motivation and KT, we would argue, is the valence associated with particular outcomes of planned KT behaviour. We refer to this in our model as incentive salience, and argue that the degree to which the individual anticipates personally-valued gains and losses associated with KT will differ markedly depending on stable-individual traits. In the forthcoming sections, we therefore put forward that individually-possessed national cultural values are likely to influence incentive salience, as well as moderate the impact of KT behaviours on KT effectiveness.

National Culture and Levels of Analysis

The term culture refers to "the collective programming of the mind, which distinguishes the members of one human group from those of another" (Hofstede, 1980, p. 24). These days, it is taken as a fundamental fact of international management that organisations are culture-bound (Hofstede, 1980)—a reality that has important consequences for practitioners and scholars. At the core of a vast majority of cross-cultural research are six national cultural dimensions identified by Hofstede (2001) and colleagues (Hofstede, Hofstede, & Minkov, 2010), which were empirically derived through a series of studies conducted with IBM half a century ago, and subsequent collaboration with Michael Bond (1984) and Michael Minkov (Hofstede et al., 2010). Data for a number of these studies were collected via the World Values Survey (World Values Survey, 2017), and each of these dimensions represents two poles of a continuum (e.g. individualism to collectivism). These dimensions are summarised in Table 1.

This study, in particular, sought to examine the influence of three of these dimensions on KT motivation. However, doing so came with inherent challenges, as many scholars maintain that Hofstede's dimensions cannot be used at differing levels of analysis. They argue that national culture exists at the macro level of

Table 1. Hofstede's national culture dimensions

Dimension	High	Low
Individualism/(Collectivism)	Society's members exhibit a preference for a loosely-knit social framework in which individuals are expected to take care of only themselves and their immediate families.	Society's members exhibit a preference for a tightly-knit framework in society in which individuals can expect their relatives or members of a particular in-group to look after them in exchange for unquestioning loyalty.
Power Distance	Society's members accept a hierarchical order in which everybody has a place, which needs no further justification.	Society's members strive to equalise the distribution of power and demand justification for inequalities of power.
Long-Term Orientation/(Short-Term Orientation)	Society encourages thrift and efforts in modern education as a way to prepare for the future.	Society prefers to maintain time-honoured traditions and norms while viewing societal change with suspicion.
Uncertainty Avoidance	Society's members feel uncomfortable with uncertainty and ambiguity. Individuals possess rigid codes of belief and are intolerant of unorthodox behaviour and ideas.	Society's members maintain a relaxed attitude about the unknowability of the future, and view practice as counting more than principles.
Masculinity/Femininity	Society exhibits a preference for achievement, heroism, assertiveness, and material rewards for success. Society at large is more competitive.	Society is more consensus-oriented, and exhibits a preference for cooperation, modesty, caring for the weak, and quality of life.
Indulgence/Restraint	Society allows relatively free gratification of basic needs and natural human drives related to enjoying life and having fun.	Society suppresses gratification of needs and regulates it by means of strict social norms.

Source: (The Hofstede Centre, 2017).

analysis, and that the dimensions proposed in Hofstede's model cannot be used to examine phenomena at the organisational or individual level (Brewer & Venaik, 2012). As such, it cannot be assumed that all members of a culture adopt a culture's values to the same extent; and thus it is inappropriate to use aggregate, ecological measures of culture in an individual-level analysis (Littrell, 2012).

In order to better examine relationships among organisational and micro-level variables that are sensitive to cultural differences, several scholars (Bearden, Money, & Nevins, 2006; Dorfman & Howell, 1988) have developed measurement instruments that are designed for use on the individual level of analysis. These redeveloped scales capture national cultural values via self-report, and have been utilised in extant research to gauge culture's impact on a range of outcomes, including leader-member exchange (Rockstuhl, Dulebohn, Ang, & Shore, 2012), voice behaviour (Botero & Van Dyne, 2009), and more. As a consequence of these new instruments' emergence, examinations of national culture on the individual level have become more common than studies on the societal level (Kirkman, Lowe, & Gibson, 2006), and thus they were utilised for the focal study of this chapter.

FOCAL STUDY

In order to assess the impact of national culture on motivation to engage in KT, this study tested for interactions between national cultural values and motivation in predicting KT outcomes via two testable models (see Figures 2 and 3). We framed our study in the context of knowledge coaching relationships by recruiting a sample of knowledge coaches and protégés in a large MNC. In the first model, we considered how national cultural values affected intrinsic motivation for knowledge coaches to transfer their knowledge. In the second model, we looked at how these values impacted upon protégés' general motivation to receive knowledge. In both models, we looked for evidence of successful transfer by capturing the perceived effectiveness of KT. In the following section, we draw upon the empirical findings and theoretical explanations from the literature previously discussed, to put forward hypotheses for testing via each of these models.

Motivational Pathways

Recent research has linked substantial differences in the effectiveness of KT to varying degrees of motivation (Hau, Kim, Lee, & Kim, 2003), and motivation of different types (Llopis & Foss, 2016). In this study, we therefore expected motivation to manifest differently, depending on whether KT was being examined from the perspective of the knowledge coach or the protégé.

When considering KT from the perspective of the knowledge coach, we suspected that intrinsic motivation would be vital, as it has been found that rewarding actors using heavy-handed, market-like incentives—such as financial reward based on total contribution to the firm's profit—does not increase motivation to transfer tacit knowledge (Osterloh & Frey, 2000). It has therefore been argued that in the

Figure 2. Knowledge coach: Hypothesised model

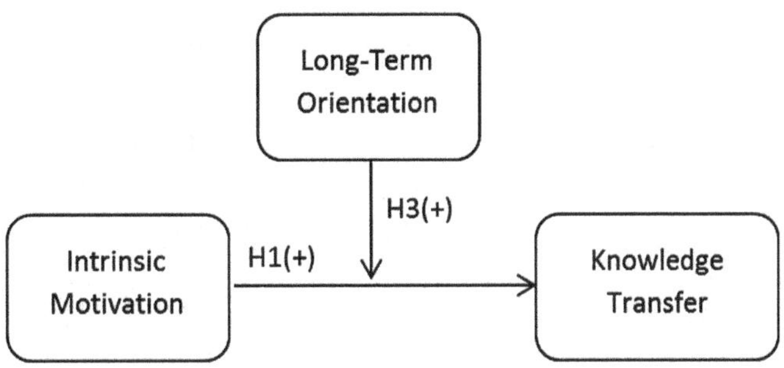

Figure 3. Protégé: Hypothesised model

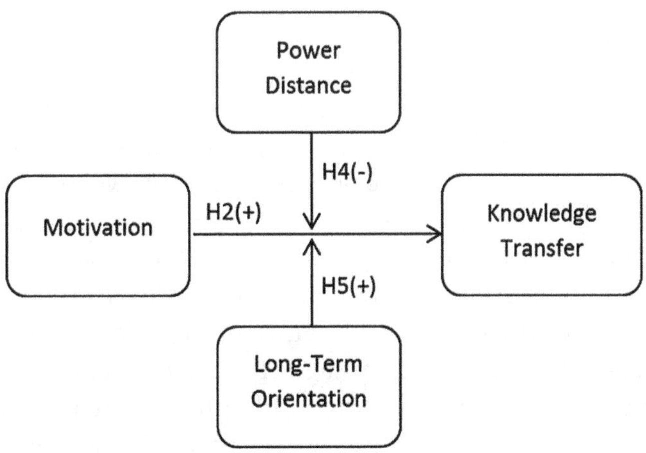

context of knowledge management, extrinsic incentives are sufficient motivators for temporary compliance, but insufficient for processes that require ongoing commitment, such as tacit KT through knowledge coaching (Hau et al., 2013). To draw on empirical findings, Minbaeva et al. (2012) found that individual-level intrinsic motivation mediated the relationship between an individual's commitment to knowledge sharing and knowledge exchanged across employee-group boundaries. Similarly, in a study conducted by Lin (2007), intrinsic motivation (operationalised as 'enjoyment in helping others') was shown to be a vital predictor of knowledge sharing intention. We therefore suspected that intrinsic motivation would be a requirement for effective KT in the model pertaining to knowledge coaches.

Hypothesis One: Higher levels of knowledge coach intrinsic motivation will be positively associated with perceived effectiveness of knowledge coaching in protégé knowledge acquisition.

In what can be an onerous, but often highly-rewarding practice, we expected that protégés would exhibit motivations to partake in knowledge coaching that differed from their knowledge coaches. As discussed, individually-possessed tacit knowledge can constitute a significant professional advantage in terms of being able to provide value for the firm (Anaza & Nowlin, 2017). At times, this knowledge may even be transferrable, providing a worker with skills that he or she can take with them beyond any given organisation, making them an attractive potential hire. For the protégé, we therefore anticipated that there would be a combination of drivers motivating them to partake in KT, which could be mapped on to the different forms of regulation described earlier (Ryan & Deci, 2000). For instance, a protégé may be under significant external pressure to acquire certain skills in the lead-up to a performance review (external regulation), or may personally recognise the value of gaining new knowledge in order perform effectively in a present or future role (identified regulation). Further, we thought it likely that upon first entering the firm, protégés would initially lack a sense of self-efficacy as they sought to adjust to their new role requirements (Jones, 1983), and would therefore be motivated to close the gap between possessed and requisite knowledge. I.e., we believed that a protégé's motivations to participate in KT would be more varied than those of the knowledge coach, and thus we captured motivation within this sample as a more holistic variable called, general motivation.

Hypothesis Two: Higher levels of protégé general motivation will be positively associated with perceived effectiveness of knowledge coaching in protégé knowledge acquisition.

Moderation Pathways

In the following sections, we put forward several key hypotheses across our two models, suggesting the presence of moderating effects. In doing so, we argue the importance of each relevant national cultural value in two ways. Firstly, we revisit our earlier notion of incentive salience (see Figure 1), and put forward the view that individuals possessing values on one pole of any given cultural dimension will perceive gains, opportunities, and pleasures associated with KT, that may not be apparent to their cultural opposite. Secondly, we argue that depending on an individual's national cultural values, he or she may be more or less inclined to utilise the strategies most suitable for transferring tacit knowledge.

We believed that knowledge coaches possessing high-long-term-orientation (LTO) values would perceive greater opportunities to experience pleasure from KT, and thus be more intrinsically motivated to invest personal resources in the process for two reasons. Firstly, high LTO individuals have been shown to exhibit organisational identification exceeding that of any other of Hofstede's cultural dimensions (Baker, Carson, & Carson, 2009). We therefore thought it likely that to those who align with their firm strongly in this way, opportunities to experience pleasure through positive work behaviours like knowledge coaching would be more salient. Secondly, being an effective knowledge coach requires continuity and a significant investment of time (Swap, Leonard, Shields, & Abrams, 2001). These are investments that those with LTO values—given their emphasis on present-day persistence in order to ensure long-term success—would find worthwhile, and inherently more attractive than someone with a short-term value orientation. To evidence empirical findings, in one study examining employees' use of an online knowledge sharing platform (Voelpel & Han, 2005), it was found that the online activity of the Chinese users almost doubled that of users in the USA, due to their high LTO; the Chinese users were more highly motivated to interact and share knowledge with others via the system.

Hypothesis Three: Higher levels of knowledge coach LTO will positively moderate the relationship between intrinsic motivation and perceived effectiveness of knowledge coaching in protégé knowledge acquisition.

A review of the knowledge management systems literature suggested that high power distance may act as an impediment to effective KT (Jacks, Wallace, & Nemati, 2012); and we believed that these findings could extend to KT on an individual level. Those high on the dimension of power distance often work most comfortably in hierarchical settings where power is unequally distributed (The Hofstede Centre, 2017). Due to the embedded acceptance of this unequal distribution of power, we anticipated that protégés high on the dimension of power distance would be less comfortable working in such close proximity to a knowledge coach, impacting negatively upon the total benefit derived from the process.

Research from the mentoring literature would suggest that when protégés feel able to freely express their opinions without fear of judgment, they will be more actively engaged in learning processes (O'Broin & Palmer, 2010). These assurances are particularly important in the context of knowledge coaching, as knowledge coaching in its most effective forms requires an open and honest dialogue between both parties. We thought that the necessity for both parties to be vulnerable and exhibit trust in one another could potentially be a barrier for protégés possessing high-power-distance values, as people's willingness to ask questions that reveal their ignorance, disagree with others in public, contradict known experts, discuss their problems, and follow others in the thread of conversation will all vary markedly across cultures (Apospori, Nikandrou, & Panayotopoulou, 2006; Wenger, McDermott, & Snyder, 2002).

Knowledge coaching requires protégés to openly make mistakes and collaborate with a superior member of the organisation. We believed that protégés exhibiting high power distance would therefore have more mental barriers to overcome when genuinely expressing their opinions, preferring to filter their verbal participation during knowledge coaching through communication approaches that convey public deference and politeness (Schermerhorn & Bond, 1997). Further, in the minds of high-power-distance individuals, the opinions of higher status individuals tend to carry greater credibility regardless of merit (Tan, Watson, & Wei, 1995), presenting a potential conflict for high-power-distance protégés. In a series of studies examining the impact of cultural differences on Russians' and Israelis' attitudes towards upward impression management, Zaidman and Drory (2001) found that the Russian sample's higher power distance made them more likely to invoke such strategies (e.g. ingratiation), while the lower-power-distance sample of Israelis tended to view such practices in a more negative light. These findings led us to believe that high-power-distance workers may place a greater importance upon how they are perceived by senior members of the organisation, leading them to say and do the 'correct' things, rather than prioritise honest responses and behaviours (i.e. ones that potentially reveal ignorance) in the process of knowledge coaching.

Hypothesis Four: Higher levels of protégé power distance will negatively moderate the relationship between general motivation and perceived effectiveness of knowledge coaching in protégé knowledge acquisition.

Those possessing collectivist values exhibit loyalty to groups and define themselves in accordance with the collective (The Hofstede Centre, 2017). It is therefore common for collectivists to experience a close sense of identification with the firm to which they belong, and the organisational members within it (Haghirian, 2010). As a consequence, we hypothesised that there would be incentives to knowledge coaching that were salient to collectivists that would not necessarily entice an individualist worker.

As mentioned previously, the choice to engage in KT is underpinned by something of a cost-benefit analysis (Cabrera & Cabrera, 2002); and studies would suggest that this balancing of potential gains and losses will pan out differently in different cultures (Voelpel & Han, 2005; Wilkesmann, Fischer, & Wilkesmann, 2009). However, those espousing collectivist values have been shown to have an inherent appreciation for the social cohesion that derives from knowledge sharing relationships (Wilkesmann et al., 2009). To illustrate, one study by Chow, Deng, and Ho (2000) found significant differences in Chinese and US nationals' openness to knowledge sharing, which was related to the two samples' differing degrees of collectivism; the Chinese sample ascribed a higher degree of importance and personal value to this process.

Another factor to consider is that acquiring requisite task-related skills will often be crucial for effective socialisation into the firm, and assimilation with its members. For instance, Jones (1983) notes that new-entrants will often seek to clarify their situational identity and evoke competence cues from others as part of this transition. To a worker with collectivist values, being effectively socialised is especially important (Kim, Cable, & Kim, 2005; Taormina & Bauer, 2000). We therefore suspected that collectivist workers would experience a heightened pressure to develop tacit knowledge in order to be accepted and viewed as competent by their colleagues; and would subsequently possess greater motivation to participate in knowledge coaching as a result.

Hypothesis Five: Higher levels of protégé collectivism will positively moderate the relationship between general motivation and perceived effectiveness of knowledge coaching in protégé knowledge acquisition.

By way of summary, the previous sections established the following arguments. Firstly, there are different motivational drivers of KT, depending on one's role in the process as either the knowledge source, or knowledge recipient. Secondly, stable-individual traits will influence the relationship between motivation and KT effectiveness. Among these, three national cultural values (LTO, power distance, and collectivism) are characteristics that have yet to be examined in the literature, but will bear a significant influence. The results of a study examining the ways in which each of these three dimensions influence KT motivation within two populations of interest (knowledge coaches and their protégés) will now be summarised. For a detailed exposition of this study's method and findings, please see Celestine and Perryer (2016).

METHOD

The sample for this study was drawn from the sales staff of a multinational provider of energy and hygiene technologies. The pools of respondents were based in Japan, Norway, Sweden, Denmark, and the UK. All members of the sample had completed the firm's 6-month coaching program, whereby district managers (DMs) teach new territory managers (TMs) on-the-job selling skills as part of the TM's induction to the organisation.

In this cross-sectional study, questionnaires in the participants' native languages were distributed to each of the samples, and responses were received from 26 DMs and 102 TMs, producing response rates of 76.47% and 64.97% respectively. The data were treated as two separate pools and were not paired. As such, the protégés (TMs) were asked to report on the extent to which they thought that being coached improved their selling skills, while the coaches (DMs) were asked about the extent to which they thought their coaching aided in the professional development of their protégés over the last 12 months. In other words, the respondents were asked to try and isolate the value of the coaching experience in their responses. All measures comprised of validated scales that were adapted from the literature (see Table 2 for a summary).

Analysis

The two models were estimated using a partial least squares (PLS) regression procedure—a particularly useful analytic technique when sample sizes are smaller than what is suitable for covariance-based approaches. Prior to PLS model estimation, analyses of descriptive statistics; an examination of the data's distributional properties; and checks for missing data, outliers, reliability, multicollinearity, and common

Table 2. Measurement instruments and sample items

Model	Construct	Source	Response Anchors	Sample Item
Knowledge coach	Intrinsic motivation	Wasko and Faraj (2000)	(1) Strongly disagree; (7) Strongly agree	*I enjoy sharing my knowledge with protégés.*
	Skill acquisition	Rentz, Shepherd, Tashchian, Dabholkar, and Ladd (2002)	(1) Not at all; (5) Considerably	*... their ability to close sales.*[1,2]
	Long-term orientation	Bearden, Money, and Nevins (2006)	(1) Strongly disagree; (7) Strongly agree	*I don't mind giving up today's fun for success in the future.*
Protégé	Motivation	Noe and Schmitt (1986); Kim-Soon and Ahmad (2012)	(1) Strongly disagree; (7) Strongly agree	*I found opportunities for learning new skills from my coach motivating.*
	Power distance	Dorfman and Howell (1988)	(1) Strongly disagree; (7) Strongly agree	*Managers should make most decisions without consulting subordinates.*
	Collectivism	Dorfman and Howell (1988)	(1) Strongly disagree; (7) Strongly agree	*Group welfare is more important than individual rewards.*

[1]Knowledge coaches were asked to report on the extent to which they felt that their coaching was responsible for their protégé/s improving their skills. The lead-in to these items was thus framed as such.

[2]This same instrument was reworded and used to measure protégés' perceptions of their own knowledge acquisition resulting from knowledge coaching.

method bias were all conducted. Following this, convergent validity was assessed using factor-loading estimations (Hair, Black, Babin, & Anderson, 2014). Once the underlying structure of the latent variables was confirmed, the paths between the independent and dependent variables were examined, and the moderating variables were introduced to test whether the national cultural values affected the relationships between the independent and dependent variables (Baron & Kenny, 1986).

RESULTS

Following an examination of demographic data, the means, standard deviations, Cronbach's alphas (CA), composite reliability (CR) estimates, average extracted variances (AVE), variance inflation factors (VIFs), and bivariate correlations were calculated for each variable in each of models. These are shown in Tables 3 and 4. By comparing these indices to accepted thresholds identified in the literature (Fornell & Larcker, 1981; Hair et al., 2014; Hair, Ringle & Sarstedt, 2011), reliability and convergent validity of the measures was confirmed. After modelling the latent variables, we also examined the indicators' loadings and cross-loadings to assess unidimensionality. If items had loadings greater than .50, this was taken to be indicative of unidimensionality (Rivard, Poirier, Raymond, & Bergeron, 1994). Given that a number of items initially loaded at levels below .50, an iterative process of removing low-loading items was undertaken until all loadings were at acceptable levels. We then examined discriminant validity between the constructs, and noted that none of the correlations exceeded the square roots of the AVE scores in either model, thus suggesting discriminant validity. Lastly, as the study relied upon self-reports collected at a single time point, a full collinearity test was conducted to test for common method bias (Kock, 2015; Kock & Lynn, 2012). As all of the VIF indices fell below 3.3, the data could be considered free of common method bias.

Prior to analysis, several control variables were included in the models—namely gender, age, education, prior work experience, and country of employment. Following this, an examination of the various hypotheses was undertaken. The moderation analyses were conducted via a two-stage approach (see Chin, Marcolin, & Newsted, 2003), and thus, variables were considered potential moderators if their inclusion in the model produced a significant increase in the R^2 values. Lastly, overall goodness of fit was assessed (Tenenhaus, Vinzi, Chatelin, & Lauro, 2005; Wetzels, Oderkerken-Schroder, & Van Oppen, 2009), suggesting the presence of large effect sizes. A summary of the results for each model are presented in Figures 4 and 5 and will now be discussed in the forthcoming section.

Table 3. Knowledge coach model: Means, standard deviations, AVE scores, composite reliability, Cronbach's alpha and correlations

Variable	M	SD	CR	CA	AVE	VIF	Sq. Rt. AVE	IM	LTO
IM	6.03	0.61	0.95	0.93	0.83	1.92	0.91		
LTO	5.23	1.14	0.86	0.75	0.67	1.54	0.82	0.01	
Sell Skill	4.01	0.57	0.85	0.80	0.55	1.83	0.72	0.33	0.15

IM = Intrinsic motivation; LTO = Long-term orientation

Table 4. Protégé model: Means, standard deviations, AVE scores, composite reliability, Cronbach's alpha and correlations

Variable	M	SD	CR	CA	AVE	VIF	Sq. Rt. AVE	Motv.	PDI	COL
Motv.	5.42	1.20	0.92	0.88	0.73	2.22	0.86			
PDI	2.73	1.12	0.83	0.69	0.62	1.07	0.79	-0.07		
Collectivism	4.32	1.25	0.87	0.70	0.77	1.21	0.88	0.27**	0.11	
Sell Skill	3.07	0.87	0.92	0.91	0.60	3.13	0.78	0.52**	0.12	0.17

** Significance at 0.01
PDI = Power distance; COL = Collectivism

Figure 4. Knowledge coach: Model results

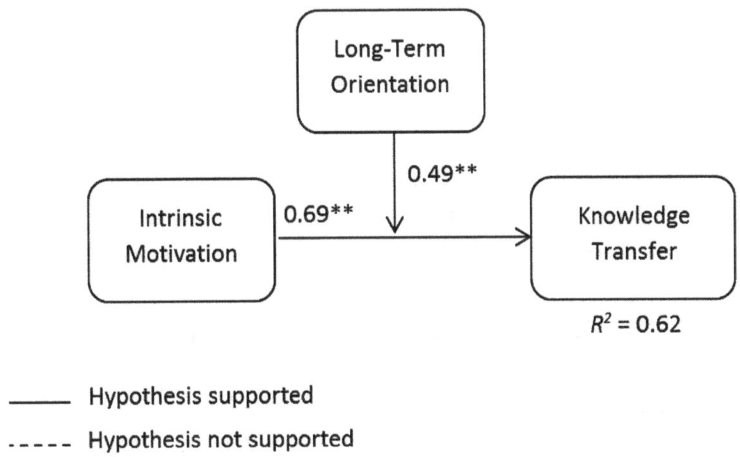

DISCUSSION

This study aimed to address identified gaps in the literature around person-level values, and their impact upon KT processes. In doing so, we sought to add to the growing body of work on micro-level mechanisms, and individual differences that facilitate the dispersion of tacit knowledge throughout the organisation, thus shedding light on the factors that can aid in the development and maintenance of strategic capabilities within the firm (Foss, 2007). As such, the influence of motivation as a primary trigger for KT between knowledge coaches and protégés was examined.

Hypothesis 1 was strongly supported by the data, suggesting that for knowledge coaching be successful, it is vital that knowledge coaches are intrinsically motivated to invest themselves fully in the process. Prior research in similar areas and on various levels of analysis served as strong reasoning to suggest that this would be the case (e.g. Minbaeva et al., 2012). It is also interesting to note that at the time of data collection, the focal firm in this study had had its induction program in place worldwide for over two years, and yet the DMs involved had never been rewarded for their coaching via any kind of extrinsic incentives. It would therefore seem that the DMs were driven to engage with the program from an internal source of motivation, suggesting that SDT may offer a sound theoretical basis for better understanding the internal drivers underpinning KT.

Figure 5. Protégé: Model results

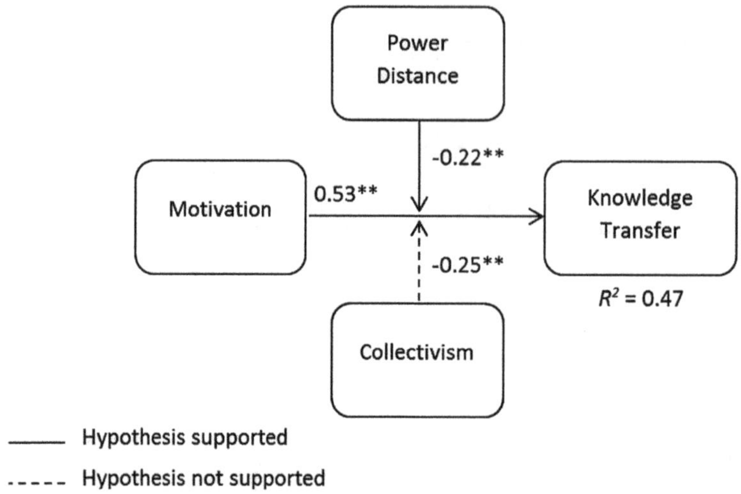

Hypothesis 2 was also supported by the data, indicating that a protégé's holistic motivation to engage in KT is crucial if the practice is to yield results. Again, we had strong reason to believe that this would be the case, as external pressures, a personal valuing of one's role with the firm, and an inherent desire to feel effective in one's environment will drive a new entrant to take advantage of training and development opportunities (Jones, 1983).

Our key contribution in this study was to uncover the impact of national cultural values on individual-level KT. We argued that these values would influence individual perceptions around the value (or lack thereof) that one would attain as a consequence of partaking in KT, resulting in the differential salience of incentives. Further, we suggested that these values could sometimes work to inspire or inhibit willingness to engage in effective knowledge coaching practices, such as guided practice or experimentation (Leonard & Swap, 2005).

In our model pertaining to knowledge coach values, LTO was found to exhibit a positive moderating effect, lending support to our third hypothesis. We suspected that the pleasures to be gained from knowledge coaching—such as the chances to make a long-lasting contribution to the firm and form lasting bonds with one's colleagues—would intrinsically motivate high-LTO individuals to engage more wholeheartedly in knowledge coaching; and indeed, these propositions were reflected in the findings of the analysis.

Meta-analytic findings from Jacks et al. (2012) informed Hypothesis 4, suggesting that high power distance would act as an impediment to KT. In line with our hypothesis, it was found that power distance negatively moderated the path between protégés' general motivation and knowledge acquisition. The findings of this study therefore suggest that a protégé's wholehearted engagement in knowledge coaching risks being derailed or blocked by other factors, such as their level of comfort when working in close collaboration with their superiors, or potentially the strength of their desire to manage impressions. As such, if individuals possessing high power distance are required to engage in communication techniques that conflict with their values and behavioural norms of deference, politeness, and agreement, additional steps may need to be taken to ensure they are comfortable and engaged in the process (Schermerhorn & Bond, 1997).

Collectivism's attenuating effect on the path between intrinsic motivation and skill acquisition in the protégé model constituted an unexpected finding—one that disconfirmed our fifth hypothesis. We therefore suspect that future research may need to tease out the specific sources of motivation underpinning the collectivist's drive to engage in knowledge coaching, so as to shed light on this unforeseen result.

Research Limitations

This study has several limitations that need to be acknowledged. Firstly, it is possible that response styles differed according to national culture, having subsequent effects on the results. For instance, the values of high power distance and masculinity have been positively associated with extreme response styles (Harzing, 2006; Johnson, Kulesa, Cho, & Shavitt, 2005). A second limitation is the study's lower than desired response rate amongst the sample of knowledge coaches. While PLS analysis is quite robust when analysing small sample sizes, the number of knowledge coach cases was too small to achieve full rank and estimate the model with the optimal level of accuracy. These findings should therefore be viewed with caution, pending future studies.

Research Implications

The findings of this study are consistent with the extant literature, indicating that motivation to engage in KT strengthens the effectiveness of KT outcomes. The question therefore remains: *What should managers do to try and enhance motivation amongst those participating in KT?* The answer: Managers should find ways to enhance KT participants' internally regulated and intrinsically motivated sources of motivation. By ensuring the presence of each of these sources of motivation, it is more likely that employees will experience an enduring willingness to partake in knowledge coaching. I.e., when an individual can personally recognise the value associated with the activity, or suspect that it will provide him or her with personal gains (e.g. a chance to develop one's coaching skills, a chance to build networks, etc.), he or she will be more likely to invest personal resources in the process. It is therefore up to managers to emphasise the gains to be had from knowledge coaching, not just from the perspective of the firm, but also the individual.

In regards to the demonstrated impact of national culture, we cautiously advise that managers understand the constructs in this study as personal values that may or may not be rooted in a particular cultural upbringing. It would therefore be unwise for managers to view these findings and come to premature conclusions about employees' KT capabilities, based on their cultural background, as sometimes individuals' cultural values will go against the grain of a society (Dorfman & Howell, 1988). Further, given that individuals tend only to apply their values when they can think of specific reasons for doing so, managers should remain conscious of whether or not their work environment encourages or discourages desirable, value-consistent behaviour that facilitates KT. For example, there would be little use in trying to hire employees with LTO values in the hope of strengthening long-term KT outcomes, when the hiring organisation heavily imposes other short-term objectives. However, if we are to assume that congruence exists between imposed objectives and desired KT outcomes, this study provides clear guidelines as to the ideal value profile of an employee who is well suited to deliver such outcomes.

FUTURE RESEARCH DIRECTIONS

We wish to use the opportunity inherent in preparing this chapter as a chance to put forward an expanded set of *three* new future research directions, which we hope to see implemented in this exciting new area of research (see Celestine & Perryer, 2016 for the original recommendations).

First, the focal study of this chapter provided a fairly high-level look at the relationships between KT, individual motivation, and national cultural values in a one-on-one KT setting. We would argue that there is opportunity to go into more depth in these areas, using methodologies and analytic techniques frequently drawn upon in other areas of management research. For instance, the recent popularity of experience sampling methods in the organisational behaviour literature presents an exciting new opportunity for the knowledge management space (see Csikszentmihalyi & Larson, 2014). Given that knowledge coaching is a lengthy process that can span over the course of several months or years, it is only to be expected that motivation will fluctuate over the course of this period. By employing an experience sampling methodology, scholars can capture these fluctuations in fine-grained detail, providing the basis for a highly rigorous, longitudinal design. Further—and returning to a consideration of national cultural values—a design such as this would allow scholars to ask questions such as: Do high-LTO knowledge coaches exhibit sustained motivation in their roles, and how does this compare to the motivational trajectories of low-LTO knowledge coaches? There may even be scope to alter such trajectories using experimental manipulation, and consider how emerging practices from the learning and development space can be drawn upon to enhance intrinsic motivation (e.g. Perryer, Celestine, Scott-Ladd, & Leighton, 2016; Statler, Roos, & Victor, 2009).

Second, future research will benefit from matching the reports of knowledge coaching effectiveness across samples of protégés and knowledge coaches. Better yet, objective indicators of knowledge coaching effectiveness obtained from firm records or formal reviews would make for an even more rigorous design. Our final major recommendation would be to further examine the notion of incentive salience, and verify its impact upon knowledge coaching motivation. In this research, we inferred that national cultural values may affect a) willingness to engage with effective coaching methods, and b) the perception that one will gain value as a consequence of engaging in the process. We would therefore recommend that future research tease apart these two mechanisms of influence.

CONCLUSION

With the rapidly changing nature of today's knowledge industries, there is a growing necessity to enhance understandings about knowledge work, and perhaps even redefine the boundaries around the traditional knowledge worker. In the future landscape of knowledge work, it is likely that individual tacit knowledge will become an exceedingly precious resource—one that enhances individual employability, as well as sustainable value for organisations. How managers approach the task of disseminating tacit knowledge within the firm will be a key determinant of future organisational success. Therefore, by taking a closer look at the stable-individual factors that stand to facilitate or inhibit the flow of knowledge within the firm, scholars can create more sophisticated models of KT effectiveness, recommend tangible courses of action for managers—such as the introduction of knowledge coaching—and ensure that the organisations of the future will be adequately equipped to ensure the continual availability of knowledge resources.

REFERENCES

Anaza, N. A., & Nowlin, E. L. (2017). What's mine is mine: A study of salesperson knowledge withholding & hoarding behavior. *Industrial Marketing Management, 64*, 14–27. doi:10.1016/j.indmarman.2017.03.007

Apospori, E., Nikandrou, I., & Panayotopoulou, L. (2006). Mentoring and women's career advancement in Greece. *Human Resource Development International, 9*(4), 509–527. doi:10.1080/13678860601032627

Baker, D. S., Carson, K. D., & Carson, P. P. (2009). An individual-level examination of the impact of cultural values on organizational identification. *The Journal of Applied Management and Entrepreneurship, 14*(2), 29–43.

Baron, R. M., & Kenny, D. A. (1986). The moderator–mediator variable distinction in social psychological research: Conceptual, strategic, and statistical considerations. *Journal of Personality and Social Psychology, 51*(6), 1173–1182. doi:10.1037/0022-3514.51.6.1173 PMID:3806354

Bearden, W. O., Money, R. B., & Nevins, J. L. (2006). A Measure of Long-Term Orientation: Development and Validation. *Academy of Marketing Science Journal, 34*(3), 456–467. doi:10.1177/0092070306286706

Becerra, M., Lunnan, R., & Huemer, L. (2008). Trustworthiness, risk, and the transfer of tacit and explicit knowledge between alliance partners. *Journal of Management Studies, 45*(4), 691–713. doi:10.1111/j.1467-6486.2008.00766.x

Bhagat, R. S., Kedia, B. L., Harveston, P. D., & Triandis, H. C. (2002). Cultural variations in the cross-border transfer of organizational knowledge: An integrative framework. *Academy of Management Review, 27*(2), 204–221.

Botero, I. C., & Van Dyne, L. (2009). Employee voice behavior interactive effects of LMX and power distance in the United States and Colombia. *Management Communication Quarterly, 23*(1), 84–104. doi:10.1177/0893318909335415

Brewer, P., & Venaik, S. (2012). On the misuse of national culture dimensions. *International Marketing Review, 29*(6), 673–683. doi:10.1108/02651331211277991

Cabrera, A., & Cabrera, E. F. (2002). Knowledge-Sharing Dilemmas. *Organization Studies, 23*(5), 687–710. doi:10.1177/0170840602235001

Cabrera, A., Collins, W. C., & Salgado, J. F. (2006). Determinants of individual engagement in knowledge sharing. *International Journal of Human Resource Management, 17*(2), 245–264. doi:10.1080/09585190500404614

Celestine, N. A., & Perryer, C. (2016). The Impact of National Cultural Values on Intrinsic Motivation to Transfer Tacit Knowledge. *International Journal of Knowledge Management, 12*(4), 1–19. doi:10.4018/IJKM.2016100101

Chilton, M. A., & Bloodgood, J. M. (2008). The Dimensions of Tacit & Explicit knowledge: A Description and Measure. *International Journal of Knowledge Management, 4*(2), 75–91. doi:10.4018/jkm.2008040106

Chin, W. W., Marcolin, B. L., & Newsted, P. R. (2003). A partial least squares latent variable modeling approach for measuring interaction effects: Results from a Monte Carlo simulation study and an electronic-mail emotion/adoption study. *Information Systems Research*, *14*(2), 189–217. doi:10.1287/isre.14.2.189.16018

Cho, N., Zheng Li, G., & Su, C. J. (2007). An empirical study on the effect of individual factors on knowledge sharing by knowledge type. *Journal of Global Business and Technology*, *3*(2), 1–15.

Chow, C. W., Deng, F. J., & Ho, J. L. (2000). The openness of knowledge sharing within organizations: A Comparative Study in the United States and the People's Republic of China. *Journal of Management Accounting Research*, *12*(1), 65–95. doi:10.2308/jmar.2000.12.1.65

Csikszentmihalyi, M., & Larson, R. (2014). Validity and reliability of the experience-sampling method. In M. Csikszentmihalyi (Ed.), *Flow and the Foundations of Positive Psychology* (pp. 35–54). Dordrecht, The Netherlands: Springer. doi:10.1007/978-94-017-9088-8_3

Dorfman, P. W., & Howell, J. P. (1988). Dimensions of National Culture and Effective Leadership Patterns: Hofstede Revisted. In R. N. Farmer (Ed.), *Advances in International Comparative Management* (Vol. 3, pp. 127–150). Philadelphia, PA: JAI Press.

Fahey, L., & Prusak, L. (1998). The Eleven Deadliest Sins of Knowledge Management. *California Management Review*, *40*(3), 265–276. doi:10.2307/41165954

Fornell, C., & Larcker, D. F. (1981). Evaluating structural equation models with unobservable variables and measurement error. *JMR, Journal of Marketing Research*, *18*(1), 39–50. doi:10.2307/3151312

Foss, N. J. (2007). The Emerging Knowledge Governance Approach: Challenges and Characteristics. *Organization*, *14*(1), 29–52. doi:10.1177/1350508407071859

Friedman, M. (2007). The methodology of positive economics. In D. M. Hausman (Ed.), *The Philosophy of Economics* (Vol. 3, pp. 145–178). Cambridge, UK: Cambridge University Press. doi:10.1017/CBO9780511819025.010

Friedman, T. L. (2005). *The world is flat: A brief history of the twenty-first century*. New York, NY: Macmillan.

Haghirian, P. (2010). *Understanding Japanese Management Practices*. New York, NY: Business Expert Press. doi:10.4128/9781606491195

Hair, J. F., Black, W., Babin, B., & Anderson, R. (2014). Multiple Regression Analysis. In J. F. Hair, W. Black, B. Babin, & R. Anderson (Eds.), *Multivariate Data Analysis* (7th ed.; pp. 151–230). Harlow, UK: Pearson Education.

Hair, J. F., Ringle, C. M., & Sarstedt, M. (2011). PLS-SEM: Indeed a silver bullet. *Journal of Marketing Theory and Practice*, *19*(2), 139–152. doi:10.2753/MTP1069-6679190202

Harzing, A. W. (2006). Response styles in cross-national survey research a 26-country study. *International Journal of Cross Cultural Management*, *6*(2), 243–266. doi:10.1177/1470595806066332

Hofstede, G., & Bond, M. H. (1984). Hofstede's culture dimensions: An independent validation using Rokeach's value survey. *Journal of Cross-Cultural Psychology, 15*(4), 417–433. doi:10.1177/0022002184015004003

Hofstede, G. H. (1980). *Culture's consequences: International Differences in Work-Related Values*. Beverly Hills, CA: Sage.

Hofstede, G. H. (2001). *Culture's consequences: Comparing values, behaviors, institutions and organizations across nations*. Beverly Hills, CA: Sage.

Hofstede, G. H., Hofstede, G. J., & Minkov, M. (2010). *Cultures and Organizations: Software of the Mind: Intercultural Cooperation and its Importance for Survival*. New York, NY: McGraw-Hill.

Jacks, T., Wallace, S., & Nemati, H. (2012). Impact of Culture on Knowledge Management: A Meta-Analysis and Framework. *Journal of Global Information Technology Management, 15*(4), 8–42. doi:10.1080/1097198X.2012.10845622

Johnson, T., Kulesa, P., Cho, Y. I., & Shavitt, S. (2005). The relation between culture and response styles evidence from 19 countries. *Journal of Cross-Cultural Psychology, 36*(2), 264–277. doi:10.1177/0022022104272905

Jones, G. R. (1983). Psychological orientation and the process of organizational socialization: An interactionist perspective. *Academy of Management Review, 8*(3), 464–474.

Kim, T. Y., Cable, D. M., & Kim, S. P. (2005). Socialization tactics, employee proactivity, and person-organization fit. *The Journal of Applied Psychology, 90*(2), 232–241. doi:10.1037/0021-9010.90.2.232 PMID:15769234

Kim-Soon, N., & Ahmad, N. (2012). Impact of Motivation to Learn on Perceived Training Transfer: Empirical Evidence from a Bank. In *Proceedings of the International Conference of Technology Management, Business and Entrepreneurship* (pp. 268-383). Melaka: ICTMBE.

Kirkman, B. L., Lowe, K. B., & Gibson, C. B. (2006). A quarter century of culture's consequences: A review of empirical research incorporating Hofstede's cultural values framework. *Journal of International Business Studies, 37*(3), 285–320. doi:10.1057/palgrave.jibs.8400202

Ko, D. G., Kirsch, L. J., & King, W. R. (2005). Antecedents of knowledge transfer from consultants to clients in enterprise system implementations. *Management Information Systems Quarterly, 29*(1), 59–85. doi:10.2307/25148668

Kock, N. (2015). Common method bias in PLS-SEM: A full collinearity assessment approach. *International Journal of e-Collaboration, 11*(4), 1–10. doi:10.4018/ijec.2015100101

Kock, N., & Lynn, G. S. (2012). Lateral collinearity and misleading results in variance-based SEM: An illustration and recommendations. *Journal of the Association for Information Systems, 13*(7), 546–580. doi:10.17705/1jais.00302

Kogut, B., & Zander, U. (1992). Knowledge of the Firm, Combinative Capabilities, and the Replication of Technology. *Organization Science, 3*(3), 383–397. doi:10.1287/orsc.3.3.383

Leonard, D., & Swap, W. (2004). Deep Smarts. *Harvard Business Review*, *82*(9), 88–97. PMID:15449858

Leonard, D., & Swap, W. (2005). *Deep Smarts: How To Cultivate And Transfer Enduring Business Wisdom*. Boston, MA: Harvard Business School Publishing.

Lin, H. F. (2007). Effects of Extrinsic and Intrinsic Motivation on Employee Knowledge Sharing Intentions. *Journal of Information Science*, *33*(2), 135–149. doi:10.1177/0165551506068174

Lin, H. F., & Lee, G. G. (2004). Perceptions of senior managers toward knowledge-sharing behaviour. *Management Decision*, *42*(1), 108–125. doi:10.1108/00251740410510181

Littrell, R. (2012). Cultural Value Dimension Theories: Hofstede – A Work in Progress. *AiB Insights*, *4*(12), 3–6.

Liyanage, C., Elhag, T., Ballal, T., & Li, Q. (2009). Knowledge communication and translation–a knowledge transfer model. *Journal of Knowledge Management*, *13*(3), 118–131. doi:10.1108/13673270910962914

Llopis, O., & Foss, N. J. (2016). Understanding the climate–knowledge sharing relation: The moderating roles of intrinsic motivation and job autonomy. *European Management Journal*, *34*(2), 135–144. doi:10.1016/j.emj.2015.11.009

Lu, L., Leung, K., & Koch, P. T. (2006). Managerial knowledge sharing: The role of individual, interpersonal, and organizational factors. *Management and Organization Review*, *2*(1), 15–41. doi:10.1111/j.1740-8784.2006.00029.x

Magnier-Watanabe, R., & Senoo, D. (2010). Shaping knowledge management: Organization and national culture. *Journal of Knowledge Management*, *14*(2), 214–227. doi:10.1108/13673271011032364

Manyika, J., Chui, M., Bughin, J., Dobbs, R., Bisson, P., & Marrs, A. (2013). *Disruptive technologies: Advances that will transform life, business, and the global economy*. Retrieved from http://www.mckinsey.com/business-functions/digital-mckinsey/our-insights/disruptive-technologies

Matzler, K., Renzl, B., Mooradian, T., von Krogh, G., & Mueller, J. (2011). Personality traits, affective commitment, documentation of knowledge, and knowledge sharing. *International Journal of Human Resource Management*, *22*(2), 296–310. doi:10.1080/09585192.2011.540156

Matzler, K., Renzl, B., Müller, J., Herting, S., & Mooradian, T. A. (2008). Personality traits and knowledge sharing. *Journal of Economic Psychology*, *29*(3), 301–313. doi:10.1016/j.joep.2007.06.004

Minbaeva, D. B., Mäkelä, K., & Rabbiosi, L. (2012). Linking HRM and Knowledge Transfer via Individual-Level Mechanisms. *Human Resource Management*, *51*(3), 387–405. doi:10.1002/hrm.21478

Noe, R. A., & Schmitt, N. (1986). The Influence of Trainee Attitudes on Training Effectiveness: Test of a Model. *Personnel Psychology*, *39*(3), 497–523. doi:10.1111/j.1744-6570.1986.tb00950.x

Nonaka, I., Umemoto, K., & Senoo, D. (1996). From Information Processing to Knowledge Creation: A Paradigm Shift in Business Management. *Technology in Society*, *18*(2), 203–218. doi:10.1016/0160-791X(96)00001-2

O'Broin, A., & Palmer, S. (2010). Exploring key aspects in the formation of coaching relationships: Initial indicators from the perspective of the coachee and the coach. *Coaching: An International Journal of Theory. Research and Practice*, *3*(2), 124–143.

Osterloh, M., & Frey, B. S. (2000). Motivation, Knowledge Transfer, and Organizational Forms. *Organization Science*, *11*(5), 538–550. doi:10.1287/orsc.11.5.538.15204

Perryer, C., Celestine, N. A., Scott-Ladd, B., & Leighton, C. (2016). Enhancing workplace motivation through gamification: Transferrable lessons from pedagogy. *International Journal of Management Education*, *14*(3), 327–335. doi:10.1016/j.ijme.2016.07.001

Rentz, J. O., Shepherd, C. D., Tashchian, A., Dabholkar, P. A., & Ladd, R. T. (2002). A measure of selling skill: Scale development and validation. *Journal of Personal Selling & Sales Management*, *22*(1), 13–21.

Rivard, S., Poirier, G., Raymond, L., & Bergeron, F. (1994). Development of a measure to assess the quality of user-developed applications. In J.F. Nunamaker, & R.H. Sprague (Eds.), *Proceedings of the Twenty-Seventh Annual Hawaii International Conference on System Sciences* (pp. 522-531). ICSS. 10.1109/HICSS.1994.323466

Rockstuhl, T., Dulebohn, J. H., Ang, S., & Shore, L. M. (2012). Leader–member exchange (LMX) and culture: A meta-analysis of correlates of LMX across 23 countries. *The Journal of Applied Psychology*, *97*(6), 1097–1130. doi:10.1037/a0029978 PMID:22985117

Ryan, R. M., & Deci, E. L. (2000). Self-Determination Theory and the Facilitation of Intrinsic Motivation, Social Development, and Well-Being. *The American Psychologist*, *55*(1), 68–78. doi:10.1037/0003-066X.55.1.68 PMID:11392867

Sarker, S., Sarker, S., Nicholson, D. B., & Joshi, K. D. (2005). Knowledge transfer in virtual systems development teams: An exploratory study of four key enablers. *IEEE Transactions on Professional Communication*, *48*(2), 201–218. doi:10.1109/TPC.2005.849650

Schermerhorn, J. R., & Bond, M. H. (1997). Cross-cultural leadership dynamics in collectivism and high power distance settings. *Leadership and Organization Development Journal*, *18*(4), 187–193. doi:10.1108/01437739710182287

Scott, J. (2000). Rational choice theory. In G. Browning, A. Halcli, & F. Webster (Eds.), *Understanding contemporary society: Theories of the present* (pp. 126–138). London: Sage Publications. doi:10.4135/9781446218310.n9

Swap, W., Leonard, D., Shields, M., & Abrams, L. (2001). Using Mentoring and Storytelling to Transfer Knowledge in the Workplace. *Journal of Management Information Systems*, *18*(1), 95–114. doi:10.1080/07421222.2001.11045668

Szulanski, G. (2000). The process of knowledge transfer: A diachronic analysis of stickiness. *Organizational Behavior and Human Decision Processes*, *82*(1), 9–27. doi:10.1006/obhd.2000.2884

Tan, B., Watson, R., & Wei, K. K. (1995). National culture and group support systems: Filtering communication to dampen power differentials. *European Journal of Information Systems*, *4*(2), 82–92. doi:10.1057/ejis.1995.10

Taormina, R. J., & Bauer, T. N. (2000). Organizational socialization in two cultures: Results from the United States and Hong Kong. *The International Journal of Organizational Analysis, 8*(3), 262–289. doi:10.1108/eb028919

Tenenhaus, M., Vinzi, V. E., Chatelin, Y. M., & Lauro, C. (2005). PLS path modeling. *Computational Statistics & Data Analysis, 48*(1), 159–205. doi:10.1016/j.csda.2004.03.005

The Hofstede Centre. (2017). National Culture. *The Hofstede Centre Website.* Retrieved August 17, 2017, from, http://geert-hofstede.com/national-culture.html

Van Wijk, R., Jansen, J. J., & Lyles, M. A. (2008). Inter-and intra-organizational knowledge transfer: A meta-analytic review and assessment of its antecedents and consequences. *Journal of Management Studies, 45*(4), 830–853. doi:10.1111/j.1467-6486.2008.00771.x

Venkitachalam, K., & Busch, P. (2012). Tacit knowledge: Review and possible research directions. *Journal of Knowledge Management, 16*(2), 357–372. doi:10.1108/13673271211218915

Vera-Munoz, S. C., Ho, J. L., & Chow, C. W. (2006). Enhancing knowledge sharing in public accounting firms. *Accounting Horizons, 20*(2), 133–155. doi:10.2308/acch.2006.20.2.133

Voelpel, S. C., & Han, Z. (2005). Managing knowledge sharing in China: The case of Siemens ShareNet. *Journal of Knowledge Management, 9*(3), 51–63. doi:10.1108/13673270510602764

Wang, S., Noe, R. A., & Wang, Z. M. (2014). Motivating knowledge sharing in knowledge management systems: A quasi–field experiment. *Journal of Management, 40*(4), 978–1009. doi:10.1177/0149206311412192

Wasko, M. M., & Faraj, S. (2000). "It is what one does": Why people participate and help others in electronic communities of practice. *The Journal of Strategic Information Systems, 9*(2), 155–173. doi:10.1016/S0963-8687(00)00045-7

Wenger, E., McDermott, R., & Snyder, W. (2002). *Cultivating Communities of Practice: A Guide to Managing Knowledge*. Boston, MA: Harvard Business School Press.

Wetzels, M., Odekerken-Schroder, G., & Van Oppen, C. (2009). Using PLS path modeling for assessing hierarchical construct models: Guidelines and empirical illustration. *Management Information Systems Quarterly, 33*(1), 177–196. doi:10.2307/20650284

Wilkesmann, U., Fischer, H., & Wilkesmann, M. (2009). Cultural characteristics of knowledge transfer. *Journal of Knowledge Management, 13*(6), 464–477. doi:10.1108/13673270910997123

World Values Survey. (2017). Data and Documentation. *World Values Survey.* Retrieved August 31, 2017, from, http://www.worldvaluessurvey.org/WVSContents.jsp

Zaidman, N., & Drory, A. (2001). Upward impression management in the work place cross-cultural analysis. *International Journal of Intercultural Relations, 25*(6), 671–690. doi:10.1016/S0147-1767(01)00031-1

KEY TERMS AND DEFINITIONS

Collectivism: Collectivist individuals value the support and security of the groups to which they belong, and will exhibit a high level of loyalty to these groups.

Extrinsic Motivation: Refers to behavior that is driven by a desire for separable outcomes (e.g., money, career advancement, ego-enhancement).

Identified Regulation: A form of extrinsic motivation whereby an individual recognizes the personal importance of a behavior and has thus accepted its regulation as his or her own.

Incentive Salience: The extent to which one perceives potential gains or value in pursuing an opportunity.

Integrated Regulation: The most autonomous form of extrinsic motivation, whereby identified regulations have been fully assimilated with the self and are fully in congruence with one's other values and needs.

Intrinsic Motivation: Refers to behavior that is driven by the inherent satisfactions of the behavior itself (e.g., interest, enjoyment).

Introjected Regulation: A type of extrinsic regulation whereby a person behaves with the aim of avoiding guilt or anxiety, or to attain ego-enhancements or pride.

Knowledge Coach: An individual who attempts to recreate their tacit knowledge in the mind of a less-experienced individual using active learning techniques.

Long-Term Orientation: Long-term oriented individuals value thrift, education, and other means to prepare for the future.

Power Distance: Individuals with high-power-distance values are comfortable in hierarchical structures and are less likely to seek justifications for inequalities of power.

Section 3
KMS Design Issues

Chapter 9
CoMIS-KMS:
An Elaborated Process Model for Transitioning MIS to KMS

Ricardo Anderson
The University of the West Indies – Mona, Jamaica

Gunjan Mansingh
The University of the West Indies – Mona, Jamaica

ABSTRACT

Modern information system environments generally contain complex computing infrastructure often mapping business processes to people and data. Data manipulation to summarize and report meaningful information is commonplace, although some of these systems and techniques have not been extensively exploited within developing countries. Even in developed economies, where resources and experience are abundant and accessible, the process for moving from information management to knowledge management is not well understood and therefore remains a challenge. In this study, a comprehensive process model that guides the transitioning of an existing information system to a knowledge management system is developed and evaluated. The results indicate that the model sufficiently represents and organizes the activities to be carried out to meet the desired outcome of converting an existing information system into a knowledge management system.

INTRODUCTION

A Knowledge management system (KMS) facilitates the creation, capture, refinement, storage, application, update and dissemination of knowledge (Alavi and Leidner, 2001; Bose, 2004). KMS may vary in sophistication and features but are generally supported by information technology. Several alternate methods and techniques have been proposed for developing knowledge based systems given observed inadequacies of traditional software development methods (Weitzel & Kerschberg, 1989; Angele et al, 1998; Schreiber, et al, 1994).

DOI: 10.4018/978-1-5225-5427-1.ch009

Although many developing countries are still at the stage of trying to use information systems to digitize, collect and manage data, having only partial automation of tasks and limited capabilities, knowledge can be of significant value and may provide for significant improvements that may allow for better decisions and an informed way forward to prosperity. With the pervasive nature of information systems across industries and in organizations, and the lagging of developing countries in exploring some of these resources to access efficiencies, research is required to provide guidance on how to access and exploit sources of potential knowledge, and harness this knowledge into existing systems towards improvement of their capabilities. The emphasis on the value of knowledge, knowledge creation and application of knowledge across all areas of endeavour necessitates the development of research to support improvements in knowledge integration in existing information systems.

The process to transition from the current state of information management to knowledge management driven by a KMS is not well articulated in the literature; a comprehensive model to guide this process does not exist. Therefore, research is needed to fill the existing gap regarding prescribed steps for the integration of knowledge into existing Information Systems (MIS) to produce a knowledge management system that is sufficient for the criteria of creation, capture, update, application and dissemination of knowledge.

This study contributes to the current literature in the prescriptive domain in the form of a process model which is developed to transition existing information system environment to a knowledge management system. The model is based on the context of a developing country environment and has been instantiated in this context. The model and its instantiation provide further contribution in design theory as well as in the descriptive domain through observation, classification and measurement (Gregor and Hevner, 2013).

In the next section, a brief review of related works is presented. This is followed by a discussion of the research methodology employed and presentation of the process model. A discussion of the model applied in two separate domains under different conditions within a developing country is then presented followed by discussion, conclusion and future work.

KNOWLEDGE ELICITATION, CODIFICATION, AND MANAGEMENT

Leonard and Sensipar (1998) define knowledge as relevant, actionable information based partially on experience. O'Dell et al (1998) suggest that knowledge is what members of an organization know about their customers, products, successes and mistakes. The authors adopt for the purpose of this study that knowledge is a 'fluid mix of framed experience, values, contextual information and expert insight that provide a framework for evaluation and incorporating new experiences and information (Davenport & Prusak, 2000). This differs significantly from information and data which are lower according to the data-information-knowledge-wisdom (DIKW) hierarchy (Rowley, 2007). Data is a collection of facts and figures, while information is organized data that provides meaning. Context, experiences, and expert insights can then be added to information to produce knowledge.

Bowman (2002) argues that knowledge can be critical to an organization's success as it can improve their capability. Examined from the capability perspective (Alavi & Leidner, 2001), knowledge enhances the ability of a firm to develop existing and new core competencies. It also influences future action which can contribute significantly to the long run survival of a firm (Darroch, 2005; Watson, 2008; Eriksson & Raven, 1996). One view of institutional knowledge capability is that this is encompassed in the firm's knowledge assets such as expertise of employees, so called knowledge documents (maps, statistics, research reports, analyses), lessons learned, data as well as policies and procedures (processes) (Freeze & Kulkarni, 2007) and how each of these assets can be used to improve or enhance the firm's capability or competencies.

Knowledge also enhances learning, supports strategy and is a source of competitive advantage (Jennex et al.,2009; Rolland, 1998; Smith et al, 2010). The importance of knowledge to the organizational success provides impetus to implement initiatives to manage knowledge to support the continued survival of the organization. The proliferation of computer-based information systems also gives rise to the need to explore how these technologies can help and facilitate the knowledge initiatives in the firm.

Polyani (1967) specified two categories of knowledge; tacit and explicit. Tacit knowledge is said to reside in one's subconscious and is not easily codified while explicit knowledge is easily codified and can be stored on machines. Further refinements of knowledge types have been articulated in the extant literature. De Jong & Ferguson-Hessler (1996) provides a description of four main types of knowledge: Situational Knowledge, Conceptual Knowledge, Procedural Knowledge and Strategic Knowledge. Alavi and Leidner (2001) suggests that knowledge can be described as know-why (causal knowledge), know-how (procedural knowledge), know-when (conditional knowledge), know-about (declarative knowledge) and know-with (relational knowledge).

Knowledge Elicitation

Knowledge considered to exist in tacit form requires explication. The type of tacit knowledge in a particular domain is critical to the planning for knowledge management as it provides for essential decisions on tools and techniques to capture, harness and apply knowledge (DeLong, 2004). There is however varying levels of difficulty in the explication of tacit knowledge. Deep tacit knowledge may require more artificial methods and is possibly the most difficult to explicate whereas implicit knowledge provides less challenges for explication (Cooke, 1994).

Knowledge also exists in data and can be elicited by the used of well-established tools and techniques for knowledge discovery in databases and data mining (KDDM). The popularity of KDDM has been spurred on by successful application to sales and marketing and by extension the retailing industry (Berry & Linoff, 1997). Over time, knowledge discovered via data mining has been used to assist with fraud detection in health-care, accounting, credit card fraud detection, among others (Ferdousi & Maeda, 2006; Sanchez et al, 2009; Phua et al, 2010; Tagaris et al, 2009).

Knowledge Management

Knowledge management in today's firms has been described as critical to the extent that it is a possible source of significant competitive advantage (Jennex et al, 2009; Ryan et al, 2012). It involves the processes that provide for the creation, storage, application and update of knowledge. Updating knowledge is particularly important since the dynamic nature of today's firm requires that outdated knowledge does not continue beyond its temporal usefulness but must be modified, or removed if new knowledge becomes available such that it invalidates the current knowledge. Dorroch (2005) posits that knowledge management plays a key role in that it provides a coordinating mechanism that supports the conversion of resources into capabilities within the firm. In addition, it is critical to the long run survival of the firm as it facilitates the development of other capabilities of the firm. In a pioneering study of firms in New Zealand, the results indicated that effectively managing knowledge was an important factor in innovativeness and the financial performance of firms (Darroch, 2005).

The SECI model proposed by Nonaka (1994) has been widely cited in the knowledge management literature as it provides an innovative theoretical framework for organizational knowledge management. Nonaka (1994) further underscores the need to facilitate knowledge creation and sharing through activities planned by the organization which will allow for the experiences to be shared and applied among stakeholders in the firm, thus propagating and advancing knowledge. More recently SMARTvision has been proposed as a knowledge management methodology that provides more detail on how initiatives should be achieved (Rubenstein-Montano et al, 2001).

Knowledge Management Systems (KMS)

A knowledge management system (KMS) is a class of information systems that are applied to managing knowledge in the organizational context. These are IT-based systems developed to support and enhance organizational processes related to knowledge management (Alavi & Leidner, 2001). KMS extend beyond traditional information systems as they provide a context within which information is coded and presented for use (Gallupe, 2001), in addition to supporting the four main knowledge management activities of knowledge creation, storage, application and update. Knowledge update is particularly important especially in a fast paced, dynamic organization (Wan, 2015). Maier (2002) notes that the key differences between a KMS and the traditional information system such as MIS include the integration of intelligent functions, a contextual combination and integration of functionality, the matching of supporting knowledge management initiatives to ensure that knowledge is integrated in processes within the organization and the dynamics of organizational learning.

Depending on the knowledge perspective, Alavi and Leidner (2001) suggest different implications for Knowledge Management Systems (KMS). The type of knowledge that exists in the organization also plays a significant role in determining the actions necessary for KM and building KMS. If the knowledge exists in experts, then capturing (knowledge elicitation) should be the focus. If the knowledge is an object then the focus should be collect, store, and share knowledge. If the knowledge is embedded in processes then the KMS should provide for improvements in the flow of the knowledge (Osei-Bryson et al, 2014). Nevo (2003) suggest three aspects as key to the design of KMS. These include 1) the network design that will support sharing of knowledge, 2) an ontological base to provide a shared language for interrogating the knowledge (or use) and 3) the meta-knowledge which is important for knowing what knowledge exists and the relevance of such to different situations that may require its use. Tauber and Schwartz (1997) note that the first generation of add-on parallel systems need to evolve into a new generation of KMS that integrates both unstructured and structured data for real-time situational assessment for decision making.

The Developing Country Context

Information Systems (IS) in many developing countries generally are either at the emergence or at the broadcast stage, thus providing very few interactive services. The few developing nations with extensive interactive, knowledge supported services typically also maintained a considerable IT and government infrastructure (Basu, 2004). No doubt, both the public and private sectors in developing countries have seen the impacts of advanced information technologies applied in the industrialized world and have sought to achieve some similar benefits from these systems based on their success in the industrialized world.

Although several implementations have failed to be successful in developing countries, the literature has several examples of local improvisations that have been successful in developing countries. These have been useful in improving the chances of success in computerization (Goonatilake et al, 2000; Silva, 2002). New, innovative and cutting-edge technology could help developing countries overcome their traditional barriers of lack of ICT infrastructure. Whilst newer versions of technology are often used to upgrade older versions in more developed environments, developing countries using older technology have the ability to leapfrog successive generations of technology to the most recent version. This often provides access to newer, efficient and often cheaper alternatives to traditional capital intensive technologies (Basu, 2004).

The current literature however notes that many failures have occurred and there is need to improve likelihood of success, even as the authors seek to transition these existing systems from primarily automation and information management, into a state for supporting knowledge acquisition, storage, application and update to reap the benefits of knowledge.

Knowledge Management Initiatives and WfMS

A workflow is a collection work items (or tasks) organized to accomplish some process, for example, processing office requisition through to settlement or managing the scientific process (Deelman et al, 2015) often established to ensure proper sequence and quality assurance. A workflow management system (WfMS) either completely or partially supports the processing of work item(s), to accomplish the objective of a group of tasks within a process. These systems usually include features for routing tasks from person to person in sequence, allowing each person to contribute before moving on in the process (Fakas & Karakostas, 1999).

Workflow systems have undoubtedly provided significant benefits to process management in many domains and is the core of many enterprise resource planning systems. WfMS, however, have not been utilized in the knowledge management domain to provide many of the studied benefits (Kammer et al, 2000; Han et al, 1998; Narendra, 2004). Having established the need for such a software solution through thorough assessment of prior research and identifying the convergence of a gap in WfMS and their use in the knowledge management domain for initiatives geared at improving knowledge capability, we develop a Knowledge Management Initiative Adaptive Workflow System (KMI-AWS) (Anderson & Mansingh, 2017). This system is primarily for managing a defined Knowledge Management Initiative process based on the activities prescribed in the CoMIS-KMS process model. The developed WfMS system was used to conduct the case study in the second domain and the results are presented in the section: Evaluation of the Process Model.

DESIGNING THE PROCESS MODEL

A process model defines a set of activities (tasks), along with the inputs and outputs for each task that is necessary to get the job done (Pressman, 2005). This work outlines a process model called CoMIS-KMS which adheres to the characteristics of traceability, effectiveness, predictability and repeatability

(Mariscal et al, 2010; Tyrrell, 2000). Both the action research (Susman and Evered, 1978) and design science (Hevner et al, 2004) methods were applied to aspects of this study.

On the basis of the five-phase cyclical process of action research: 1) diagnosing, 2) action planning, 3) action taking 4) evaluating 5) specifying learning (Susman and Evered, 1978), the process of diagnosis began. First, the diagnostic stage involved a collaborative analysis of the situation by the researcher and the subjects of the research. This led to the identification of a gap in the use of potential knowledge that could be of significant help to improve the delivery services in the social welfare domain.

Having further determined that there is need for action to maximize knowledge application for improved services delivery in the domain, the researchers needed a method for carrying out this action. The action required a step-by-step process for transitioning the current information system for integrated knowledge application. This then required action planning to drive the process of transitioning the current environment.

Guided by the design science methodology proposed in (Hevner et al, 2004), the need and relevance of such process was established based on existing literature related to issues integrating knowledge in existing systems, resulting in a KMS. The authors conducted a review of literature on management information systems (MIS), knowledge management and knowledge management systems to identify the state of the content in this area. The design search process guideline (Hevner et al, 2004) then started with initial version of the process model which benefited significantly from the domain context in which it was designed to be used. Working with the subject matter experts of the domain, and having constant communication of the planned activities, feedback was provided by them on the steps to be performed and what is required for the planned activities, identifying the expected outcomes.

The literature provided important guidance on the important activities for successful knowledge management and key metrics for evaluation of knowledge management systems. Based on the requirements of KMS the authors began to identify the tasks that would be necessary to develop a KMS particularly in this domain that would support the capture, storage, application and update of knowledge. The importance integration of knowledge within the existing systems was articulated as critical (Tauber and Schwartz, 2011) and therefore an important requirement was a knowledge integration architecture. The activities enlisted were reviewed and based on task dependencies, natural groupings emerged. Tasks were arranged to ensure dependencies are first completed after which tasks with common dependencies were grouped to create each phase as defined.

The techniques applied and the results at each stage coupled with inputs from the subject matter experts were used to change, re-order and otherwise add or refine activities as the process was being developed and applied. Therefore, the model benefited from feedback as it progressed; this informed the final version presented as CoMIS-KMS. The final artefact developed is a process model which specifies steps segmented in phases and cluster necessary activities, giving a sequence for execution. Each phase has a set of inputs and outputs. Outputs give an indication of the achievements necessary to proceed to the next phase. The CoMIS-KMS model is discussed in detail in the next section.

The taking of action provided for evaluating the artefact within the domain via the case study observational method as prescribed in the Design Evaluation guideline (Hevner et al, 2004). The collaborative change experiment was conducted by applying the CoMIS-KMS model. The effects of the CoMIS-KMS application became visible through the changes that were viewed based on the new application of knowledge to impact the processes in the domain. Further, in line with the fifth aspect of action research-Specify learning, an extensive review was done that specifies how the action taken benefits the domain both in the specific case and through analysis of other potential cases of social welfare systems

(Anderson et al, 2014). This specification of learning through expanded review and analysis supports the notion that action research merges research and praxis thus producing exceedingly relevant research findings (Baskerville & Wood-Harper, 1996).

THE COMIS-KMS PROCESS MODEL

The CoMIS-KMS Model has six phases arranged based on a set sequence. The sequence is derived from the outputs that are produced, as necessary for the next phase of activities and therefore, though there are overlaps, the subsequent phase cannot be completed without completing the prior phase. Each phase of the model has associated activities which depend on several techniques that exist in the literature. The model can be applied iteratively since the context for evolution may allow for discrete processes or domain segments and their supporting systems to be transitioned independently to support a group of the organization's processes rather than applying the model at once to the entire enterprise. This can be done especially in context where the knowledge assets pertain to independent areas, functions and contexts such that they are not applicable across the entire domain, context or organization. The following details the phases of the model and provides descriptions of the activities grouped in each phase.

Figure 1. The CoMIS KMS model

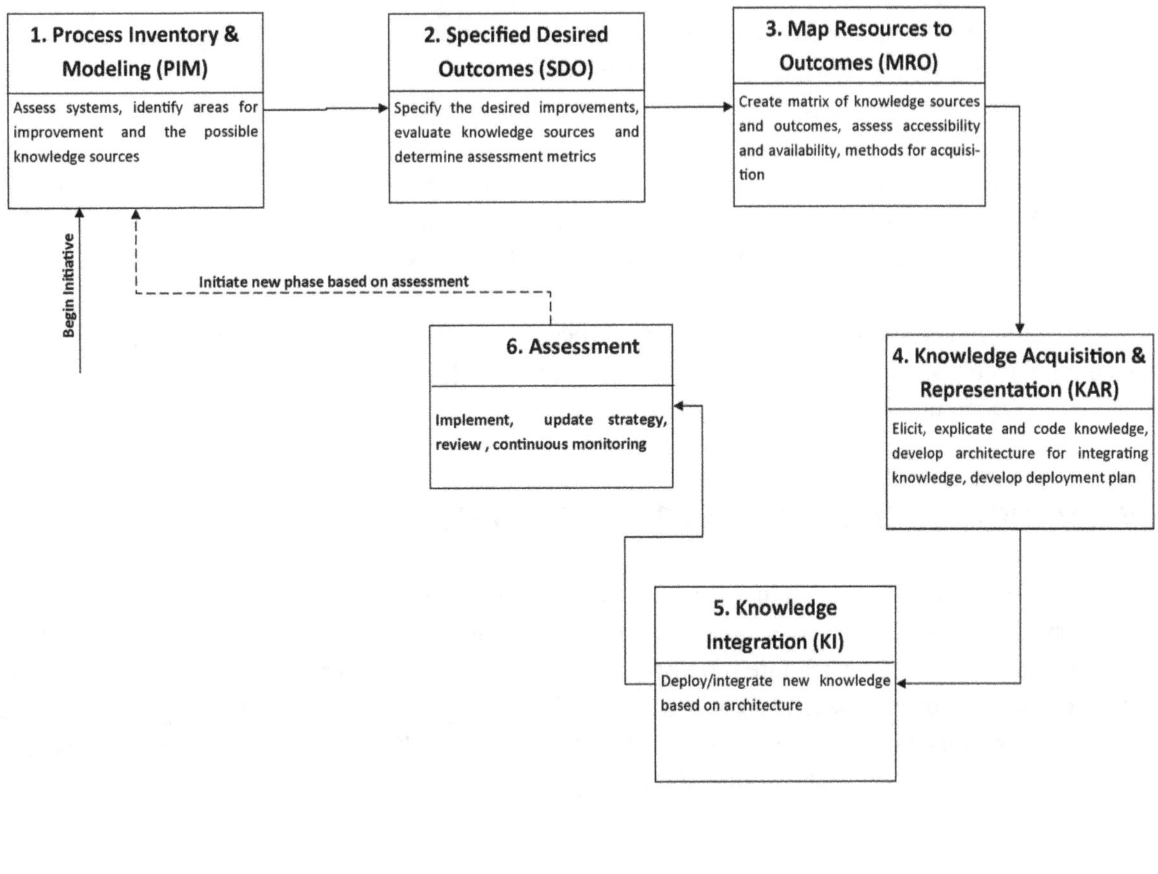

- **Process Inventory and Modelling (PIM):** In this phase the main aim is to assess the business information systems and processes and identify possible areas for improvement with corresponding knowledge sources that may be useful in such improvements. This stage can progress relatively quickly if systems and processes are already well documented. The aim is to identify the processes that exist as well as any possible knowledge assets and have a comprehensive inventory.
- **Specify Desired Outcomes (SDO):** In this phase, given the review of outputs from the PIM, decisions on desired improvements and areas of focus for Knowledge Management are decided. The main aim is to set some objectives for the initiatives to be undertaken and use the identified assets/sources from the PIM phase to create a matrix of supporting resources with an evaluation of these assets/sources. This phase also prescribes that techniques be identified for knowledge explication if necessary.
- **Map Resources to Outcomes (MRO):** In this phase, given the evaluation of knowledge sources and the outcomes specified in the SDO phase, a matrix which represents mapping of resources that may support those outcomes is derived and validated. This phase primarily focuses on linking the available, evaluated knowledge sources and assets to the outcomes specified such that there is clarity on what resources will be used and the way it will be used in order to meet the desired outcome.
- **Knowledge Acquisition and Representation (KAR):** At this stage, given the resources identified, mappings and techniques are used to acquire/elicit and organize knowledge. This process will be driven by the nature of the assets/sources selected from the previous phase. The knowledge discovered here must be coded and an appropriate integration mechanism developed so that it can guide the deployment and continuous update and application of the knowledge.
- **Knowledge Integration:** This phase focuses on deploying the knowledge resources into the current environment. The deployment plan and architecture is activated and testing is done.
- **Assessment:** At this stage, the knowledge update and supporting knowledge strategies are activated to ensure the right knowledge is available, that the outcomes specified are realized and changes are made where necessary which may include the application of this model again. The key assessment activities in this phase include firstly assessing the implementation based on KMS metrics and the desired outcomes. After that assessment, depending on the outcome, improvements are planned which may require another iteration of the CoMIS-KMS.

This model can be applied to impact different processes within the domain and may be applied based on unmet objectives as indicated by performance measures. Therefore, each process/context or department may be done in one iteration through the process model.

EVALUATION OF THE COMIS-KMS PROCESS MODEL

The CoMIS-KMS model was evaluated in two separate domains: social services and education using the observational method via a case study (Hevner, 2004). In the case study of social services domain, a Conditional Cash Transfer Welfare programme called *PATH* was the subject of the evaluation, while in the education domain, a secondary school (*HS*) was the subject of the study. The application of the model in the secondary school was augmented by the use of the KMI-AWS (Anderson & Mansingh, 2017). This section discusses the results of applying the CoMIS-KMS model to both domains.

Phase 1: Process Inventory and Modelling (PIM)

The main activities in this phase of the model are: 1. Review of business processes - this should produce a list of the key processes and an overview of the possible sources of knowledge and 2. Conduct a Gap Analysis - this should highlight areas for strengthening.

Business Process Review

In PATH, the key business processes were well defined and documented in process flow diagrams and were easily elaborated by the programme staff. Whilst HS did not have well documented processes, interaction with key stakeholders and subject matter experts provided easily elaborated business processes. Table 1 gives a high-level listing of the key business processes in both domains.

Processes were reviewed with key stakeholders in both domains, identifying the data sources, purpose, use and impact within their operations. Table 2 gives a summarized list.

Gap Analysis

PATH

The programme currently struggles with verification and subsequent recertification due to significant resource constraints. The number of social workers available is insufficient and hence a significant backlog of months is experienced between application and verification, thus the start of the benefit disbursement is delayed. Once disbursement begins, monitoring the beneficiaries' adherence to contidionalities is necessary. The application, verification and recertification processes accumulate vast amounts of data about living conditions and the behaviours and progress of beneficiaries over the period that they

Table 1. Business processes: PATH process descriptions

Domain	Process Name	Description
PATH	Targeting	Method by which potential beneficiaries can access the programme.
	Assessment	Review of application details submitted by target group members.
	Verification	Social worker visits the address indicated as residence for applicant and reviews the information submitted on the application and verifies details.
	Re-certification	Periodic verification.
	Monitoring	Review of beneficiary data for compliance with programme guidelines.
HS	Teaching and Assessment	Lesson planning, delivery and administration of assignments to assess student's progress.
	Support and Co-Curricular	Enrolment, monitoring and performance in non-academic activities
	Staff Assessment	In-class assessment of teaching, end of period comprehensive assessment with school leaders
	Student Registration	Enrolment and monitoring of new and returning students for each period.
	Incident Management	Recording and action taken to monitor student behaviours
	Attendance management	Recording and evaluating student attendance.

Table 2. Data/information sources and usage

Domain	Data/Information Source Description
PATH	• Performance Analysis Reports generated internally and by external assessments to determine overall operations of PATH • Operational Reports – reports generated as outputs from internal processes • Agency Reports – structured and unstructured logs of beneficiaries access to services
HS	• Academic Department Reports – submitted monthly by each department head • Termly student reports – provided for parents at the end of a term • Attendance profiles and behaviour records – log of attendance and behaviour incidents • School Board report – produce monthly for the school board with summary of the status of the school • Student health and welfare records • Ministry of Education reports – summaries of all relevant school operation activities

remain on the programme. This accumulated data can be a source of knowledge providing insights into social behaviours, economic patterns, school and health-care services access and trends in agencies that provide these services. There are benefits to be derived, especially in the specific processes of application verification and proactive monitoring of at risk households, such as continued support in meeting the conditionalities.

HS

Several areas were identified for improvement in the school. These included desired improvements in lesson planning and delivery, overall student academic achievement to reach the target of 70 percent achieving average of 70% or over, analyses to determine the effectiveness of support systems in the performance of students and improving systems for identification and intervention for students who are at risk.

The KMI-AWS was used extensively in this phase for HS to manage the documents for review and recording the gaps which produced an automated gap analysis report.

Phase 2: Specify Desired Outcomes (SDO)

This phase requires critical review of the outputs from the previous phase *(PIM)* with the subject matter experts and managers of the programme for them to communicate which areas should be prioritized for strengthening.

Desired Outcomes

In consultation with the stakeholders and subject matter experts, the set of desired outcome and specific indicators of successful achievement were specified for both domains. In PATH verification and proactive monitoring were prioritised as areas for improvement. In HS, identification of performance and behaviour indicators and improving lesson planning were prioritized.

The desired outcomes for each domain is summarised in Table 3.

To support the outcomes specified, the authors evaluated the identified knowledge sources identified in the *PIM* phase. The assessment was based on accessibility and availability of the knowledge sources. The knowledge sources from each domain that met the accessibility and availability criteria were selected

Table 3. Summary of desired outcomes for both domains

Desired Outcomes	
PATH	**HS**
An integrated system that will allow for aspects of the verification process to be automated or informed by knowledge.	Provide a way to determine students who are predisposed to good academic performance or deviance.
Reduction in the time taken to complete verification of potential beneficiaries.	Provide details based on school historical data, to determine the most important factors that contribute to good academic achievement.
Reduced need to manually verify each potential beneficiary.	Provide a mechanism for staff members to share lessons, best practices and insights into teaching methods.

for use in the next phase. This activity of e valuation and filtering available knowledge sources for use in the next phase was done by the KMI-AWS in HS. In addition, the KMI-AWS was used to record prioritization scoring for the desired outcomes in HS.

Phase 3: Map Resources to Outcome (MRO)

This phase requires further review and assessment of the potential knowledge sources previously identified and plotting these to the outcomes specified in the *SDO* phase.

In both domains, with the existence of a Management Information System which has mainly structured data that captured the key operational activities, this was one of the main potential sources of knowledge selected for exploitation towards achieving the desired outcomes. To complete this phase, subject matter experts in both domains guided us in locating and further examining relevant data sources (which may contain knowledge) which were then mapped to the outcomes from the *SDO* phase. Table 4 provides a summary of the outcomes and the potential knowledge sources for each domain.

Phase 4: Knowledge Acquisition and Representation (KAR)

Having completed the mapping of knowledge sources to outcomes in the previous phase, this phase focuses on determining the knowledge elicitation methods, applying these methods to acquire knowledge and determine how the architecture for integrating the acquired knowledge.

KAR in PATH

The initial tasks in this phase included discussion with domain experts, process tracing and observation. This process provided better understanding of the critical steps in the verification process and their in-

Table 4. Selected potential knowledge sources

Supporting Knowledge Sources	
HS	**PATH**
School MIS, Guidance & Health Records, Term Reports, Teacher appraisals, Faculty Expertise by subject area	Domain Experts, Operational Reports, Process maps, Beneficiary MIS Data

teraction with the verification result. It was also necessary to discuss the data and some report findings with the programme management team. Based on the knowledge sources that are mapped to the processes that will be supported by knowledge, the relevant subsets of data were identified from the BMIS.

In order to better understand the application verification processes, the data in the BMIS that was collected during the application process were examined with the help of domain experts at PATH who guided the understanding of the data. The completed data selection and preparation resulted in 56 data fields that were then formatted for input to our mining tool SAS eMiner for application of algorithms for extracting patterns. Decision tree (DT) induction was used since the objective was to use the results for predictive purposes; future verification of applicants. Approximately 156 thousand applicant records were included in the data set. The data was partitioned into training, validation and test sets based on a 50:30:20 ratio. The decision tree model produced 257 English rules that explained the data. This new knowledge must now be integrated into the environment for use based on the desired outcomes.

HS

The knowledge sources selected for each desired outcome provided the basis for selecting the techniques for elicitation/acquisition of knowledge, codification where necessary and the deployment for use, hence, improving the knowledge capability of the institution.

The School MIS stores profile data of students in structured format. Given that the outcome was both to determine and be able to use the determination for predictive modelling, the techniques applied to support desired outcomes include association rule mining, regression analysis, interviews, questionnaires, process tracing/observation and document review and analysis. Data preparation for each of the selected techniques required a thorough analysis of the existing data intermediary steps to properly format the data.

Several further utility scripts were written using R, PHP and Python to better understand the data, discover missing data and for basic descriptive statistics. The data were formatted for import into the analysis tools using utility scripts which generated the required dimensions. Forty-three (43) data dimensions were used to develop various models using R, both for association rules mining. Applying variations of datasets, with model assessment criteria, the best models were selected. This resulted in approximately 89 rules that had predictive or explanatory significance.

In applying multiple linear regression, the data were formatted in several dimensions for initial modeling. Several field combinations were used for preliminary exploration. Eighteen models were built using different dimensions of the data. The results were evaluated for variable significance in predicting the value of student overall average. The evaluation resulted in only twelve of the models being selected based on t-values, adjusted R-squared, F-statistic and p-value.

In relation to improving lesson planning and delivery, several techniques were used to elicit best practices from teachers who were high achievers.

Interviews and document review: Teachers were interviewed about best practices and contributors to their outstanding student achievements. The teachers were asked to complete a questionnaire which was sent in digital format, this provided preliminary information that was used to structure the interview and recording sessions. The teachers were selected by the school leaders. The selection was based on the institution's determination of outstanding teachers from which others could learn of how they achieved their successes with students. Recording was then done and edited. Lesson plans are an important component of the best practices that may assist teachers in subject specific areas. Therefore, sample lesson plans were collected for review with stakeholders to evaluate and decide which aspects of lessons could benefit others who teach similar subjects. Having concluded this review, a determination was made that accessibility and access to lesson plans and recorded interviews could help knowledge sharing, since one teacher may learn from all/any aspects of lessons or experience of others.

Having conducted interviews, document review, administered the questionnaires and recorded responses towards eliciting expertise and accessing already existing codified knowledge in the form of lesson plans, the activity to determine how the knowledge should be deployed is now required.

Given the need to integrate knowledge into the current MIS to allow automated knowledge application, an integration and deployment architecture was required. Studying the architecture of the MIS in both domains, and seeking to meet the metrics that were specified for the desired outcomes, an initial architecture was designed to guide the integration of the newly discovered knowledge. After several iterations and discussion with the technical leads and subject matter experts, the KIISA architecture (Anderson and Mansingh, 2015) depicted in figure 2 was finalized for knowledge integration.

Deployment Plan

The deployment of the acquired knowledge is guided by the integration architecture depicted in Figure 2. The coded knowledge is stored in the knowledge repository and applied by the knowledge service.

Through this process two fundamental functions of a KMS are achieved; application and sharing of knowledge. The KIISA architecture (Anderson and Mansingh, 2015) provides for the capture of knowledge offline from data and expertise that can be added to the knowledge repository, thus fulfilling the KMS requirement of knowledge capture. This new knowledge can serve as an update to previous knowledge or for addition of newly discovered knowledge.

The BMIS represents the current MIS used to manage the operation in PATH and is substituted for the School MIS in HS. It is intended that knowledge discovery will be an ongoing process where new knowledge is added to the knowledge service through a plugin architecture. This can be done in the KDDM software utilities.

Figure 2. Knowledge integrated information system architecture (kiisa)

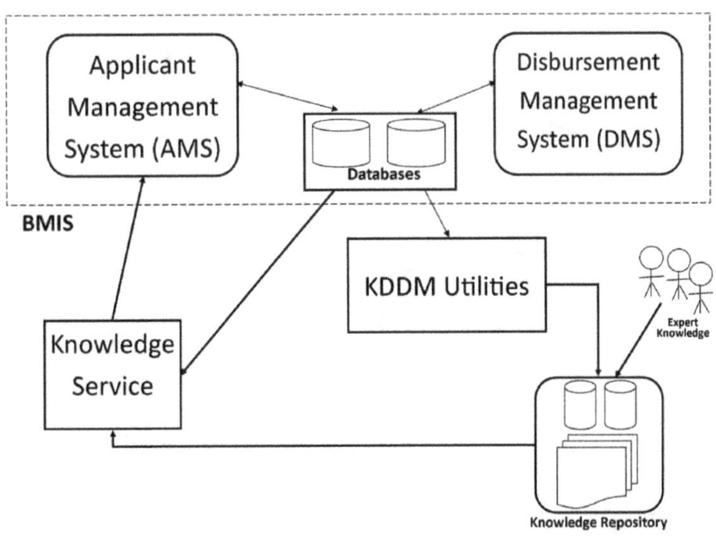

The best practices for improving teacher lesson planning and delivery will be deployed in a separate KMS which manages the current workflow of lesson planning and assessment, and will provide an indexed, searchable web based repository of best practices and lesson plans.

Phase 5: Knowledge Integration (KI)

This stage depends on the knowledge produced from the *KAR* phase and the deployment plan. The key activities in this phase include the development of tools to integrate the newly acquired knowledge and testing this new knowledge in the extended environment.

PATH: The main tool developed in this phase was an expert system which used the *English Rules* that came from knowledge acquisition in the previous phase. This knowledge was coded into production rules and were then used to implement the expert system (ES) using the Clips Engine. Figure 3 depicts the expert system components implemented as the knowledge service as seen in the KIISA architecture (Figure 2).

In Figure 3, the User Interaction console allows an end user to evaluate an application and receive results of the inference engine. This combines the facts that are extracted from the live application database stored in relational tables (BMIS). It extracts and loads the data, appropriately translated to produce the required fields and values that are added to the working memory of the inference engine in real-time with the rules. Once inferencing is complete, the result is presented in the User Interaction Console. This architecture provides for a semi-automated extraction of relational data using a schema aware data extraction and conversion module that inserts facts related to the application currently being evaluated to determine verification result.

Figure 3. Expert system components

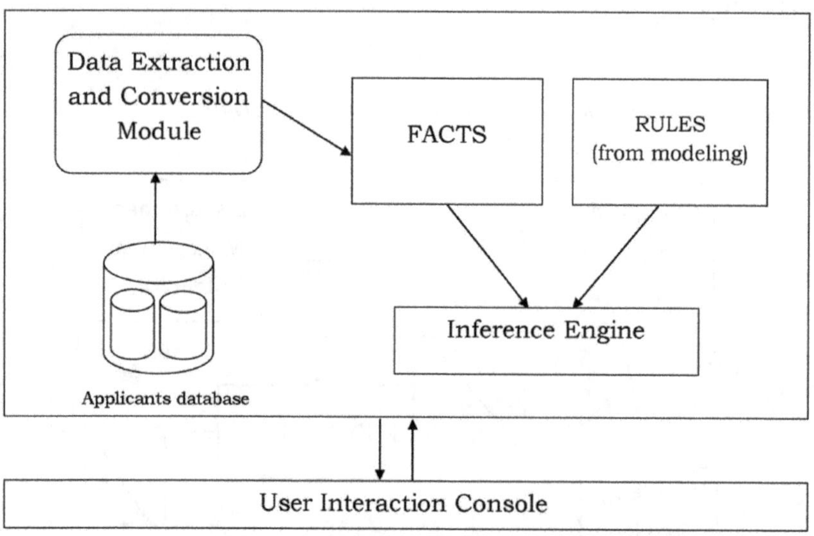

HS

Based on the deployment plan developed in the prior phase this phase involves the development and integration of software components with the various interfaces in an intranet based solution to allow the storage, sharing and application of knowledge. There are 2 major application components in the deployment: 1) Intelligent Analyzer, 2) Lesson Plan Manager).

Intelligent Analyzer (depicted in Figure 4) allows the user to access models from the results phase. For each option in the system, the appropriate model which was the result of the knowledge acquisition is can be invoked with interpretation. This becomes useful since this interprets the models using a context aware set of expressions to produce explanations to the user. Therefore, for all the options except the Profile Student a specific model, is attached and the system reads that model and uses the model to provide easily understood interpretation of that knowledge.

The *Profile Student* option (Figure 4) represents the deployment of the association rules. This feature uses the selected frequent pattern to assess any student and make a prediction of their likely performance level. Thus, when an ID number is entered in this section, a connection is made to the student data to fetch and transform the student combined profile which is then compared with the association rules. This inferencing returns the rules with the best match based on support and lift values defined in the underlying model assessment. These implementations represent the deployed knowledge, appropriately integrated based on the KIISA architecture.

The lesson plan manager deploys the knowledge about best practices and was implemented as a web-based tool. Sample screes from the tool are depicted in Figure 5 and 6.

Figure 4. Intelligent analyzer sample screen

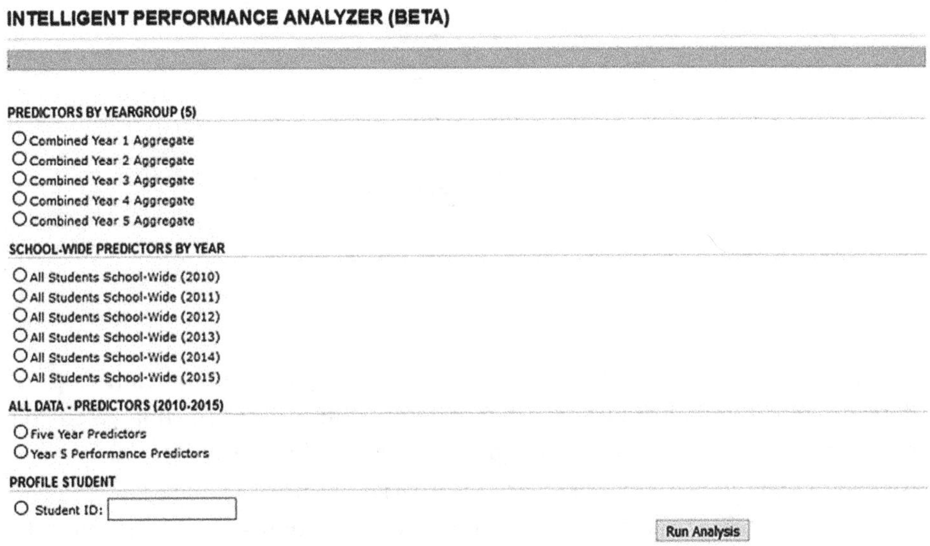

Figure 5. Lesson plan manager: Review lesson

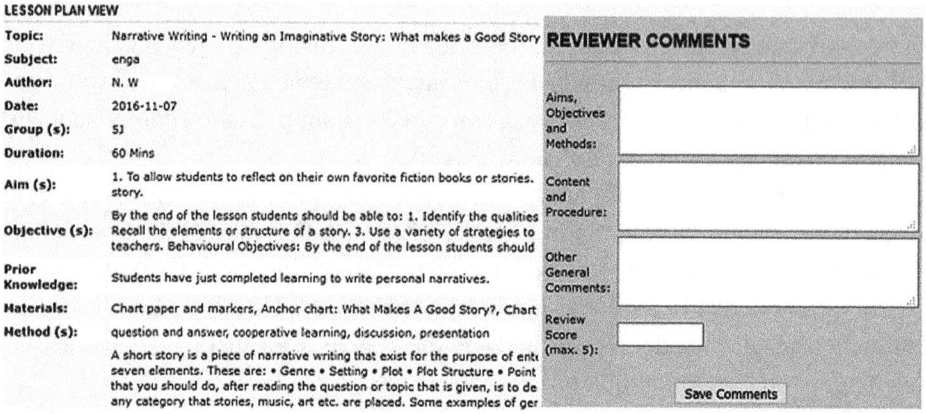

Figure 6. Lesson plan manager: Search repository

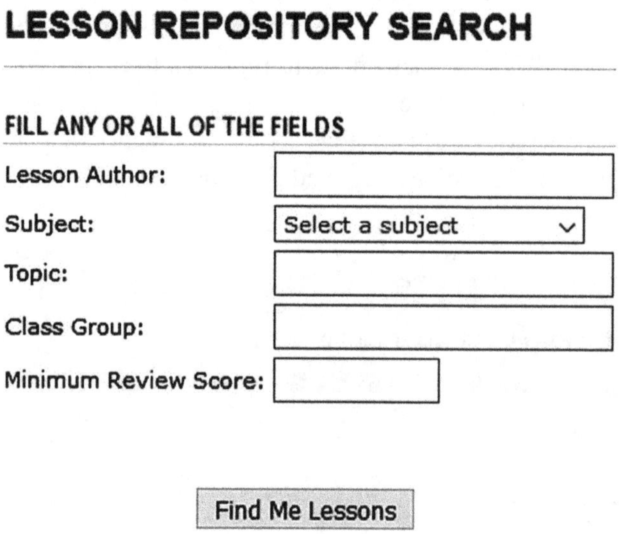

Assessment

This phase requires a review of the activities and their results against the outcomes specified in each case study domain. This will inform how to move forward by improving the implemented system and/or integrating more knowledge to support other processes and objectives.

The assessment is done in two areas: (1) using the metrics specified by the desired outcomes and (2) using the requirements for KMS classification. Table 5 summarizes the assessment results related to KMS.

The assessment of each dimension prescribed for KMS shows that our final product from the CoMIS-KMS process model produced a system consistent with the requirements for classification as KMS. Table 6 outlines the assessment of the deployment in relation to each of the desired outcomes.

Table 5. KMS assessment of deployed system

KMS Metric	Assessment of the DEPLOYMENTS	
	PATH	HS
Knowledge Dissemination / Sharing	Due to the integrated architecture that was used to deploy the knowledge, the programme managers have ready access to the knowledge in their decision making related to verification of potential beneficiaries.	Based on the integrated architecture that was used to deploy the knowledge, school leaders have online access to the knowledge in their decision making related key factors that impact student performance, student profile and likely performance level as provided in the Intelligent Analyzer.
Knowledge Application	The deployment of the expert system integrated as the knowledge service allows the application of knowledge directly in determining the verification status of an applicant to the CCT programme.	The deployment of the decision support expert system integrated as the knowledge service allows the application of knowledge directly in profiling students, identifying potential factors that influence student outcomes and to what extent, in addition the lesson objects have content that may be used to improve lesson planning for other subject teachers.
Knowledge Storage	This is provided by the repository which stores knowledge acquired from data and experts.	The knowledge acquired is stored for use in two ways: 1) As models from regression analysis and association rules and 2) the Lesson plan manager which stores knowledge objects for retrieval and use.
Knowledge Capture / Update	The update of knowledge is provided for in the deployed architecture through easy modifications to the knowledge repository. The current deployment however does not explicitly provide for finding, linking and integrating sources of knowledge. This will be addressed in future work.	The update of knowledge is provided for in the deployed architecture through easy modifications to the knowledge repository in the case of the Intelligent Analyzer. The Lesson Plan Manager integrates into the current lesson planning process in the institution which ensure weekly updates.
Knowledge Creation	The KDDM utilities provide for the application of techniques to acquire/create new knowledge for integration and application in the current environment. Additional processes can be supported in the current system.	The KDDM utilities provide for the application of techniques to acquire/create new knowledge for integration and application in the current environment. The fact that the Lesson Plan Manager now manages lesson planning, each object becomes a new knowledge asset created by teachers each week of the academic year.

The result of the assessment done validates the CoMIS-KMS process model can be used to convert an existing information system to a knowledge management system. The next section discusses the implications of these results.

DISCUSSION

The CoMIS-KMS process model guides the evolution of an existing information system to a knowledge management system. The defined phases in the CoMIS-KMS model guided the identification of key processes, the supporting data sets, knowledge resources and the critical decisions supported by information systems in the respective environments. These resources were aligned to the specific outcomes required by the domain managers. The acquisition, representation and integration of the discovered knowledge then followed based on the prescribed steps in the CoMIS-KMS process model. This work included the application of several elicitation techniques and algorithms to extract knowledge that were then deployed as an extension of existing system to include the newly discovered knowledge. The integration was guided by the KIISA architecture which we also developed. The evolved systems provide

Table 6. Assessment of desired outcomes

Domain	Outcome Metric	Assessment of Deployment
PATH	Automatic verification of potential beneficiaries by system	This is achieved through the expert system which connects to the live data in the BMIS extracts and converts the data and conduct inferencing to produce a verification result.
	Verification result available to end user of the BMIS system without use of auxiliary tools	Verification result is returned to the user automatically in the BMIS system once it is produced and written to the user record in the BMIS database.
	Accuracy of verification result exceeds 50 percent	The test of the expert system produced 94 percent accuracy on evaluation of 156 thousand application records after implementation.
	Actual time taken to complete verification	Verification results returned within 10 seconds of requesting evaluation.
	Number of beneficiaries that still require manual verification	Some special conditions require manual verification. These are imposed by external stakeholders and also the need to double check system results.
HS	Ability to identify key contributors to academic performance in order to make decisions on how to improve on these key factors for greater student achievement.	This has been deployed in the intelligent analyzer. Stakeholders are able to identify the impact of subject performance, co-curricular involvement, disciplinary issues, attendance etc on the student's academic achievement. This makes it possible for comparisons and decisions on resource allocation for improvements. The reporting feature also provides simulations on possible estimated improvements.
	Ability to compare current students based on identified achievement factors.	
	The materials are accessible beyond the time of employment of the staff member.	This was deployed in the lesson planning manager allowing teachers to upload knowledge objects in the form of lesson plans or best practices. This system also integrates with the current process to automate the inclusion of new lessons into an indexed and searchable repository.
	Lesson plans, best practices can be shared among staff members through an accessible mechanism not limited to person-to-person interaction.	

knowledge that assists in improved decision quality and speed. Further, the use of knowledge also impacts the revision of operations for more efficient resource allocation, cost reduction and quality improvements in core operational processes.

The implementation of the KMI-AWS was aimed at improving the execution of the tasks associated the KMI in HS based on observations from PATH regarding lack of process visibility, task routing and accountability for activities. The implementation and deployment of KMI-AWS significantly improved the management of the activities in the case of *HS*. The observed benefits are consistent with prior observations of WfMS in other domains (Kammer et al, 2000; Han et al, 1998; Narendra, 2004).

CONCLUSION AND FUTURE WORK

This study demonstrates the development and evaluation of the CoMIS-KMS process model to guide the evolution of information system into knowledge management system in two distinct domains within a developing country context. The KIISA architecture used for integration extends the current operational software to provide integrated use of knowledge within current automated operations.

Based on the metrics provided for assessment, the outcomes specified in phase 2 *(SDO)* of the CoMIS-KMS model, the assessment in Table 6 shows that each indicator related to the desired outcomes was appropriately addressed having instantiated the model in the domains. In addition, the integrated systems that is based on the KIISA architecture meet the criteria specified as necessary and sufficient to be classified as KMS. Given these results, this work demonstrates the successful application of the CoMIS-KMS model to give the desired result of transitioning MIS to KMS.

Future work will include the use of the model in additional domains to further improve the activities and steps if necessary such that it can become more generally applicable.

REFERENCES

Alavi, M., & Leidner, D. E. (2001). Review: Knowledge management and knowledge management systems: Conceptual foundations and research issues. *Management Information Systems Quarterly*, *25*(1), 107–136. doi:10.2307/3250961

Anderson, R., Gordon, S., & Mansingh, G. (2014). An assessment of the potential impacts of knowledge-driven decision support in social welfare. In *Global Humanitarian Technology Conference (GHTC)* (pp. 499-505). IEEE. 10.1109/GHTC.2014.6970329

Anderson, R., & Mansingh, G. (2015). KIISA: Towards a Reference Architecture for Integrated Knowledge Services. In Intelligent Systems' 2014 (pp. 247-257). Springer International Publishing.

Anderson, R., & Mansingh, G. (2017). A Workflow Management System for Knowledge Management Initiatives. *Proceedings of the 23rd Americas Conference on Information Systems, AMCIS 2017.*

Angele, J., Fensel, D., Landes, D., & Studer, R. (1998). Developing knowledge-based systems with mike. *Automated Software Engineering*, *5*(4), 389–418. doi:10.1023/A:1008653328901

Baskerville, R. L., & Wood-Harper, A. T. (1996). A critical perspective on action research as a method for information systems research. *Journal of Information Technology*, *11*(3), 235–246. doi:10.1080/026839696345289

Basu, S. (2004). E-government and developing countries: An overview. *International Review of Law Computers & Technology*, *18*(1), 109–132. doi:10.1080/13600860410001674779

Berry, M. J., & Linoff, G. (1997). *Data mining techniques: for marketing, sales, and customer support.* John Wiley & Sons, Inc.

Bose, R. (2004). Knowledge management metrics. *Industrial Management & Data Systems*, *104*(6), 457–468. doi:10.1108/02635570410543771

Bowman, B. J. (2002). Building knowledge management systems. *Information Systems Management*, *19*(3), 32–40. doi:10.1201/1078/43201.19.3.20020601/37168.5

Cooke, N. J. (1994). Varieties of knowledge elicitation techniques. *International Journal of Human-Computer Studies*, *41*(6), 801–849. doi:10.1006/ijhc.1994.1083

Darroch, J. (2005). Knowledge management, innovation and firm performance. *Journal of Knowledge Management*, *9*(3), 101–115. doi:10.1108/13673270510602809

Davenport, T. H., & Prusak, L. (1998). *Working knowledge: How organizations manage what they know*. Harvard Business Press.

De Jong, T., & Ferguson-Hessler, M. G. (1996). Types and qualities of knowledge. *Educational Psychologist*, *31*(2), 105–113. doi:10.120715326985ep3102_2

DeLong, D. W. (2004). *Lost knowledge: Confronting the threat of an aging workforce*. Oxford University Press. doi:10.1093/acprof:oso/9780195170979.001.0001

Eriksson, I., & Raven, A. (1996). *Gaining competitive advantage through shared knowledge creation: in search of a new design theory for strategic information systems*. Academic Press.

Ferdousi, Z., & Maeda, A. (2006, September). Anomaly detection using unsupervised profiling method in time series data. ADBIS Research Communications.

Freeze, R. D., & Kulkarni, U. (2007). Knowledge management capability: Defining knowledge assets. *Journal of Knowledge Management*, *11*(6), 94–109. doi:10.1108/13673270710832190

Gallupe, B. (2001). Knowledge management systems: Surveying the landscape. *International Journal of Management Reviews*, *3*(1), 61–77. doi:10.1111/1468-2370.00054

Goonatilake, L., Maizza-Neto, O., & Jayawardene, P. (2000). Enhancing enterprise competitiveness in developing countries through the promotion of management information and benchmarking tools. In *Proceedings of the 6th International Working Conference of IFIP WG 9.4: Information Flows, Local Improvisations and Work Practices*. STS Conferences.

Gregor, S., & Hevner, A. R. (2013). Positioning and presenting design science research for maximum impact. *Management Information Systems Quarterly*, *37*(2), 337–355. doi:10.25300/MISQ/2013/37.2.01

Hevner, A. R., March, S. T., Park, J., & Ram, S. (2004). Design science in information systems research. *Management Information Systems Quarterly*, *28*(1), 75–105. doi:10.2307/25148625

Hult, M., & Lennung, S. A. (1980). Towards a definition of action research: A note and bibliography. *Journal of Management Studies*, *17*(2), 241–250. doi:10.1111/j.1467-6486.1980.tb00087.x

Jennex, M. E., Smolnik, S., & Croasdell, D. T. (2009). Towards a consensus knowledge management success definition. *Vine*, *39*(2), 174–188. doi:10.1108/03055720910988878

Leonard, D., & Sensiper, S. (1998). The role of tacit knowledge in group innovation. *California Management Review*, *40*(3), 112–132. doi:10.2307/41165946

Maier, R. (2002). *Knowledge management systems: Information and communication technologies for knowledge management*. Berlin: Springer. doi:10.1007/978-3-662-04380-6

Mariscal, G., Marban, O., & Fernandez, C. (2010). A survey of data mining and knowledge discovery process models and methodologies. *The Knowledge Engineering Review*, *25*(02), 137–166. doi:10.1017/S0269888910000032

Nevo, D. (2003). *Developing effective knowledge management systems*. Academic Press.

O'Dell, C., & Grayson, C. J. (1998). If only we knew what we know: Identification and transfer of internal best practices. *California Management Review*, *40*(3), 154–174. doi:10.2307/41165948

Osei-Bryson, K. M., Mansingh, G., & Rao, L. (2014). Understanding and applying knowledge management and knowledge management systems in developing countries: Some conceptual foundations. In Knowledge Management for Development (pp. 1-15). Springer US.

Phua, C., Lee, V., Smith, K., & Gayler, R. (2010). *A comprehensive survey of data mining-based fraud detection research*. arXiv preprint arXiv:1009.6119

Pressman, R. S. (2005). *Software engineering: a practitioner's approach*. Palgrave Macmillan.

Rolland, C. (1998, June). A comprehensive view of process engineering. In *Advanced Information Systems Engineering* (pp. 1–24). Springer Berlin Heidelberg. doi:10.1007/BFb0054216

Rowley, J. E. (2007). The wisdom hierarchy: Representations of the DIKW hierarchy. *Journal of Information Science*, *33*(2), 163–180. doi:10.1177/0165551506070706

Rubenstein-Montano, B., Liebowitz, J., Buchwalter, J., McCaw, D., Newman, B., & Rebeck, K. (2001). Smartvision: A knowledge-management methodology. *Journal of Knowledge Management*, *5*(4), 300–310. doi:10.1108/13673270110411724

Ryan, S., Harden, G., Ibragimova, B., & Windsor, J. (2012). The business value of knowledge management. *AMCIS 2012 Proceedings*.

Sánchez, D., Vila, M. A., Cerda, L., & Serrano, J. M. (2009). Association rules applied to credit card fraud detection. *Expert Systems with Applications*, *36*(2), 3630–3640. doi:10.1016/j.eswa.2008.02.001

Schreiber, G. (2000). *Knowledge engineering and management: the CommonKADS methodology*. MIT Press.

Schreiber, G., Wielinga, B., de Hoog, R., Akkermans, H., & Van de Velde, W. (1994). CommonKADS: A comprehensive methodology for kbs development. *IEEE Expert*, *9*(6), 28–37. doi:10.1109/64.363263

Silva, L. O. (2002). Outsourcing as an improvisation: A case study in Latin America. *The Information Society*, *18*(2), 129–138. doi:10.1080/01972240290075057

Smith, T. A., Mills, A. M., & Dion, P. (2010). Linking business strategy and knowledge management capabilities for organizational effectiveness. *International Journal of Knowledge Management*, *6*(3), 22–43. doi:10.4018/jkm.2010070102

Susman, G. I., & Evered, R. D. (1978). An assessment of the scientific merits of action research. *Administrative Science Quarterly*, *23*(4), 582–603. doi:10.2307/2392581

Tagaris, A., Konnis, G., Benetou, X., Dimakopoulos, T., Kassis, K., Athanasiadis, N., & Koutsouris, D. (2009, November). Integrated web services platform for the facilitation of fraud detection in health care e-government services. In *Information Technology and Applications in Biomedicine, 2009. ITAB 2009. 9th International Conference on* (pp. 1-4). IEEE. 10.1109/ITAB.2009.5394355

Tauber, D., & Schwartz, D. G. (2011). *Integrating Knowledge Management with the Systems Analysis Process*. Academic Press.

Tyrrell, S. (2000). The many dimensions of the software process. *Crossroads*, *6*(4), 22–26. doi:10.1145/333424.333435

Wan, Z., Haggerty, N., & Wang, Y. (2015). Individual Level Knowledge Transfer in Virtual Settings: A Review and Synthesis. *International Journal of Knowledge Management*, *11*(2), 29–61. doi:10.4018/IJKM.2015040103

Watson, R. T. (2008). *Data management, databases and organizations*. John Wiley & Sons.

Weitzel, J. R., & Kerschberg, L. (1989). Developing knowledge-based systems: Reorganizing the system development life cycle. *Communications of the ACM*, *32*(4), 482–488. doi:10.1145/63334.63340

Chapter 10
Knowledge Management Strategy Implementation Through Knowledge Ambassadors

Annabeth Aagaard
Aarhus University, Denmark

ABSTRACT

The aim of this chapter is to explore and explain how knowledge management strategies are implemented effectively across divisions through the application of change agents in the shape of so-called "knowledge ambassadors." Knowledge ambassadors can play an active role in supporting knowledge sharing and effective integration of knowledge management strategies. The chapter is based on a three-year longitudinal participative case study of the implementation of a knowledge management strategy across an organization and its seven geographically dispersed divisions. The results of the study reveal different steps and key activities in effective implementation of a knowledge management strategy and how knowledge ambassadors play a vital role in effective integration of knowledge management tools, continuous support of knowledge sharing practices, and in developing a knowledge culture within the organization.

INTRODUCTION

Knowledge Management (KM) has become a key factor in the performance of organizations (Perez-Soltero et al., 2015) and several organizations have established some sort of knowledge management strategy in an attempt to structure and support knowledge sharing across the internal and external organization. Yet, in practice many companies fail in ensuring a continuous focus on knowledge sharing, the full integration of knowledge management strategies and on continuously developing a knowledge culture. Oftentimes the 'news value' of the knowledge management strategy and the knowledge management initiatives wear off, when operations and daily assignments take over the employees' focus.

DOI: 10.4018/978-1-5225-5427-1.ch010

The challenges of overcoming the obstacles to effective strategy integration and execution have been discussed theoretically and empirically by various authors (Giles, 1991; Pellegrinelli & Bowman, 1994; Al-Ghamdi, 1998; Hrebiniak, 2005, 2006; Kaplan & Norton, 2006; Wheelen & Hunger, 2006; Brenes, Mena & Molina, 2008; Crittenden & Crittenden, 2008; Cater & Pucko, 2010). According to Hrebiniak (2005) all these obstacles relate to one of four groups: the change management context, the organizational culture context, the organizational power structure context and the leadership context. Change agents have been suggested as one of the key tools in ensuring more efficient implementation of changes and in ensuring integration of strategies by a number of authors (Ginsberg & Abrahamson, 1991; Burgelman, 1991; Armenakisi & Fredenberger, 1995; Pitt, McAulay & Sims, 2002; Birkenshaw, Hamel & Mol, 2010). Yet, the majority of the research performed on change agents, focuses on leaders as change agents (Yukl, 2002; Lines, 2007; Westover, 2010) and external change agents such as consultants (Armenakisi & Fredenberger, 1995; Greenwood, Hinings & Suddaby, 2002) and not on employees, the internal change agents (Birkenshaw, Hamel & Mol, 2008; Shanker & Sayeed, 2012). Furthermore, change agents have not been studied within the knowledge management literature or in terms of knowledge management strategy implementation. Therefore, the concept of 'knowledge ambassadors' will be introduced and explored in this chapter. Knowledge ambassadors serve as change agents in implementing changes through knowledge management strategies integration and in sustaining the organization's continuous change and development towards effective knowledge sharing in a knowledge company. In the following a framework for this examination will be presented including a mapping of the key aspects of knowledge and knowledge management, knowledge management strategies and -implementation through the use of change agents in the form of knowledge ambassadors.

Identifying the Concepts of Knowledge and Knowledge Management

The nature of knowledge has been described as 'justified true belief' (Nonaka & Takeuchi, 1995) and as 'a fluid of framed experience, values, contextual information and expert insight that provides a framework for evaluating and incorporating new experiences and information' (Davenport & Prusak, 1998). Moreover, Bollinger and Smith (2001) describe knowledge as 'the understanding, awareness, or familiarity acquired through study, investigation, observation, or experience over the course of time'; that is, knowledge is an individual's interpretation of information based on personal experiences, skills and competencies. Managing knowledge well can develop new opportunities, creating value for customers, obtaining competitive advantages or improving performance (Lloria, 2008). According Dayan and Evans (2006) the activities of KM include knowledge capture, documentation, retrieval and reuse, creation, transfer and sharing of its knowledge assets integrated in its operational and business processes. Earl (2001) suggests that knowledge management can be viewed from seven dimensions with their focuses as follows: (1) system: technology; (2) cartographic: maps; (3) engineering: processes; (4) commercial: income; (5) organizational: networks; (6) spatial: space; and (7) strategic: mindset. Alavi and Leidner (1999, 2001) concluded in their study that the perceived organizational benefits of KM scheme could be shown in two primary dimensions: process improvement (e.g., enhancing communication, reducing problem-solving time, better serving the clients and providing accountability) and organizational outcomes (e.g., cost reduction, increased sales, personnel reduction and higher profitability).

Recent empirical effort has found support for the direct impact of knowledge management on company performance, yet we should take the positive results with some cautions (Yeoh & Roth, 1999; Decarolis & Dees, 1999; Tanriverdi &Venkatraman, 2005). A simplistic conclusion of these studies could mistakenly lead managers to think that the more learning the better and the more knowledge the better. However, knowledge management encompasses much more than technologies that facilitate knowledge storage and sharing. According to Alavi and Leidner (2001) people and the culture of the workplace constitute the key drivers that ultimately determine the success or failure of knowledge management. Rubenstein-Montano et al., (2001) also stress that emphasis on technologies forces a too narrow view of that may affect the growth and diffusion of knowledge management in a company. Thus, technical and organizational initiatives should be aligned and integrated to provide the optimal infrastructure supporting knowledge management implementation. As stressed by Zheng, Yang, and McLean (2010) knowledge management plays a potentially mediating role in connecting organizational context and strategy with organizational effectiveness. While an integrated infrastructure tends to enhance knowledge management capabilities to exploit and explore knowledge, it does not insure that a company is making the best investment of its knowledge or that a company is managing its knowledge in the right way (Zack, 1999a).

This raises the questions of how a company should determine which knowledge effort is appropriate, what determines the success of a KM activity and how it the success measured and evaluated. Defining when a knowledge management (KM) project or initiative is successful is difficult and several different KM success model can be applied (Jennex & Olfman, 2006). The issue of measuring the impact of knowledge management has been discussed in literature by several researchers (e.g. Choy, Yew & Lin, 2006; Kulkarni, Ravindran & Freeze, 2007; Jennex 2008, 2012). Jennex, Smolnik, and Croasdel (2012) emphasize that KM success is measured in four dimensions: impact on business processes, impact on knowledge management strategy, leadership/management support, and knowledge content. The results of the survey reveals that the more successful a knowledge management project/initiative is perceived to be the more of the tested measures will be observed and more likely the organization sees success in all four dimensions. This chapter applies definition of knowledge management success by Jennex, Smolnik, Croasdell (2013):

KM success is a multidimensional concept. It is defined by capturing the right knowledge, getting the right knowledge to the right user, and using this knowledge to improve organizational and/or individual performance. KM success is measured by means of the dimensions: impact on business processes, impact on KM strategy, leadership/management support, and knowledge content (Jennex, Smolnik, Croasdell, 2009, 2013, p. 34).

Mapping and Implementing Knowledge Management Strategy

The main context for determining the activities of knowledge management in a company is it's competitive strategy (Earl, 2001; Desouza & Awazu, 2005). The competitive strategy of an organization provides the 'blueprint' for predicting the need of future knowledge resources and determines the activities of knowledge management in effective support of the company's mission and objective. However, the link between business strategy and knowledge management, although often talked about, has been widely ignored in practice (Davenport et al., 1998; Ruggles, 1998). And this is where knowledge strategies come into the picture as the missing link. Even though knowledge may be considered the key strategic resource to gain and sustain competitive advantage, an organization must still articulate its strategy intent among its competitors and identify the knowledge strategy to execute its intended strategy (Zack, 1999a; Teece, 2000; Dunford, 2000). Several studies have examined the relationship of the strategy in managing knowledge and company performance (Gold et al., 2001; Lee & Choi, 2003; Darroch, 2005;

Tanriverdi, 2005; Kim et al., 2014; Wang et al., 2016). In a survey conducted of 301 organizations, the results suggest that knowledge management fully mediates the impact of organizational culture on organizational effectiveness, and partially mediates the impact of organizational structure and strategy on organizational effectiveness (Zheng, Yang, and McLean, 2010). Another study by Imran et al. (2016) of 206 responses obtained from the banking sector of Pakistan stated that KM strategies have an indirect effect on successful organizational change through organizational learning and change readiness.

In this chapter a knowledge management strategy refers to the development and application of knowledge assets through the capabilities of knowledge processes, including knowledge creation, sharing, application, and integration and may be defined as the set of strategic choices that direct and shape the organization's learning process and subsequently determine the company's knowledge resources (Zack, 1999a). Whereas, knowledge management refers to the portfolio of procedures and techniques used to get the most from a company's knowledge assets (Teece, 2000). While knowledge management strategies deal with structural and technical management issues, knowledge strategies deal with business outcomes and support for competitive advantage. The knowledge management strategy taxonomy revolves around the management of knowledge rather than its application for competitive purposes (Denford & Chan, 2011) and in the present study the focus is on the development and integration of a knowledge management strategy. Bettiol et al. (2012) mention that KM strategies serves as a map to guide organizations towards more efficiency and to essentially improve all KM processes such as creation, codification, transfer and reuse. An appropriate KM strategy enables a firm to create, acquire, access, and leverage knowledge in a timely manner, thereby resulting in better performance (Gray & Meister, 2004). KM strategy can be divided into two dimensions, internal- and external-oriented strategy (Hughes et al., 2007). The internal-oriented strategy focuses on creating, sharing and accumulating knowledge inside an organization while the external-oriented strategy tends to learn, imitate and transfer knowledge beyond the boundary of a firm (Choi et al., 2008; Zack, 1999).

Previous research on knowledge management implementation has primarily focused on technological aspect in supporting the management process of knowledge exploitation and exploration. Insufficient attention has been paid to the organizational and people aspects, such as culture, commitment, and reward system. Swan et al. (2000) argue that an information technology it-based approach to knowledge management potentially attempts to codify tacit knowledge, but this neglects the fact that some tacit knowledge cannot be easily expressed and is probably difficult to codify. Therefore, organizational, and people management practices need to be considered in facilitating the effective use of knowledge management technology (Zack, 1999b). Several researchers have elaborated the basic concept to discuss the possible implementation approaches for practice. E.g. Hansen et al. (1999) studied the KM practices of companies in several industries and found that the companies did not take a uniform approach to manage their knowledge. Bixler (2002) argued that a key perspective to developing an enterprise knowledge strategy lies in determining the right tools to support knowledge management. Masa'deh and Shannak (2012) emphasize that KM strategy plays an important role in supporting the creation, transfer and application of knowledge inside and outside the workplace.

The Key Roles of Knowledge Ambassadors in KM Strategy Implementation

Knowledge ambassadors in shape of KM change agents are examined in this chapter as a key tool to support KM. Interest in the theory of planned change has created the new professional role of the "change agent." According to Pitt, McAulay and Sims (2002) effective strategic change depends on an evolving, relevant and timely agenda and on the contributions made by internal change agents who adopt 'agenda playmaking' roles and act as ambassadors for change. Management can legitimize these playmaking activities by promoting organizational norms and rewards whose aim is to align personal and organizational interests and to encourage diverse, yet inclusive contributions (Pitt, McAulay & Sims, 2002). Westover (2010) emphasize that the change agent's job is to develop a climate for planned change by overcoming resistances and rallying forces for positive growth.

Birkenshaw, Hamel and Mol (2008) define two types of change agents: internal and external change agents. The internal change agents are employees of the focal organization and external change agents are not, which, in turn, implies that internal change agents will typically have superior knowledge and networks inside the organization and greater accountability for delivering results than their external counterparts (p. 840). The present study focuses on the role of internal change agents (internal knowledge ambassadors) in effective knowledge sharing and the integration of a knowledge management strategy. The internal knowledge ambassador acts as a change agent but may also be described using the literature and definitions of champions. Although the term "champion", first suggested by Schon (1963), is now used widely in academic literature (Higgins, 1990), there are ambiguities in both scope and coverage in this usage (Howell and Higgins, 1990; Markham et al., 1998). While Tushman and Nadler (1986) describe champions as individuals who emerge informally in organizations, and make a contribution to innovations by actively and enthusiastically promoting their progress over time, Howell and Higgins (1990) refer to them as individuals who are innovative, who are prone to take risks, and who exhibit a transformational leadership style. In contrast, Chakrabarti (1974) identifies the characteristics of champions as technical competence, knowledge about the company, knowledge about the market, drive and aggressiveness, and political astuteness.

In the study by Gupta, Cadeaux and Dubelaar (2006) the concept of 'implementation champions' is introduced, and they are described to display most of the following characteristics: "(a) they are intimately familiar with the aims of the new venture; (b) they possess the required skill and experience with the technology; (c) they distill creative ideas from information sources and then enthusiastically promote them within the organization; (d) they are able to coordinate the planning and implementing activities for a business model using the new technology; and (e) they are able to maintain discipline within their own department, within other departments, and with third parties who are essential in providing resources to support the launch of the new venture (p. 551)". This definition in many ways characterizes the roles, characteristics and assignments of the 'knowledge ambassador' in implementation KM strategies.

Apart from supporting the implementation of the KM strategy, the knowledge ambassadors also play a pivotal role in facilitating knowledge sharing across the organization. In explaining these assignments I apply Nonaka's knowledge creation theory (Nonaka, 1991, 1994; Nonaka & Takeuchi, 1995; Nonaka & Toyama, 2003). Most knowledge exists in the form of experiences that people have made, or, in other words, as tacit knowledge (Polanyi, 1966), which is characterized by the difficulty to express in a spoken or written form and in the transfer of this knowledge to other people. Yet, a successful creation and sharing of knowledge requires a transfer of this knowledge and to describe how a transfer of tacit knowledge can occur, Nonaka refers to these four processes: socialization, externalization, combination, and internalization and refers to them as the knowledge spiral. The knowledge spiral process has received critique by e.g. Gourlay, 2006, but will be applied here in explaining the roles and assignments of the knowledge ambassadors in supporting knowledge sharing and knowledge management in the integration of a knowledge management strategy.

In the process of socialization people share their experience and in this process tacit knowledge can only be transferred from one individual to the other through immediate experience, observation and imitation. The knowledge acquired by the observer as a result of this transfer is tacit knowledge as well, which is why a process of externalization is necessary. Externalization constitutes the articulation of tacit knowledge and the translation of tacit knowledge into explicit knowledge. When knowledge has become explicit, it can be combined with other knowledge in order to develop new knowledge within the organization. This leads to the process of combination that refers to the re-arrangement of explicit knowledge though e.g. new structures and which is facilitated by interpersonal knowledge sharing. Finally, explicit knowledge will, once again, be converted into tacit knowledge through a process of *internalization*, which is an individual process. Internalization may often occur as a result of 'learning by doing' (Kimmerle, Cress and Held, 2010).

A Participative Study of KM Strategy Implementation

A participative case study was carried out in the exploration of KM strategy implementation through knowledge ambassadors. This methodology is described as a mode of case study research that involves the participant in all the phases of the research process. Reilly (2010, p. 658) explains participative research "as research that is fully collaborative and emphasizes deliberate participation, contribution, empowerment and emancipation of all relevant parties in actively examining some issue, which participants experience as problematic. An outcome of the research process is to change and improve the situation under investigation". However, participative research and action research do share some similarities, which oftentimes cause confusion. E.g. both participative research and action research work through a cyclical process of consciously and deliberately: (a) planning, (b) taking action, (c) evaluating that action and (d) leading to further planning (Moss, 2007). Baskerville (1997) underlines that the prospect of a participative case study being confused with action research is a simple error in properly describing the research method and stresses: "action research is more rigorous, more difficult, and longer in duration than participative case studies. For example, an action research report that fails to discuss (at least in an appendix) the client-system infrastructure, the collaborative nature of the research team, the iterative theory development (especially theory failure and modification) may be open to the challenge that the method has been erroneously described, and is indeed a participative case study" (Baskerville, 1997, p. 42).

The participative case study approach has been selected in this study to provide real-life knowledge of the actual process and experienced effects of enhancing knowledge sharing in practice through the development and integration of a knowledge management strategy and through the usage of knowledge ambassadors. The summarize is written on the basis of a three-year longitudinal case study on how to support knowledge sharing through the implementation of a knowledge management strategy and the application of knowledge ambassadors, which was conducted across seven divisions of the Danish Road Directorate. The public case company, the Danish Road Directorate, was selected as the main case company as it represents the challenges of many established large sized public organizations in improving knowledge sharing across geographically dispersed divisions. The Danish Road Directorate is responsible for the state-owned roads in Denmark, which consist of highways, a number of main roads and many of its bridges. The present study has focused on the construction division that is represented in seven different locations in Denmark. The division does not conduct the construction assignments, which are tendered to private suppliers, but attend to the planning of the constructions.

The practical process of carrying out the participative case study of this chapter is captured in the process model below, which is inspired by Moss et al. (2007). The process model is applied as a guide and overview of the process in exploring the chapter's research question and depicts the four main activities: diagnosing and reviewing KM practice, designing the KM strategy, integrating the KM strategy and evaluating the process and effects/results, which were carried out as part of the participative case study. The arrows between the four activities including the text, "L&R" represent the learning and reflections obtained by the participative researcher during and after each activity. The participative study process was performed through the four steps presented in model 1, and the findings were a result of carrying out these process steps and the learning and reflections made during and after each step, which is why the findings are presented as part of these four strategy process steps.

Identifying KM Issues and KM Practice

The first group of activities constitutes the diagnosis of the problem areas to address in the participative study of the case company. As in most organizations the management group would normally have defined the key focus areas of the strategy themselves. In an attempt to avoid the typical challenges of management not having full insight of the day-to-day knowledge sharing challenges a different diagnosis approach was established. First, the company's knowledge sharing activity was monitored through observations. Then the derived findings of the observations were applied in formulating an interview guide

Figure 1. The participative research process
Inspired by Moss (2007: 297).

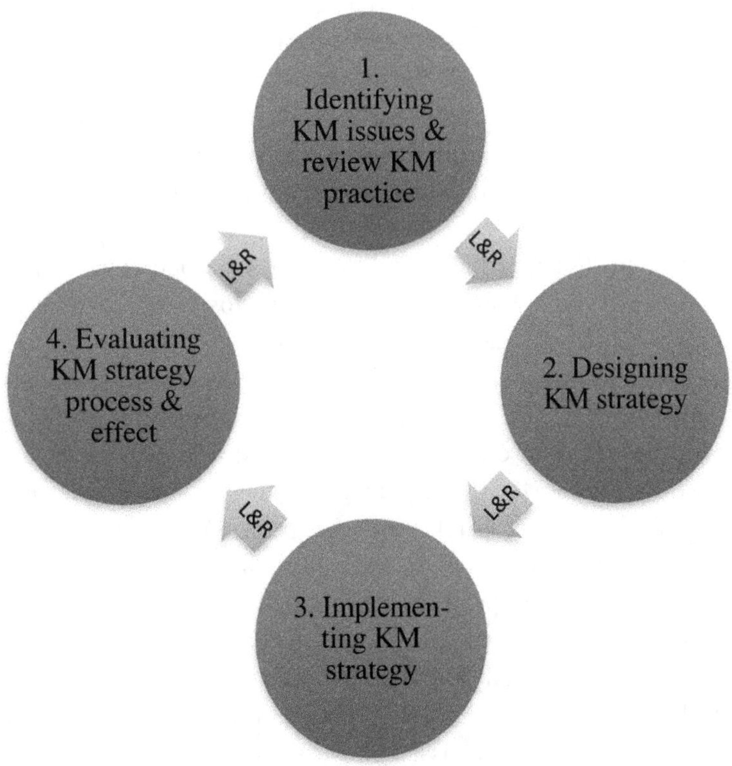

to be utilized in identification of the key issues of knowledge management across the different divisions and organizational levels. Finally, seven semi-structured focus group interview sessions were conducted with representatives of the organization's seven divisions and different employee groups (administrative personnel, technicians, specialists, project managers, department managers) with an even mix of male and female informants when possible and of employees of different age groups and with low to high levels of seniority (from ½ - 25 years of company experience). These criteria were included, as the authors had an assumption about potential differences in knowledge sharing activity among employees on different organizational levels and locations, gender and seniority.

The findings revealed this assumption to be correct to some extent. It appeared that the focus group informants from the seven different divisions shared a number of generic knowledge management concerns (e.g. a need for more emphasis on developing a corporate knowledge culture, more storytelling about great knowledge sharing cross-company, an improvement of the existing knowledge management system, more focus on knowledge leadership than just knowledge management). Yet, the informants from each division also presented specific knowledge management challenges that originated from the specific operations and collaborations that each division was engaged in (e.g. regarding specific approval processes, necessary communications and placement of responsibility when going from one project stage to the next etc.). There were no significant differences in the answers between female and male informant. Yet, there appeared some differences in focus areas addressed by informant of shorter versus longer seniority, where the former emphasized education, mentoring, development opportunities and company communication more than the latter group of informants, which is not that surprising as newcomers are seeking more knowledge and training to be able to perform the tasks represented in their job description.

The derived data from the focus group interviews were categorized and analyzed through the application of the contemporary method, meaning categorization (Kvale & Brinkmann, 2008: 203), where similar statements were grouped in categories of the key development areas as presented by the informants. Seven meaning categories were derived from this process: 1. Knowledge culture, 2. Knowledge leadership, 3. Knowledge strategy, 4. Knowledge management systems, 5. Knowledge sharing between employees, 6. Knowledge sharing with external partners, 7. Communications about knowledge sharing. These meaning categories were explained in a short report presented to management and employees, and later applied in the development of the company's knowledge management strategy targeting the identified development areas.

Key findings from participative activities step 1 revealed the importance of inviting the employees into the identification phase of strategy making. First, the employees in the focus interviews stated that they felt much more involved in the strategy process, where one informant explained it this way: "we are normally never involved in strategy making only in the implementation, which can be confusing as management does not always address the day-to-day problems that we as employees experience". Second, the management group admitted that some of the issues raised were new to them and would therefore not have been integrated in the knowledge management strategy, had the management group designed it themselves. The lack of management insights into the specific knowledge problems may often result in too generic knowledge management strategies and actions, and with this process, the actions planned were much more specific targeting the specific knowledge challenges which differed between the organization's seven divisions each division and each employee group, which one of the (project manager) informants commented on: "I can actually tell from the report that management are made aware of the knowledge sharing challenges that I experience in my work, which is comforting to know". The people that where later selected as knowledge ambassadors participated in the seven focus groups carried out and they distinguished themselves by playing an active role in identifying the key knowledge management issue to be addressed in the company's knowledge management strategy.

Planning Actions: Designing a KM Strategy

The second group of activities includes the mapping of the key knowledge focus areas and the development of a knowledge management strategy in facilitation of more efficient knowledge management across the seven divisions of the case company. The process of planning actions constituted a management workshop in which the top management group discussed and prioritized the identified knowledge management issues identified through the observations and the meaning categories (Kvale & Brinkman, 2008) derived from the focus group interviews performed in participative activities step 1. This process was facilitated by the participative researcher where the seven meaning categories were reduced and combined into four strategic focus areas: 1) Knowledge culture, 2) Knowledge management, 3) Internal knowledge handling and 4) External knowledge sharing, including specific corporate activities and activities that each division should work with.

The learnings from this workshop revealed a necessity for identification of knowledge types and a formulation of the company's core knowledge and competences for the organization to know which knowledge to prioritize in the company's knowledge management activities. Three knowledge types were identified: written knowledge to be codified and stored in the company's extensive knowledge management system, verbal/tacit knowledge to be share through the organization's formal specialist networks and finally competences, which cannot be transferred to others directly only indirectly through mentoring and improved overviews of where the company's specialist competences are located. This joint definition was important for the entire process and 'knowledge' had become a 'blurry' concept.

The mapping of core knowledge and competences became a much more complex and political assignment than first assumed, as management with this process wanted to address the issue, that not all knowledge should be 'core' to the company or in-house. It took ten months to get an elaborate mapping of the key knowledge categories representing knowledge necessary to perform all the operations and tasks carried out within the company. And with this mapping management decided to outsource a number of knowledge areas, where the general knowledge of the area was kept in-house, but where the specialist knowledge was sought among existing and new external partners. The process was both fruitful and painful for the organization, but necessary, as budget cuts and higher delivery requirements from the Danish State had forced the organization to focus on core knowledge areas and management to prioritize these areas with resources to avoid spreading resources too thin.

The corporate knowledge management strategy was presented on a joint knowledge management workshop across the organization, where each focus area and the related activities were discussed and received input from all the representatives of the seven divisions attending the workshop. The feedback was collected and included and a revised corporate knowledge management strategy was mailed out to each division after the workshop. Hereafter each of the seven divisional managers was requested to plan and carry out their own workshops with the employees of their respective divisions in identification of the concrete activities to be carried out in support of knowledge sharing and knowledge management in the specific divisions and its departments. The managers were asked to consult their ideas for the workshop program and the knowledge activities with key knowledge personnel as a way of 'testing' the activities among representatives of the daily working routines. In this context, the key knowledge personnel is represented by the knowledge ambassadors and key specialists.

Taking Action: Implementing KM Strategy

The third group of activities included the implementation of the corporate knowledge management strategy throughout the seven divisions of the organization. This process included the divisional workshops mentioned above and initiation of corporate cross-company activities. Several activities were set in motion almost immediately by the top management group as the identified issues were either pressing or problem areas, which had needed attention for some time. This observation revealed that a knowledge management strategy implementation process may also provide the 'umbrella' and the process for starting corrections and elimination of existing problems, both relating to processes, organizational structures, management etc., which have not found the right opportunity or frame for execution.

The implementation of the knowledge management strategy in practice were carried out by the divisions themselves, but observed by the researcher and supervised to some extent. Yet, it was decided by top management that the divisions should carry out this process themselves to ensure commitment, empowerment and ownership. As a new initiative in ensuring effective strategy implementation the researcher introduced the management group to the concept of 'knowledge ambassadors' in the strategy integration process, which was accepted by management. Thereafter a rather complex process took place of defining profiles, roles, assignments of the knowledge ambassadors and actually assigning the proper persons to the assignments. A key learning from this process and an observation made revealed the necessity of management's definition of these roles and assignments, as the ambassadors later became almost sparring partners to many managers in relations to knowledge management issues in their units. Thus knowing the boundaries of responsibility between manager and ambassador became an important factor in the successful collaboration between managers vs. knowledge managers.

Seven knowledge ambassadors were selected and assigned to the task of supporting an effective integration of the knowledge management strategy by help communicating the strategy in the different units, assisting colleagues with knowledge management related issues (e.g. how to apply the updated/new knowledge management system, how to find the proper knowledge network or specialist, and in help guiding knowledge and collaborations in ways that enhanced efficiency and quality etc.). These knowledge ambassadors had been detected during the identification and strategy making process as key knowledge personnel taking an active role in the process. They were characterized by having a large network across disciplines, high people and communication skills, and a medium level of seniority. In addition, they were acknowledged for their expertise and were well known across the organization due to their involvement in various projects across the divisions of the organization.

In the strategy implementation process they functioned as day-to-day change agents, helping the execution of the changes, which the knowledge management strategy implementation required. In addition, they provided very valuable information about status quo and actually knowledge issues etc. Furthermore, they appeared to help stimulate the knowledge spiral (Nonaka, 1991) by enhancing the level of particularly socialisation, externalization and combination of knowledge through how they shared knowledge and communicated about new knowledge (storytelling) and who possessed specific knowledge and core competences and through the facilitation of forums for dialogues and informal networks.

Evaluating Actions: Evaluating KM Strategy Process and Effects

The fourth group of activities, evaluation of the process and effects, constitutes a collection and analysis of the process and the monitored effects of the participative study. The evaluation data were collected through focus group interviews sessions with 3-4 representatives of each of the seven divisions. One key finding revealed the effect of the knowledge ambassadors in the evaluation process. On the two first focus group evaluation interviews the assigned knowledge ambassadors could not attend. This resulted in a rather poor dialogues and almost one-way 'conversions'. Based on the two interviews, it was decided that the next five evaluation focus group interviews were only to be carried out if the knowledge ambassador could attend. The effect of their presence was a much more vivid and open dialogue and discussion of the entire process and the effects and the issues of the knowledge strategy process as experienced personally by each of the focus group informants. It appeared as though the knowledge ambassador acted as a mediator and a team facilitator in generating an open-hearted dialogue.

The findings from the evaluation process revealed that "the key roles that the knowledge ambassadors had played in making the knowledge management strategy come alive" as stated by one informant. In all the seven focus group interviews 'the knowledge ambassadors' was mentioned as the key success factor of the strategy integration. Another informant stated: "without the knowledge ambassadors, there would have been no real strategy integration". In addition, the informants also stressed the need for a continuous development process after the knowledge strategy had been implemented. In response to these findings management decided to continue the operations of the knowledge managers and established a so-called "Knowledge ambassador team" including the appointed knowledge ambassadors from the seven divisions. The knowledge ambassador team was appointed the responsibility and given the resources to carry on the continuous development activities in support of corporate and divisional knowledge management across the seven divisions after the knowledge management strategy implementation process. This was done in an attempt to keep momentum and the focus on effective knowledge sharing as well as knowledge development across the organization.

CONCLUSION AND MANAGERIAL IMPLICATIONS

This chapter's topic of knowledge ambassadors in KM strategy implementation is explored through a longitudinal case study. However, performing case studies over longer periods of time entail a common issue related to the challenges of studying a moving target. Time affects the study at hand through the many internal and external factors influencing the object of study. The issue of time is particularly pervasive in participative and action oriented studies, as the influences of time and external factors on the obtained effects of the actions are difficult to identify and isolate. Isolation is not an option as we are dealing with reality and a study of reality. Abrupt events or occurrences may change behaviors and have an effect on the outcome and your findings as a researcher. This issue was witnessed in practice during the research study as the organization underwent a radical organizational change and an increase in employee turnover as result of this. These changes altered the focus and motivation of the participants, which further influenced the initiatives and the results of the planned intervention activities.

Yet, the study findings revealed that the implementation process of the corporate KM strategy and the divisional KM activities indicate that the continuous communication of progress, results and learning between the divisions and top management is an essential element in enhancing the effects, as knowledge and best practices were shared between the divisions. A number of informants in the evaluation had stressed that lack of communication of the progress of the process had given them the impression that the process was not successful. Continuous communication of progress is therefore essential in building the organization's experience of success. Furthermore inviting divisions for joint presentations and discussions of actions and results, where the divisions can meet and collaborate on similar activities to enhance cross functional and divisional knowledge sharing and collaboration is critical and should have been provided more often during the participative research study. Some division arranged for these collaborative arrangements themselves and carried out joint activities with other divisions with challenges similar to their own. Continuously monitoring and storytelling across the organization also proved critical as a way of keeping focus and in attaining positive effects of the development process.

Another finding from the process, which may not be that surprising, but still central in a KM strategy implementation study, is that a strategy implementation process is never better than the manager who performs it. In the study it became apparent that the specific leadership and process skills of the individual manager had a direct effect on the success of the process and the evaluated effects. The way to counteract this issue and a suggestion for improvements would be to provide more guidelines, strategy process training and tool boxes and for all managers, as it appeared that the less skilled managers did not readily acknowledge their own weaknesses. However, tools and training can only do so much, as personal motivation and leadership skills differ, so variations in results may be expected.

It was also evident in the study that managers act as role models in the knowledge management strategy process and their enthusiasm affects the behaviors of their employees. Thus, the leaders who encouraged employee ownership and participated actively in development activities and ensured the open dialogue and knowledge sharing of the process and the obtained effects, also had the most passionate and participating employees, who gladly spend extra hours and made time for knowledge sharing in their daily work. Thus, making sure that all managers involved are motivated, feel committed and understand the concepts and basics of the organization's knowledge management (e.g. the organization's definitions of knowledge management and core knowledge and competences and the knowledge sharing systems and tools etc.) is essential. Hence, a joint training program for all managers could potentially have improved the level of commitment, the quality and the results of the knowledge management and knowledge sharing actions.

The overall conclusion of study and this chapter is that it is possible to actively improve a company's knowledge sharing and knowledge management strategy integration through the application of knowledge ambassadors. The key roles of the knowledge ambassadors in effective strategy implementation may be summarized in four key role activities as addressed in the findings:

1. The knowledge ambassadors have the insights into the daily working routines and that provides them with the ability to identify which daily knowledge problems to address in ensuring effective knowledge sharing across the company.
2. The knowledge ambassadors act as sparring partners to the divisional managers in designing and carrying out strategy workshops and in ensuring employee involvement as they have networks across different disciplines of the company.
3. The knowledge ambassadors have the relations and people skills to ensure effective integration of the knowledge strategy through employee involvement and open dialogues with colleagues.
4. The knowledge ambassadors have the role of mediators and have the respect and trust of colleagues to ensure open dialogues and evaluations of strategy implementation processes and of the attained results.

The case company had not previously worked with knowledge management strategy implementation. Yet, several factors affected the strategy outcome in a positive way. Among other the high involvement from management and the CEO's high priority of the project and the timing with several pressing organizational questions and developments, which fit the project perfectly. The exact same benefits and advantages from knowledge management strategy implementation may therefore not be expected for all companies applying the same method, although this research study present a number of key considerations to be made in planning and carrying out an effective strategy development and implementation process.

The implications of the study suggest knowledge ambassadors as key facilitators of successful KM strategy implementation. In practice this implies that companies should establish more formal teams of employees (ambassadors) in securing valid employee involvement and effective integration of KM strategies. Furthermore the findings indicate that the KM ambassador teams should not be dissolved after the KM strategy has been implemented, but should be continued to ensure the long-term integration, application and continuous development of knowledge management practices. Based on the findings and the presented key learnings from this four-phase research process the following managerial implications for support of effective KM strategy implementation are suggested:

- Involve employees from all organizational levels and subdivisions to ensure that the knowledge management strategy addresses actual knowledge challenges and involves and empowers employees in identifying the real issues.
- Establish a joint and clear definition of what knowledge and knowledge management is within the organization and which core knowledge areas and core competences the organization wants to prioritize and focus their resources on to avoid spreading knowledge management resources too thin.
- Integrate knowledge ambassadors as part of the knowledge strategy process to ensure that implementation and integration is actively and continuously supported be a selected group of people with a main focus on knowledge strategy integration and on facilitating knowledge sharing and knowledge management on a daily basis
- Let the knowledge strategy facilitate concrete opportunities and actual forums for cross divisional and cross functional discussions, collaborations and knowledge sharing – as it is not a lack of interest, but lack of time and opportunity that holds back the daily focus and support of knowledge sharing activity.
- Communicate status of strategy implementation continuously and preferably through storytelling with actual stories of how the strategic initiatives have helped improve specific areas, tasks and results within the daily routines of the organization.
- Set up a task force of employee representatives in collaboration with managers after the strategy implementation process to ensure the continuous development, monitoring, communication and evaluation of knowledge management activities and effects - and as a way of keeping up steam.

The chapter is based on a single case study of a public organization. An interesting further research would be to conduct a cross-industrial study to reveal the potential differences and similarities across industries (public and private) in the ways knowledge management strategies are designed and implemented. Another interesting further research suggestion would be to carry out a longitudinal study of the qualitative and quantitative effects from integrating knowledge ambassadors in support knowledge management. This way the long-term effects could be evaluated of implementing knowledge ambassadors as part of knowledge management strategies and in the daily knowledge sharing practice across organizations and their geographically dispersed divisions.

REFERENCES

Al-Ghamdi, S. M. (1998). Obstacles to successful implementation of strategic decisions: The British experience. European Business Review, 98(6), 322–327. doi:10.1108/09555349810241590 doi:10.1108/09555349810241590

Alavi, M., & Leidner, D. E. (1999). Knowledge management systems: Issues, challenges, and benefits. Communications of the AIS, 1(7), 1–37.

Alavi, M., & Leidner, D. E. (2001). Review: knowledge management and knowledge management systems: conceptual foundations and research issues. Management Information Systems Quarterly, 25(1), 107–146. doi:10.2307/3250961 doi:10.2307/3250961

Baskerville, R. L. (1997). Distinguishing action research from participative case studies. Journal of Systems and Information Technology, 1(1), 24–43. doi:10.1108/13287269780000733 doi:10.1108/13287269780000733

Bettiol, M., Di Maria, E., & Grandinetti, R. (2012). Codification and Creativity: Knowledge Management Strategies in KIBS. Journal of Knowledge Management, 16(4), 550–562. doi:10.1108/13673271211246130 doi:10.1108/13673271211246130

Birkenshaw, J., Hamel, G., & Mol, M. J. (2008). Management Innovation. Academy of Management Review, 33(4), 825–845. doi:10.5465/AMR.2008.34421969 doi:10.5465/AMR.2008.34421969

Bixler, C. (2002). Codification, personalization, and integration. KMWorld, 11(3), 18.

Bollinger, A. S., & Smith, R. D. (2001). Managing organizational knowledge as a strategic asset. Journal of Knowledge Management, 5(1), 1–8. doi:10.1108/13673270110384365 doi:10.1108/13673270110384365

Brenes, E. R., Mena, M., & Molina, G. E. (2008). Key success factors for strategy implementation in Latin America. Journal of Business Research, 61(6), 590–598. doi:10.1016/j.jbusres.2007.06.033 doi:10.1016/j.jbusres.2007.06.033

Burgelman, R. A. (1991). Intraorganizational ecology of strategy making and organizational adaptation: Theory and field research. Organization Science, 2(3), 239–262. doi:10.1287/orsc.2.3.239 doi:10.1287/orsc.2.3.239

Cater, T., & Pucko, D. (2010). Factors of effective strategy implementation: Empirical evidence from Slovenian business practice. Journal for East European Management Studies, 3, 207–236.

Chakrabarti, A. K. (1974). The role of champion in product innovation. California Management Review, 17(Winter), 58–62. doi:10.2307/41164561 doi:10.2307/41164561

Choi, B., Poon, S. K., & Davis, J. G. (2008). Effects of knowledge management strategy on organizational performance: A complementarity theory-based approach. Omega International Journal of Management Science, 36(2), 235–251. doi:10.1016/j.omega.2006.06.007 doi:10.1016/j.omega.2006.06.007

Choy, C. S., Yew, K. V., & Lin, B. (2006). Criteria for measuring KM performance outcomes in organisations. Industrial Management & Data Systems, 106(7), 917–936. doi:10.1108/02635570610688850 doi:10.1108/02635570610688850

Crittenden, V. L., & Crittenden, W. F. (2008). Building a capable organization: The eight levers of strategy implementation. Business Horizons, 51(4), 301–309. doi:10.1016/j.bushor.2008.02.003 doi:10.1016/j.bushor.2008.02.003

Darroch, J. (2005). Knowledge management, innovation, and firm performance. Journal of Knowledge Management, 9(3), 101–115. doi:10.1108/13673270510602809 doi:10.1108/13673270510602809

Davenport, T. H., & Prusak, L. (1998). Working Knowledge: How Organizations Manage What They Know. Boston, MA: Harvard Business School Press.

Dayan, R., & Evans, S. (2006). KM your way to CMMI. Journal of Knowledge Management, 10(1), 69–80. doi:10.1108/13673270610650111 doi:10.1108/13673270610650111

Decarolis, D. M., & Deeds, D. (1999). The impact on stocks and flows of organizational knowledge on firm performance: An empirical investigation of the biotechnology industry. Strategic Management Journal, 20(10), 953–968. doi:10.1002/(SICI)1097-0266(199910)20:10<953::AID-SMJ59>3.0.CO;2-3 doi:10.1002/(SICI)1097-0266(199910)20:10<953::AID-SMJ593.0.CO;2-3

Denford, J. S., & Chan, Y. E. (2011). Knowledge strategy typologies: Defining dimensions and relationships. Knowledge Management Research and Practice, 9(2), 102–119. doi:10.1057/kmrp.2011.7 doi:10.1057/kmrp.2011.7

Desouza, K. C., & Awazu, Y. (2005). Maintaining knowledge management systems: A strategic imperative. Journal of the American Society for Information Science and Technology, 56(7), 765–768. doi:10.1002/asi.20149 doi:10.1002/asi.20149

Dunford, R. (2000). Key challenges in the search for the effective management of knowledge in management consulting firms. Journal of Knowledge Management, 4(4), 295–302. doi:10.1108/13673270010379849 doi:10.1108/13673270010379849

Earl, M. (2001). Knowledge management strategies: Towards a taxonomy. Journal of Management Information Systems, 18(1), 215–233. doi:10.1080/07421222.2001.11045670 doi:10.1080/07421222.2001.11045670

Giles, W. D. (1991). Making strategy work. Long Range Planning, 24(5), 75–91. doi:10.1016/0024-6301(91)90254-L doi:10.1016/0024-6301(91)90254-L

Ginsberg, A., & Abrahamson, E. (1991). Champion of change and strategic shifts: The role of internal and external change advocates. Journal of Management Studies, 28(2), 173–190. doi:10.1111/j.1467-6486.1991.tb00276.x doi:10.1111/j.1467-6486.1991.tb00276.x

Gold, A. H., Malhotra, A., & Segars, A. H. (2001). Knowledge management: An organizational capabilities perspective. Journal of Management Information Systems, 18(1), 185–214. doi:10.1080/07421222.2001.11045669 doi:10.1080/07421222.2001.11045669

Gray, P. H., & Meister, D. B. (2004). Knowledge sourcing effectiveness. Management Science, 50(6), 821–834. doi:10.1287/mnsc.1030.0192 doi:10.1287/mnsc.1030.0192

Greenwood, R., Hinings, C. R., & Suddaby, R. (2002). Theorizing change: The role of professional associations in the transformation of institutionalized fields. Academy of Management Journal, 45(1), 58–80. doi:10.2307/3069285 doi:10.2307/3069285

Griffin, A., & Page, A. L. (1993). Introduction to Action Research. Thousand Oaks, CA: Sage.

Gupta, S., Cadeaux, J., & Dubelaar, C. (2006). Uncovering multiple champion roles in implementing new-technology ventures. Journal of Business Research, 59(5), 549–563. doi:10.1016/j.jbusres.2006.01.003 doi:10.1016/j.jbusres.2006.01.003

Hansen, T. M., Nohria, N., & Tierney, T. (1999, March). What's your strategy for managing knowledge. Harvard Business Review, 106–116. PubMed PMID:10387767

Higgins, A. (1990). Champions of technological innovation. Administrative Science Quarterly, 35(2), 317–342. doi:10.2307/2393393 doi:10.2307/2393393

Howell, J. M., & Higgins, C. A. (1990). Champions of technological innovations. Administrative Science Quarterly, 35(2), 317–341. doi:10.2307/2393393 doi:10.2307/2393393

Hrebiniak, L. G. (2005). Making strategy work: Leading effective execution and change. Upper Saddle River, NJ: Pearson Education.

Hrebiniak, L. G. (2006). Obstacles to effective strategy implementation. Organizational Dynamics, 35(1), 12–31. doi:10.1016/j.orgdyn.2005.12.001 doi:10.1016/j.orgdyn.2005.12.001

Hughes, M., Hughes, P., & Morgan, R. E. (2007). Exploitative learning and entrepreneurial orientation alignment in emerging young firms: Implications for market and response performance. British Journal of Management, 18(4), 359–375. doi:10.1111/j.1467-8551.2007.00519.x doi:10.1111/j.1467-8551.2007.00519.x

Imran, M. K., Rehman, C. A., Aslam, U., & Bilal, A. R. (2016). What's organization knowledge management strategy for successful change implementation? Journal of Organizational Change Management, 29(7), 1097–1117. doi:10.1108/JOCM-07-2015-0130 doi:10.1108/JOCM-07-2015-0130

Jennex, M. E., & Olfman, L. (2006). A Model of Knowledge Management Success. International Journal of Knowledge Management, 2(3), 51–68. doi:10.4018/jkm.2006070104 doi:10.4018/jkm.2006070104

Jennex, M. E. (2008). Impacts from Using Knowledge: A Longitudinal Study from a Nuclear Power Plant. International Journal of Knowledge Management, 4(1), 51–64. doi:10.4018/jkm.2008010105 doi:10.4018/jkm.2008010105

Jennex, M. E., Smolnik, S., & Croasdell, D. (2012) Where to Look for Knowledge Management Success. In *45th Hawaii International Conference on System Sciences, HICSS45*. IEEE Computer Society. doi:10.1109/HICSS.2012.64110.1109/HICSS.2012.641

Jennex, M. E., Smolnik, S., & Croasdell, D. (2013). The Search for Knowledge Management Success. Global Perspectives on Engineering Management, 2(2), 34–44.

Jennex, M. E., Smolnik, S., & Croasdell, D. T. (2009). Towards a Consensus Knowledge Management Success Definition. VINE. The Journal of Information and Knowledge Management Systems, 39(2), 174–188.

Kaplan, R. S., & Norton, D. P. (2006). How to implement a new strategy without disrupting your organization. Harvard Business Review, 84(3), 100–109. PubMed PMID:16515159

Kim, T. H., Lee, J. N., Chun, J. U., & Benbasat, I. (2014). Understanding the effect of knowledge management strategies on knowledge management performance: A contingency perspective. Information & Management, 51(4), 398–416. doi:10.1016/j.im.2014.03.001 doi:10.1016/j.im.2014.03.001

Kimmerle, J., Cress, U., & Held, C. (2010). The interplay between individual and collective knowledge: Technologies for organizational learning and knowledge building. Knowledge Management Research and Practice, 8(1), 33–44. doi:10.1057/kmrp.2009.36 doi:10.1057/kmrp.2009.36

Koch, T., & Harrington, A. (1998). Reconceptualizing rigour: The case for reexivity. Journal of Advanced Nursing, 28(4), 882–890. doi:10.1046/j.1365-2648.1998.00725.x PubMed doi:10.1046/j.1365-2648.1998.00725.x PMID:9829678

Kulkarni, U. R., Ravindran, S., & Freeze, R. (2007). A Knowledge Management Success Model: Theoretical Development and Empirical Validation. Journal of Management Information Systems, 23(3), 309–347. doi:10.2753/MIS0742-1222230311 doi:10.2753/MIS0742-1222230311

Kvale, S., & Brinkmann, S. (2008). InterViews – Learning the Craft of Qualitative Research Interviewing (2nd ed.). SAGE Publications.

Lee, H., & Choi, B. (2003). Knowledge management enablers, processes, and organizational performance: An integrative view and empirical examination. Journal of Management Information Systems, 20(1), 179–228. doi:10.1080/07421222.2003.11045756 doi:10.1080/07421222.2003.11045756

Lines, R. (2007). Using Power to Install Strategy: The Relationships between Expert Power, Position Power, Influence Tactics and Implementation Success. Journal of Change Management, 7(2), 143–170. doi:10.1080/14697010701531657 doi:10.1080/14697010701531657

Lloria, M. B. (2008). A review of the main approaches to knowledge management. Knowledge Management Research and Practice, 6(1), 77–89. doi:10.1057/palgrave.kmrp.8500164 doi:10.1057/palgrave.kmrp.8500164

Markham, S. K. (1998). A longitudinal examination of how champions influence others to support their projects. Journal of Product Innovation Management, 15(6), 490–504. doi:10.1016/S0737-6782(98)00031-9 doi:10.1016/S0737-6782(98)00031-9

Masa'deh, R., & Shannak, R. O. (2012). Intermediary Effects of Knowledge Management Strategy and Learning Orientation on Strategic Alignment and Firm Performance. *Research Journal of International Studies*, 112-128.

Moss, Q.Z., Alho, J., & Alexander, K. (2007). Performance measurement action research. *Journal of Facilities Management, 5*(4), 290-300.

Nonaka, I., & Takeuchi, H. (1995). The Knowledge-creating Company: How Japanese Companies Create the Dynamics of Innovation. New York: Oxford University Press.

Pellegrinelli, S., & Bowman, C. (1994). Implementing strategy through projects. Long Range Planning, 27(4), 125–132. doi:10.1016/0024-6301(94)90062-0 doi:10.1016/0024-6301(94)90062-0

Perez-Soltero, A., Zavala-Guerrero, A. G., Barcelo-Valenzuela, M., Sanchez-Schmitz, G., & Meroño-Cerdan, A. L. (2015). A Methodology for the Development and Implementation of Knowledge Management Strategy in a Mexican SME Trading Company. Journal of Knowledge Management, 13(2), 25–44.

Pitt, M., McAulay, L., & Sims, D. (2002). Promoting strategic change: 'playmaker' roles in organizational agenda formation. Strategic Change, 11(3), 155–172. doi:10.1002/jsc.586 doi:10.1002/jsc.586

Reilly, R. C. (2010). Participatory case study. In Encyclopedia of case study research (pp. 658–660). Thousand Oaks, CA: Sage.

Rubenstein-Montano, B., Liebowitz, J., Buchwalter, J., McCaw, D., Newman, B., & Rebeck, K. (2001). A system thinking framework for knowledge management. Decision Support Systems, 31(1), 5–16. doi:10.1016/S0167-9236(00)00116-0 doi:10.1016/S0167-9236(00)00116-0

Ruggles, R. (1998). The state of the notion: Knowledge management in practice. California Management Review, 40(3), 80–89. doi:10.2307/41165944 doi:10.2307/41165944

Schon, D. A. (1963). Champions for radical new inventions. Harvard Business Review, (March–April): 77–86.

Shanker, M., & Sayeed, O. B. (2012). Role of Transformational Leaders as Change Agents: Leveraging Effects on Organizational Climate. Indian Journal of Industrial Relations, 47(3), 470–484.

Swan, J., Newell, S., & Robertson, M. (2000). Knowledge management when will people management enter the debate? In R.H. Sprague (Ed.), *Proceedings of the 33rd Hawaii International Conference on System Sciences* (pp. 3026–3035). Los Alamitos, CA: IEEE Computer Society Press. doi:10.1109/HICSS.2000.92670910.1109/HICSS.2000.926709

Tanriverdi, H., & Venkatraman, N. (2005). Knowledge relatedness and performance of multibusiness firms. Strategic Management Journal, 26(2), 97–119. doi:10.1002/smj.435 doi:10.1002mj.435

Tanriverdi, H. (2005). Information technology relatedness, knowledge management capability, and performance of multibusiness firms. Management Information Systems Quarterly, 29(2), 311–334. doi:10.2307/25148681 doi:10.2307/25148681

Teece, D. J. (2000). Strategies for managing knowledge assets: The role of firm structure and industrial context. Long Range Planning, 33(4), 35–54. doi:10.1016/S0024-6301(99)00117-X doi:10.1016/S0024-6301(99)00117-X

Tushman, M. L., & Nadler, D. (1986). Organizing for innovation. California Management Review, 28(3), 74–92. doi:10.2307/41165203 doi:10.2307/41165203

Wang, Z., Wang, N., Cao, J., & Ye, X. (2016). The impact of intellectual capital – knowledge management strategy fit on firm performance. Management Decision, 54(8), 1861–1885. doi:10.1108/MD-06-2015-0231 doi:10.1108/MD-06-2015-0231

Westover, J. H. (2010). Managing Organizational Change: Change Agent Strategies and Techniques to Successfully Managing the Dynamics of Stability and Change in Organizations. International Journal of Management and Innovation, 2(1), 45–50.

Wheelen, T. L., & Hunger, D. J. (2006). Concepts in strategic management and business policy. Upper Saddle River, NJ: Pearson – Prentice Hall.

Yeoh, P., & Roth, K. (1999). An empirical analysis of sustained advantage in the U.S. pharmaceutical industry: Impact of firm resources and capabilities. Strategic Management Journal, 20(7), 637–653. doi:10.1002/(SICI)1097-0266(199907)20:7<637::AID-SMJ42>3.0.CO;2-Z doi:10.1002/(SICI)1097-0266(199907)20:7<637::AID-SMJ423.0.CO;2-Z

Yukl, G. A. (2002). Leadership in Organizations (5th ed.). Upper Saddle River, NJ: Prentice Hall.

Zack, M. H. (1999a). Developing a knowledge strategy. California Management Review, 41(3), 125–145. doi:10.2307/41166000 doi:10.2307/41166000

Zack, M. H. (1999b). Managing codified knowledge. Sloan Management Review, 40(4), 45–58.

Zheng, W., Yang, B., & McLean, G. N. (2010). Linking organizational culture, structure, strategy, and organizational effectiveness: Mediating role of knowledge management. Journal of Business Research, 63(7), 763–771. doi:10.1016/j.jbusres.2009.06.005 doi:10.1016/j.jbusres.2009.06.005

Chapter 11
Measuring the Relationship Among Learning Enablers and IT Project Success

Donald Stuart McKay
Ashford University, USA

Timothy J. Ellis
Nova Southeastern University, USA

ABSTRACT

Knowledge enablers exist at the organizational and project levels. There is, however, no meaningful means to measure organizational or project knowledge sharing. The need to understand the elements that enable this flow of knowledge is dramatically evidenced in information technology organizations in which insufficient knowledge sharing leads to intellectual capital loss, rework, skills deterioration, and repeated mistakes that increase project costs or failures. The goal of this chapter is to describe the relationships between knowledge sharing processes at the organizational—organizational learning enablers (OLEs)—and project levels—project learning enablers (PLEs)—with project success variables (PSVs).

BACKGROUND

IT Projects continue to fail for many of the same reasons that they did 30 years ago (Cerpa & Verner, 2009). These failures lead to economic consequences. For example, companies spent millions of dollars on failed ERP implementations (Wu, Ong, Hsu, 2008). In the United States, the cost of failed IT projects amounts to $63 billion (McCafferty, 2010). Citing Panorama Consulting, Jeng and Dunk (2013) reported that 59% of ERP implementations cost more than anticipated. One interviewee, in Reich (2007) opined that project knowledge issues cost 10% of the total amount of a $60 million IT project. A failed hospital IT implementation cost $13 million and wasted six years of effort (Gauld, 2007). Customers concluded that too many of their IT projects fail (Ballou, Belardo, & Pazer, 2010)

DOI: 10.4018/978-1-5225-5427-1.ch011

Outside of the IT industry there are also project failures due to inadequate use of existing knowledge. Fortune 500 companies lose billions of dollars every year because their employees do not share knowledge (Marouf & Khalil, 2015). The International Association of Engineering Insurers reported that greater than €570 million in losses occurred in 18 tunneling projects globally between 1994 and 2005 (Cárdenas, Al-Jibouri, Halman, van de Linde, Kaalberg, 2014). A significant number of the tunneling failures happened because available knowledge was not used (Cárdenas, et al. 2014). In developing countries, a lack of project knowledge management impeded the sustainability of "reproductive health development" initiatives which, if corrected, could potentially improve the skills and knowledge of health care professionals and the quality of services to the nation (Dumrak, Baroudi, Hadjinicolaou, 2017).

The scope of the problem is significant. The magnitude of IT expenditures, lost benefits during the period of delay (Banker and Kemerer, 1992), forgone value when projects fail or under deliver, and employee impact combined suggest a large problem. Small and large organizations have failed to "effectively mine lessons learned" from projects leading to lost opportunities (Larson & Gray, 2014, p. 522). In a very meaningful sense, "these dismal findings can be traced to poor organizational learning mechanisms in software organizations" (Desouza, Dinsøyr, & Awazu, 2005, p. 204). Project teams are not learning lessons from other teams and this contributes to higher project costs (Hanisch, Lindner, & Mueller, & Wald 2009). Vital knowledge from prior projects is lost and not passed on to subsequent project teams (Jugdev, 2012). In short, failure to share knowledge is a key reason that IT projects fail (Nemani, 2012).

Knowledge frequently does not flow among project teams (Ajmal & Koskinen, 2008; Newell, Bresnen, Edelman, Scarbrough, & Swan 2006; Owen, Burstein, & Mitchell, 2004; Petter & Randolph, 2009; von Zedtwitz, 2003). Organizational failures to extract and apply project lessons learned are widespread (Newell & Edelman, 2008). Knowledge is neither captured nor shared with future project teams (Handzic & Durmic, 2015). Since knowledge exists at both the organizational and project levels, barriers to knowledge flow can exist at the organizational or project level (Ajmal & Koskinen, 2008; Crossnan, Lane, & White, 1999; Keegan & Turner, 2001; Nonaka, von Krogh, & Voelpel, 2006). Meaningful means to measure organizational or project level knowledge enablers do not appear to exist.

When knowledge does not flow among project teams within an IT organization resources are wasted. New project teams 'reinvent the wheel' as opposed to learning from prior projects (Newell, et al., (2006). Some projects repeat errors for years because learning from previous projects did not occur (Ajmal & Koskinen, 2008). Furthermore, companies experience waste in the form of lost potential to build employee skills (von Zedtwitz, 2003). Project implementation or process change management is adversely affected when knowledge transfer is ineffective (Alkhuraiji, Liu, Oderanti, Annansingh, & Pan, 2014). Thus, when project teams do not share lessons learned, poor solutions are duplicated, mistakes repeated, and knowledge of good procedures lost, leading to rework and missed opportunities (Owen, et al., 2004; Petter & Randolph, 2009).

THE UNDERLYING PROBLEM

IT leaders often do not make it a priority to share lessons learned among project teams. Managers may not understand the value derived from sharing lessons among project teams. For example, a knowledge manager facing a challenge of convincing senior management on the value of KM exclaimed: "My bosses want to see how KM implementation improves the ROI [return on investment] of the company, and how am I going to convince them since it is hard to measure KM using dollars and cents?" (Choy,

Yew, & Lin, 2006, p. 930). KM leaders need evidence that justifies KM investments (Jennex, Smolnic, & Croasdell, n.d.). In addition, IT staff resist efforts to capture and share lessons learned (Jugdev, 2012). In short, IT leaders fail to reuse "knowledge to improve organizational effectiveness by providing appropriate knowledge to those that need it when it is needed." (Jennex, Smolnik, & Croasdell, 2009, p. 185). Knowledge management success is not being achieved and IT leaders may not realize the cost of this over sight.

Although attempts have been made to solve the problem using IT, success is limited. For example, project managers at the National Aeronautics and Space Administration (NASA) did not use available technology to access lessons learned because many felt the system was too onerous (The United States General Accounting Office (GAO), 2002). Project managers did not even use knowledge management systems that were acknowledged to be easy to use and accessible, because of distractions from other work (Newell, et al., 2006).

Research has demonstrated the value of KM. Ganguly, Mostashari, and Mansouri (2011) theorized a proposed set of metrics to measure KM inputs and outputs related to organizational success. Some of the important output metrics included patent value / the amount spent on KM per year; new products; and operational cost/efficiency improvements. The value of the metrics were established using a hypothetical case. Nevertheless, organizations have made insufficient attempts to affect cultural changes and processes to share lessons between projects (Ajmal & Koskinen, 2008). Culture, as in the case of NASA, can impede sharing lessons between projects (GAO, 2002), and increased global competition may erode social bonds between people and organizations, making it difficult to learn lessons and benefit from them in the future (Keegan & Turner, 2001). Lacking culture and processes, IT solutions have been ineffective.

Organizations have not implemented the culture and processes for various reasons. Lack of time is a significant barrier because customers demand timely responses to their requests (Keegan & Turner, 2001). Organizations thus prioritized new business opportunities over learning lessons from previous projects. Lack of time is often a barrier to sharing lessons between projects because schedules and budgets do not make room for learning (Disterer, 2002). In addition, managers quickly redeploy teams to other IT projects around the globe (Disterer, 2002). Ironically, bureaucracy interferes with true learning exacerbating time constraints (von Zedtwitz, 2003). Managers may not set time aside to share lessons learned because the project end-dates must be met (Newell, et al. 2006).

It can be painful to review problems in a prior project. Quoting Boddie (1987) Disterer noted "the postmortem experience is much like a losing football team watching a game film. It's not comfortable, but if the team pays attention to its mistakes, it can perform better the next time it plays" (p. 516). Team members find it difficult to reflect (von Zedtwitz, 2003). Poor communication and a reluctance to blame others also impede learning lessons from projects (von Zedtwitz, 2003). Knowledge sharing barriers may fall into three major categories: cultural issues; perceptions and fear; and structural barriers (Barratt-Pugh, Kennett, & Bahn (2013). Indeed, one participant in a recent study of Swedish Military stated that the common view of lessons learned as an evaluation defeats the true desire to save lives (Dessne, 2013). The very nature of projects as temporary structures and high personnel mobility also impedes project learning (Sareminia, Shamizanjani, Mousakhani, & Manian, 2016). It appears that organizations need to overcome barriers to implement the foundation for culture and process, but they may not understand the relationship between organizational learning, project learning and project success.

THE UNDERLYING FOUNDATION

Knowledge Management and Project Management

Learning and knowing are ongoing activities. Learning can be enabled or hindered (Dessne, 2013). One answer to the problem of lost knowledge is to better understand the enablers of knowledge flows at the organizational and project layers, and the impact of these enablers on project success. If a positive relationship exists among organizational-level and project-level learning enablers and project success, then a better understanding of the value of promoting organizational learning initiatives and project learning within IT organizations would ensue.

Knowledge Management (KM) creates value out of intangible assets (Liebowitz & Megolugbe, 2003) by enabling an organization to capture, store, transfer and retrieve knowledge, ensuring effective utilization (Alavi & Leidner, 2001) in order to provide employees with an understanding of why, how, and what to accomplish (Ebert & DeMan, 2008). "KM is what the organization does with respect to identifying and using knowledge" (Dittes, Smolnik, Jennex, & Croasdell, 2016, p. 52). Project management entails the knowledge and specific skills to realize business value from projects (Project Management Institute, 2013, p. 5); Shenhar & Dvir, 2007a). Organizations initiate temporary organizations better known as projects to achieve specific objectives. Projects are unique because they bring together people from multiple functions to work outside of normal routines (Hanisch, et al., 2009). Project work is also often unpredictable and complex (Shenhar & Dvir, 2007a). Primary functions of the project manager include management of knowledge within the team and with other teams to achieve success (Pretorius & Steyn, 2005). The program executive at the enterprise level facilitates enterprise-wide knowledge sharing, knowledge reuse, training, benchmarking, and capturing lessons (Owen, 2006). Thus, KM within project-based organization is a specialized area that entails learning at the organizational and project layers within project-based entities.

Organizational Learning Enablers

Organizational learning enablers (OLEs) relate to the systems and processes at the corporate level that facilitate individual learning. Organizations can impede or promote learning (Lakshman, 2009). Organizational policies can cause project teams to focus more on applying historical information rather than first understanding the relevance of the lessons for the emerging project (Haas & Hansen, 2005). Furthermore, when IT organizations face complex challenges and uncertainty, lessons learned should be evaluated in the context of how they may apply in the future (Intezari & Pauleen, 2017). The organization's culture, systems and procedures, as well as IT, enable knowledge transfer between projects (Karlsen & Gottschalk, 2004). Organizational climate contributes to a willingness to share knowledge (Ranasinghe & Dharmadasa, 2013). Scovetta and Ellis (2015), through empirical research demonstrated the impact of leadership power constructs on KM success. Improving KM success leads to project success. Senior management support for an organizational KM system promotes project management success (Zqikael, Levin, & Rad, 2008). Ranasinghe and Dharmadasa also found in their survey that knowledge sharing was more likely to occur when senior managers and peers supported the behaviors. Knowledge management enablers include the infrastructure such as organizational culture and IT systems (Jeng and Dunk, 2013). An ontology study for project knowledge management identified three high level categorizations: people, process, and technology (Sareminia, et al., 2016). Within the people categorization leadership and culture were identified as the main sub-categories of a knowledge management infrastructure for project-based organizations. Organizations that implement communities of practice can strengthen knowledge sharing (Davel, Du Toit, & Mearns, 2017). Empowering a work force through "participative knowledge management" can strengthen a quality management information system that helps the work force to better solve problems that in turn can help reduce manufacturing defects (Campatelli, Richter, & Stocker, 2016, p. 47). Finally, incentives and rewards enable knowledge sharing (Marouf & Khalil, 2015). Winter and Chaves (2017), using action-based research, implemented a process supported by a wiki to manage lessons learned in a medium sized IT organization. The wiki contributed to the development of a collaborative environment. Thus, OLEs such as culture, processes, systems, tools, policies, incentives, and leadership impact for better or worse the relationship between learning in project-based organizations and project success.

Project Learning Enablers

Project Learning Enablers (PLEs) are the project processes and activities that mature teams conduct to capture, store, and transfer lessons learned, and emerging project teams conduct to access, evaluate, and decide which lessons to apply. Lessons learned as knowledge gained from experience are important and relevant (Garon, 2006). For example, the Space Project Management Lessons Learned initiative enabled organizations to plan and manage future projects better. Project lessons came from previous or current projects and support improvement in future project management (Garon, 2006). Another information system captured annotations and artifacts that helped project leaders make decisions that future project managers could access (González-Rojas, Pedraza-Garcia, Correal, & Beltrán, 2016). "Project learning practices involve each project undertaking regular project reviews and maintaining project documentation" (Newell & Edelman, 2008). Lessons learned can be the most important contribution of the project closure process (Larson & Gray, 2014). The value of post-project reviews comes from the flow of lessons learned to future projects and the organization (Anbari, Carayannis, & Voetsch, 2008). Post-project reviews provide a structured means to capture lessons learned for the benefit of future project teams (von Zedtwitz, 2003). The combination of these ideas suggests a concept that can be labeled PLEs.

Project Success Variables

Project success variables (PSVs) are indicators of project success. Projects can be evaluated based on meeting schedule and delivering within budget (Anantatmula & Kanung, 2008; Anbari, et al., 2008; Karlsen & Gottschalk, 2004; Shenhar & Dvir, 2007a). One may measure project efficiency based on evaluating cost and time performance (Shenhar & Dvir, 2007b). Project success may also be evaluated based on the quality of the product in that it meets stated requirements, contains few defects, is maintainable (Banker & Kemerer, 1992; Pall, 1987; PMI, 2013), and demonstrates quality in the form of conformance to requirements, effective communication of requirements, and delivery without defects (Pall, 1987). PMI relates quality to the degree that the product delivers to specifications (PMI, 2013, p. 556). Project Success has also been gauged based on user satisfaction (Anbari, et al., 2008; Shenhar & Dvir, 2007a) and on the business benefits delivered. Project success should encompass accomplishment of greater business value (PMI, 2013, p. 15). Another view of project success suggests that it is multidimensional and aligned with an organization's strategies (Shenhar, Milosevic, Dvir, & Thamhain, 2007, p. 13). In short, project managers should strive to deliver measureable organizational benefits.

Organizational benefits can refer to strategic advantage, financial returns, market position and impact on growth (Shenhar & Dvir, 2007a).

Capturing and Evaluating Lessons Learned

It is important to execute project learning well (Newell & Edelman, 2008). Lessons can be captured after milestones or at project completion but the analysis helps only if the insights contribute to future project team endeavors (Desouza, et al., 2005). Project review meetings need to be conducted properly and effectively. For example, a trained facilitator should lead meetings and team members should prepare for the meetings (von Zedtwitz, 2003). Knowledge useful in one situation may not be useful in another (Haas & Hansen, 2005; Intezari & Pauleen, 2017). The emerging project team has an obligation to evaluate and judge the relevance of lessons learned for the new project. It is important to understand the quality level or effectiveness of KM (Holsapple & Wu, 2008). In this research project success provides a means to evaluate the effectiveness or quality of OLEs and PLEs.

THE UNDERLYING QUESTIONS

It is important to explore the enablers of knowledge within an IT organization by: 1) identifying those enablers at the organizational level that impact knowledge sharing among IT project teams; 2) identifying those enablers at the unit or project team level that impact knowledge sharing; and 3) establishing the impact those enablers have on a tangible measure of successful transfer of knowledge, in this case IT project success.

The theoretical framework (depicted in Figure 1) is based on the expected interaction of the OLEs, PLEs, and PSVs.

The theoretical framework outlines the relationship between OLEs, PLEs, and PSVs and leads to three questions:

1. What constitutes OLEs, PLEs, and PSVs?

Figure 1. Theoretical framework: Relationship among OLEs, PLEs, and PSVs

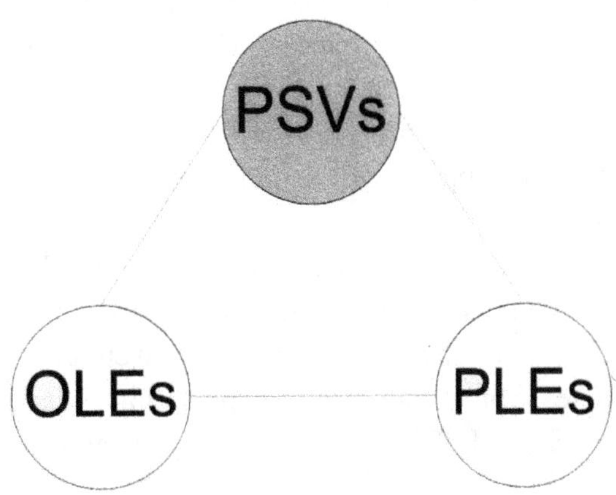

2. What relationships exist in IT organizations among OLEs, PLEs, and PSVs
3. What portion of project success can be attributed to OLEs and PLEs?

The importance of deriving answers to this question is established in the literature. Handzic and Durmic (2015) suggest further research to comprehend the value of knowledge transfer as it relates to project success and Turner, Zimmerman, and Allen (2012) and Terzieva and Morabito (2016) call for research on knowledge transfer among teams. Alkhuraiji, et al. (2014) suggest quantitative research to evaluate knowledge networks. Jennex, et al. (2016) established a key mini-track at the Hawaii International Conference on System Sciences within the Knowledge, Innovation, and Entreprenurial Systems track that calls for research related to measuring the value of knowledge management. Andres (2011) suggested that team learning needs further examination by team type. Henry, McCray, Purvis, and Roberts (2007) called for research to guide project managers to utilize lessons learned from prior experience. Anbari, et al. (2008) invited research that encouraged teams to develop lessons learned and emerging teams to use the information. Newell, et al. (2006) suggested that researchers explore how organizations can generate project-level learning. Helping leaders to understand the relationship among OLEs, PLEs and PSVs may result in efforts to promote organizational and project learning. Newell and Edelman (2008) opined that the majority of KM research has focused on the supply side or development of lessons learned. Newell and Edelman balanced their research between supply and demand. This research focuses on the demand for knowledge.

EXPLORING THE RELATIONSHIP AMONG OLES, PLES, AND PSVS

Content Analysis

Seventy-nine peer reviewed KM articles were analyzed to formulate the set of recognized OLEs, PLEs, and PSVs. A six-step process guided the analysis (Creswell, 2003). Data was organized and prepared (step 1); reviewed multiple times (step 2); analyzed and characterized (step 3); coded and described (step 4); charted and represented (step 5); and interpreted (step 6). This analysis identified the major ideas through synonyms and understanding of relationships with other terms (Coakes & Coakes, 2009). Elements were categorized and the number of times concepts were mentioned was counted (Heisig, 2009; Lee, Shin, & Lee, 2011). In order to organize and prepare the data (step 1) a purposeful sample was drawn using databases in Table 1 (Creswell, 2005). A spreadsheet enabled the authors to record the identification, quotation, citation, and research type for each element. The recorded data was listed in three separate tables; one each for OLEs, PLEs, and PSVs.

Data included "sentences, paragraphs, or themes" (Tesch, 1990, p. 79) that suggested useful organizational learning and project learning approaches. In addition, data was extracted that defined measures for project performance.

The protocol to capture the data (step 1 continued) involved taking short notes or quoting the sources using a table that captured the note or quote and a citation (Creswell, 2003). For each data element, the quote or short note and the citation (authors, year, and page) were captured. In this research most of the data consisted of direct quotes. Sometimes a short explanation expanded on a quote or captured the essence of a concept. OLEs, PLEs, and PSVs, emerged from the captured data.

Table 1. Databases

Database Name
ABI/Inform Complete-ProQuest
ACM Digital Library
IEEE Computer Society Digital Library
Computers and Applied Sciences Complete - EBSCO host
Academic OneFile - Gale Cengage Learning
Applied Science and Technology Full Text - Wilson Web
Emerald Management ejournals - Emerald Group
IBI Global Science Direct - Elsevier
Dissertations and Theses - ProQuest

References to the culture, processes, systems, tools, policies, and leadership that affected organizational learning suggested an OLE. For example, "it is necessary for project-based organizations to develop an organizational culture that coordinates and facilitates knowledge transfer" (Ajmal & Koskinen, 2008) suggested an OLE. If an article referred to activities that emerging project teams conducted to access, evaluate, or decide which lessons to apply, a PLE was suggested. In addition, references to methods to capture, store, and transfer lessons learned also suggested a PLE. For example, "organizations that routinely schedule and hold post-project reviews can continually refine the conduct and improve the benefits of these exercises" (Anbari, et al., 2008). If an author concluded that project success should be measured based on certain dimensions or metrics, a PSV was suggested. For example, "meeting customer expectations, final quality, and successful mitigation of all identified risks and threats to project completion" (Anbari, et al., 2008, p. 637).

The process step mentioned above was repeated (step 2) twice by the authors. This work often validated the first review. However, during the second and third reviews there were times the authors added or deleted data. Also, the authors continually checked articles to ensure that the quote reflected the researcher's context.

The authors analyzed and categorized the data (step 3) by iteratively developing a theme and then grouping data within a theme. For example, several researchers suggested that there be an environment of trust and support within an organization's culture to facilitate knowledge sharing. This led to a single group designation labeled as "trusting and supportive culture."

The authors assigned identification codes to each element (step 4). The coding process involved several iterations (Creswell, 2005; Tesch, 1990). For example, a data element could initially appear to be an OLE identified as OLE1. Upon reexamination, OLE1 could become a PLE with an id PLE27.

In this research, findings were represented in tables (step 5). Finally, the interpretation (step 6) led to groups of learning practices that were translated into survey questions.

Survey Development and Validation

The survey was developed and validated using the Delphi Process, as outlined by Turoff and Hiltz (1996) and adapted by Yousuf (2007), Skulmoski, et al., 2007, and Harlow (2008). The Delphi implementation in the current study consisted of three steps. First the goal was developed. Second, the team-member selection criteria were established and members were invited. Third, the team executed the process.

The team's goal was to reach consensus that the survey would be an effective tool to answer the research question. Consensus was determined by the statistical average including each team member's response, as suggested by Yousuf (2007). In the current study, consensus was defined as the mean for each survey question equaled four or better and no individual question received a score of two or less.

Before the Delphi team could begin its work, the qualifications for team membership and team size had to be established (Step 2). Based upon the methods followed in a similar study (Landaeta, 2008), the team for the current study was composed of people experienced in one of three domains: KM, project management, or survey development. The Project Management Institute's requirements (Petter & Randolph, 2009) that certificate candidates must have three years experience served as an operational definition for "experience" for all three domains. In addition to the requirement for experience in one of the three domains, the following criteria were also used to identify qualified candidates: communication skills, academic achievements, and practical experience. Ultimately, 10 candidates agreed to participate.

Once the team was formed, the Delphi process was divided into rounds conducted via an anonymous online modality. In preparation for the first round, participants received: a description of the research; a short description of the Delphi process; a draft survey derived from the OLEs, PLEs, and PSVs identified in the content analysis phase; and finally a questionnaire about the survey. In round one, the participants were not asked to quantitatively rate the survey but rather were asked open-ended question concerning the capacity of the survey questions to appropriately measure OLEs, PLEs, and PSVs. The team members were asked to describe how questions they deemed deficient may be improved.

In preparation for round 2 the researcher prepared a return comment matrix, a revised survey, and a questionnaire that included quantitative ratings for each survey question. Each team member's comment and the author's reply were returned to the team, allowing individual participants to confirm that their comments were included and action taken. The researcher sent the aggregated comments, a revised survey based on those comments, and the question-rating questionnaire back to the team to commence round 2. The team members provided both textual and numeric feedback to the researcher on each survey question.

In preparation for rounds 3 and 4, a return comment matrix similar to the one developed subsequent to round 1 was developed. This document, however, contained both textual comments (if any) and numeric ratings for each survey question. The researcher sent the following information to the team members to start the round: the aggregated comments for each question, the mean rating for each question, the distribution of ratings for each question, a revised survey based on those comments, and the question-rating questionnaire. As mentioned above, prior to the start of the validation process consensus was operationally defined as each survey question having a mean rating of four or better and no individual question receiving a score of two or less. Consensus on the survey's validity was reached in round 4.

The survey instrument utilized a five-point Likert scale for all questions except the first two questions for PSVs addressing budget and schedule performance. The five-point scale included "strongly agree," "agree," "neither agree nor disagree," "disagree," or "strongly disagree." The three-point scale used for budget and schedule questions included "agree," "neither agree nor disagree," and "disagree.

Using a Pilot to Check Survey Reliability

A test-retest pilot survey was conducted after the Delphi group reached consensus. Pearson's r has been used to conduct a test-retest correlation (Leech, Barrett, & Morgan, 2011). The test-retest in this study was two-tailed striving for significance at $\rho<0.5$. A test-retest involves calculation of correlation coefficients to compare two sets of responses (Lindbergh, 2009). Test-retest correlations were moderate to highly positive with r >.70, ρ<0.5 (Lindbergh, 2009).

Cronbach's alpha was used to test for the internal consistency of the results in the pilot and general surveys (Creswell, 2005). The ideal range was between 0.7 and 0.9 (Leech, et al., 2011). A Cronbach's alpha above .9 suggests that redundant questions may be in the survey (Leech, et al., 2011).

Population and Sample

The population ideally covered all IT projects in the United States. As this population was too large, a working population or a sub-set of the population was used (Rea & Parker, 2005). In this research the sample was drawn from ZoomInfo. (ZoomInfo, 2010). ZoomInfo's database contained approximately 5,000 names of managers in IT organizations in the United States with 1,000 or more employees. The employee information included had been updated within the last 18 months.

Rea & Parker (2005) provided a formula that was used to derive minimum acceptable sample size. Based upon initial assumptions that the confidence interval = 95%, standard deviation = 2, margin of error = 0.25 and working population = 4,400, a sample size of 233 was initially derived.

Based on the actual survey results, the minimum acceptable sample size was recalculated. Data from the surveys received indicated that the highest standard deviation for any of the 32 questions related to the OLEs, PLEs, and PSVs was actually 1.199, not the assumed 2.0. When this change from the assumed standard deviation was entered into formula (Ajmal & Koskinen, 2008), the results reduced the required minimum sample size to 87 respondents. In the survey, 101 IT managers responded producing 97 completed surveys.

Researchers have used similar sample sizes in research related to KM in project environments. For research on factors affecting knowledge transfer in IT projects a sample of 68 respondents for a survey instrument that included 51 questions was used that included a similar scale to this research (1 to 5) (Karlsen & Gottschalk, 2004). In a study on the relationship between improving project management and use of KM the sample size was 99 respondents for a survey containing 43 questions (Lierni & Ribière, 2008). Finally, in research that evaluated knowledge transfer across projects used a sample of 46 respondents (one per project) to answer 48 questions (Landaeta, 2008). Thus, it was decided that 97 respondents were adequate to complete this study.

Statistical Analysis

Statistical methods were an essential component of correlational studies (Creswell, 2005). The statistical analysis would prove successful if the results enabled an answer to the research question in a manner that could withstand peer review. Data analysis involved two tasks to answer the research question. The first task was to describe the data to gain a broad understanding of the information. The second task was to answer the research question by use of inferential statistics.

Within the first step, the project data was evaluated noting the frequency distribution, central tendency, variability, and ranking (Creswell, 2005; Rea & Parker, 2005). The descriptive data was used to identify any unusual issues and provide a sense of lessons that could be learned from the survey (Rea & Parker, 2005).

The second task addressed the second and third research questions. Pearson's Product Moment Correlation (two-tailed) was used to correlate the variables in his study (Jugdev, 2007). "Experts differ on the interpretation of the strength of the correlation amongst the variables. As result Table 2 was developed to enable interpretation using a common lexicon in the top row. Very low and very high are not shown.

Upon completing Pearson's Product Moment a multiple regression was conducted to determine what portion of project success could be attributed to OLEs and PLEs. An ANOVA was used to test the significance of each of the independent variables (OLE and PLE) in the model derived from the multiple regression.

RESULTS

Survey Instrument

Table 3 provides an abbreviated version of each question of the survey validated by the Delphi team. The survey contained questions for 12 OLEs, 11 PLEs, and nine PSVs.

Administration: Sample Description

The sample for the pilot test came from a convenience sample of 15 IT managers with experience in large corporations. The pilot group was asked to take the survey twice with an interval of two weeks between the surveys (Lindbergh, 2009). However, three people in Pilot 2 took the survey about four weeks after the initial pilot.

In this research, the general survey introduction was sent to 4,986 people and was posted on a website (ASAPM, 2012). A letter was then sent to 3,340 potential respondents. Addresses were not available for all people in the initial working population and the initial population. Following up on the letter campaign, two email reminders were sent. From these efforts 101 people responded, producing 97 usable responses.

Reliability: Pilot Survey

A pilot survey was conducted to determine the reliability of the survey. The participants came from a convenience sample of 15 experienced IT managers. The test-retest showed a positive correlation between pilot 1 and pilot 2. The PSVs, OLEs, and PLEs had correlations of 0.919 (.01 sig.), 0.727 (.01 sig.), and 0.57 (.05 sig.) respectively.

Table 2. Expert interpretation of relationships

Researchers	Low	Medium	High
Creswell (2005)	.20- .35	.35–.65	.66- .85
Jugdev (2007)	.20- .40	.40-.70	.70-.90
Leech, et al. (2011)	.10-.30	.30-.50	.50-.70
Gray & Kinnear (2012)	.10-.30	.30-.50	.50-1.0

Table 3. Sample survey questions

Type	Variable Description	No. of Articles
PSV	Budget was within a tolerable variance	11
PSV	Schedule was within a tolerable variance	11
PSV	Delivery was within final customer specifications	3
PSV	Quality (bugs, user interface, maintainability, reliable data, and smooth implementation)	9
PSV	Delivered measureable organizational benefits (strategic value, financial returns, market share, stronger brand, and future capabilities)	9
PSV	Customer satisfaction base on objective feedback	8
PSV	Communications between customer and the team	Delphi
PSV	Change control process in place	Delphi
PSV	Mitigated all risks	Delphi
OLE	Trusting and supportive culture enables knowledge sharing	28
OLE	Senior management support	23
OLE	Sufficient resources to support knowledge sharing	11
OLE	Training in knowledge sharing	13
OLE	Access to information systems that facilitate knowledge sharing	36
OLE	Can locate an expert without knowing name or location (expert locator)	9
OLE	Sufficient time in the schedule for knowledge sharing	8
OLE	Requirement exists to do post-project reviews	Delphi
OLE	Process exists to facilitate learning among teams.	26
OLE	Incentives used to encourage knowledge sharing	19
OLE	Organizational structure facilitates knowledge sharing	14
OLE	Staff actively share knowledge through personal communication (Communities of practice, get-togethers, and other social settings).	18
PLE	Last team benefitted from post-project reviews completed within the same organization.	12
PLE	Individual team member used lessons brought from earlier projects within the same organization	4
PLE	Brought right skills and experience to the team	8
PLE	Team networked inside and outside of organization to gain knowledge	6
PLE	Lessons learned were disseminated during the kick-off meeting	5
PLE	Outside resources enabled knowledge sharing among teams	5
PLE	Information systems facilitated knowledge sharing	12
PLE	Team actually located an expert without knowing name or location using directory or IT system	Delphi
PLE	Team evaluated lessons learned by other teams for appropriate use in current project	9
PLE	Applied lessons learned by other projects	11
PLE	Team captured lessons learned from its experience	Delphi

Cronbach's Alpha was calculated from the pilot data for the PSVs, OLEs, and PLEs. N is double the number of questions because both Pilot 1 and Pilot 2 were included in the results as shown in Table 4.

It was determined that the survey was reliable, and thus no changes were made to the survey confirmed by the Delphi team.

Reliability: General Survey

Cronbach's Alpha was also calculated for the general survey as shown in Table 5. All Cronbach's alpha studies fell between 0.8 and 0.9; the ideal range.

Relationship Among Constructs

Descriptive statistics were captured in Table 6.

Table 7 indicated a positive correlation amongst the three constructs significant at the 0.01 level (**).

Using the common interpretation across the top of Table 3, the results may be interpreted for each expert and combined in Table 8. This permits a judgment to be made about the strength of relationships in qualitative terms which most experts may accept.

The multiple regression model is illustrated in Table 9. The table shows that 30% of project success was related to OLEs and PLEs.

ANOVA results showed that the regression was statistically significant ($f = 20.727$, $p = 0.000$). OLEs taken separately were also statistically significant ($\beta = 0.394$, $t = 3.251$, $p = 0.002$). PLEs were not statistically significant ($\beta = 0.200$, $t = 1.654$, $p = 0.101$).

Table 4. Cronbach's alpha for pilot

Variable	Cronbach's Alpha	N
PSVs	.860	18
OLEs	.894	24
PLEs	.889	22

Table 5. Cronbach's alpha for general survey

Variable	Cronbach's Alpha	N
PSVs	.802	9
OLEs	.887	12
PLEs	.862	11

Table 6. Descriptive statistics

Variable	N	Min.	Max.	Mean	Std. Dev.
PSVs	97	2.44	4.78	3.72	0.56
OLEs	97	1.33	4.50	3.09	0.76
PLEs	97	1.30	4.73	3.25	0.69

Table 7. Correlations among PSVs, OLEs, and PLEs

	PSV	OLE	PLE
PSV	1	.537**	.474**
OLE	.537**	1	.705**
PLE	.474**	.705**	1

Table 8. Interpretation of correlations

Researcher	OLEs – PSVS 0.537	PLEs – PSVs .474	OLEs – PLEs 0.705
Creswell (2005)	Medium	Medium	High
Jugdev (2007)	Medium	Medium	High
Leech, et al. (2011)	High	Medium	High
Gray & Kinnear (2012)	High	Medium	Very High
Finding	Medium/ High	Medium	High

Table 9. Multiple regression results

R	R^2	Adjusted R^2	Std. Error of the Estimate
.553	.306	.291	.47530

CONCLUSION

This research extended prior research by delineating the variables at the organizational and project layers within IT organizations based on the foundations of KM. Nonaka, et al. (2006) theorized that knowledge assets could be used at the organizational and project layers. The research brought together a common set of specific organizational and project enablers that may lead to increased project success in IT organizations.

Gartner Group (2015) estimated that around $3.7 trillion world-wide was spent on information technology. In this research sample, 65% of the projects appear to be large serving customers throughout an organization or multiple organizations. Yet while the magnitude of IT expenditures and their impact on society and organizations is large, it rests with individual senior IT leaders and project leads to prioritize learning for each IT organization. The results of this study presented a justification for IT leaders to further explore the potential of KM in their organizations.

This research demonstrated a positive and significant correlation among organizational learning, project learning, and project success in IT organizations. The correlations among the variables were significant at the .01 level. The multiple regression indicated that the combination of OLEs and PLEs accounted for approximately 30% of a project's success. Thus, learning may be a critical success factor to enable project success in IT organizations.

Yet this research has limitations. Originally, a goal of this research was to achieve 233 respondents using a conservative standard deviation of two. This goal was not achieved. Using the sample sizing formula again based on the highest actual standard deviation the acceptable sample was revised to 87 respondents. In this study 101 IT managers responded, producing 97 valid responses. In addition, the test-retest for PLEs at 0.57 was less than 0.70 recommended by experts.

The aforementioned limitation may have led to the finding that PLEs did not contribute significantly to the multiple regression in this study. The methods employed for the content analysis, and the Delphi teamwork may also be factors. The content analysis was based upon 220 citations in support of OLEs but only 83 citations for PLEs. During the Delphi team process it was somewhat more difficult to reach consensus on the PLEs. This suggests that more research is necessary to define PLEs and understand their use.

The research confirms that OLEs are an important framework for learning in IT organizations. For example, if an organization lacks trust, leadership, and incentives it is less likely that project teams will implement PLEs. Knowledge management to-date has primarily been focused on the organizational level. Another part of the problem maybe that different terms are used to describe lessons learned in project settings (Jugdev, 2012). In addition, project teams often determine that the project has succeeded if the project is delivered to the scope, on-time, and within budget. While these traditional goals are important knowledge management can enable project-based organizations to attain strategic goals such as brand image, learn new technologies, and strengthen strategic partnerships (Solli-Sæther, Karlsen, & van Oorschot, 2015). Taking the time to capture and share lessons learned may help project teams to better align their efforts to strategic objectives. There appears to be a significant opportunity to improve knowledge of learning at the unit or project layers.

Organizations and their employees may not have implemented PLEs because organizational leaders consider the time or costs of capturing and storing lessons learned too high. A study at Volvo Technology indicated that employees used knowledge management solutions that used the least amount of time (Johansson, Moehler, & Vahidi, 2013). Exploratory research through interviews and surveys of 12 project management professionals in different industries suggests efficient KM can lead to improved profitability (Terzieva & Morabito, 2016). Researchers are now considering time use and costs in their design of KM solutions supporting project teams. González-Rojas, et al. (2016) stated that they plan to further automate their system because the cost of the current knowledge management system currently requires a dedicated resource. Organizations may also consider assigning a smaller number of project team members to enable knowledge sharing to reduce investments in knowledge transfer (Taveres & Schneck de Paula Pessoa, 2014). For manufacturing ramp-up projects a knowledge management approach was designed to enable knowledge sharing "while not being too resource intensive" (BuBwolder, Burgahn, Hübner, & Werker, 2016, p. 49). Web 2.0 technologies offer an affordable alternative for developing lessons learned applications (Chaves & Veronese, 2014). Knowledge management systems could be more cost effective using mobile and cloud technologies (Ferrada, Núñez, Neyem, Serpell, Sepúlveda, 2016). Thus, while use of PLEs can potentially improve knowledge sharing leading to greater project success there may be a need to deliver more cost effective and budget friendly solutions in order to gain acceptance in IT organizations. The finding of a positive relationship amongst OLEs, PLEs, and PSVs and the significant contribution of learning to project success is similar to other research findings. In a survey, 114 project practitioners expressed a belief that knowledge management processes can improve project management (Al-Zayyat, Al-Khaldi, Tadros, & Al-Edwan, 2010). In a survey of 220 mid-level managers, knowledge management correlated with innovation and operational performance (Al-Hakin & Hassan, 2013). In yet another survey of 187 ERP system users in the footwear and apparel industry, it was determined that knowledge creation predicts ERP implementation success (Jeng & Dunk, 2013). A relationship was found between knowledge transfer amongst project teams, their consultants, and users which in turn correlated with user perceptions of system quality (r = .45) and user benefits (r = .53) (Lee, et al., 2011). A causal relationship was found between systems integration project success and team member knowledge (Hong, Kim, Kim, & Leem, 2008). Also, a relationship was found between KM practices and project management (Lierni & Ribière, 2008). In other research it was found that the combination of traditional project management practices and KM enabled schedule and budget predictability (Henry, et al., 2007). Also, research concluded that culture and leadership, organization and processes, and information systems correlated with KM effectiveness in project-based organizations (Lindner & Wald, 2011). Knowledge leadership enables knowledge sharing based on empirical research. Specifically,178 construction engineers through a survey validated the importance of knowledge leadership (Zhang & Cheng, 2015). Positive relationships between KM and success in project-based organizations demonstrated by the literature validated the results of this research.

Jennex, Smolnick, and Croasdell (2012) found in the their survey that KM success measures included business process impact, KM strategy, leadership support, and knowledge content which tends to support major OLEs developed in the content analysis and validated by the Delphi team. A process to support knowledge was mentioned in 23 articles and leadership support in 20 articles suggests that the major OLEs are based on KM success that in turn relates to project success

Specific cases illustrated a relationship between project success and learning. Duke Engineering and Services reported that applying lessons learned for projects in which generators were replaced at power stations achieved good results (Terrell, 2000). Using lessons learned the company was able to reduce the critical path of the emerging project by 33% while accomplishing 27% more work. In another specific endeavor an IT KM system and a process enabled research and development projects over six years to shorten project lead times (Hirai, Uchida, & Fujinami, 2007). Alcatel-Lucent developed a KM process and reported that 89% of the sales and marketing forces considered the tool an important for their jobs (Ebert & DeMann, 2008). The company also uncovered 40% of all defects sooner in the process enabling a cost savings of 30%. NPTC, India's largest power company, illustrated that competiveness and sustainability were linked to knowledge management processes (Goel, Rana, & Rastogi, 2010).

Fong (2003) theorized that some repetition of processes improved learning prospects among projects. Fully 66% of the respondents, in this research, worked on projects in which the organization had prior experience. Another 28% worked on projects new to the company suggesting that an emphasis on external networking, benchmarking, and alliances may be helpful to improve project success. Thus, 94% of IT projects may benefit from knowledge flow among projects.

The demographics, strength of the correlation among OLEs, PLEs, and PSVs, and the relatively high proportion of project success that can be attributed to learning suggest that IT organizations have an opportunity to improve project success using KM.

PRACTICAL IMPLICATIONS

IT organizations should consider implementing or strengthening their KM initiatives. This research developed a set of OLEs and PLEs some or all of which may enable managers to define a program that meets the needs of an organization. Both survey respondents and the literature emphasized factors such as trust, senior management leadership, process management, and personal communication suggesting that organizational emphasis on the OLEs may be a good way to begin implementation of KM in IT organizations. This could begin to bring about the cultural shift necessary to become a learning organization (Garon, 2006). Furthermore, implementation of OLEs in an organization may help leaders take a holistic view of the company's strategies and learning infrastructure (Chronéer & Backlund, 2015). Organizations may also consider methods to evaluate the effectiveness of their KM program. The survey instrument could be tailored to individual organizational needs initially to define gaps and later to assess continuous improvement towards becoming a learning organization.

FUTURE RESEARCH

KM related to projects is an emerging field of study that offers exciting opportunities for further research. More research is needed at the project level to define and measure the relationship with project success. Researchers may continue to explore the value of project knowledge management through models that managers may use to develop cost/benefit studies and ongoing assessments (Desouza, et al., 2005; Jennex, et al., 2012). Models may also involve non-financial metrics that have an impact on product quality, service reliability, customer service, productivity, and so on. This research also provides a basis for action research through implementation of OLEs and PLEs in an IT organization.

PLEs are not being extensively used. More research to investigate methods to measure and reduce the costs and time invested in knowledge sharing among project teams could lead to improved use of KM processes in project-based organizations.

Future research may also consider means to measure knowledge sharing value across organizational and national boundaries including external stakeholders such as suppliers, customers, regulators, communities, and others (Alkhuraiji, et al., 2014; Mueller, 2015). Elements of leadership, related cultural considerations, and collaboration among citizens of different nations pose research opportunities to execute and measure the impact of knowledge sharing internationally (Scovetta & Ellis, 2015). Tavres and Schneck de Paula Pessoa (2014) conducted a survey to understand the flow of knowledge along the supply chain. The value of knowledge sharing among organizations in a common business network presents interesting challenges as suppliers compete to serve similar customers who are in turn competing with each other. (McKay & Edick, 2017).

IT Project teams need to capture tacit and explicit knowledge from their customers in order to develop system requirements and plan systems development. Knowledge sharing will become more important as IT project work becomes more complex requiring multiple project teams to work simultaneously to support over arching programs. It will become more important to share tacit knowledge as IT projects support users in many countries and grow in complexity. For example, storytelling and observation in the work place will need to be integrated within a project-based organization's KM framework (Bretschneider & Zogaj, 2016). IT project development teams use agile project management techniques with increasing frequency that requires effective and timely knowledge sharing. More research can enable business leaders to define and measure the role and contribution of KM in support of an increasingly complex and dynamic IT project environment.

REFERENCES

Ajmal, M. M., & Koskinen, U. K. (2008). Knowledge transfer in project-based organizations: An organization culture perspective. *Project Management Journal*, *39*(1), 7–15. doi:10.1002/pmj.20031

Al-Hakim, L. A. Y., & Hassan, S. (2013). Knowledge management strategies, innovation, and operational performance: *An empirical study of the Iraqi MTS. Journal of Advances in Management Research*, *10*(1), 58–71. doi:10.1108/09727981311327767

Al-Zayyat, A. N., Al-Khaldi, F., Tadros, I., & Al-Edwan, G. (2010). The effect of knowledge management processes on project management. *Journal of IBIMA Business Review*, *2010*, 1–6.

Alavi, M., & Leidner, D. E. (2001). Review: KM and KM systems: Conceptual foundations and research issues. *Management Information Systems Quarterly*, *25*(1), 107–136. doi:10.2307/3250961

Alkhuraiji, A., Liu, S., Oderanti, F. O., Annansingh, F., & Pan, J. (2014). Knowledge network modeling to support decision-making for strategic intervention in IT project-oriented change management. *Journal of Decision Systems*, *23*(3), 285–302. doi:10.1080/12460125.2014.886499

Anantatmula, V., & Kanung, S. (2008). Role of IT and KM in improving project management performance. *The Journal of Information and KM Systems*, *38*(3), 357–369. doi:10.1108/03055720810904862

Anbari, F. T., Carayannis, E. G., & Voetsch, R. J. (2008). Post-project reviews as a key project management competence. *Technovation*, *28*(10), 633–643. doi:10.1016/j.technovation.2007.12.001

Andres, H. P. (2011). Team learning and reflexivity in technology-mediated collaboration. *International Journal of Knowledge Management*, *7*(4), 22–36. doi:10.4018/jkm.2011100102

ASAPM. (2012). Please participate in this survey: In this survey learning best practices and project success. *ASAPM February 2012 Newsletter*.

Ballou, D. P., Belardo, S., & Pazer, H. L. (2010). A project staffing model to enhance the effectiveness of knowledge transfer in the requirements planning phase for multi-project environments. *International Journal of Knowledge Management*, *6*(2), 1–21. doi:10.4018/jkm.2010040101

Banker, R., & Kemerer, C. F. (1992). Performance evaluation metrics for information systems development: A principal-agent model. *Information Systems Research*, *3*(4), 379–400. doi:10.1287/isre.3.4.379

Barratt-Pugh, L., Kennett, P., & Bahn, S. (2013). Managing knowledge: The critical role of culture and ownership as mediator of systems. *International Journal of Knowledge Management*, *9*(2), 20–37. doi:10.4018/jkm.2013040102

Bretschneider, U., & Zogaj, S. (2016). Exploring strategies for capturing customer's tacit knowledge in customer integration methods. *International Journal of Knowledge Management*, *12*(2), 1–19. doi:10.4018/IJKM.2016040101

BuBwolder, P., Burgahn, F., Hübner, M., & Werker, M. (2016). *Classification of company-specific influence factors as part of knowledge management system for ramp-up projects.* Paper presented at the 3rd ICRM 2016 International Conference on Ramp-Up Management. doi: 10.1016/j.procir.2016.05.042

Campatelli, G., Richter, A., & Stocker, A. (2016). Participative knowledge management to empower manufacturing workers. *International Journal of Knowledge Management, 12*(4), 37–50. doi:10.4018/IJKM.2016100103

Cárdenas, I. C., Al-Jibouri, S. S. H., Halman, J. I. M., van de Linde, W., & Kaalberg, F. (2014). Using prior risk-related knowledge to support risk management decisions: Lessons learnt from a tunneling project. *Risk Analysis, 34*(10), 1923–1943. doi:10.1111/risa.12213 PMID:24842516

Cerpa, N., & Verner, J. M. (2009). Why did your projects fail? *Communications of the ACM, 52*(12), 130–134. doi:10.1145/1610252.1610286

Chaves, M. S., & Veronese, G. S. (2014). A proposal to manage lessons learned in projects: web 2.0 technologies to promote innovation. International Journal of Innovation, 2(1).

Choy, C. S., Yew, K. Y., & Lin, B. (2006). Criteria for measuring KM performance outcomes in organization. *Industrial Management & Data Systems, 106*(7), 917–936. doi:10.1108/02635570610688850

Chronéer, D., & Backlund, F. (2015). A holistic view on learning in project-based organizations. *Project Management Journal, 46*(3), 61–74. doi:10.1002/pmj.21503

Coakes, E., & Coakes, J. (2009). A meta-analysis of the direction and state of sociotechnical research in a range of disciplines: For practioners and academics. *International Journal of Sociotechnology and Knowledge Development, 1*(1), 1–52. doi:10.4018/jskd.2009010101

Creswell, J. W. (2003). *Research Design: Qualitative, Quantitative, and Mixed Methods Approaches* (2nd ed.). Thousand Oaks, CA: Sage Publications.

Creswell, J. W. (2005). Educational Research: Planning, Conducting, and Evaluating Quantitative and Qualitative Research (2nd ed.). Upper Saddle Back River, NJ: Pearson and Merrill Prentice Hall.

Crossnan, M. M., Lane, H. W., & White, R. E. (1999). An organizational learning framework: From intuition to institution. *Academy of Management Review, 24*(3), 522–537.

Davel, R., Du Toit, S. A., & Mearns, M. M. (2017). Understanding knowledge networks through social network analysis. *International Journal of Knowledge Management, 13*(2), 1–17. doi:10.4018/IJKM.2017040101

Desouza, K. C., Dingsøyr, T., & Awazu, Y. (2005). Experiences with conducting project postmortems: Reports versus stories. *Software Process Improvement and Practice, 10*(2), 203–215. doi:10.1002pip.224

Dessne, K. (2013). Formality and informality: Learning in relationships in an organisation. *International Journal of Knowledge Management, 9*(4), 17–32. doi:10.4018/ijkm.2013100102

Disterer, G. (2002). Management of project knowledge and experiences. *Journal of Management Information Systems, 6*(5), 512–520. doi:10.1108/13673270210450450

Dittes, S., Smolnik, S., Jennex, M. E., & Croasdell, D. T. (2016). Eleven years of the knowledge management track at HICSS: An overview. *International Journal of Knowledge Management, 12*(4), 51–61. doi:10.4018/IJKM.2016100104

Dumrak, J., Baroudi, B., & Hadjinicolaou, N. (2017). Exploring the association between project management knowledge areas and sustainable outcomes. *Journal of Engineering*, *182*, 157–164. doi:10.1016/j.proeng.2017.03.152

Ebert, C., & De Man, J. (2008). Effectively utilizing project, product, and process knowledge. *Information and Software Technology*, *50*(6), 579–594. doi:10.1016/j.infsof.2007.06.007

Ferrada, X., Núñez, D., Neyem, A., Serpell, A., & Sepúlveda, M. (2016). A lessons-learned system for construction project management: A preliminary application. *Journal of Social and Behavioral Sciences*, *226*, 302–309. doi:10.1016/jsbspro.2016.06.192

Fong, P. S. W. (2003). Knowledge creation in multidisciplinary project teams: An empirical study of the processes and their dynamic relationships. *International Journal of Project Management*, *21*(7), 479–486. doi:10.1016/S0263-7863(03)00047-4

Ganguly, A., Mostashari, A., & Mansouri, M. (2011). Measuring knowledge management/knowledge sharing efficiency and effectiveness in enterprise networks. *International Journal of Knowledge Management*, *7*(4), 37–54. doi:10.4018/jkm.2011100103

Garon, S. (2006). Space project management lessons learned: A powerful tool for success. *Journal of Knowledge Management*, *10*(2), 103–112. doi:10.1108/13673270610656665

Gartner Group. (2015). *Gartner Worldwide Spending Forecast*. Retrieved February 2, 2015, from http://www.gartner.com/technology/research/it-spending-forecast

Gauld, R. (2007). Public sector information system project failures: Lessons from a New Zealand hospital organization. *Government Information Quarterly*, *24*(1), 102–114. doi:10.1016/j.giq.2006.02.010

Goel, A., Rana, G., & Rastogi, R. (2010). Case study: Knowledge management as a process to develop sustainable competitive advantage. *South Asian Journal of Management*, *17*(3), 104–116.

González-Rojas, O., Pedraza-Garcia, G., Correal, D., & Beltrán, G. (2016). MONO+KM: Knowledge management in collaborative project development, *Lng. University of Bogotá*, *20*(2), 267–302. doi:10.11144/Javeriana.iyu20-2.mkkm

Gray, C. D., & Kinnear, P. R. (2012). *IBM SPSS Statistics 19 Made Simple*. New York, NY: Psychology Press.

Haas, M. R., & Hansen, M. T. (2005). When using knowledge can hurt performance: The value of organizational capabilities in a management consulting company. *Strategic Management Journal*, *26*(1), 1–24. doi:10.1002mj.429

Handzic, M., & Durmic, N. (2015). Knowledge management, intellectual capital and project management: Connecting the dots. *Electronic Journal of Knowledge Management, 13*(10), 51-61.

Hanisch, B., Lindner, F., Mueller, A., & Wald, A. (2009). KM in project environments. *Journal of Knowledge Management*, *13*(4), 148–160. doi:10.1108/13673270910971897 PMID:19622123

Harlow, H. (2008). The effect of tacit knowledge on firm performance. *Journal of Knowledge Management*, *12*(1), 148–163. doi:10.1108/13673270810852458

Heisig, P. (2009). Harmonisation of KM - comparing160KM frameworks around the globe. *Journal of Knowledge Management*, *13*(4), 4–30. doi:10.1108/13673270910971798

Henry, R. M., McCray, G. E., Purvis, R. L., & Roberts, T. L. (2007). Exploiting organizational knowledge in developing IS project cost and schedule estimates: An empirical study. *Information & Management*, *44*(6), 598–612. doi:10.1016/j.im.2007.06.002

Hirai, C., Uchida, Y., & Fujinami, T. (2007). A KM system for dynamic organizational knowledge circulation. *International Journal of IT & Decision Making*, *6*(3), 509–522. doi:10.1142/S0219622007002617

Holsapple, C. W., & Wu, J. (2008). In search of a missing link. *KM Research and Practice, 2008*(6), 31-40.

Hong, H., Kim, J.-S., Kim, T., & Leem, B.-H. (2008). The effect of knowledge on system integration project performance. *Industrial Management & Data Systems*, *108*(3), 385–404. doi:10.1108/02635570810858787

Intezari, A., & Pauleen, D. J. (2017). The past-present-future conundrum: Extending time-bound knowledge. *International Journal of Knowledge Management*, *13*(1), 1–15. doi:10.4018/IJKM.2017010101

Jeng, D. J.-F., & Dunk, N. (2013). Knowledge management enablers and knowledge creation in ERP system success. *International Journal of Electronic Business Management*, *11*(1), 49–59.

Jennex, M. E., Smolnic, S., & Croasdell, D. T. (n.d.). *Knowledge value, success and performance measurements minitrack*. Retrieved on 8/24/2017 from http://hicss.hawaii.edu/tracks-51/knowledge-innovation-and-entrepreneurial-systems

Jennex, M. E., Smolnik, S., & Croasdell, D. (2012). *Where to look for Knowledge Management success.* Paper presented at the 2012 Hawaii International Conference on Systems Sciences, Maui, HI. 10.1109/HICSS.2012.641

Jennex, M. E., Smolnik, S., & Croasdell, D. T. (2009). Towards a consensus knowledge management success definition. *The Journal of Information and Knowledge Management Systems*, *39*(2), 174–188. doi:10.1108/03055720910988878

Johansson, T., & Moehler, R. C., & Vahidi. (2013, March). *Knowledge sharing strategies for project knowledge management in the automotive sector*. Paper presented at the 26th IPMA (International Project Management Association), World Congress, Crete, Greece. 10.1016/j.sbspro.2013.03.018

Jugdev, K. (2007). Closing the circle: The knowledge management spiral of project management. *International Journal of Knowledge Management Studies*, *1*(3/4), 423–441. doi:10.1504/IJKMS.2007.012533

Jugdev, K. (2012). Learning from lessons learned: Project management research program. *American Journal of Economics and Business Administration*, *4*(1), 13–22. doi:10.3844/ajebasp.2012.13.22

Karlsen, J. T., & Gottschalk, P. (2004). Factors affecting knowledge transfer in IT projects. *Engineering Management Journal*, *16*(1), 3–10. doi:10.1080/10429247.2004.11415233

Keegan, A., & Turner, J. R. (2001). Quantity versus quality in project-based learning practices. *Management Learning*, *32*(1), 77–98. doi:10.1177/1350507601321006

Lakshman, C. (2009). Organizational knowledge leadership; An empirical examination of KM by top executive leaders. *Leadership and Organization Development Journal*, *30*(4), 338–364. doi:10.1108/01437730910961676

Landaeta, R. E. (2008). Evaluating benefits and challenges of knowledge transfer across projects. *Engineering Management Journal*, *20*(1), 29–38. doi:10.1080/10429247.2008.11431753

Larson, E. W., & Gray, C. F. (2014). *Project management: The managerial process* (6th ed.). New York: McGraw Hill Education.

Lee, S., Shin, B., & Lee, H. (2011). Boundary spanning role of the IS development team in consultant-partnered projects: Knowledge management perspective. *International Journal of Knowledge Management*, *7*(2), 22–48. doi:10.4018/jkm.2011040102

Leech, N. L., Barrett, K. C., & Morgan, G. A. (2011). *IBM SPSS for Intermediate Statistics* (4th ed.). New York: Routledge.

Liebowitz, J., & Megbolugbe, I. (2003). A set of frameworks to aid the project manager in conceptualizing and implementing knowledge management initiatives. *International Journal of Project Management*, *21*(3), 189–198. doi:10.1016/S0263-7863(02)00093-5

Lierni, P. C., & Ribière, V. M. (2008). The relationship between improving the management of projects and the use of KM. *The Journal of Information and Knowledge Management Systems*, *38*(1), 133–146.

Lindbergh, L. B. (2009). *The relationship between project manager perceived capability, organizational culture, and project outcome*. Capella University.

Lindner, F., & Wald, A. (2011). Success factors of knowledge management in temporary organizations. *International Journal of Project Management*, *29*(7), 877–888. doi:10.1016/j.ijproman.2010.09.003

Marouf, L. N., & Khalil, O. M. (2015). The influence of individual characteristics on knowledge sharing practices, enablers, and barriers in a project management context. *International Journal of Knowledge Management*, *11*(1), 1–27. doi:10.4018/IJKM.2015010101

McCafferty, D. (Producer). (2010). *Why IT projects fail*. Retrieved from http://www.cioinsight.com/c/a/IT-Management/Why-IT-Projects-Fail-762340/

McKay, D., & Edick, K. (2017, May 23). Global supply chains have changed – Here's how to adapt [Web log post]. Retrieved from https://www.ashford.edu/online-degrees/business/global-supply-chains-have-changed-heres-how-to-adapt

Mueller, J. (2015). Formal and informal practices and knowledge sharing between project teams and enacted cultural characteristics. *Project Management Journal*, *46*(1), 53–68. doi:10.1002/pmj.21471

Nemani, R. (Producer). (2012). *Making your IT project a success*. Retrieved from http://resources.css.edu/academics/cis/graduate/docs/PMWorkshop20120305.pdf

Newell, S., Bresnen, M., Edelman, L., Scarbrough, H., & Swan, J. (2006). Sharing knowledge across projects: Limits to ICT-led project review practices. *Management Learning*, *37*(2), 167–185. doi:10.1177/1350507606063441

Newell, S., & Edelman, L. F. (2006). Developing a dynamic project learning and cross-project learning capability: Synthesizing two perspectives. *IT Journal, 18*, 567–591.

Nonaka, I., von Krogh, G., & Voelpel, S. (2006). Organizational knowledge creation theory: Evolution paths and future advances. *Organization Studies, 27*(8), 1179–1208. doi:10.1177/0170840606066312

Owen, J. (2006). Integrating KM with programme management. *International Journal of Knowledge Management, 2*(1), 41–57. doi:10.4018/jkm.2006010104

Owen, J., Burstein, F., & Mitchell, S. (2004). Knowledge reuse and transfer in a project management environment. *Journal of IT Cases and Applications, 6*(4), 21–35.

Pall, G. A. (1987). *Quality process management.* Englewood Cliffs, NJ: Prentice-Hall.

Petter, S., & Randolph, A. B. (2009). Developing soft skills to manage user expectations in IT projects: Knowledge reuse among IT project managers. *Project Management Journal, 40*(4), 45–59. doi:10.1002/pmj.20130

PMI. (2013). A guide to the project management body of knowledge (PMBOK Guide, 5th ed.). Project Management Institute.

Pretorius, C. J., & Steyn, H. (2005). KM in project environments. *South African Journal of Business Management, 36*(3), 41–50.

Ranasinghe, S. B., & Dharmadasa, P. (2013). Intention to knowledge sharing: From planned behavior and pyschological needs perspectives. *International Journal of Knowledge Management, 9*(4), 33–50. doi:10.4018/ijkm.2013100103

Rea, L. M., & Parker, R. A. (2005). *Designing and Conducting Survey Research: A Comprehensive Guide* (3rd ed.). San Francisco: Jossey-Bass.

Reich, B. H. (2007). Managing knowledge and learning in IT projects: A conceptual framework and guidelines for practice. *Project Management Journal, 38*(2), 5–17. doi:10.1080/1058053080215210

Sareminia, S., Shamizanjani, M., Mousakhani, M., & Manian, A. (2016). Project knowledge management: An ontological view. *Knowledge Management & E-Learning: An International Journal, 8*(2), 292–316.

Scovetta, V., & Ellis, T. J. (2015). Leadership social power as a component of KMS success. *International Journal of Knowledge Management, 11*(2), 1–14. doi:10.4018/IJKM.2015040101

Shenhar, A. J., & Dvir, D. (2007a). *Reinventing project management.* Boston: Harvard Business School Press.

Shenhar, A. J., & Dvir, D. (2007b). Project management research-The challenge and opportunity. *Project Management Journal, 38*(2), 93–99.

Shenhar, A. J., Milosevic, D., Dvir, D., & Thamhain, H. (2007). *Linking project management to business strategy.* Newtown Square, PA: Project Management Institute.

Skulmoski, G. J., Hartman, F. T., & Krahn, J. (2007). The Delphi method for graduate research. *Journal of Information Technology Education, 6*, 1–21. doi:10.28945/199

Solli-Sæther, H., Karlsen, J. T., & van Oorschot, K. (2015). Strategic and cultural knowledge sharing barriers in project networks. *Project Management Journal, 46*(3), 49–60. doi:10.1002/pmj.21501

Tavares, E., & Schneck de Paula Pessoa, M. (2014). Technological innovation projects: Proposal for an integrative model between project management and knowledge management in a customer-supplier perspective. *Journal of Information Systems and Technology Management, 11*(1), 105–130. doi:10.4301/S1807-17752014000100007

Terrell, M. S. (2000, January). Implement a lessons learned process that works. *Hydrocarbon Processing*, 97–100.

Terzieva, M., & Vincenzo, M. (2016). Learning from experience: The project team is the key. *Business Systems Research, 7*(1), 1–15. doi:10.1515/bsrj-2016-0001

Tesch, R. (1990). *Qualitative research.* New York: The Falmer Press.

Turner, J. R., Zimmerman, T., & Allen, J. M. (2012). Teams as a sub-process for knowledge management. *Journal of Knowledge Management, 16*(6), 963–977. doi:10.1108/13673271211276227

Turoff, M. & Hiltz, S.R. (1996). Computer based Delphi processes. *Gazing into the oracle: The Delphu method and its application to social policy and public health*, 56-85.

United States General Accounting Office. (2002). *NASA: Better Mechanisms Needed for Sharing Lessons Learned.* Washington, DC: United States General Accounting Office.

Von Zedtwitz, M. (2003). Post-project reviews in R&D. *Research Technology Management, 46*(5), 43–49.

Winter, R., & Chaves, M. S. (2017). Innovation in the management of lessons learned in an IT project with the adoption of social media. *International Journal of Innovation, 5*(2), 156–170. doi:10.5585/iji.v5i2.155

Wu, L.-C., Ong, C.-S., & Hsu, Y.-W. (2008). Active ERP implementation management: A real options perspective. *Journal of Systems and Software, 81*(6), 1039–1050. doi:10.1016/j.jss.2007.10.004

Yousuf, M. I. (2007). Using experts' opinions through Delphi technique. *Practical Assessment, Research & Evaluation, 12*(4), 1–8.

ZoomInfo. (2010). *Home.* Retrieved September 1, 2010, from http://www.zoominfo.com/search

Zqikael, O., Levin, G., & Rad, P. F. (2008). Top management support - The project friendly organization. *Cost Engineering (Morgantown, W. Va.), 50*(9), 22–30.

Chapter 12
Practical Relevance of Management Research:
The Role of Doctoral Program Graduates

Madora Moshonsky
Business Development Bank of Canada, Canada

Alexander Serenko
Lakehead University, Canada

Nick Bontis
McMaster University, Canada

ABSTRACT

Recently, a number of academics and practitioners have questioned the relevance and practical impact of management research. This study, based on an analysis of interviews with 20 doctoral program graduates, demonstrates that such claims are not fully warranted. Instead, academic research reaches practitioners because graduates of doctoral business programs act as knowledge-transfer intermediaries that aggregate, summarize, communicate, and implement findings reported in academic publications. Demand for evidence-based knowledge in the practitioner's environment determines his or her probability of applying academic knowledge. However, not all academic knowledge is perceived as useful by practitioners, and limited access to academic literature is a major impediment to the application of scholarly findings in practice. The practitioners' connection with academia after graduation influences their probability of using academic literature. Academic journals also have the potential to disseminate scholarly knowledge beyond the academic world.

INTRODUCTION

The debate about the perceived irrelevance of academic business research dates back to the 1980s when scholars, practitioners, and public officials started criticizing scholars for placing priority on scientific rigor over relevance to industry (Bennis & O'Toole, 2005; Knights, 2008; Van de Ven & Johnson, 2006).

DOI: 10.4018/978-1-5225-5427-1.ch012

The disconnect between academics and practitioners has been deemed "the Great Divide" because the theoretical contributions of researchers are rarely implemented in practice (Rynes, Bartunek, & Daft, 2001). The very value and relevance of academic research has been called into question as a result of the perceived lack of applicability and generalizability of academic knowledge (Benjamin & O'Reilly, 2011). For example, the utilization of academic research on a regular basis by human resource managers is below one percent (Rynes et al., 2001), and information systems professionals are generally unaware of academic research in their field (Pearson, Pearson, & Shim, 2005). As a result, a flurry of papers has been published which reflects on this divide between academia and practice (Jennex, 2001; Rottman, 2008; Simmons et al., 2001; Starkey & Madan, 2001). Literature suggests that the "academic-practitioner gap" exists in virtually all social science disciplines including management (Bartunek & McKenzie, 2018), and researchers should focus on the actual value of their research by asking unique questions, addressing problems that others cannot address, and consolidating their discipline's fragmented body of knowledge (Hassan, 2014; Hassan & Mathiassen, 2018). At the same time, all of these works clearly emphasize the importance of transferring academic knowledge to practice (Straub & Ang, 2011).

There are several factors that justify the importance of the transfer of academic knowledge to practice. First, in the current knowledge-based economy, organizations must utilize recent and relevant knowledge in their decision-making to remain competitive (Parent, Roy, & St-Jaques, 2007). Second, the volume of scientific research of a nation is positively correlated with its overall wealth (King, 2004; Rousseau & Rousseau, 1998). This correlation, however, becomes even stronger when a larger proportion of scientific discoveries reach practitioners. Third, the application of academic research has been shown to increase an organization's sales and productivity (Fontana, Geuna, & Matt, 2006). Fourth, empirical evidence suggests a positive relationship between the commercialization of academic findings and organizational performance levels (Susanty et al., 2011). In order to ensure the success of an academic discipline, it must have an impact on the state of both theory and practice (Jennex & Olfman, 2005, 2006). At the same time, managers see value in academic findings as long as these may help them improve their efficiency and effectiveness (Heisig et al., 2016; Lönnqvist, 2017). Therefore, calls have been made for studies that examine possible transfer methods of evidence-based knowledge to practitioners (Rousseau & McCarthy, 2007).

Knowledge refers to people's ability to act based on the information they possess (Berger & Luckmann, 1966). Knowledge provides a justification and motivation to alter decisions (Hannabuss, 2001). Accordingly, industry practitioners require knowledge to develop and implement action strategies. Therefore, academic knowledge is only relevant to industry if it motivates practitioners to take action inspired by its content (Taylor, Artman, & Woelfer, 2012). However, prior investigations have identified a glaring gap between the knowledge presented in scholarly publications and the use of this knowledge. Booker, Bontis, and Serenko (2008) studied how knowledge management professionals access and utilize academic research in their daily work. They found that, while practitioners value academic research, it is the accessibility of this research that produces the detachment. This accessibility refers to the receiver's (i.e., the practitioner's) ability to effectively consume knowledge. Simmons et al. (2001) also established that the process of knowledge transfer mostly fails on the side of the receiver. Serenko, Bontis, and Hull (2011) argued that future research examining the transfer of academic knowledge to practice should focus on knowledge transfer mechanisms, specifically indirect knowledge transfer channels. Direct channels of knowledge transfer work when an individual accesses, understands, and executes the knowledge directly from the source (i.e., from an academic publication) (Almond, 2001). Unfortunately, direct knowledge dissemination channels are underutilized by practitioners. In contrast, knowledge is transferred through indirect channels when the knowledge is modified and/or distributed to the end user by an intermediary that converts this knowledge into a format that may be easily comprehended by busy practitioners (Nohria & Eccles, 1998).

Direct knowledge dissemination channels are ineffective because academic works are usually targeted at other academics, reviewers, and editors. They are written in complicated language, contain jargon, present advanced statistical techniques, have abstract ideas and theories, and assume the reader's familiarity with previous literature and academic research in general. Thus, the accessibility of academic publications is a major barrier for the transfer of academic research to practice because practitioners often lack the academic training which is required to read and understand academic works. At the same time, graduates of doctoral business programs who are employed in the non-academic sector (e.g., managers who hold a Ph.D. in business) are fully qualified to read academic publications and use academic findings in their decision making. This study attempts to contribute to the knowledge base by exploring whether doctoral business program graduates who work in industry are knowledge ambassadors acting as an indirect channel of knowledge transfer between academics and practitioners. Particularly, the purpose of this study is to explore whether doctoral business program graduates who enter the non-academic workforce acquire, utilize, and disseminate academic knowledge in their daily decision making.

THEORETICAL BACKGROUND

The Rigor-Relevance Debate in the Management Discipline

Throughout the history of science, creating, communicating, and utilizing authenticated knowledge have been recurrent purposes of academic institutions (McLuhan, 1962; Roberts & Skeat, 1983; Saenger, 1975). Presently, the creation and dissemination of scholarly knowledge is a common component of mission statements for universities, and research is considered one of the most important activities of faculty members (Jagodinski, 2008; Serenko, Bontis, & Moshonsky, 2012). However, there is debate about whether the academic institution is fulfilling its self-expressed mission. Khurana (2007) empirically studied business schools and concluded that the top U.S. schools have lost focus on the mission of the legitimization of management and are motivated by self-interest instead of knowledge creation. Pfeffer and Fong (2002) support this critique of business schools and state that the applicability of both education and research has fallen.

The role of the university often changes depending on the stakeholder consulted. The academic researcher has been viewed as the primary stakeholder for academic output. Spender (2005) argues that management research has been driven by "a search for legitimacy, ownership and the control of management knowledge" (p. 1283) by these academic researchers. Policy makers' aspiration for universities is to enhance the quality of education and supply of useful knowledge for society as a whole in response to the rising demand for knowledge-intensive products and solutions (Yusuf, 2008). Policy makers also have the added pressure of addressing the demands of funding agencies to ensure continued funding. In many practitioners' view, the purpose of the university is to be a generator of knowledge for the development of industry (Feldman, 1994). Where the commercialization of knowledge has become a focus for some institutions, collaboration with practice has become routine (Hitt, 1998; Van Aken, 2005).

Another threat to the reputation of management scholars as knowledge producers comes from business consultants, who became popular in the 1980s. Consultants often act as independent advisors to various organizations and deliver new knowledge to legitimize their profession (Alvarez, 1998; Scarbrough, 2002; Suddaby & Greenwood, 2001). From the perspective of their customers, consultants infuse new ideas, offer solutions, implement technologies, and assist in decision-making processes. In this, their actions resemble those that are supposed to be done by academics, who – instead of consultants – should be considered the creators and disseminators of knowledge.

This topic raises the question: what type of knowledge should academics pursue – fashionable or fundamental (Abrahamson, 1991; Weick, 2001)? Scarbrough (2002) defined fashionable knowledge as "knowledge that has been diffused, but which has not been institutionalized" (p. 89). There is an argument on both sides of the debate as to what type of knowledge should be created by academics. On the one hand, there are views that academics should be encouraged to tailor management research to the practitioner audience, which reverses the proper relationship between academia and industry (Knights, 2008). On the other hand, there are arguments that researchers should focus on fundamental issues that constructively criticize industry practices and do what is best for the development of science in general (Starkey & Madan, 2001). The problem, however, is that to the practical world, management research is viewed as not generalizable and as lacking the potential for practical implementation (Jacob, 2001). According to Rynes et al. (2001), organizations tend to ignore numerous research findings, solutions, and strategies provided by academia.

The management discipline has two goals to balance – scholarly rigour and social usefulness (Benbasat & Zmud, 1999; Hodgkinson & Starkey, 2011). Consequently, it is important to understand the mechanisms that contribute to the widening of the Great Divide in order to sustain advancement in the management discipline. If the gap between the stakeholders continues to exist, it may impact the justification of the role of the academic researcher and even the sustainability of academia (Starkey & Madan, 2001). Business schools must become more responsive in addressing practical considerations; otherwise, practitioners will access substitute suppliers of knowledge.

In summary, policy makers are searching for methods of knowledge transfer from academia to practice in order to address the increasing need for relevant knowledge, to promote innovation, and to encourage competitiveness (Hanberger & Schild, 2004). This process begins with the mobility of university-educated students who develop an association with academia and later enter the labour force (Fleming & Frenken, 2007), often as business consultants. According to Bramwell and Wolfe (2008), former students act as intermediaries between industry and academia to transfer academic research to practitioners, establish future research directions, and improve curricula. However, the extent and manner in which these graduates transfer academic knowledge to practice requires further study. In line with this school of thought, this study will strive to gain an understanding of the possible function of Ph.D. graduates as an intermediary between academia and practice. The purpose is to understand what hinders the transfer of academic knowledge to practice and whether doctoral business program graduates who are employed in private and public non-academic organizations may bridge this gap.

Knowledge Transfer

Knowledge transfer has become one of the most important strategic organizational tools (Christensen, 2003). It is the key concept that all successful managers are aware of and apply in their daily work (Cavusgil, Calantone, & Zhao, 2003). When knowledge is allowed to flow within an organization, it enables organizational learning (Lahti & Beyerlein, 2000). When people face an opportunity or a problem, they require accessible knowledge to make the required modifications to their behavior (Liyanage et al., 2009).

The value of knowledge lies in its ability to help managers undertake better actions and improve their decision making (Davenport & Prusak, 1998).

There are various theories explaining knowledge transfer and how knowledge is communicated from one individual to another. In the past, knowledge was considered an object which could simply be passed from one person to another without regard for the surrounding context (Parent et al., 2007). It was also assumed that knowledge transfer was a hierarchical, top to bottom interaction where the receiver of the knowledge was a passive actor (Roling, 1992). However, this traditional model has been criticized for its linear perspective which ignores context and exchanges between the two participants. Instead, the knowledge transfer process is bi-directional, and it mostly fails on the receiver's side (Simmons et al., 2001; Szulanski, 1996). Therefore, the receivers cannot be passive entities that are bestowed with knowledge from a source. Instead, they must be active problem-solvers who generate their own knowledge base (Hutchison & Huberman, 1994). Knowledge transfer is a result of the interaction within a dyadic relationship (Knights & Scarbrough, 2010). The newer, process-based, knowledge transfer models are of the social constructivist perspective, which assumes that knowledge has an individual meaning to different people based on their experiences (Parent et al., 2007). Process-based models take into account the environment in which the knowledge is transferred and applied. In an organizational context, the legitimacy of new knowledge is validated against the organization's culture (Roling, 1992), and received knowledge is adapted to fit the receiver's individual situation (Foss & Pedersen, 2002).

Knowledge transfer processes may also be understood from the perspective of the theory of communication, which focuses on the behaviors exhibited during interaction between the source and receiver (Baxter & Braithwaite, 2008; Dillard, 1990; Giles, 2008; Hewes & Planalp, 1987; Wilson, 1997). The Goals-Plans-Action Theory illustrates that the source can influence the receiver based on his or her behavior during the transfer process (Dillard, 2008). The Uncertainty Reduction Theory states that the primary goal of an individual's communication is to decrease uncertainty (Bylund, Peterson, & Cameron, 2012). This allows one to better predict the actions of others and the outcomes of different situations. The Communication Accommodation Theory posits that individuals accommodate their communication approaches based on their desire to either converge and match, or diverge and differentiate, from the other person's style (Bylund et al., 2012). Usually, the individual who is perceived as possessing the least power in the relationship will do the accommodating (Giles, 2008).

With respect to the transfer of academic knowledge to practice, the aforementioned theories underline the importance of understanding the relationship between the participants and the environment in which the exchange takes place. In terms of this study, it means paying attention to the practitioners' previous and current relationships with the academic sector, the communication channels through which they access academic literature, the cultures of their organizations, and their communication styles.

The Knowledge Transfer Model

Knowledge transfer occurs over a variety of mediums through direct or indirect methods. A direct channel means that the receiver accesses the material written by the creators of the academic knowledge through mediums including journals, books, and conference proceedings. However, practitioners are rarely directly exposed to or utilize current academic material (Pearson et al., 2005). Rather, these practitioners access knowledge through indirect channels where the knowledge is transformed by an intermediary into an accessible format that is applicable to the receiver's environment (Nohria & Eccles, 1998). Understanding and identifying effective indirect channels, therefore, is key to conveying academic research to practitioners (Booker, Bontis, & Serenko, 2012).

For example, medical patients avoid information they believe themselves to be unqualified to consume and, instead, defer to the information provided by their healthcare providers as authorities (Baxter & Braithwaite, 2008). Similarly, a practitioner who does not possess a Ph.D. can indirectly access academic material by communicating with practitioners holding Ph.Ds., thus using them as channels for knowledge transfer. This indirect channel occurs when non-Ph.D.-holding individuals are exposed to academic theory through the Ph.D. graduates who have the ability to synthesize and communicate the initially inaccessible knowledge. Accordingly, Zolnik (2010) demonstrates that post-doctoral fellows play a vital role in technology transfer and the dissemination of scholarly knowledge beyond the boundaries of academia.

In order to explore the dissemination of knowledge, the present study adapted the process-based model of knowledge transfer proposed by Liyanage et al. (2009) (see Figure 1). This model conforms to the notion that knowledge is not an object which can be passed in static form from one person to another because it is through the process of interaction that an individual attaches new meaning to the knowledge environment (Parent et al., 2007). This model depicts a process which occurs on different levels of the organization. Both the source and the receiver of the knowledge have to actively engage in the knowledge transfer process and possess the necessary capabilities for the receiver to be able to effectively gain new knowledge and be able to act upon it. Each step in the knowledge transfer process must be completed before proceeding onto the next one. If all of these steps are not completed, then the process of knowledge transfer cannot occur, and the recipient's behavior will not be impacted by the knowledge. As expressed by Knights and Scarbrough (2010), the need for such a model is emphasized by the constant debate in this field.

The model assumes that there exists a body of scholarly knowledge which the recipient needs to be aware of in order to consume it. The process of knowledge transfer begins with the recipient identifying what kind of knowledge is required to solve a particular problem. Therefore, the receiver must be able to correctly assess the situation and the surrounding environment. The receiver must next acquire the necessary academic knowledge – knowledge that is unprocessed and may not be applied immediately or as is. After this, scholarly knowledge is transformed into new, applied knowledge that builds on the recipient's existing knowledge, skills, or capabilities. The integration of this knowledge involves adapting the knowledge to the situation and environment at hand and making it useful in a given context. In the next stage, this knowledge is applied to the current problem in an actionable strategy. Last, this knowledge should be transferred from the receiver to other organizational members and retained by not only the receiver but also his or her organization because newly acquired knowledge should have a lasting impact on as many stakeholders as possible.

This process-based model of knowledge transfer is applied in the present study to explore whether doctoral business program graduates who are employed in the non-academic workforce acquire, utilize, and disseminate academic knowledge in their daily decision making.

METHODOLOGY

Twenty semi-structured phone and Skype interviews with practitioners who possess a Ph.D. were conducted during 2012-2013. These practitioners obtained their doctorates in Canada in business, management, or a management-related discipline. The rationale for using doctoral degree holders relates to the previously

Figure 1. The process-based model of knowledge transfer
Adapted from Liyanage et al. (2009).

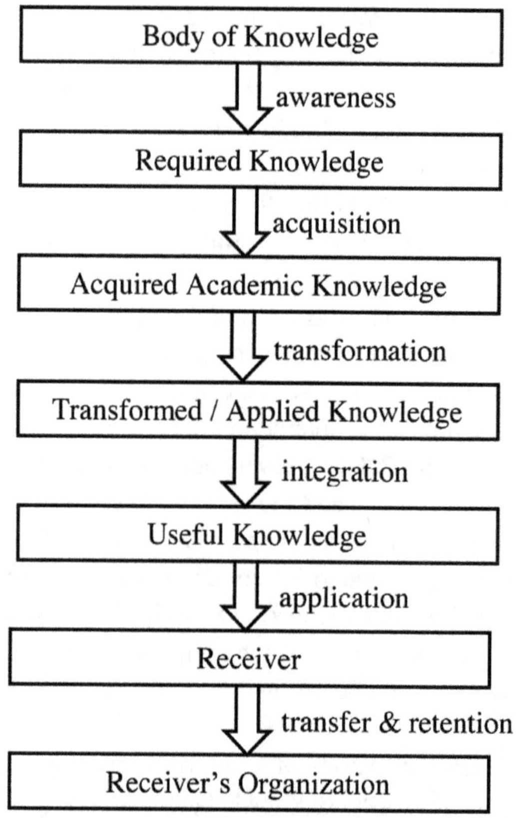

identified problem of the accessibility of academic research. Past studies revealed that academic papers are inaccessible to most practitioners due to jargon, length, writing style, and complicated statistics. Scholarly papers also mostly contain theoretical recommendations that need to be converted to practical application (Booker et al., 2008; Serenko, et al., 2012). Additionally, most business practitioners are unaware of the existence of scholarly publications. By studying individuals who are equipped with the necessary skills and experience to utilize academic material, this study acts to negate the inaccessibility issue. Therefore, this study focuses on alternative explanations for the gap in the transfer of academic knowledge to practice.

The interview questions were designed to follow the process of knowledge transfer and explore how the participants progressed through the various stages of the proposed model as described in Figure 1 (see Appendix I). The interview protocol was subjected to face validity assessment by consulting a group of business faculty members to address concerns of ambiguity and social desirability, and various adjustments to the questions were made. When respondents did not offer comprehensive answers, several probing follow-ups were used.

Participants were recruited in two ways: Internet searches and referrals (i.e., snowballing). The first method was initially used to identify individuals through a Google search of each person listed on a Canadian university's published graduate list from an applicable Ph.D. program. Graduates who worked in industry were identified through this process. If the individual's contact information was not available online, his or her dissertation supervisor was contacted to assist in reaching the graduate. The second method refers to approaching the individuals who were mentioned by the previous interviewees. Overall, the participants included ten consultants, six government employees, three investment managers, and one employee of a private company (20 in total). The participants were initially told that the study was focusing on their use of various sources of knowledge, not specifically on their use of academic research. This misdirection was important in order to eliminate social desirability bias (Crowne & Marlowe, 1960; Fisher, 1993).

The interviews were analyzed using content analysis, which is a systematic process of analyzing written, verbal, or visual content (Cole, 1988). Content analysis is the most appropriate method for analyzing the data from interviews because of the allowed flexibility in the research design (Harwood & Garry, 2003). This technique allows for a continual re-evaluation of categories established from existing theoretical models. Additionally, since content analysis provides a formalized analysis procedure, it facilitates the comparison of the different subjective viewpoints (Flick, 2002). More specifically, deductive content analysis was used to test previous knowledge established in the knowledge transfer field. Deductive content analysis structures the analysis around an earlier theory and moves from the general to the specific (Burns & Grove, 2005; Elo & Kyngas, 2008).

The data was analyzed to determine the underlying relationships among an individual's characteristics, values, experiences, and environment with his or her demand, valuation, and use of knowledge. The raw data from the interviews was transformed into manageable content categories based on systematic coding (Weber, 1990). The codes used to analyze the data for this study were developed based on the process-based model of knowledge transfer (Figure 1), aspects of holistic knowledge transfer, and known antecedents for and barriers to knowledge transfer (Dey, 1993). While traveling the dynamic path of deductive content analysis, the codes were continually re-evaluated and transformed as the analysis progressed. The researchers then returned to existing theory which might explain observed phenomena and further direct the analysis. This check of reliability also involved an additional review of the data to ensure the material was analyzed properly.

Theory triangulation was achieved by incorporating multiple theories to strengthen the credibility of the findings (Erlandson et al., 1993). The differing perspectives of each participant in the study also contributed to data triangulation. The data analysis process was facilitated through the use of a qualitative data analysis program NVivo. Nvivo was used to organize and analyze the content from the interviews through queries, visualization, and report generation.

Findings

Eleven participants were female and nine were male. The participants graduated with their respective Ph.D. degrees between 1991 and 2010 with the average year being 2005 (i.e., seven years before the interview date). Figure 2 shows that the participants possessed a wide range of degrees.

Knowledge Awareness

Knowledge awareness was focused on the knowledge that the practitioners believed they required to perform their job. Additionally, the study explored whether these practitioners perceived academic knowledge to be a necessary source. Participants described (a) what they believed was the necessary

Figure 2. Ph.D. degrees of participants

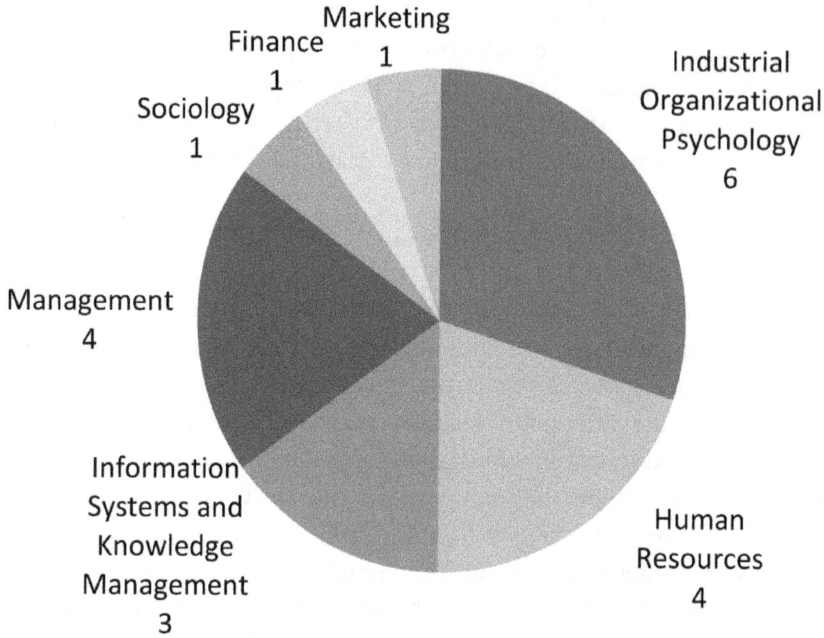

knowledge to search for in order to perform their job-related duties and (b) how they accessed it. Four categories of necessary knowledge emerged (Figure 3).

Often, participants stated they require a variety of knowledge:

I tend to think of it as I need a background in industrial organizational psychology, background in clinical accounting psychology, so literature you can think of it as… background in the business literature and also really understanding the actual clients that I'm working with.

Figure 3. Perceived knowledge requirement

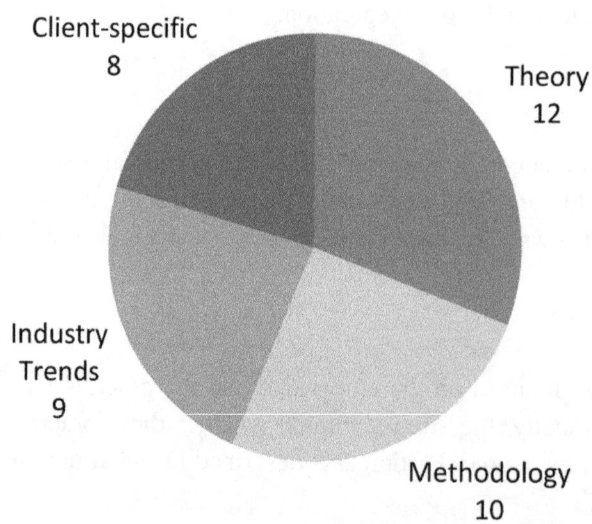

Out of the twenty participants, nine subscribed to an alert system which notified them of new knowledge from a source (Figure 4). Interestingly, when asked, several respondents who did not receive alerts said they had not thought of that but would use such notifications in the future. The following excerpt provides an example of a participant's use of alert tools:

So there [are] four journals that they, that the Academy, produces, and every month I get an email that tells me the titles of all of the articles in each of the journals. So you can very quickly scan, the, basically the index of that month's publication and see if there is something that is of particular interest to your area.

Additionally, the participants described how they decided which source was required (Figure 5). The responses were coded in the following manner: 1) situational (i.e., the participant decides which source to access based on the nature of the problem he or she was encountering), 2) timely (i.e., the source which provides the quickest answer), 3) internal experience (i.e., the past experience of the individuals themselves and of their colleagues), 4) audience (i.e., the group that the practitioner will present the new knowledge to), and 5) reliability (i.e., how consistent the source is with providing accurate, proven information).

In terms of discrepancies between consultants and government employees, government employees were slightly more likely to mention theory and methodology as required knowledge. While several consultants decided which source is required based on their audience, none of the government participants did.

Knowledge Acquisition

Figure 6 presents knowledge sources that practitioners access on a regular basis. Academic journals were mentioned most frequently, with eleven practitioners stating they access this source. Non-academic journals and knowledge obtained during their Ph.D. programs were the next highest accessed sources with nine instances each. On average, the practitioners would consult with three different sources.

Figure 4. Use of alert tools

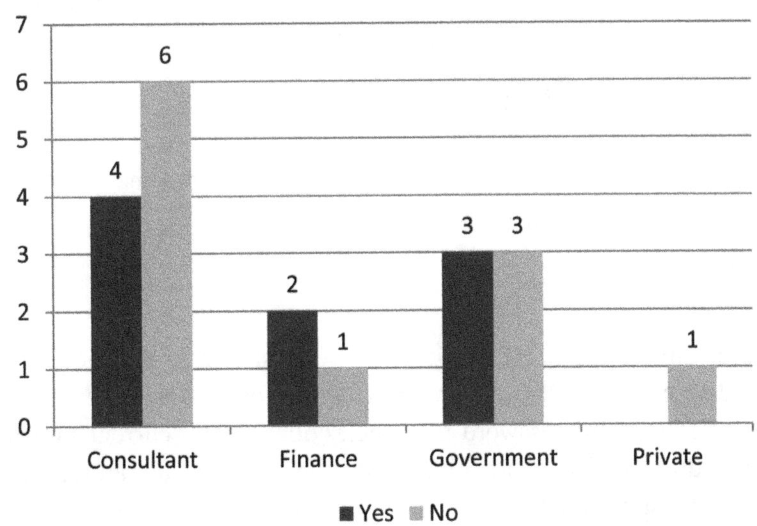

Figure 5. Decision criteria

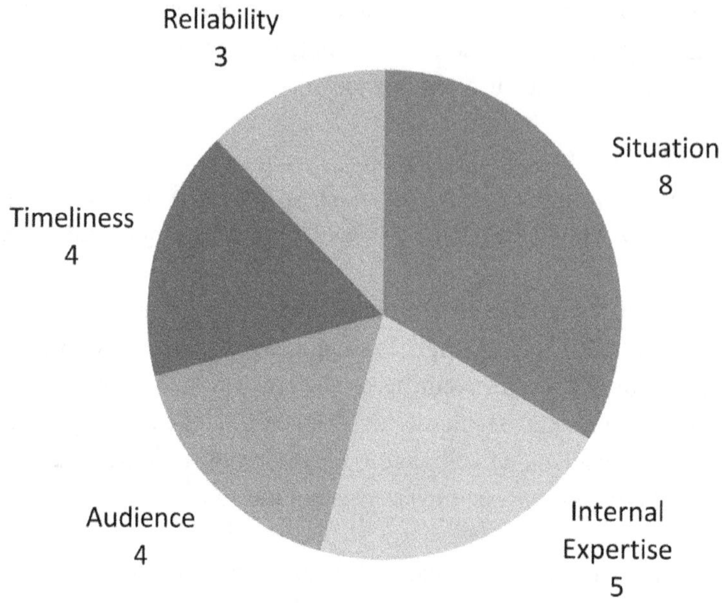

Figure 6. Knowledge sources accessed

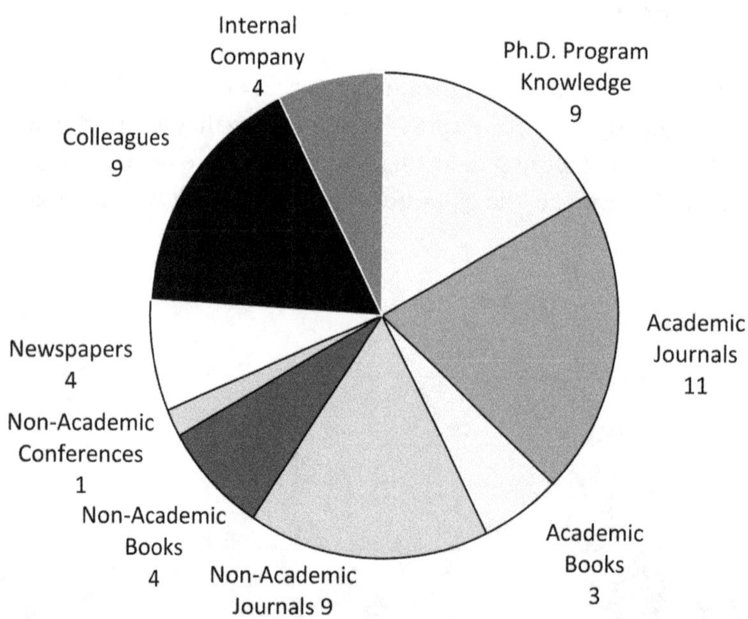

In focusing on academic research, the participants were asked how they access academic findings. Nine participants had access to academic literature at work, four through their Ph.D.-granting institution, and two obtained a login and password from their colleagues or current students (legally or not). Regrettably, four individuals said they didn't have access to academic literature, and, if they wanted to, they would have to pay for it personally:

Well, it's not easily available, like Harvard Business Review would be a good source and I'll read it from time to time. It's expensive, and I'm at the early stages of my business, and being able to buy academic, peer-reviewed journal articles is not something that I can afford. But, if it was free, I would certainly look at it.

Nine practitioners said they only accessed academic literature when they encountered a problem that required academic knowledge. Five responded that they accessed academic literature on a regular basis. Because two participants never accessed academic literature, they were excluded from this question. Additionally, responses were not obtained for four participants. While many practitioners expressed their desire to access academic literature, most could not do so due to budgetary or time constraints. Nevertheless, most respondents did access academic literature either regularly or intermittently. While government employees all had access to academic literature at work, the only consultant with access to academic literature at work owned his or her own practice. Consultants had the largest barrier to accessing academic literature because they had to rely on their colleagues who had access or they had to pay for it personally.

Knowledge Transformation

The knowledge transformation stage in the knowledge transfer process involves converting the newly accessed academic knowledge into a useable form for its intended consumers. The participants were asked if, when they accessed academic literature, it contributed to the development of their knowledge base. It was assumed that, if the participants were able to transform and adapt the new knowledge, they improved their ability to make better managerial decisions. If the newly acquired knowledge was obsolete or irrelevant, no transformation processes would take place.

Eleven participants stated that they gained new knowledge when they accessed academic literature. Four said that sometimes it was new and sometimes it was not. Three respondents mentioned that academic literature did not contribute to their knowledge base. Two respondents did not comment. Overall, a majority commented that academic knowledge was valuable and worth adapting.

Those who did not find academic knowledge worth transforming commented on the obsolescence of and lack of novelty in academic findings:

Academic publications, unless they are still working papers, they are a bit dated because it takes a while until they get published. So it's not really new information.

My bias is that 90% percent of academic literature out there is rehashing and not contributing anything new to the field. And so I try to keep up with it, but I'd say only 10% of it is providing me with new knowledge.

Additionally, the practitioners were asked if accessing academic literature had improved their skills or capabilities. Ten responded affirmatively. Three said that this occurred sometimes, and five said that it did not. One participant said that he or she gained new knowledge from academic literature but that it did not enhance his or her capabilities. At the same time, it may potentially improve his or her understanding of the situation, underlying factors, and related concepts that may be applied in the future or in different contexts:

I guess my abilities are not influenced by academic research; my understanding of a particular area is what's improved with the academic research. So it doesn't necessarily change how I do things; it might change how I understand something.

There are, again, a few differences between the responses from consultants and government employees. Five of the six government employees answered that new knowledge is gained by accessing academic literature, and one said that this occurs "sometimes." In comparison with the government employees' responses, of the nine consulting responses, four said yes, three said sometimes, and two said no.

Knowledge Integration

The knowledge integration portion of the interview studied how the practitioners found academic research fit into their work environment by addressing organizational needs. The participants were asked to describe the general usefulness, applicability, and relevance of academic research in performing job-related duties. Five participants stated that academic literature is unconditionally relevant and applicable to their work responsibilities. For example, one person commented on how important academic literature is to his or her responsibilities:

Totally relevant and necessary, I could not do my job without academic sources.

Six replied that academic literature is relevant in creating the foundation for their knowledge but not for implementable recommendations. Three did find relevance and usefulness in academic knowledge but noticed that it requires extensive transformation in order to be applied. Six participants replied that it was rare that they were able to apply academic knowledge because of its lack of relevance or usefulness. Many answered that the applicability and relevance of academic research is based on certain conditions, which included language, sample type, scope, and intended audience. For example, one participant expressed how academic research required conversion:

It's applicable, but I would have to take the time to convert it into something...so it's applicable at a conceptual level, less at a pragmatic level. If I just want to analyze the why and the how, from an analytical perspective, it's very helpful, but less so if I need something that I can use right away.

In a comparison between the ten consultant participants with the six government participants, only one consultant said that academic knowledge is relevant and applicable to his or her working environment, whereas three government employees said yes.

Knowledge Application

Knowledge application occurs when the useful knowledge is utilized to address the current problem the practitioner has encountered. In this stage in the knowledge transfer process, the practitioners acted upon the new knowledge they obtained. The participants were asked how frequently they applied in practice their findings from academic literature, and the responses were categorized as: 1) regularly, 2) sometimes, 3) rarely, and 4) never (Figure 7). Overall, a majority applied the body of academic knowledge in practical settings. For example, a practitioner expressed his or her enthusiasm as follows:

Figure 7. Practical application of academic knowledge

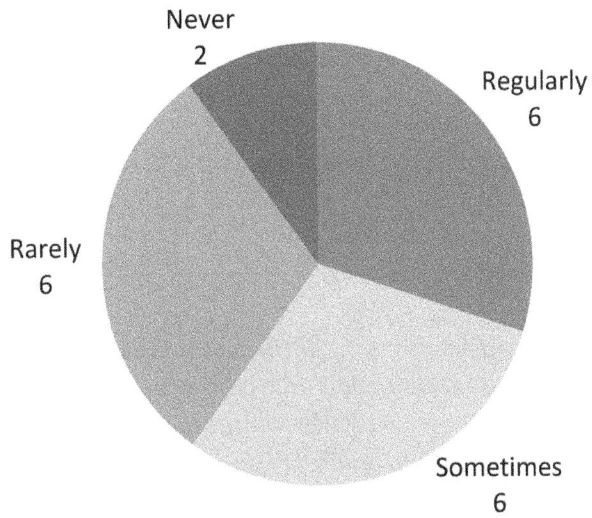

I am the type of person who, if somebody gives me something that I find interesting ... I will spend the next three weeks telling everybody about it and trying to apply it all over the place, which I'm kind of known for.

The participants were also asked if their company or clients recognized the value in applying academic research to solve managerial problems. The responses were less optimistic (Figure 8).

It was observed that an employee's organization or clients dramatically influenced the probability of applying scientific findings for decision making. If the organizational culture, job settings, fellow colleagues, and policies valued academic knowledge and encouraged its application for decision making,

Figure 8. Client or company values applied academic knowledge

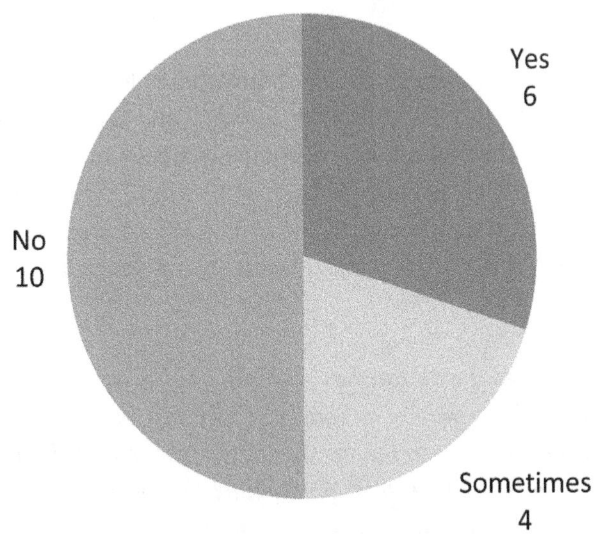

practitioners were more likely to use it. For example, one participant explained how his or her clients would appreciate the application of academic knowledge because it increases the validity of the recommendations made:

I think there is an appreciation of the face validity that that would bring to the advice that you provide. In terms of that information is valid and we can believe that that is true.

However, an opposite effect was also observed. For instance, an interviewee who never used academic literature described the deterrents he or she faced:

A couple of things, number one the demands of the job in the private sector: it's about productivity; you're not measured by how many articles you are going to read... you are measured by how many hours can you... have you been on a client. So based on how you are measured, your behavior changes and so, if the culture of the firm doesn't support you to read all of the academic articles to see what ideas and what is the latest thinking... there's not many firms that pay you to do that. So by default then you don't do it.

Another consultant stated that his or her use of academic knowledge could have an opposite effect in his or her organization:

Nobody would ever [care about the inclusion of academic knowledge]; in fact, you could reduce your credibility if you heavily reference your findings.

Yet again, there is a stark contrast between the answers provided by the consultant and government participants. Only two consultants (out of ten) versus three government employees (out of six) affirmed that they apply academic knowledge on a regular basis. The majority of consultants revealed that they rarely applied academic knowledge. Additionally, there was a difference in whether or not the participants believed that their company or clients value the application of academic knowledge because most consultants stated that it was not valued.

Knowledge Transfer and Retention

Knowledge transfer and retention ensures that the academic knowledge acquired and utilized by practitioners is embedded in the organization for future consultation and action. For knowledge to be retained, it must have a lasting impact on the individual's or company's behavior. The participants were asked if they see themselves as translators of academic research by making academic knowledge usable for those who would not be able to access it themselves. Most responded affirmatively (Figure 9). For instance, one participant discussed how he or she transfers academic knowledge and the challenges that must be overcome in this effort:

I'm using it, and people know it, but I will translate. I would use it by translating it, and I would definitely not hesitate to use what I have learned here as long as I can make it understandable and useful for my colleagues. I will give you an example: in the academic environment in literature, we're talking about commitment and then we're talking about this topic [that] has been highly researched. And it gets to very more detailed terminology like employee commitment, supervisor commitment... I don't see myself

Figure 9. Transfer of academic knowledge

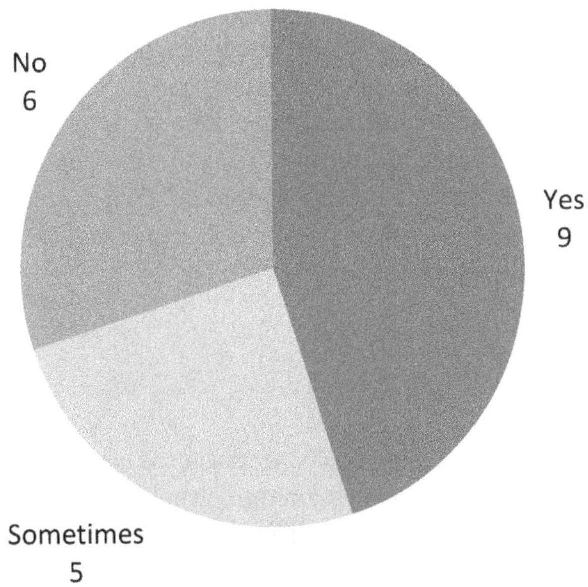

talking about these definitions here, because people won't follow me. So I need to stay broad, talking about commitment, but not going with all the refinements that we can have in the literature. This is where I will lose them. I need to use this knowledge but not go as deeply as academic knowledge will do.

Most participants viewed themselves as translators of academic knowledge to those who were not able to consume it by means of the direct knowledge dissemination channels. By providing a higher level of abstraction, they believed that educational differences did not create an unbridgeable gap:

It's not that they weren't intelligent; they were just maybe a different level.

The majority of government employees replied that they transfer academic knowledge to their colleagues who lack the necessary education. In contrast, few consultants believed they successfully perform this function.

Other Factors

During data analysis, two factors were discovered that affected the probability of practitioners using academic research. The first one pertains to the practitioners' connection with academia. Only seven of the twenty practitioners replied that they had been published in peer-reviewed journals and conference proceedings after obtaining their Ph.D. Almost all of those who published also reported that they regularly use academic knowledge. Nine participants had also taught at a college or university since graduation, and they were more likely to use academic research than those who had not taught. Thus, a connection with academia, which may be continued through publishing in peer-reviewed outlets and teaching, increases the likelihood of an individual using academic literature.

The second factor relates to the degree to which academic knowledge is valued in the practitioner's environment. It was observed that the audience to which the practitioner presented his or her recommendations greatly impacted the sources the practitioner employed. If the audience requires evidence-based knowledge, then academic research is more likely to be used. One of the strongest indicators of the probability of an individual accessing and applying academic literature is whether or not his or her clients or company values academic knowledge. If the client or company does not value academic research, the likelihood of the practitioner referencing the material declines. One reason why this relationship could exist is that, if the participant's employer does not value academic content, then it will not pay for the practitioner to have access to this material.

IMPLICATIONS

The overall purpose of this study was to investigate the transfer of academic knowledge to practice. Particularly, it explored whether doctoral business program graduates who are employed in the non-academic sector act as an indirect knowledge dissemination channel by acquiring, utilizing, and disseminating the academic body of knowledge in their daily practice. Based on the findings, several implications are offered.

Implication #1: *The Previous Claims That Academic Research Does Not Influence Decision Making of Industry Practitioners are Not Fully Warranted*

Previously, a number of scholars, practitioners, and policy makers expressed their concern that academic research in the business discipline does not reach industry professionals, remains unimplemented in practical settings, and has no impact on the state of practice (Bennis & O'Toole, 2005). In other words, they stated that the transfer of academic knowledge to practice does not occur. The present study offers empirical evidence that refutes this claim. It was found that the academic body of knowledge does have an impact on the behavior, operations, and decisions of practitioners when it is delivered through appropriate channels.

Implication #2: *Graduates of Doctoral Business Programs Act as Knowledge Transfer Intermediaries That Aggregate, Summarize, Communicate, and Implement Findings Reported in Academic Publications*

According to the direct knowledge transfer channel, practitioners are supposed to access, read, and use the body of knowledge existing in peer-reviewed publications. This channel, however, is rarely utilized because practitioners generally lack the necessary training required to consume academic knowledge directly from its source. In contrast, academic knowledge may be successfully transferred to practice by means of indirect dissemination channels which aggregate, summarize, communicate, and convert academic knowledge into a format that may be easily comprehended by busy practitioners. This study demonstrated that doctoral business program graduates fulfill this role relatively successfully. Many of them not only use academic knowledge in practice but also deliver it to other members of their organization and clients who, in turn, will retain it and use in the future.

Implication #3: *Academic Journals Have a Potential to Disseminate Scholarly Knowledge Beyond the Academic World.*

Peer-reviewed journals have served as a primary tool for the promotion and development of science for centuries (de Vaujany, Walsh, & Mitev, 2011; Greco et al., 2006; Merton & Sztompka, 1996). They help scholars communicate their discoveries to a wider audience, retain copyright over their intellectual contributions, ensure the quality of publications by means of peer-reviewed processes, establish the direction of scientific progress, and preserve knowledge for future generations. This also applies to all types of business disciplines, including knowledge management (Serenko, 2013; Serenko & Dumay, 2015). This study demonstrates that, in addition to the critical mission above, academic journals play an important role in disseminating and preserving scholarly knowledge not only in the academic environment but also in practical settings. In this study, it was observed that many doctoral-degree-holding practitioners still access academic journals to acquire new knowledge. Thus, scholarly outlets may potentially promote academic knowledge far beyond the boundaries of scholarly institutions. For this, appropriate knowledge dissemination mechanisms should be further established.

Implication #4: *The Principles of the Dissemination of Academic Knowledge in the Practitioner-Oriented Environment Should Become a Component of Student Training in Doctoral Business Programs*

Doctoral business program graduates can act as valuable knowledge distribution channels that can enhance the effectiveness and efficiency of an organization. During their doctoral degree training, they should be prepared with the skills and experience necessary to act as intermediaries that promote the benefits of academic literature in the practitioner-oriented environment. Because the knowledge that practitioners receive during their Ph.D. training is applied outside of academia, all doctoral students should be taught the principles of scholarly knowledge dissemination. In a similar vein, De Meyer (2013) argues that doctoral education in business should include more interdisciplinary work and be targeted at a wider selection of career paths.

The ability of Ph.D. program graduates to translate and promote the latest scientific discoveries beyond the academic world has been already well-recognized in the field of medicine. For example, several Ph.D. programs in science have already introduced specialized curricula for this purpose (Smith, Jarrett, & Bierer, 2013). Many leading healthcare organizations institutionalized the position of knowledge broker whose primary goal is to bridge the gap between research and practice (Meyer, 2010; Ward, House, & Hamer, 2009). Thus, there are exemplars that the developers of doctoral business program curricula may follow.

Implication #5: *Demand for evidence-based knowledge in the practitioner's environment determines his or her probability of applying academic knowledge.*

One of the strongest indicators of the probability of an individual accessing and applying academic literature is whether or not his or her employer or client values academic knowledge. If the company or client does not value academic research, the likelihood of the practitioner referencing the academic material declines. One reason why this relationship exists is that, if the participant's employer does not value academic content, then it will not pay for the practitioner to have access to this material. This implication is somewhat similar to the findings of Siegel, Waldman, and Atwater (2004) who argue that cultural misunderstandings inhibited agreement between academics and practitioners. Some reasons why academic content is not seen as valuable included a lack of relevance, the scope of the research, and the complex language used. Additionally, as stated by many of the participants, it took considerable time and effort to convert academic knowledge into a form that is consumable by other users. This is partially attributed to the absorptive capacity and the responsive capacity of the organization as identified in the capacity-based model of knowledge transfer (Cohen & Levinthal, 1990; Parent et al., 2007). The absence of these capacities inhibits knowledge transfer within an organization. There is no motivation for these practitioners to undertake this effort if it is not appreciated or recognized as valuable. Therefore, managers should change their attitude towards the use of credible scientific evidence in order to promote the transfer of academic knowledge to industry.

Implication #6: *Not All Academic Knowledge is Perceived as Useful by Practitioners*

On the one hand, many practitioners found academic knowledge applicable to their business practice. On the other hand, the usefulness of academic knowledge is sometimes limited. For example, academic theories, principles, and recommendations are often very abstract. Practitioners, however, need to put the abstract ideas expressed in academic sources into the context relevant to their organization and current situation. Another issue is the timeliness of academic findings. It takes years from the beginning of a scholarly investigation until it appears in print and reaches its target audience. As a result, practitioners frequently perceive academic findings as somewhat obsolete. The same finding was also reported in previous empirical investigations of the practical relevance of academic business research (Thomas & Tymon, 1982; Varadarajan, 2003). Nevertheless, many practitioners are still able to benefit from academic discoveries reported in peer-reviewed sources, but too often the search process for applicable knowledge becomes time consuming, which in turn negatively affects the perception of usefulness of academic output.

Implication #7: *Limited Access to Academic Literature is a Major Impediment to the Application of Scholarly Findings in Practice*

In this study, a major difference between consultants and government employees was observed: the latter acquired, transformed, integrated, applied, and transferred academic knowledge to a greater extent. The issue, however, was not in the differences in the sectors (private vs. public), job titles, or tasks performed by both groups of practitioners. The problem lay in the accessibility of academic literature – very few consultants had access to subscription-based academic publications whereas most government employees did. Thus, access plays a key role in the distribution of scholarly knowledge beyond the academic world.

Implication #8: *The Practitioners' Connection With Academia Upon Graduation is Linked to Their Probability of Using Academic Literature*

Those practitioners who kept in touch with the academic environment through publications or engaging in university or college teaching also maintained their academic connection by accessing and employing academic knowledge in workplace. This finding may be explained from the perspective of the social identity theory (Tajfel & Turner, 1979), which explains how an individual's self-concept can predict behavior. In-group preference appears where individuals favour the group they identify with, such as the academic community (Sachdev & Bourhis, 1987). This in-group preference for academic community formed continued appreciation for scientific knowledge. As a result, practitioners who continued publishing scholarly works after graduation or taught at an academic institution were far more likely to use academic research than those who did not publish or teach courses. Therefore, they were more likely to self-identify as a member of the academic society while practitioners who did not have this connection might have seen academics as an out-group. Future research should further explore the role of academic ties of university graduates with the academic environment and their use of scholarly research in workplace.

Implication #9: *Alert Tools are an Important Mechanism Helping Practitioners Stay Abreast of Relevant Academic Knowledge*

Time pressure is one of the most important components of the contemporary workplace environment. Unlike full-time academics, practitioners are not paid for continually digesting and staying up-to-date with the latest scientific discoveries. They cannot dedicate time to continually browsing new issues of peer-reviewed outlets and accumulating new knowledge in case they may need it in the future. Alert tools, which inform the subscriber about new publications that match specific criteria, were employed by almost half of the interviewed practitioners. Interestingly, many had not heard about the advantage of alert tools before and indicated they would use them in the future. Therefore, electronic alert tools may serve as an important mechanism to help busy practitioners keep track of the relevant knowledge in their respective areas.

CONCLUSION

The results of this study support and further develop current literature in the realm of knowledge transfer. The adapted and extended process-based model of knowledge transfer incorporates relevant theory regarding knowledge and how it should impact the behavior of the receiver. Additionally, the findings of this study contribute to understanding indirect knowledge dissemination channels and how intermediaries process knowledge for the consumption by others. Particularly, this study empirically demonstrates that doctoral program graduates who join the non-academic sector after graduation promote the dissemination of academic knowledge.

The results suggest that there must be a greater demand and appreciation for evidence-based knowledge. Current organizational cultures outside the public sector are not conducive to a practitioner accessing and applying academic literature. The findings presented in this study enhance our understanding of the factors that affect a doctoral business program graduate's likelihood of acquiring, utilizing, and disseminating academic knowledge. Organizations employing doctoral degree holders should consider providing them with access to academic literature, which may improve their decision-making. Most importantly, these individuals may act as knowledge ambassadors to deliver academic knowledge to their colleagues and present it in an appropriate format. This, in turn, may improve overall organizational performance. Particularly, this is an important issue for consulting companies, which, in contrast to public organizations, rarely provide their employees with access to academic material.

Recently, the role of the academic institution and its responsibility for the accessibility and distribution of academic research was questioned. One important finding from this study is that the characteristics of academic publications created a barrier for many practitioners. These limiting features include the narrow scope, limited generalizability, complex language, and the sample population used in such publications. Yet, if academics were to present their research to accommodate practitioners, they would never be published in academic journals. Because journal publications are a significant aspect of academics' performance evaluation, it doesn't make sense for them to do this. Therefore, the required form and preferred function of academic articles are apparently not well aligned between the stakeholders. While some participants believed these institutions should be changing to address what industry values, others argued this is not so and that such accommodation should not be the function of universities. This echoes the debate in academic circles. However, it is unclear if an academic institution or academic journal can be sustainable if it does not fill industry's need for knowledge: can there be a self-sufficient industry with academics publishing solely for themselves? With consultants increasingly being viewed as a viable alternative for academic knowledge dissemination, this is becoming an urgent matter for policy makers.

Considering the increasing competitiveness of the academic job market around the world, more doctoral business program graduates will join the non-academic sector in the future. Therefore, they need to receive not only theoretical but also applied knowledge during their training. Particularly, an ability to convert academic findings to actionable items should be strongly emphasized. For this, changes to the academic curricula are required at both institutional and national levels.

ACKNOWLEDGMENT

This study is kindly supported by the research grant 864-07-0181 from the Social Sciences and Humanities Research Council of Canada (SSHRC).

REFERENCES

Abrahamson, E. (1991). Managerial fads and fashions: The diffusion and rejection of innovations. *Academy of Management Review*, *16*(3), 586–612.

Almond, N. (2001). The transportability and dissemination of knowledge: A surveyor's perspective. *Property Management*, *19*(2), 124–135. doi:10.1108/02637470110387849

Alvarez, C. (1998). Making the case for hiring a consultant. *Clinical Nurse Specialist CNS*, *12*(2), 52–65. doi:10.1097/00002800-199803000-00003 PMID:9708106

Bartunek, J. M., & McKenzie, J. (Eds.). (2018). *Academic-practitioner relationships: Developments, complexities and opportunities*. Routledge.

Baxter, L. A., & Braithwaite, D. O. (Eds.). (2008). *Engaging theories in interpersonal communication: multiple perspectives*. Los Angeles, CA: Sage. doi:10.4135/9781483329529

Benbasat, I., & Zmud, R. W. (1999). Empirical research in information systems: The practice of relevance. *Management Information Systems Quarterly*, *23*(1), 3–16. doi:10.2307/249403

Benjamin, B., & O'Reilly, C. (2011). Becoming a leader: Early career challenges faced by MBA graduates. *Academy of Management Learning & Education*, *10*(3), 452–472. doi:10.5465/amle.2011.0002

Bennis, W. G., & O'Toole, J. (2005). How business schools lost their way. *Harvard Business Review*, *83*(5), 96–104. PMID:15929407

Berger, P. L., & Luckmann, T. (1966). *The social construction of reality: A treatise on the sociology of knowledge*. New York, NY: Anchor Books.

Booker, L., Bontis, N., & Serenko, A. (2008). The relevance of knowledge management and intellectual capital research. *Knowledge and Process Management*, *15*(4), 235–246. doi:10.1002/kpm.314

Booker, L., Bontis, N., & Serenko, A. (2012). Evidence-based management and academic research relevance. *Knowledge and Process Management*, *19*(3), 121–130. doi:10.1002/kpm.1392

Bramwell, A., & Wolfe, D. A. (2008). Universities and regional economic development: The entrepreneurial University of Waterloo. *Research Policy*, *37*(8), 1175–1187. doi:10.1016/j.respol.2008.04.016

Burns, N., & Grove, S. K. (2005). *The practice of nursing research: Conduct, critique & utilization*. St Louis, MO: Elsevier.

Bylund, C. L., Peterson, E. B., & Cameron, K. A. (2012). A practitioner's guide to interpersonal communication theory: An overview and exploration of selected theories. *Patient Education and Counseling*, *87*(3), 261–267. doi:10.1016/j.pec.2011.10.006 PMID:22112396

Cavusgil, S. T., Calantone, R. J., & Zhao, Y. (2003). Tacit knowledge transfer and firm innovation capability. *Journal of Business and Industrial Marketing*, *18*(1), 6–22. doi:10.1108/08858620310458615

Christensen, P. H. (2003). *Knowledge sharing - time sensitive and push-pull strategies in a non-hype organization*. Copenhagen: Copenhagen Business School.

Cohen, W. M., & Levinthal, D. A. (1990). Absorptive capacity: A new perspective on learning and innovation. *Administrative Science Quarterly*, *35*(1), 128–152. doi:10.2307/2393553

Cole, F. L. (1988). Content analysis: Process and application. *Clinical Nurse Specialist CNS*, *2*(1), 53–57. doi:10.1097/00002800-198800210-00025 PMID:3349413

Crowne, D. P., & Marlowe, D. (1960). A new scale of social desirability independent of psychopathology. *Journal of Consulting Psychology*, *24*(4), 349–354. doi:10.1037/h0047358 PMID:13813058

Davenport, T. H., & Prusak, C. (1998). *Working Knowledge*. Boston: Harvard Business School Press.

De Meyer, A. (2013). The future of doctoral education in business administration. *Journal of Management Development*, *32*(5), 477–486. doi:10.1108/02621711311328264

de Vaujany, F.-X., Walsh, I., & Mitev, N. (2011). An historically grounded critical analysis of research articles in IS. *European Journal of Information Systems*, *20*(4), 395–417. doi:10.1057/ejis.2011.13

Dey, I. (1993). *Qualitative Data Analysis. A user-friendly guide for social scientists*. London: Routledge. doi:10.4324/9780203412497

Dillard, J. P. (1990). A goal-driven model of interpersonal influence. In J. P. Dillard (Ed.), *Seeking compliance: The production of interpersonal influence messages*. Scottsdale, AZ: Gorsuch Scarisbrick.

Dillard, J. P. (2008). Goals-plans-action theory of message production: Making influence messages. In L. A. Baxter & D. O. Braithwaite (Eds.), *Engaging theories in interpersonal communication: Multiple perspectives* (pp. 65–76). Los Angeles, CA: Sage. doi:10.4135/9781483329529.n5

Elo, S., & Kyngas, H. (2008). The qualitative content analysis process. *Journal of Advanced Nursing, 61*(1), 107–115. doi:10.1111/j.1365-2648.2007.04569.x PMID:18352969

Erlandson, D. A., Harris, E. L., Skipper, B. L., & Allen, S. D. (1993). *Doing naturalistic inquiry: A guide to methods*. Newbury Park, CA: Sage Publications.

Feldman, M. P. (1994). The university and economic development: The case of the Johns Hopkins University and Baltimore. *Economic Development Quarterly, 8*(1), 67–76. doi:10.1177/089124249400800105

Fisher, R. J. (1993). Social desirability bias and the validity of indirect questioning. *The Journal of Consumer Research, 20*(2), 303–315. doi:10.1086/209351

Fleming, L., & Frenken, K. (2007). The evolution of inventor networks in the silicon valley and Boston regions. *Advances in Complex Systems, 10*(1), 53–68. doi:10.1142/S0219525907000921

Flick, U. (2002). An introduction to qualitative research (2nd ed.). London: Sage Publications.

Fontana, R., Geuna, A., & Matt, M. (2006). Factors affecting university - industry R&D projects: The importance of searching, screening, and signalling. *Research Policy, 35*(2), 309–323. doi:10.1016/j.respol.2005.12.001

Foss, N. J., & Pedersen, T. (2002). Transferring knowledge in MNCs: The role of sources of subsidiary knowledge and organizational context. *Journal of International Management, 8*(1), 49–67. doi:10.1016/S1075-4253(01)00054-0

Giles, H. (2008). Communication accommodation theory. In L. A. Baxter & D. O. Braithwaite (Eds.), *Engaging theories in interpersonal communication: Multiple perspectives* (pp. 161–173). Los Angeles, CA: Sage. doi:10.4135/9781483329529.n12

Greco, A. N., Wharton, R. M., Estelami, H., & Jones, R. F. (2006). The state of Scholarly Journal Publishing: 1981 – 2000. *Journal of Scholarly Publishing, 37*(3), 155–214. doi:10.3138/jsp.37.3.155

Hanberger, A., & Schild, I. (2004). Strategies to evaluate a university-industry knowledge-exchange programme. *Evaluation, 10*(4), 475–492. doi:10.1177/1356389004050217

Hannabuss, S. (2001). A wider view of knowledge. *Library Management, 22*(8), 357–363. doi:10.1108/01435120110406273

Harwood, T. G., & Garry, T. (2003). An overview of content analysis. *The Marketing Review, 3*(4), 479–498. doi:10.1362/146934703771910080

Hassan, N. R. (2014). Value of IS research: Is there a crisis? *Communications of the Association for Information Systems, 34*, 801-816.

Hassan, N. R., & Mathiassen, L. (2018). Distilling a body of knowledge for information systems development. *Information Systems Journal.*

Heisig, P., Suraj, O. A., Kianto, A., Kemboi, C., Arrau, G. P., & Easa, N. F. (2016). Knowledge management and business performance: Global experts' views on future research needs. *Journal of Knowledge Management*, *20*(6), 1169–1198. doi:10.1108/JKM-12-2015-0521

Hewes, D. E., & Planalp, S. (1987). The individual's place in communication science. In C. R. Berger & S. H. Chaffer (Eds.), *Handbook of communication science*. Newbury Park, CA: Sage.

Hitt, M. (1998). Twenty first century organizations? Business firms, business schools, and the academy. *Academy of Management Review*, *23*(2), 218–225.

Hodgkinson, G. P., & Starkey, K. (2011). Not simply returning to the same answer over and over again: Reframing relevance. *British Journal of Management*, *22*(3), 355–369. doi:10.1111/j.1467-8551.2011.00757.x

Hutchison, J., & Huberman, M. (1994). Knowledge dissemination and use in science and mathematics education: A literature review. *Journal of Science Education and Technology*, *3*(1), 27–47. doi:10.1007/BF01575814

Jacob, M. (2001). Managing the institutionalisation of Mode 2 knowledge production. *Science Studies*, *14*(2), 83–100.

Jagodinski, C. M. (2008). The university press in North America: A brief history. *Journal of Scholarly Publishing*, *40*(1), 1–20. doi:10.3138/jsp.40.1.1

Jennex, M. E. (2001). Research relevance - you get what you reward. *Communications of the Association for Information Systems, 6*, 49-52.

Jennex, M. E., & Olfman, L. (2005). Assessing knowledge management success. *International Journal of Knowledge Management*, *1*(2), 33–49. doi:10.4018/jkm.2005040104

Jennex, M. E., & Olfman, L. (2006). A model of knowledge management success. *International Journal of Knowledge Management*, *2*(3), 51–68. doi:10.4018/jkm.2006070104

Khurana, R. (2007). *From higher aims to hired hands: The social transformation of American business schools and the unfulfilled promise of management as a profession*. Princeton, NJ: Princeton University Press.

King, D. A. (2004). The scientific impact of nations: What different countries get for their research spending. *Nature*, *430*(6997), 311–316. doi:10.1038/430311a PMID:15254529

Knights, D. (2008). Myopic rhetorics: Reflecting epistemologically and ethically on the demand for relevance in organizational and management research. *Academy of Management Learning & Education*, *7*(4), 537–552. doi:10.5465/AMLE.2008.35882194

Knights, D., & Scarbrough, H. (2010). In search of relevance: Perspectives on the contribution of academic-practitioner networks. *Organization Studies*, *31*(9), 1287–1309. doi:10.1177/0170840610374398

Lahti, R. K., & Beyerlein, M. M. (2000). Knowledge transfer and management consulting: A look at "the firm". *Business Horizons*, *43*(1), 65–74. doi:10.1016/S0007-6813(00)87389-9

Liyanage, C., Elhag, T., Ballal, T., & Li, Q. (2009). Knowledge communication and translation - a knowledge transfer model. *Journal of Knowledge Management*, *13*(3), 118–131. doi:10.1108/13673270910962914

Lönnqvist, A. (2017). Embedded knowledge management: Towards improved managerial relevance. *Knowledge Management Research and Practice*, *15*(2), 184–191. doi:10.105741275-017-0053-y

McLuhan, M. (1962). *The Gutenberg Galaxy: The making of typographic man*. Toronto, Canada: University of Toronto Press.

Merton, R. K., & Sztompka, P. (1996). *On social structure and science*. Chicago, IL: University of Chicago Press.

Meyer, M. (2010). The rise of the knowledge broker. *Science Communication*, *32*(1), 118–127. doi:10.1177/1075547009359797

Nohria, N., & Eccles, R. G. (1998). Where does management knowledge come from? In J. L. Alvarez (Ed.), *The diffusion and consumption of business knowledge* (pp. 278–304). doi:10.1007/978-1-349-25899-4_13

Parent, R., Roy, M., & St-Jaques, D. (2007). A systems-based dynamic knowledge transfer capacity model. *Journal of Knowledge Management*, *11*(6), 81–93. doi:10.1108/13673270710832181

Pearson, J. M., Pearson, A., & Shim, J. P. (2005). The relevancy of information systems research: The practitioner's view. *Information Resources Management Journal*, *18*(3), 50–67. doi:10.4018/irmj.2005070104

Pfeffer, J., & Fong, C. T. (2002). The end of business schools? Less success than meets the eye. *Academy of Management Learning & Education*, *1*(1), 78–95. doi:10.5465/AMLE.2002.7373679

Roberts, C. H., & Skeat, T. C. (1983). *The birth of the codex*. London: Oxford University Press.

Roling, N. G. (1992). The emergence of knowledge systems thinking: A changing perception of relationships among innovation, knowledge process and configuration. *Knowledge and Policy*, *5*(1), 42–64. doi:10.1007/BF02692791

Rottman, J. W. (2008). Successful knowledge transfer within offshore supplier networks: A case study exploring social capital in strategic alliances. *Journal of Information Technology*, *23*(1), 31–43. doi:10.1057/palgrave.jit.2000127

Rousseau, D. M., & McCarthy, S. (2007). Educating managers from an evidence-based perspective. *Academy of Management Learning & Education*, *6*(1), 84–101. doi:10.5465/AMLE.2007.24401705

Rousseau, S., & Rousseau, R. (1998). The scientific wealth of European nations: Taking effectiveness into account. *Scientometrics*, *42*(1), 75–87. doi:10.1007/BF02465013

Rynes, S. L., Bartunek, J. M., & Daft, R. L. (2001). Across the great divide: Knowledge creation and transfer between practitioners and academics. *Academy of Management Journal*, *44*(2), 350–355.

Sachdev, I., & Bourhis, R. (1987). Status differentials and intergroup behaviour. *European Journal of Social Psychology*, *17*(3), 277–293. doi:10.1002/ejsp.2420170304

Saenger, P. (1975). Colard Mansion and the evolution of the printed book. *The Library Quarterly*, *45*(4), 405–418. doi:10.1086/620426

Scarbrough, H. (2002). The role of intermediary groups in shaping management fashion. *International Studies of Management & Organization*, *32*(4), 87–103.

Serenko, A. (2013). Meta-analysis of scientometric research of knowledge management: Discovering the identity of the discipline. *Journal of Knowledge Management*, *17*(5), 773–812. doi:10.1108/JKM-05-2013-0166

Serenko, A., Bontis, N., & Hull, E. (2011). Practical relevance of knowledge management and intellectual capital scholarly research: Books as knowledge translation agents. *Knowledge and Process Management*, *18*(1), 1–9. doi:10.1002/kpm.363

Serenko, A., Bontis, N., & Moshonsky, M. (2012). Books as a knowledge translation mechanism: Citation analysis and author survey. *Journal of Knowledge Management*, *16*(3), 495–511. doi:10.1108/13673271211238797

Serenko, A., & Dumay, J. C. (2015). Citation classics published in knowledge management journals. Part I: Articles and their characteristics. *Journal of Knowledge Management*, *19*(2), 401–431. doi:10.1108/JKM-06-2014-0220

Siegel, D. S., Waldman, D. A., Atwater, L. E., & Link, A. N. (2004). Toward a model of effective transfer of scientific knowledge from academicians to practitioners: Qualitative evidence from the commercialization of university technologies. *Journal of Engineering Technology*, *21*(1), 115–142.

Simmons, P. G., Dawley, D., Richie, W., & Anthony, W. (2001). An exploratory examination of the knowledge transfer of strategic management concepts from the academic environment to practicing managers. *Journal of Managerial Issues*, *13*(3), 360–376.

Smith, C. L., Jarrett, M., & Bierer, S. B. (2013). Integrating clinical medicine into biomedical graduate education to promote translational research: Strategies from two new PhD programs. *Academic Medicine*, *88*(1), 137–143. doi:10.1097/ACM.0b013e31827644ef PMID:23165264

Spender, J. (2005). Speaking about management education: Some history of the search for academic legitimacy and the ownership and control of management knowledge. *Management Decision*, *42*(10), 1282–1292. doi:10.1108/00251740510634868

Starkey, K., & Madan, P. (2001). Bridging the relevance gap: Aligning stakeholders in the future of management research. *British Journal of Management*, *12*(S1), S3–S26. doi:10.1111/1467-8551.12.s1.2

Straub, D., & Ang, S. (2011). Editor's comments: Rigor and relevance in IS research: Redefining the debate and a call for future research. *Management Information Systems Quarterly*, *35*(1), iii–xi. doi:10.2307/23043485

Suddaby, R., & Greenwood, R. (2001). Colonizing knowledge: Commodification as a dynamic of jurisdictional expansion in professional service firms. *Human Relations*, *54*(7), 933–953. doi:10.1177/0018726701547007

Susanty, A., Puspitasari, D., Puspitasari, N. B., & Ninditarini, M. R. (2011). *Preliminary study of key success factors for effective knowledge transfer in SMEs Batik*. 2011 International Conference on Industrial Engineering and Operations Management, Kuala Lumpur, Malaysia.

Szulanski, G. (1996). Exploring internal stickiness: Impediments to the transfer of best practice within the firm. *Strategic Management Journal*, *17*(1), 27–43. doi:10.1002mj.4250171105

Tajfel, H., & Turner, J. C. (1979). An integrative theory of intergroup conflict. In W. G. Austin & S. Worchel (Eds.), *The social psychology of intergroup relations*. Monterey, CA: Brooks.

Taylor, H., Artman, E., & Woelfer, J. P. (2012). Information technology project risk management: Bridging the gap between research and practice. *Journal of Information Technology*, *27*(1), 17–34. doi:10.1057/jit.2011.29

Thomas, K. W., & Tymon, W. G. Jr. (1982). Necessary properties of relevant research: Lessons from recent criticisms of the Organizational Sciences. *Academy of Management Review*, *7*(3), 345–352.

Van Aken, J. E. (2005). Management research as a design science: Articulating the research products of Mode 2 knowledge production in management. *British Journal of Management*, *16*(1), 19–36. doi:10.1111/j.1467-8551.2005.00437.x

Van de Ven, A. H., & Johnson, P. E. (2006). Knowledge for theory and practice. *Academy of Management Review*, *31*(4), 802–821. doi:10.5465/AMR.2006.22527385

Varadarajan, P. R. (2003). Musing on relevance and rigor of scholarly research in marketing. *Journal of the Academy of Marketing Science*, *31*(4), 368–376. doi:10.1177/0092070303258240

Ward, V., House, A., & Hamer, S. (2009). Knowledge brokering: The missing link in the evidence to action chain? *Evidence & Policy: A Journal of Research. Debate and Practice*, *5*(3), 267–279. doi:10.1332/174426409X463811

Weber, R. P. (1990). *Basic content analysis*. Newbury Park, CA: Sage Publications. doi:10.4135/9781412983488

Weick, K. E. (2001). Gapping the relevance bridge: Fashion meets fundamentals in management research. *British Journal of Management*, *12*(s1), 71–76. doi:10.1111/1467-8551.12.s1.9

Wilson, S. R. (1997). Developing theories of persuasive message production: The next generation. In J. O. Greene (Ed.), *Advances in communication theory* (pp. 15–43). Hillsdale, NJ: Lawrence Erlbaum.

Yusuf, S. (2008). Intermediating knowledge exchange between universities and businesses. *Research Policy*, *37*(8), 1167–1174. doi:10.1016/j.respol.2008.04.011

Zolnik, E. J. (2010). The role of postdoctoral fellows in technology transfer: Evidence from the National Capital Region of the USA. *International Journal of Knowledge-Based Development*, *1*(3), 158–175. doi:10.1504/IJKBD.2010.035657

KEY TERMS AND DEFINITIONS

Knowledge Management: A systematic process of discovering, creating, sharing, retaining, and applying knowledge within and among organizations.

Knowledge Management Discipline: A multidisciplinary scholarly field building upon discoveries from various reference disciplines, including organizational behavior, human resources, strategic management, psychology, information systems/technologies, computer science, finance, etc. Its goal is to develop effective and efficient knowledge management principles.

Knowledge Transfer: A process by which translated academic knowledge reaches industry professionals.

Knowledge Translation: A process by which knowledge documented in academic sources, such as peer-reviewed journals, conference proceedings, and scholarly books, is converted to the format suitable for busy practitioners lacking rigorous academic training.

Research Relevance: A line of inquiry in order to understand the practical impact of academic findings.

APPENDIX: INTERVIEW QUESTIONS

Demographic Question Set #1

- What is your current job title and responsibilities? In what industry?
- If you were to describe the type of knowledge do you create on an average day, how would you do it?

Knowledge Awareness

Question #1: What sort of knowledge do you seek on an average day?
Question #2: How do you decide what source of knowledge to choose to solve each managerial problem? In other words, does your selection of knowledge sources depend on the nature of the problem?
Probe #1: What way of gaining new knowledge have you found most beneficial?
Question #3: Do you use specific tools to make you aware of new knowledge that is available to be consumed?

Knowledge Acquisition

Question #1: In your daily work you probably come across situations when you need credible sources of knowledge. Please, list in order of importance, sources of knowledge you use in your daily work.
Probe #1: Do you use knowledge that you learned in your PhD program? Your MBA/Management program?
Probe #2: Do you use academic journals, practitioner journals, books, or the Internet?
Which journal titles do you use most frequently?
Do you have access to academic research? How do you access it?
Do you read academic literature regularly or do you search for a particular topic only when you need it?
If you don't use academic publications, why not?
Do you attend academic/practitioner conferences? If yes, which ones? If not, why not?
Probe #3: Do you frequently use knowledge from your colleagues?

Knowledge Transformation

Question #1: With respect to your job, do you gain new knowledge by accessing academic literature? Does this new knowledge improve any of your existing skills or capabilities?

Knowledge Integration

Question #1: Can you comment on the general usefulness, applicability, and relevance of academic knowledge in aiding you in your daily work?

Knowledge Application

Question #1: How frequently do you utilize academic knowledge to benefit you and your organization?
Question #2: Can you give me a few examples when you applied academic knowledge in your daily work? What were the outcomes and benefits for you and the company?
Question #3: Have your company or clients ever recognized the value of applying knowledge from academic publications? If yes, can you offer an example of this?

Knowledge Transfer and Retention

Question #1: In your work, do you communicate or transfer your academic knowledge to your colleagues who don't have a PhD degree? Can you see yourself as a translator of academic knowledge?

Demographic Question Set #2

- When did you graduate from your PhD program?
- After getting your PhD, have you taught part-time or full-time at a college or university?
- What area was your PhD in?
- Have you published any works after obtaining your PhD?

Note: the list above represents a starting point to solicit the necessary answers. Additional questions and probes were used when needed to ensure that each point was adequately covered.

Section 4
KM and Security

Chapter 13
Integrating IS Security With Knowledge Management:
What Can Knowledge Management Learn From IS Security Vice Versa?

Murray Eugene Jennex
San Diego State University, USA

Alexandra Durcikova
Oklahoma University, USA

ABSTRACT

Knowledge management focuses on capturing and sharing knowledge. Because of this, KM researchers tend to focus on issues related to knowledge capture, storage, and sharing. However, because knowledge is valuable, it is a target needing to be protected. This chapter posits that KM researchers and practitioners also need to think about security and explores how important security skills are to KM practitioners and researchers. A literature review was performed to determine how much attention is paid by KM researchers to knowledge security. Additionally, KM job postings were examined to determine if security skills are considered important by those hiring KM practitioners. Next, a survey was prepared for exploring security attitudes of KM practitioners as an area of future research. Finally, future research areas for IS security are proposed that can greatly benefit from lessons learned in the areas of both knowledge sharing and knowledge sourcing.

INTRODUCTION

News reports have been reporting on several security breaches that either compromised the confidentiality of information and data stored in knowledge management systems (KMSs), availability of this knowledge, or its confidentiality. An example of the former is the Target breached with approximately in December 2013 when 40 million credit and debit card accounts exposed and financial and personnel data on up to 70 million customers accessed (Weiss and Miller, 2015). An example of the compromise of the

DOI: 10.4018/978-1-5225-5427-1.ch013

availability is the WannaCry ransomware that affected more than ten thousand computers in 36 hospitals and prevented medical staff from accessing vital patient information (Perlroth & Sanger, 2017). Finally, a recent spear-phishing prank (Tapper, 2017) that targeted some White House officials during summer of 2017 led to the disclosure of sensitive private information thus jeopardizing the confidentiality of it. While these examples my not be directly linked to a breach of a KMS, the magnitude and fall out from these massive security breaches also raises questions on the security of the knowledge stored in KMS. Information System (IS) security is about protecting IS assets, networks, data, information, computers, and applications by restricting access to the assets and preventing unauthorized modification or destruction. Knowledge management (KM) focuses on sharing and transferring knowledge from knowledge providers to knowledge users. It is not intuitive that security and KM are related as they have contradictory objectives: KM is about providing access to knowledge, while IS Security is about restricting access to knowledge. However, it is our position that KM and IS Security are complementary. While KM can be used to improve security performance, training and awareness (San Nicolas-Rocca, Schooley, & Spears, 2014), this chapter primarily focuses on the use of security in KM and touches on what IS Security can learn from KM. Knowledge has value and items of value are targets of theft and attack. This chapter posits that KM does not have close enough links with IS Security. It is posited that this is evidenced by not only a lack of research literature addressing the integration of KM and IS Security but also a lack of interest in integrating KM and IS Security by KM practitioners. To investigate the links between KM and IT Security this chapter performs a review of the KM research literature with respect to IS Security. Additionally, to assess how KM practitioners' value IS Security skills and capabilities KM job postings are analyzed to determine what skills and capabilities are desired in new KM position hires. Nest, we explore KM practitioner attitudes with respect to the role of IT Security in KM an exploratory survey is generated and presented in this chapter. Finally, before discussion, we touch upon papers that have already applied lessons learned from KM and KMS research into IS Security research.

The value of this chapter is in providing insight into perceptions and attitudes with respect to integrating IS Security into KM. The concern is that there is too little integration and that KM practitioners and researchers need to put more effort into creating secure KM and KMS. We believe this is necessary given the threat level in our cyber environment. As cyber threat is growing so is the cost associated with a breach. The Ponemon Cost of Cybercrime Report shows that the cost of data breeches has risen to an average cost of $8.9 million per breech in 2012, a 6% increase from 2011 (note that this is for the organizations in their survey) (Ponemon, 2012). The 2013 Ponemon Cost of Cybercrime report shows that the average cost per record breached rose from $188.00 in 2012 (Ponemon, 2013) to $2.3 billion USD annually according to the FBI (McCabe, 2016). In addition, the financial cost of a security breach, there is also the good will of all customers that is affected by it. Why then, as the case of Target breach illustrates, even if evidence of a breach is at hands, responsible managers don't take action? This chapter proposes that the lack of security awareness and knowledge by non-security practitioners, those practitioners responsible for sales, operations, and the performance of the organization; leads to poor security decision making. Our proposal is also supported by increased regulatory pressure on organizations to improve security awareness of their decision makers (Thune, 2017).

Finally, how are cyber-attacks carried out? While there is a variety of means, the most costly come from malicious insiders, web based attacks, and phishing attacks (Riecicky & Spichiger, 2012) with advanced persistent threats (APT) coming from state sponsored espionage (McWhorter, 2013) and sophisticated cybercriminal rings (Sophos, 2013). In addition, movement towards cloud and mobile technologies increases the reach of all of these threats. The cloud is increasing the use of service based products such as Software as a Service (SaaS), Infrastructure as a Service (IaaS), and Platform as a Service (PaaS). While these cloud services provide cost savings to organizations they also tend to increase security concerns (Kajiyama, Jennex, Addo, & Paolini, 2013). Finally, mobile technologies are easy to use and are widely adopted. However, there are many attacks focused on these devices and bring your own device (BYOD) policies are increasing the risk to the organization Sophos, 2013. In conclusion, there are persistent cyber threats to the knowledge relied upon by our organizations. We posit that it is a responsibility of KM researchers and practitioners to develop secure KM so that these critical knowledge assets can be stored, accessed, and utilized.

RESEARCH APPROACH

This is exploratory research focused on determining if KM and IS Security are integrated well enough to protect knowledge assets from the persistent threats of disclosure, modification, and destruction. Two data gathering methods are used and a third is piloted. The first is a literature review of the research literature. Google Scholar and the AIS eLibrary were used to search for KM papers that also discuss security. Relevant papers were reviewed and the security aspects summarized. Conclusions were reached based on the numbers of relevant papers found and the security concepts discussed.

The second data gathering method was to find current KM job postings. Google was used to search job posting sites such as monster.com, carrerbuilder.com, and glassdoor.com as well as independent job postings found on the web. 50 job postings were posted during the spring of 2013. The 50 job postings were divided into two groups, KM managers and KM technicians. Additionally, the postings were grouped by those related to defense activities and those related to civilian business. Each of the two groups, KM management and KM technician, were analyzed by coding the job requirements into a set of overall job requirements ranked by how often each was mentioned. This process was repeated in August 2014 to determine if there is an impact from the Target breach on KM job requirements. 38 job postings were found, 18 were KM manager postings, 20 were KM technician postings. Additionally, the 38 postings represented 23 different companies with 33 postings being for the United States, 3 for Canada, and 2 for the United Kingdom. Finally, the process was repeated one more time in September, 2017 to determine if there was an impact from the breaches and ransomware attacks from 2016 and 2017. 20 job postings representing the first page of a google search for KM job descriptions. An unexpected finding was the identification of two standard KM job descriptions.

The third data gathering method was to generate and pilot a survey for gathering opinion data on how important IS Security is to KM practitioners. The survey was generated based on general IS Security attitude surveys and included demographic items. The survey was distributed using the web and posting to KM practitioner groups and listservers. The results are considered a pilot study as the response was limited, only 13 were received. Survey results are analyzed using descriptive statistics.

LITERATURE REVIEW

The KM research literature was reviewed with respect to incorporating security into KM. Three types of papers were found: 1) KM papers that recognized the need for security for KM to be successful; 2) KM papers that applied specific IS Security technologies to solving specific KM issues; and 3) KM papers that incorporated risk management techniques. These are summarized below.

Security as Part of KM Success

How to be successful with KM? Initial research focused on the identification of critical success factors (CSFs) for initiating KM/KMS and the application of these CSFs into KM success models (Note: CSFs are areas in which satisfactory results ensure successful competitive performance and are the minimum key factors that an organization must have or do in order to achieve some goal (Alazmi & Zairi, 2003). Davenport, DeLong, and Beers (1998) identify four objectives for knowledge-based projects: create knowledge repositories, improve knowledge access, enhance knowledge environments, and manage knowledge as an asset. KM projects are successful when there is a growth in resources attached to the project, a growth in knowledge content, a likelihood that a project would survive without the support of a particular individual or two, and some evidence of financial return. Jennex and Olfman (2005) surveyed the literature on KM and KM project success to generate a list of KM CSFs. The following list of CSFs is ordered by the popularity of the CSF in the literature with the most commonly mentioned CSF listed first and the least mentioned CSF last:

- A Knowledge Strategy that identifies users, sources, processes, storage strategy, knowledge and links to knowledge for the KMS.
- Motivation and Commitment of users including incentives and training
- Integrated Technical Infrastructure including networks, databases/repositories, computers, software, KMS experts
- An organizational culture and structure that supports learning and the sharing and use of knowledge
- A common enterprise wide knowledge structure that is clearly articulated and easily understood
- Senior Management support including allocation of resources, leadership, and providing training
- Learning Organization
- There is a clear goal and purpose for the KMS
- Measures are established to assess the impacts of the KMS and the use of knowledge as well as verifying that the right knowledge is being captured
- The search, retrieval, and visualization functions of the KMS support easy knowledge use
- Work processes are designed that incorporate knowledge capture and use
- Security/protection of knowledge

The last CSF focuses on security and protection of knowledge. This has been incorporated into KM success models including Jennex and Olfman (2006) who modified the DeLone and McLean's (2003) IS Success Model to incorporate the KM CSFs listed above into a KM success model. Additionally, Lindsey (2002) proposed a KM effectiveness model based on combining Organizational Capability Perspective theory Gold, et al., 2001 and Contingency Perspective Theory (Irma Becerra-Fernandez, 2001) and included a protection construct.

In addition to the KM success models several other authors have identified security as a CSF. Holsapple and Joshi (2000) identified control as an influence on KM success and included security as a part of the control construct. Security was defined as the administrative and technical controls used to prevent unauthorized disclosure, modification, or destruction/obsolescence of knowledge.

Malhotra (2004) identified security and protection of knowledge in knowledge management systems, KMS, as an inhibitor to the sharing of knowledge and a cause for why KMS fail.

Alavi and Leidner (1999) identified as a concern of managers when considering implementing a KMS, the protection of data on the web.

Wipawayngkool (2009) discussed the need for KMS to be compliant with IS Security policies and practices. This was considered necessary for organizations to be competitive on an international basis.

Ilvonen, Jusilla, Kärkkäinen, and Paivarint (2015), Ilvonen, Alnne, Helander, and Vayrunen (2016), Ilvonen, Jussila, Karkkainen & Paivarinta (2016), and Sarigianni, Thallmann, and Manhart (Sarigianni, Thalmann, & Manhart, 2016) focus on how to protect shared knowledge either within the organization or outside of organizations where knowledge is shared via social media. Spears and San Nicolas-Rocca (Spears & San Nicolas-Rocca, 2016) suggest that one way to overcome potential knowledge loss due to security reasons is to build information security capacity skills and offer a case study from the health and human services sector that handle very sensitive client information. Jennex and Durcikova (2014) highlight that KM practitioners and researchers need security skills in order to be able to protect organizational knowledge. Finally, Saha, Paramaswaran, Chakraborty, and Mahanti (2013) offer a formal analysis of fraud when it comes improper to knowledge sharing.

To summarize, there is a small but growing body of literature that recognizes that successful KM requires the integration of IS Security even though this integration may result in lesser knowledge usability, making it more difficult to transfer knowledge. Note that usability as used in this paper refers to ease of use construct as defined by Davis (1989) in the Technology Acceptance Model (TAM). This integration is best stated as KM being compliant with the security goals and policies of the organization and IS Security best practices and standards. The implication is that the design and implementation of KMS should include IS Security requirements but in a thoughtful and balanced approach that minimizes the impact of IS Security on usability of the KMS

Security as Reflected in Risk Management in KM

There is also a small body of literature that looks at the integration of KM and IS Security through the application of risk management to KM. KM has been described as a selective capture and use of knowledge Jennex (M. E. Jennex, 2005) and KM success has been described as getting the right knowledge to the right people at the right time (M. E. Jennex, Smolnik, & Croasdell, 2012). Both of these activities have risk where risk is a measure of the likelihood and consequence of something bad happening(Stoneburner, gogunen, & Feringa, 2007); or in these cases the likelihood and probability of not capturing the right knowledge or getting it to the right people at the right time. Authors integrating risk management into KM are summarized below:

Jennex (2006, 2013) explored the risk of forgetting and losing knowledge through loss of experts and lack of persistence of knowledge storage media. The most notable example is the inability of the National Aeronautical and Space Administration (NASA) to return to the moon quickly due to the loss of knowledge on how to build moon mission spacecraft. The proposal is to use risk management in developing a KM strategy to identify what knowledge is to be captured and how it is to be stored.

Ermine (Ermine, 2013) discussed the knowledge loss risk associated with retiring workers. He proposed three levels of knowledge loss: knowledge gap, knowledge loss, and knowledge gap with the knowledge loss level being based on the amount of knowledge lost and the prediction that the upcoming surge in retirements will cause a knowledge crash in many organizations. He proposes that KM can be used to manage this risk. Jennex (2009, 2014) and Jennex and Durcikova (2013) also investigated the upcoming retirement surge as well as knowledge losses due to other departing workers and used risk management to create a framework for prioritizing knowledge assets so that organizations can allocate resources as appropriate and use KM to capture critical knowledge before it is lost. This allows organizations and KM managers to manage the risk of losing knowledge to departing knowledge workers by identifying where critical knowledge resides and the likelihood of that knowledge source leaving.

Lambe (Lambe, 2013) identifies four types of knowledge risk: knowledge continuity, knowledge acquisition, knowledge outsourcing, and knowledge articulation. These are risks related to an organization not being able to obtain or maintain critical knowledge resources and it is proposed that KM strategy needs to include a focus on these risks. Riecicky and Spichiger (2012) addressed these risks and used risk management to create a framework for valuing and managing knowledge assets. This allows organizations and KM managers to make better decisions on how to allocate resources to manage all KM assets including, hardware, software, people, and data.

Jennex and Zyngier (2007) proposed using McCumber's cube (2004) as a framework for analyzing a KM initiative/KMS. The goal is to use this analysis to provide a risk based approach to identifying the security policies and controls for securing KM/KMS.

According to Schinagl, Paans, and Schoon (2016) leveraging IT risk management techniques and the usage of security standards and certification can reduce the risk of knowledge loss. Genre-based assessment of information and knowledge security risk can add additional safeguards to knowledge loss because it identifies organizational communication patterns through which organizational knowledge is shared (Padyab, Päivärinta, & Harnesk, 2014).

Knowledge loss not only occurs through improper sharing but also because of departing employees. Jennex and Durcikova (2013) offer a methodology of knowledge loss risk assessment that prioritizes efforts within an organization to capture knowledge from departing employees.

To summarize, there is a small but growing body of literature that is applying risk based approaches to managing KM to be successful by addressing the risks of losing critical knowledge, not capturing the right knowledge, not getting it to the right people, and not getting it to them at the right time. The current body of literature is not sufficient to address these risks but is a start in integrating IS Security risk management techniques into KM.

IS Security Used to Address Issues in KM

This is a small body of literature that is focused on addressing specific security issues in KM. Particularly are the issues of access control, secure communication, and secure storage. Upadhyaya, Rao, and Padmanabhan (2006) discuss the components of a KMS. These include secure languages such as security-assertion markup languages and secure knowledge query and manipulation language for secure communication; circles of trust where two or more organizations share supplier/customer authentication information (also for secure communication); and digital rights management and secure content management for access control.

Thuraisingham and Nonmonotonic (1992) focused on database design to create secure database management systems and addressing the secure storage issue. Boella and van der Torre (2006) identified the access control policies for managing knowledge sharing in virtual KM communities. Randeree (Randeree, 2005) investigated access control versus the need to share medical data and information.

Cannataro and Talia (Cannataro & Talia, 2003) discuss the security requirements for creating secure parallel and distributed knowledge discovery systems, PDKD, called a knowledge grid.

Finally, Neville, Powell, and Panteli (2003) propose a research model and agenda for investigating the relationships between IS Security components and knowledge. The model maps factors for and barriers to security to factors for and barriers to knowledge. This allows researchers to identify issues that can be mapped into KM processes.

To summarize, KM researchers have begun a small body of literature that is starting to address the areas of concern as outlined in the McCumber cube (McCumber, 2004).

SECURITY IN KM JOB POSTINGS

The second step of this research was to see what KM practitioners were looking for in new hires. We were particularly looking to see if security skills and/or knowledge were considered important. To determine this, 50 KM job postings were analyzed by coding the job requirements in order to determine the most asked for skills and knowledge. The 50 job postings were grouped into two categories, KM manager jobs (21 postings) and KM technician jobs (29 postings). The process was repeated for 38 job postings (18 KM manager and 20 KM technician) collected at the end of August 2014. Table 1 lists the main job criteria (more than one job posting listing the criteria) for the KM Manager group. Seventeen criteria were identified from the first 50 job postings with possessing a Bachelor's degree and having good written and oral skills being the top criteria. An additional four criteria were identified from the second 18 postings. What is most interesting for this chapter is that two security related criteria are listed. Possessing an active security clearance (secret or top secret were requested) appeared on 7 out of 21 postings (the seventh most requested criteria). Understanding organizational security standards/policies or possess a security certification was tied for the eleventh most requested criteria (4 listings).

Table 2 lists the main job criteria (more than one job posting listing the criteria) for the KM technician group. Nineteen criteria were identified from the job postings with possessing a Bachelor's degree and having technical experience being the top criteria. What is most interesting for this chapter is that two security related criteria are listed. Possessing an active security clearance (secret or top secret were requested) appeared on 9 out of 29 postings (the fourth most requested criteria). Understanding organizational security standards/policies or possess a security certification was tied for the thirteenth most requested criteria (2 listings each).

Prior to analyzing the KM job postings, it was hypothesized that there would be little to no security requirements listed. This is only partly true. Security clearances are an access control mechanism that appeared on 16 of the first 50 postings but on none of the second 38 postings. This was surprising at first until a review of the postings found that eleven postings were specifically for defense related KM positions (defense related positions required access to classified materials in order to manage the knowledge) while none of the second 38 postings were defense related (although 6 were financial related). This implies that really only five out of 39 non-defense postings requested a clearance of some type and so is not as impressive.

What was interesting with the second 38 postings was the identification of requirements of maintaining discretion and being familiar with regulatory/legislative requirements. This may be indicative of a realization of the need to meet regulatory based security requirements, such as in the financial industry, and a general need to not spread sensitive information.

Table 1. KM manager job requirements #first 21/# second 18

Job Skill/Knowledge Criteria	Number Listing
BA or BS degree	18/15
proven oral and written skills	18/18
experience with project management/PMP	12/10
experience with knowledge management	12/16
advanced degree	9/4
experience promoting & providing technical support for collaborative tools including SharePoint 2010, web conferring tools and other Web 2.0 technologies (Communities of Interest, wikis, blogs, forums, etc.)	8/5
Must possess active security clearance	7/0
Knowledge of the SDLC	6/3
Able/willing to travel in the U.S./internationally	6/3
Ability to work independently and as part of a team	5/9
experience with information management	4/3
ITIL cert	4/2
understand organizational information security standards/security certs	4/0
experience with business process improvement	3/5
CKM/other Certification	3/1
experience with using/implementing social media	2/4
experience in Enterprise IT	2/2
Experience with business/business strategy	/8
Experience in finance	/4
Experience with CRM/SCM	/2
Familiar with regulatory/legislative requirements	/3

We fully expected to see security requirements in the sample of 20 2017 KM job postings due to the flurry of security issues in 2016 and 2017. We were pleased that the two standard KM descriptions included security requirements:

- The KM Specialist generic job description included the requirement to be able to encrypt the data/information/knowledge once it is stored in digital format so that it cannot be manipulated, destroyed, or otherwise get infected by virus, trojan, malware, etc.
- The Knowledge Manager generic job description included the requirement to act as the organisation's data protection officer by managing data protection policies and the organization's response to data protection breaches. Additional duties include awareness training and handling ad hoc requests for data sharing advice, and data protection audits.

Table 2. KM technician job requirements # first 29/# second 20

Job Skill/Knowledge Criteria	Number Listing
B.A/B.S. degree.	18/17
technical experience	18/20
Communication Skills (written and oral)	12/20
security clearance	9/0
proficient: HTML, XHTML, CSS, Microsoft SharePoint 2007/2010, Microsoft Office 2007/2010, Adobe Acrobat Professional.	8/14
KM project experience/PMP	6/9
Experience in KM Systems	4/10
U.S. Citizen.	3/0
web portal design techniques: SharePoint portal 2007/2010; maintenance, permissions management, and web environment design.	3/1
experience in Enterprise IT	3/3
Advanced degree	3/11
ITIL Cert	3/1
Change Management/Business Readiness	2/2
Training Delivery Expert	2/0
Training Development Expert	2/3
JavaScript, SQL, and ASP.NET (VB and/or C#), Microsoft SQL Server, Microsoft SharePoint Designer, Adobe Acrobat Professional (8.0+), Adobe Photoshop (CS3+).	2/3
certifications in A+/Net+, Security+ GAIC GSEC	2/0
understand organizational information security standards	2/0
experience with multiple countries/cultures	2/0
Maintain discretion when necessary	/2
Collaborative, team player	/16
Familiar with regulatory/legislative requirements	/1

Unfortunately, the actual job descriptions did not support the generic requirements. Of the 20 job descriptions, 11 were for non-defense related positions and 9 were for defense related positions. All the defense related positions required a security clearance while 3 added the following two requirements:

- Shall understand applicable Privacy, Security and DoD Information Assurance directives that apply to information systems.
- Shall have a comprehensive knowledge of Navy and DoD regulations and protocols for information dissemination.

None of the 11 non-defense related positions had clearance or other security requirements. The conclusion after looking at the 2017 sample is still that KM job positions are wanting in requiring security knowledge.

SURVEY OF KM PRACTITIONER SECURITY ATTITUDES

An online exploratory survey was designed to gather data on KM practitioners' attitudes with respect to the role of IT Security in KM. The participants were KM professions who subscribed to a KM listserv. A call to fill out a survey on IT Security and KM was posted on this listerv and a reminder was sent out one week after the initial call for participation. The survey was open for 14 days and we received 13 usable responses.

Scales for this survey were adapted from DeSouza (2007) and all items were measure on a five and seven-point Liker scale anchored on 1 (strongly disagree) and 5 (strongly agree). The survey focused on three distinct areas of KM security: (1) organizational believes about security, (2) preventative measures to secure knowledge assets, and (3) KM professionals' opinions about KM security.

The response rate to the survey was 48% (13 out of 27 people that viewed the survey filled it out). The average age of respondents was 48 years, and 9 were males. Our subjects worked for various industries, including IT, telecommunications, nonprofit, utility; the number of employees in these organizations were from 100 to 100,000. These KM professionals were employed at their respective organizations on an average for 13 years, and as KM professionals on an average of 5 years. These organizations used a variety of KM tools including SharePoint. The results to our survey are in Table 3.

The results of this survey suggest that majority of the companies consider knowledge to be an asset (items 1-4) but also suggests that the organizations they work for are not that concerned with loosing or potential damage to the knowledge they created (items 5-7). This might be reason that they are not that concerned with creating a knowledge security framework nor they have a contingency plan in place (item 8-10). While companies do background checks on employees, they are not too worried about other ways of securing the knowledge in their organizations (items 12-18). While KM professions do consider about security issues (item 19) it seems that organizational leaders are not too concerned about security issues around KM (item 20).

We also asked respondents to insert comments about KM and security. Interestingly, five respondents indicated that they their organization dealt with a security issue in their KM initiative and several mentioned that the main issue that KM security professionals face is the "… ability to balance security issues with accessibility issues" (a KM security specialist).

KNOWLEDGE SHARING AND SOURCING RESEARCH

Lot of IS Security research focuses on improving compliance with organizational security policy and on strategies to diminish cyber threat mostly through phishing email detection. Knowledge management techniques including motivators to knowledge sharing and sourcing can be of help in this area because researchers have been studying what motivates employees to share and source knowledge for decades. For example, Burns, Roberts, Posey, Bennett, and Courtney (2015) suggest that proper motivation can

Table 3. Survey results

My Organization... (7 Point Likert Scale)	
1. has a way of identifying intellectual assets	5.93
2. considers knowledge one of the most important assets in my organization.	5.40
3. is concerned with knowledge being lost.	5.33
4. is concerned with knowledge being stolen.	5.73
5. is concerned with knowledge being destroyed.	4.33
6. believes that security skills are necessary for a knowledge management professional.	4.47
7. is concerned with knowledge access rather than securing knowledge.	4.33
8. has a knowledge security framework in place.	4.00
9. has a contingency plan in place in case a knowledge breach would occur.	3.80
10. monitors access to knowledge.	4.33
11. experienced a knowledge breach in the past.	4.47
My Organization Employs the Following Preventative Measure to Secure the Knowledge Assets (7 point Likert Scale)	
12. Background checks on employees	5.38
13. Counter-intelligence	3.00
14. Incentive schemes for employees	3.77
15. Educating employees	5.54
16. Securing electronic channels	5.23
17. Securing duplication and storage of knowledge	4.58
18. Securing the application of knowledge	4.23
Tell Us About Your Opinions on Secure Knowledge Management (5 Point Likert Scale)	
19. I consider security issues in my KM initiative(s)	3.46
20. I am asked by organizational leaders about security in my KM initiative(s)	2.77
21. I have had to deal with a security issue in my KM initiative(s)	2.23

improve the success of security education, training, and awareness (SETA) programs. KM research that focuses on knowledge sharing does it through investigation of the role of knowledge management systems (KMSs). Lessons learned from this stream of research are especially applicable because some KMSs support fast changing environment, such as tracking phishing attacks.

In addition, KM literature states that personal acknowledge and personal benefits do indeed increase motivation to contribute to a KMS (Constant, Sproull, & Kiesler, 1996). Jensen, Durcikova, and Wright (Jensen, Durcikova, & Wright, 2017) found that a message board that shows contribution activity of each employee to a KMS in real time and offers a real time validation of these posts (e.g., IT Security department checks whether the message submitted is indeed a phishing message), it creases not only motivation to report phishing attacks abut also increases knowledge sharing about these attacks. In addition, they found that validation or acknowledge of submission of knowledge by itself does not increase motivation or actual submission of knowledge about phishing attacks; both acknowledgement and validation must be present.

Similarly, Safa and Von Solms (Safa & Von Solms, 2016) argue that earning a reputation, and gaining promotion as an extrinsic motivation and curiosity satisfaction as an intrinsic motivation have positive effects on employees attitudes toward sharing information security knowledge. Their model relied on the motivation theory and the theory of planned behavior. In addition, Flores, Antonsen, and Ekstedt (2014) found that information security governance (organizational structure and business-based IS management) sets the platform to establish knowledge sharing about information security where coordinating information security processes mediate this relationship in addition to national culture.

These are only few examples of where IS Security research benefited from what we know about knowledge sharing and sourcing. More research is needed that focuses on the intersection of KM and individual and organizational information security (Soomro, Shah, & Ahmed, 2016).

DISCUSSION

Ultimately, security is about controlling access to data, information, and knowledge while KM is about getting needed data, information, and knowledge to the right people at the right time. Given the competitive value this data, information, and knowledge brings to the organizations possessing it, it is only natural that organizations would want to take steps to protect these assets. The review of the academic research literature found that there is a small body of literature that is addressing the key issues of showing the linkage between IS Security and KM success; applying risk management approaches managing KM activities, and applying Information Security technologies to addressing issues such as access control, secure communications, and secure storage. This is a good finding but the research needs to continue. 2012 was the year of Big Data, 2013 saw the Target breach, 2014 onwards may well be start the years of the Persistent Threat. Cybercrime, cyber espionage, cyber war, and cyber terrorism. Additionally, the development and adoption of cloud and mobile technologies all lead to the volatility of security approaches, making this an area needing further research. Interestingly enough Information Security practitioners are investigating the use of KM as a way improving overall Information Security through the sharing of attack methods and defense approaches.

While the findings in the KM research literature were better than expected, the findings in KM practitioner job postings were as expected, and to a degree disappointing. The findings reflect that KM practitioners still totally focus on knowledge capture, storage and sharing and that Information Security is an afterthought. Only just over 10% of the non-defense related job postings had requirements for understanding organizational security standards or policies and only two KM technician postings requested Information Security certifications. The survey results mimic this suggesting that the focus is on capturing and sharing knowledge and much less focus is paid on security access to knowledge. Organizations that understand the value of their knowledge would want their KM personnel to be able to protect it. It is expected that all KM managers should be familiar with organizational Information Security standards and policies. It is understandable if organizations were to create new KM technician positions focused on secure KM but this study found only one posting that fit this: Knowledge Management Secured Messaging Specialist. Since there were 11 postings focused on defense organizations it is surprising to note this specialist posting was for a commercial bank.

CONCLUSION

The goal of this exploratory study was to survey the area of Information Security within the KM literature and practice. Before discussing potential areas of future research, we will tend to the limitations of this study.

Limitations

The following are the limitations of this study: small sample size in terms of survey responses and job postings. The chief limitation is with respect to KM practitioner job requirements. 108 job postings from three-time periods were examined and analyzed. While 108 postings is a significant number, only seven of these postings were for jobs that were outside the United States (and these were for Canada and the United Kingdom, two countries very similar to the United States).

What is not a limitation is the breadth of the postings. While 101 are for United States positions, these 101 postings reflect 88 different organizations ranging from consultants to banks to health care to construction to automotive to defense and finally to information services. This is a wide range of organizations that does support generalizing the findings to all United States organizations and does offer a window into the how organizations consider security and knowledge management.

Areas for Future Research

This chapter identifies many areas for future research. The first is to perform a quantitative study with respect to KM practitioner and researcher attitudes towards security. The pilot survey presented in this study does provide an instrument that can be used to collect this data.

Specifically, KM in cloud and mobile environments needs much attention. There are key questions as to who owns data, information, and knowledge stored in the cloud and produced by multi-organizational teams. Access control schemes for large numbers of knowledge objects and knowledge users need to be developed.

KM governance needs to be developed to include Information Security management issues. KM models need to include security as a critical success factor and access control and classification as critical KM processes.

Additionally, how to add skills such as risk assessment and management and secure communications and storage into KM practitioner skill sets needs to be explored. KM certifications need to include security knowledge requirements. Organizations need to adopt those certifications that incorporate security as a KM skill requirement.

The conclusion of this chapter is that the protection of knowledge artifacts used in KM is not well enough developed and understood. KM governance and management needs to integrate organization Information Security standards and policies into their KM strategy. This can be achieved through collaboration between Information Security practitioners and KM practitioners. Innovations such as cloud and mobile technologies are improving the ability of organizations to capture and make available knowledge to those that need it at the time they need it. However, organizations need to balance the ease and speed of knowledge sharing with controls that ensure only those authorized to have access to that knowledge are the ones getting that knowledge; and governance that ensures that those who created the artifacts maintain their ownership. Additionally, many KM researchers focus on identifying and defeating barriers to knowledge transfer/sharing. These same researchers need to be aware of Information Security access controls and so ensure that overcoming barriers to knowledge transfer/sharing do not invalidate needed access control schemes or make it more difficult to design and implement a access control paradigm on what is normally considered a critical organizational asset.

REFERENCES

Alavi, M., & Leidner, D. E. (1999). Knowledge management systems: Issues, challenges, and benefits. *Communications of the AIS*, *1*(2es), 1.

Alazmi, M., & Zairi, M. (2003). Knowledge management critical success factors. *Total Quality Management & Business Excellence*, *14*(2), 199–204. doi:10.1080/1478336032000051386

Boella, G., & Van Der Torre, L. (2006). Security policies for sharing knowledge in virtual communities. *IEEE Transactions on Systems, Man, and Cybernetics. Part A, Systems and Humans*, *36*(3), 439–450. doi:10.1109/TSMCA.2006.871793

Burns, A. J., Roberts, T. L., Posey, C., Bennett, R. J., & Courtney, J. F. (2015). *Assessing the Role of Security Education, Training, and Awareness on Insiders' Security-Related Behavior: An Expectancy Theory Approach.* Paper presented at the HICSS. 10.1109/HICSS.2015.471

Cannataro, M., & Talia, D. (2003). The Knowledge Grid. *Communications of the AIS*, *46*(1), 89–93.

Constant, D., Sproull, L., & Kiesler, S. (1996). The kindness of strangers: The usefulness of electronic weak ties for technical advice. *Organization Science*, *7*(2), 119–135. doi:10.1287/orsc.7.2.119

Davenport, T. H., De Long, D. W., & Beers, M. C. (1998, Winter). Successful Knowledge Management Projects. *Sloan Management Review*, 40–57.

Davis, F. D. (1989). Perceived usefulness, perceived ease of use, and user acceptance of information technology. *Management Information Systems Quarterly*, *13*(3), 319–340. doi:10.2307/249008

Desouza, K. C. (2007). *Managing knowledge security: strategies for protecting your company's intellectual assets*. Kogan Page Publishers.

Ermine, J. L. (2013). Knowledge Management as Knowledge Risk Prevention. *iKNOW, the Magazine for Knowledge Workers, 3*(1), 8-11.

Flores, W. R., Antonsen, E., & Ekstedt, M. (2014). Information security knowledge sharing in organizations: Investigating the effect of behavioral information security governance and national culture. *Computers & Security*, *43*, 90–110. doi:10.1016/j.cose.2014.03.004

Holsapple, C. W., & Joshi, K. D. (2000). An investigation of factors that influence the management of knowledge in organizations. *The Journal of Strategic Information Systems*, *9*(2), 235–261. doi:10.1016/S0963-8687(00)00046-9

Ilvonen, I., Alanne, A., Helander, N., & Vayrynen, H. (2016). *Knowledge Sharing and Knowledge Security in Finnish Companies*. Paper presented at the HICSS. 10.1109/HICSS.2016.499

Ilvonen, I., Jussila, J., Karkkainen, H., & Paivarinta, T. (2015). *Knowledge Security Risk Management in Contemporary Companies -- Toward a Proactive Approach*. Paper presented at the HICSS.

Ilvonen, I., Jussila, J., Karkkainen, H., & Paivarinta, T. (2016). *Knowledge Security Risk Management in Contemporary Companies -- Toward a Proactive Approach*. Paper presented at the HICSS.

Irma Becerra-Fernandez, R. S. (2001). Organizational knowledge management: A contingency perspective. *Journal of Management Information Systems*, *18*(1), 23–55. doi:10.1080/07421222.2001.11045676

Jennex, M., & Olfman, L. (2005). Assessing knowledge management success. *International Journal of Knowledge Management*, *1*(2), 33–49. doi:10.4018/jkm.2005040104

Jennex, M., & Olfman, L. (2006). A Model of Knowledge Management Success. *International Journal of Knowledge Management*, *2*(3), 51–68. doi:10.4018/jkm.2006070104

Jennex, M. E. (2005). What is Knowledge Management? *International Journal of Knowledge Management*, *1*(4), i–iv.

Jennex, M. E. (2006). Why We Can't Return to the Moon: The Need for Knowledge Management. *International Journal of Knowledge Management*, *2*(1), i–iv.

Jennex, M. E. (2009). *Assessing Knowledge Loss Risk*. Paper presented at the AMCIS 2015.

Jennex, M. E. (2013). Knowledge Management: The Risk of Forgetting. *iKNOW, the Magazine for Knowledge Workers*, *3*(1), 4-7.

Jennex, M. E. (2014). A Proposed Method for Assessing Knowledge Loss Risk with Departing Personnel. *VINE: The Journal of Information and Knowledge Management Systems*, *44*(2), 185–209. doi:10.1108/VINE-07-2012-0028

Jennex, M. E., & Durcikova, A. (2013). *Assessing knowledge loss risk*. Paper presented at the System Sciences (HICSS), 2013 46th Hawaii International Conference on. 10.1109/HICSS.2013.103

Jennex, M. E., & Durcikova, A. (2014). *Integrating IS Security with Knowledge Management: Are We Doing Enough to Thwart the Persistent Threat?* Paper presented at the HICSS. 10.1109/HICSS.2014.429

Jennex, M. E., Smolnik, S., & Croasdell, D. (2012). *Where to look for knowledge management success*. Paper presented at the System Science (HICSS), 2012 45th Hawaii International Conference on. 10.1109/HICSS.2012.641

Jennex, M. E., & Zyngier, S. (2007). Security as a contributor to knowledge management success. *Information Systems Frontiers*, *9*(5), 493–504. doi:10.100710796-007-9053-4

Jensen, M., Durcikova, A., & Wright, R. (2017). Combating Phishing Attacks: A Knowledge Management Approach. *Proceedings of the 50th Hawaii International Conference on System Sciences*. 10.24251/HICSS.2017.520

Kajiyama, T., Jennex, M. E., Addo, T., & Paolini, C. (2013). *Cloud Computing Security: How Risks And Threats Are Affecting Cloud Adoption Decisions*. Paper presented at the 12th Security Conference.

Lambe, P. (2013). Four Types of Knowledge risk. *iKNOW, the Magazine for Knowledge Workers, 3*(1), 21-24.

Lindsey, K. (2002). Measuring knowledge management effectiveness: A task-contingent organizational capabilities perspective. *AMCIS 2002 Proceedings*, 285.

Malhotra, Y. (2004). *Why knowledge management systems fail: enablers and constraints of knowledge management in human enterprises. In Handbook on Knowledge Management 1* (pp. 577–599). Springer.

McCabe, J. (2016). *FBI Warns of Dramatic Increase in Business E-Mail Scams*. Retrieved 4/4/2016, 2017, from https://www.fbi.gov/contact-us/field-offices/phoenix/news/press-releases/fbi-warns-of-dramatic-increase-in-business-e-mail-scams

McCumber, J. (2004). *Assessing and managing security risk in IT systems: A structured methodology*. CRC Press. doi:10.1201/9780203490426

McWhorter, D. (2013). *Mandiant Exposes APT1 – One of China's Cyber Espionage Units & Releases 3,000 Indicators*. Retrieved 9/5/2017, 2017, from https://www.fireeye.com/blog/threat-research/2013/02/mandiant-exposes-apt1-chinas-cyber-espionage-units.html

Neville, K., Powell, P., & Panteli, N. (2003). Knowledge and security. *AMCIS 2003 Proceedings*, 262.

Padyab, A. M., Päivärinta, T., & Harnesk, D. (2014). *Genre-Based Assessment of Information and Knowledge Security Risks*. Paper presented at the HICSS. 10.1109/HICSS.2014.428

Perlroth, N., & Sanger, D. E. (2017). *In Computer Attacks, Clues Point to Frequent Culprit: North Korea*. Retrieved 9/5/2017, 2017, from https://www.nytimes.com/2017/05/15/us/nsa-hacking-shadow-brokers.html?mcubz=1

Ponemon, L. (2012). *2012 Cost of Cyber Crime Study*. Ponemon Institute.

Ponemon, L. (2013). *2013 Cost of Data Breach: Global Analysis*. Ponemon Institute.

Randeree, E. (2005). Secure health knowledge: balancing security, privacy and access. *AMCIS 2005 Proceedings*, 287.

Riecicky, V., & Spichiger, A. (2012). Knowledge asset risk management framework. *Human Resource Management Review*, *2*(4), 38–45.

Safa, N. S., & Von Solms, R. (2016). An information security knowledge sharing model in organizations. *Computers in Human Behavior*, *57*, 442–451. doi:10.1016/j.chb.2015.12.037

Saha, P., Parameswaran, N., Chakraborty, B., & Mahanti, A. (2013). *A Formal Analysis of Fraud in Banking Sector*. Paper presented at the System Sciences (HICSS), 2013 46th Hawaii International Conference on. 10.1109/HICSS.2013.36

San Nicolas-Rocca, T., Schooley, B., & Spears, J. L. (2014). Exploring the effect of knowledge transfer practices on user compliance to is security practices. *International Journal of Knowledge Management, 10*(2), 62–78. doi:10.4018/ijkm.2014040105

Sarigianni, C., Thalmann, S., & Manhart, M. (2016). *Protecting Knowledge in the Financial Sector: An Analysis of Knowledge Risks Arising from Social Media*. Paper presented at the HICSS.

Schinagl, S., Paans, R., & Schoon, K. (2016). *The Revival of Ancient Information Security Models, Insight in Risks and Selection of Measures*. Paper presented at the System Sciences (HICSS), 2016 49th Hawaii International Conference on. 10.1109/HICSS.2016.501

Soomro, Z. A., Shah, M. H., & Ahmed, J. (2016). Information security management needs more holistic approach: A literature review. *International Journal of Information Management, 36*(2), 215–225. doi:10.1016/j.ijinfomgt.2015.11.009

Sophos. (2013). *Security Threat Report 2013*. Author.

Spears, J. L., & San Nicolas-Rocca, T. (2016). *Information Security Capacity Building in Community-Based Organizations: Examining the Effects of Knowledge Transfer*. Paper presented at the HICSS. 10.1109/HICSS.2016.498

Stoneburner, G., Gogunen, A., & Feringa, A. (2007). NIST Special Publication 800-30, Risk Management Guide for Information Technology Systems. United States National Institute of Standards and Technology.

Tapper, J. (2017). *White House Officials Tricked by Email Prankster*. Retrieved 9/5/2017, 2017, from http://www.cnn.com/2017/07/31/politics/white-house-officials-tricked-by-email-prankster/index.html

Thune, J. (2017). *Senate Committee on Commerce, Science, and Transportation*. Retrieved from https://www.govtrack.us/congress/committees/SSCM

Thuraisingham, V., & Nonmonotonic, A. (1992). *Typed Multilevel Logic for Multilevel Secure Data / Knowledge Base Management Systems - II*. Bedford, MA: The MITRE Corporation.

Upadhyava, S., Rao, H. R., & Padmanabhan, G. (2006). Secure Knowledge Management. In D. G. Schwartz (Ed.), *Encyclopedia of Knowledge Management* (pp. 795–801). Idea Group Inc. doi:10.4018/978-1-59140-573-3.ch104

Weiss, N. E., & Miller, R. S. (2015). *The Target and Other Financial Data Breaches: Frequently Asked Questions*. Congressional Research Service. Retrieved on September 9, 2017 from https://fas.org/sgp/crs/misc/R43496.pdf

Wipawayangkool, K. (2009). Information security compliances and knowledge management capabilities in international diversification. *AMCIS 2009 Proceedings*, 604.

Chapter 14
Knowledge Management in Support of Enterprise Risk Management

Eduardo Rodriguez
University of Wisconsin – Stevens Point, USA

John S. Edwards
Aston University, UK

ABSTRACT

This chapter examines possible relationships between knowledge management constructs related to knowledge sharing and two risk management concepts: perceived quality of risk control and perceived value of enterprise risk management. From a literature review, relationships with eight knowledge management variables covering people, process, and technology aspects were hypothesized. The results showed that the perceived quality of risk control is significantly associated with four knowledge management variables: perceived quality of risk knowledge sharing, perceived quality of communication among people, web channel functionality, and risk management information system functionality. However, the relationships of the knowledge management variables to the perceived value of enterprise risk management are not significant. The authors conclude that better knowledge management is associated with better risk control, but that more effort needs to be made to break down organizational silos in order to support true enterprise risk management.

INTRODUCTION

The separation between Knowledge Management (KM) and Risk Management (RM) is part of current organizational reality. The aim of this research is to study how KM concepts may help improve RM and help to turn it into true Enterprise Risk Management (ERM). It builds on previous work on knowledge-related constructs within RM (Rodriguez & Edwards, 2010). ERM implementation has evolved and studies show that organizations have expectations of improving the ERM process to perform better. Callahan and Soileau (2017) identified that organizations with ERM process in higher maturity levels have

DOI: 10.4018/978-1-5225-5427-1.ch014

a better operational performance than their peers in the market. Fraser and Simkins in 2007 presented the misconceptions of ERM that reduced the implementation of the ERM process, in Fraser & Simkins (2016) presented solutions to the challenges and misconceptions. These solutions are based on the ERM best practices study. These solutions are associated with knowledge management practices where knowledge transfer in policy and integration of ERM in strategic planning, committees conversations, performance evaluation and development of common language around ERM terms.

Bromiley et al. (2015) indicated the importance of studying several aspects of ERM form the management perspective mainly on ERM adoption and capacity to implement ERM in organizations. In this article, we consider the relationships between KM variables related to knowledge sharing, and two RM variables: perceived quality of risk control (representing the operational level of RM) and perceived value of the ERM implementation (representing the strategic level). Crouhy, Galai, and Mark (2001) indicate the need for risk systems to control risk at individual and enterprise level.

This article begins with the identification of events that have affected the financial services industry and that indicate the need for better management of risk management knowledge. The succeeding sections introduce relevant concepts of risk management and knowledge management, present the research model that comprises eight hypotheses, and describe the analysis of the results of two regression models that were used to test them. The final sections discuss the findings and seek to interpret their meaning.

The Context of Financial Services

The financial crisis of recent years has raised many questions about the performance of financial institutions in response to adverse events. There are doubts about their capacity to execute the three knowledge components of the management of risk: use of models, use of technology and leveraging on people (Beasley, Bronson, & Hancock, 2009; Champion, 2009; Taleb, Goldstein, & Spitznagel, 2009). The development of financial services organizations is not only going in the direction of knowledge-based service quality but also in the development of metrics for sustainable appropriate performance. On the one hand, Paul et al. (2015) pointed out the importance of products knowledge in the quality of services in banking. On the other hand, O'Brien and Szerszeń (2017) shows that the review of risk metrics (VAR) is part of the activities to develop by banks in order to improve and use the knowledge of the banks performance and financial positions in the markets.

Financial institutions are information and knowledge organizations (Fourie & Shilawa, 2004). Risk is one of the principal business issues a financial institution must deal with. To manage risk "is frequently not a problem of a lack of information, but rather a lack of knowledge with which to interpret its meaning" (Marshall, Prusak, & Shpilberg, 1996, p. 82). Knowledge reduces uncertainty (Nonaka, 1991) and therefore, knowledge reduces risk (Dickinson, 2001). However, it is not clear how knowledge is organized in, and provides support to, financial institutions in order to deal with uncertainty and risk.

The performance of financial institutions is affected by the management of wide risk exposure represented by an offer that includes more products and services than in the past. Financial institutions thus need to evolve from a risk management process based on silos of risk analysis towards Enterprise Risk Management (ERM) which is a dynamic risk management process across the company (Dickinson, 2001). To transform RM into ERM is a strategic step in managing risk, but is essentially good risk management practice (Lam, 2003) with an holistic view.

As KM also needs to take an holistic view of the organization (Edwards, 2009), it therefore seems reasonable to suppose that these two disciplines, when working together and complementing one another, can better handle the risks affecting financial organizations.

THEORETICAL BACKGROUND

This section presents concepts relating to managing risk (risk management processes, the differences between RM and ERM, and a description of the risk management system) and managing knowledge (knowledge management processes and knowledge management systems). It goes on to consider the small amount of existing work combining KM and RM.

Risk Management in Financial Institutions

For financial institutions, RM is "the overall process that a financial institution follows to define a business strategy, to identify the risks to which it is exposed, to quantify those risks and to understand and control the nature of the risks it faces" (Cumming & Hirtle, 2001, p. 2) and a "collection of processes, people and systems aligned for the purpose of measuring, managing, monitoring, and controlling risk exposure." (Levine, 2004, p. 31).

The RM processes as described by Brown (2001) are shown in Table 1. Brown's categorization is based on types of action, presenting the activities involved in the RM processes by risk type. His model indicates that these actions are related to: managing the cost of risks; developing alerts; organizing measurement systems that can produce results that can be used by management to make decisions; and defining protection and solutions. The actions are carried out across the organization as a whole through a governance structure, controls and people development. A crucial element that Brown identifies is the need to support and develop the people involved, as only they can perform the RM actions and interpret the outcomes.

As mentioned above, the main difference between RM and ERM is in the enterprise strategic view of risk analysis for the whole organization as opposed to RM's "silo view". A more detailed list of the differences between RM and ERM is presented in Table 2:

Table 1. Risk management processes

RM Process	Description
Risk identification	This process refers to the group of actions developed in the organization to classify and map risks that can affect the organization in their current and expected business conditions
Risk measurement	Quantification and assessment of risk are important actions in RM; particularly, the capacity to provide evaluation of the impact, frequency and severity of risks in the business operation.
Risk monitoring	These sets of actions represent the capacity to follow up on what has been designed for managing risks.
Risk control	This represents the capacity to assure the adequacy of the RM actions, such as risk mitigation, risk transfer and, in general, risk alignment to the policies and strategy.
Risk application	Policies and solutions for the business processes and the required conditions to keep risk effects under control

Brown, 2001.

Table 2. Comparison between risk management and enterprise risk management

Risk Management (RM)	Enterprise Risk Management (ERM)	
• Silo, individual view of risk • Specific risk analysis • Tactic orientation • Related to control and minimization • Organization specific, department or business unit. Concentrated on business events • Disaggregated methods for risk analysis • Responsibility on the functional managers • Performance evaluation concentrated on the particular problem solved • Protection of adverse financial effects of bad events. Earnings volatility protection from the source • Reactive • Specific control on section or division expenditures • Individual risk analysis • The priority is in the portfolio and individual sources	• Global, holistic view of risk • Risk analysis across the organization • Strategic orientation • Related to competitiveness • Individuals, business units and the complete organization. Corporate view • Aggregated methods • Governance/stakeholders responsibility • Risk performance evaluation enterprise wide and based on risk • Organization stability protection. Decision making process based on risk • Proactive • Reviews and reduction of duplication of risk management expenditures • Interdependent risk analysis • Priority can be in portfolio structure, assets modification, strategic movements	

Meulbroek 2002; Lam 2003; Cumming and Hirtle 2001; Dickinson 2001.

The Risk Management Information System (RMIS)

Risk management processes require the support of a risk management information system (RMIS). RMISs need to have an open data architecture based on Internet standards, and a flexible design to enable new RM processes to be supported by appropriate risk workflows (Levine, 2004). However, it has been argued that existing RMISs present more of a challenge to the development of ERM proper than a solution (Lee & Lam, 2007).

McNally (2013) described the new framework for risk control proposed by COSO (the Committee of Sponsoring Organizations of the Treadway Commission) and indicates the need to manage information and communication as one of the key elements of the control system. Using relevant information, and internal and external communication capacity stand alongside developing means to evaluate and communicate deficiencies across the organization, and performing independent evaluations for processes, business units and areas in the organization structure. In summary, to integrate ERM with the management control systems in order to achieve strategic goals.

According to Hsu, Backhouse, and Silva (2013), there is limited research on understanding the role of IT as a support in operational risk management; in particular, IT can be a source of new risks. There is no clear road to understanding how managers decide to take a particular course of action in managing operational risk. Their findings from a case study of a financial organization show the value of knowledge sharing as a risk mitigant in operational risk management, and that organizational awareness of the importance of proper IT adoption is required in order to avoid new risks in the organization.

Zhao et al. (2015) indicate the use of aligning expert knowledge in ERM with the design of Decision Support Systems to develop action plans. The proposed action plans are the means to improve ERM implementation.

Knowledge and Knowledge Management

Wiig (2004, p. 335) indicates that "knowledge consists of truths and beliefs, perspectives and concepts, judgements and expectations, methodologies and know-how." There are many other definitions of knowledge. As respondents in our study were not presented with a definition of knowledge, but rather left to answer according to what they thought knowledge meant, we will not discuss its definition further here.

The same applies to knowledge management, which Wiig (1997, p. 3) defines as: "... the systematic, explicit and deliberate building, renewal, and application of knowledge to maximize an enterprise's knowledge–related effectiveness and returns from its knowledge assets." Again, we will not discuss the definition further here, but we do need to consider in more detail what KM consists of.

Heisig (2009) analysed no fewer than 160 KM frameworks, of which 117 included a list of processes or activities (p.9). In descending order of frequency, these were: share knowledge; create knowledge; use knowledge; store knowledge; identify knowledge; acquire knowledge. Ribière and Walter (2013) also found that knowledge sharing was the most common keyword in ten years of research papers on KM. We thus feel that it is reasonable to regard knowledge sharing as perhaps the most fundamental activity in KM. In the ERM context risk knowledge sharing is basic for risk policies implementation and control. ERM is a setting with a technical and analytic orientation where the study of Oliveira et al. (2015) can fit. They reported that in a technology organization setting and present a view about using scales to measure knowledge sharing behaviour and the relationship with innovation. They found that the scale change is not showing differences in the results and that tacit knowledge is shared more than explicit knowledge.

Supporting the KM processes requires a Knowledge Management System (KMS). Alavi and Leidner (2001, p. 114) identified KMSs as the "kind of information systems applied to managing organizational knowledge." However, a KMS is not just technology-oriented; it also has to include the social and cultural components of KM (Davenport & Prusak, 1998). Thus the KMS elements can be summarized as people interactions with both technology and processes (Edwards, 2009).

Knowledge Management and Risk Management

Very few articles attempt to combine KM and RM. There are specific risk management process that have been integrated to the knowledge creation process in combination with the analytics process. Saha et al. (2016) presented a review of the loan granting process using knowledge-based risk assessment approach. The approach combines the expert opinion and the use of statistical learning. In those that do, typically the approach has been more on identifying risks of knowledge management practice rather than applying KM to RM. Massingham (2010, p. 464) describes "knowledge risk management", the application of risk management principles, mainly risk scoring approaches, to manage risks relating to knowledge (especially its loss) in a case study from the Australian Department of Defence. Thus Massingham is looking more at how a KM metric can be an indicator of RM improvement than at how KM processes and systems can improve RM. Trkman and Desouza (2012) also use the phrase "knowledge risk management" to refer to the risks associated with knowledge management, for example that too much sharing of knowledge can adversely impact competitive advantage. They categorize these knowledge risks, and examine how these risks affect knowledge transfer. The other interaction between RM and KM in the literature is at a higher level. It builds on the consideration of RM as part of security in KM programs developed by Jennex and Zyngier (2007), to cover the governance of KM programs and the risk factors affecting a KM program (e.g. Zyngier, 2008; Zyngier & Burstein, 2012; Zyngier & Venkitachalam, 2011).

Most of the prior work on KM in financial institutions addresses either strategy (e.g. Shaw, Hall, Edwards, & Baker, 2007; Spies, Clayton, & Noormohammadian, 2005), both of which take the holistic view, or specific aspects such as customer knowledge (e.g. Värlander, 2008) or knowledge mapping (Fourie & Shilawa, 2004) without explicit mention of risk management. Recent exceptions are related to operational risk. The paper by Hsu et al. (2013) which identified knowledge sharing as a valuable factor for managing operational risk. This helps justify knowledge sharing as the area within KM whose influence on RM we shall study first. The same as Luburic (2016) who identify the importance of the process of learning and development of knowledge to identify and manage operational risk.

RESEARCH MODEL AND HYPOTHESES

The hypotheses are expressed as the existence of relationships between independent variables describing knowledge sharing activities, and the dependent variables perceived quality of risk control, and perceived value of ERM implementation. The reason for using the two dependent variables is to capture the differentiation between RM practice, viewing separate risk areas risk by risk, and ERM's integral, holistic view of risk across the organization (Francis & Paladino, 2008). We are measuring perceptions of performance of RM as a process, rather than measures of RM outcome performance.

The hypotheses (see Figure 1) are divided into three groups: people, process and technology (Edwards, 2009).

Figure 1. Research model

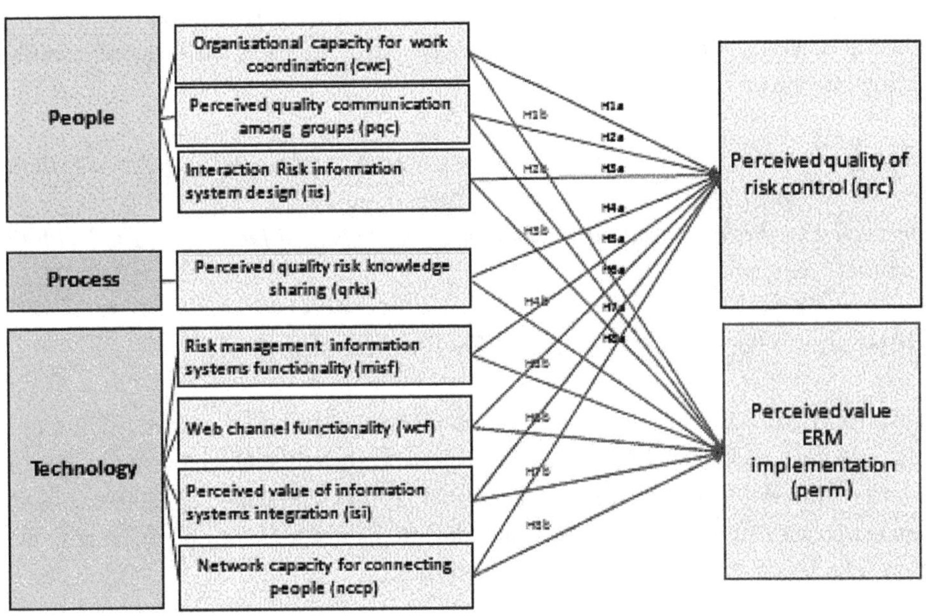

People Hypotheses

Three hypotheses were formulated regarding the effect of people variables: organizational capacity for work coordination, perceived quality of communication among people, and people's interaction for risk information system design.

Organizational Capacity for Work Coordination (*cwc*)

Work coordination refers to the assignment of responsibilities and accountabilities between the participants in a business initiative. Ideally, organizational capacity for work coordination in risk management would be such that all the people who are involved in a risk process, project or assignment can achieve the goals by a synchronized development (Banham, 2004; Meulbroek, 2002).

The hypotheses formulated were:

H1a: *Organizational capacity for work coordination is positively associated with the perceived quality of risk control.*

H1b: *Organizational capacity for work coordination is positively associated with the perceived value of ERM implementation.*

Perceived Quality of Communication Among People (*pqc*)

Knowledge sharing and effective communication depends on the overlap and amalgamation of knowledge bases among people (Te'eni, 2006). The flow of information for risk knowledge in an ERM context needs communication capacity between the different groups in risk management (Uzzi & Lancaster, 2003). A clear relationship between communication and risk knowledge sharing has not yet been identified. We chose the following "positive" hypotheses:

H2a: *The perceived quality of communication among people is positively associated with the perceived quality of risk control.*

H2b: *The perceived quality of communication among people is positively associated with the perceived value of ERM implementation.*

People's Interaction for Risk Information System Design (*iis*)

The design of an RMIS needs to encompass requirements from different risk areas, including both people interaction in each area, and the creation of collective risk knowledge which can contribute to the evolution of RM into ERM. Majchrzak, Beath, and Lim (2005) and Uchupalanan (2000) explain the value of collaboration in accomplishing information system design, facilitated learning management and support system design. The hypotheses were formulated as follows:

H3a: *People's interaction for risk information system design is positively associated with the perceived quality of risk control.*
H3b: *People's interaction for risk information system design is positively associated with the perceived value of the ERM implementation.*

Process Hypothesis

Perceived Quality of Risk Knowledge Sharing (*qrks*)

Shariq and Vendelo (2006, p. 833) state that: "When people solve complex problems, they bring knowledge and experience to the situation, and they engage in problem solving: they create, use, and share tacit knowledge." Improvement in knowledge sharing develops capacities inside the organization. RM might be influenced by attributes such as work satisfaction and capacity to share knowledge without a limitation on the number of people sharing. Organizational silos could adversely influence knowledge sharing, and business units can require assistance in knowing how to transfer their practical experiences to other units. In RM, the knowledge sharing process is not clearly identified. The hypotheses formulated were:

H4a: *The perceived quality of risk knowledge sharing is positively associated with the perceived quality of risk control.*
H4b: *The perceived quality of risk knowledge sharing is positively associated with the perceived value of the ERM implementation.*

Technology Hypotheses

The four KM variables in this group were: risk management information systems functionality, web channel functionality, perceived value of information systems integration, and quality of the network capacity for connecting people (in the organization).

Risk Management Information Systems Functionality (*misf*)

Functionality is identified as the capacity that the system has for answering the needs of the user (O'Brien, 1996). Functionality of information systems is an attribute that organizations as a whole and users look for in order to perform their activities. In general, the information systems design needs to provide the support required for RM/ERM, especially how to achieve goals of compliance with the new market conditions. Ideally, that functionality should support and facilitate knowledge sharing. Thus the functionality of a risk management information system might have influence on risk control and ERM. The following hypotheses were formulated:

H5a: *The risk management information system functionality is positively associated with the perceived quality of risk control.*
H5b: *The risk management information system functionality is positively associated with the perceived value of ERM implementation.*

Web Channel Functionality (*wcf*)

One specific knowledge sharing channel is the web channel used in order to improve communication capacity. Financial institutions have different technologies at their disposal to deliver their services to their internal and external customers. However, it seems that having independent intranets, putting emphasis on IT for knowledge sharing and having a reduced flow of KM processes through network systems all reduce knowledge sharing (Swan, Newell, Scarborough, & Hislop, 1999). The issue is about the contribution of the use of web channel support to connect people, to support access to data, processes, applications and in general to the generation of consistency in all the dimensions of risk across the organization. Nevertheless, the hypotheses were again formulated "positively" as:

H6a: *The web channel functionality is positively associated with the perceived quality of risk control.*
H6b: *The web channel functionality is positively associated with the perceived value of ERM implementation.*

Perceived Value of Information Systems Integration (*isi*)

Complex, large organizations and multi-divisional businesses require the design of an enterprise architecture in order to support and co-ordinate access to data and systems (Hobday, Davies, & Prencipe, 2005). Data should be in formats that are accessible and usable, and of sufficient quality to share and adopt in business processes. This research regards this transformation as a key part of KM: information systems have to be more efficient, effective, and integrated in order to help people to make more complex decisions. The hypotheses formulated were:

H7a: *The perceived value of information systems integration is positively associated with the perceived quality of risk control.*
H7b: *The perceived value of information systems integration is positively associated with the perceived value of ERM implementation.*

Quality of the Network Capacity for Connecting People (*nccp*)

Effective KMS implementation needs the identification of stakeholders and the association of their different types of knowledge (Lehaney, Clarke, Coakes, & Jack, 2004). Additionally, Earl (2001) introduces the networking capacity for connecting people in the organization as a success factor for KM implementation, often through a "Yellow Pages" knowledge directory approach. In risk control and ERM implementation, it could be an advantage to interrelate technological, methodological and business factors. Thus, the hypotheses formulated were:

H8a: *The quality of the network capacity for connecting people is positively associated with the perceived quality of risk control.*
H8b: *The quality of the network capacity for connecting people is positively associated with the perceived value of the ERM implementation.*

RESEARCH METHOD AND HYPOTHESES

The research takes a positivist approach: building a relationship model based on existing literature and on the experience of one of the authors in the sector, formulating hypotheses and testing them using statistical modelling (Babbie, 1998). The initial questions in the survey covered four demographic variables: risk management area of work, risk management process on which most time is spent, length of time in the current position and length of experience in risk management, followed by the actual item questions. The lack of previous studies combining KM and RM meant it was necessary to develop scales for the variables concerned that were appropriate for the respondents in the industry.

The survey was applied to a random sample of full-time employees in RM in financial institutions, in two phases. First, a sample of RM employees of financial services organizations' headquarters were contacted and (if willing) the questionnaire was administered face-to-face. This was used to test the questionnaire.

There were no substantive changes after the testing phase, only minor changes in the wording of some items, to avoid technical KM terms in the item formulation and assume greater literacy of RM people in computer and web related topics.

For the second phase, an electronic web-based questionnaire was constructed (Aaker, Day, & Kumar, 1998). The final list of items is shown in Table 3 (space does not permit a full justification of each item here). An email invitation to participate in the survey was sent to members of the Professional Risk Managers International Association or the Risk Management Association who worked for financial institutions. Given this population's literacy in computer systems, the use of the web was not expected to bias the results.

The unit of analysis is thus an employee who is involved in RM activities in any RM process in a financial institution. All groups of RM employees are exposed to KM activities in a similar way given the centralization and corporate level decision processes that policies and strategy definition require in the sector.

The changes after the first phase were sufficiently small that it was judged reasonable to include the responses from this phase in the final analysis. The survey was thus distributed to a total of 620 full-time employees in RM in financial institutions. In total, 121 usable responses were received, giving a response rate of 19.5%, 19 of the responses being face-to-face. The population was based world-wide, although more than 50% were from North America.

Measurement, Reliability, and Validity

All 53 substantive items in the survey were rated on the same Likert scale: 1 strongly disagree, 2 disagree, 3 neutral, 4 agree and 5 strongly agree. The Likert scale is used so that respondents evaluate their agreement with, preference for and attitude towards the statements for each item (Aaker et al., 1998). Values for the variables were then derived from the item scores. An important issue in aggregating item scores was not to assume that simple addition of the item scores (i.e. equal weight) would be appropriate (Alfares & Duffuaa, 2008). For each observation, the original value x was transformed to a new value given by z-score $(x-\mu)/\sigma$, (Bohrnstedt & Knoke, 1982) where μ is the mean and σ the standard deviation. This transformation allows for the combination of items with a different mean and standard deviation.

Table 3. Research variables and items used for their construction

Ftable Variable	Items (in the Order in Which They Appeared on the Survey)	Attributes and Reliabillity
Perceived value of information systems integration (isi)	1. The same standards are used 2. A common data structure is used 3. A common data-warehouse is used 4. A common user interface is used 5. A common report system is used 6. A common application access is used	Independent. Score index of six items, each measured on a 5-point scale (Cronbach's alpha 0.89)
Organizational capacity for work coordination (cwc)	1. The organization encourages interdisciplinary work 2. The organization encourages interdepartmental work 3. There are good web based collaboration tools 4. People are willing to work with multiple groups 5. There are guiding principles for working with different groups 6. There are standards for using collaboration tools	Independent. Score index of six items, each measured on a 5-point scale (Cronbach's alpha 0.80)
Perceived quality of risk knowledge sharing (qrks)	1. People are willing to share risk knowledge 2. The availability of documentation is good 3. The access to experience is good 4. There is an appropriate environment to discuss results interdepartmentally 5. There is an appropriate environment for the creation of shared solutions	Independent. Score index of five items, each measured on a 5-point scale (Cronbach's alpha 0.79)
Perceived quality of risk control (qrc)	1. The risk mitigation tools are good 2. The risk assessment process is good 3. The risk transfer process is good 4. The risk product evaluation is good 5. The risk aggregation analysis is good	Dependent. Score index of five items, each measured on a 5-point scale (Cronbach's alpha 0.86)
Risk Management Information systems functionality (misf)	1. The systems provide support to the risk modeling process 2. The systems provide access to experience in risk analysis 3. The systems provide adequate data management support 4. The systems provide capacity to improve work flow 5. The systems provide capacity to work with multiple groups on a project	Independent. Score index of five items, each measured on a 5-point scale (Cronbach's alpha 0.88)
Quality of network capacity for connecting people (nccp)	1. There is an enterprise portal structure supporting interdepartmental work 2. There are collaboration tools easily available 3. People use web based workspaces for working on projects 4. Solutions are created because of multidepartment work 5. Sharing my work with others is easy	Independent. Score index of five items, each measured on a 5-point scale (Cronbach's alpha 0.86)
Perceived quality of communication among people (pqc)	1. The communication between the Risk Management groups is good 2. The communication within my Risk Management group is good 3. The communication environment fosters the interchange of different points of view 4. There is a good capacity to get conclusions easily during meetings 5. The communication environment promotes team work	Independent. Score index of five items, each measured on a 5-point scale (Cronbach's alpha 0.88)
People's interaction for risk information system design(iis)	1. Perceived quality of people interactions in the ERMIS design	Independent. This is a single item
Web channel functionality (wcf)	1. The Risk Management Intranet provides access to collaboration tools 2. The Risk Management Intranet provides access to all applications used in risk management 3. The Risk Management Intranet provides access to the proper data 4. The Risk Management Intranet facilitates interaction in problem solving process 5. The Risk Management Intranet supports communication among risk management people 6. The Risk Management Intranet supports risk management controls	Independent. Score index of six items, each measured on a 5-point scale (Cronbach's alpha 0.92)
Perceived value of ERM (perm)	1. ERM improves collaboration 2. ERM promotes our experience sharing 3. ERM reduces the number of times we reinvent the wheel 4. ERM improves the quality of data 5. ERM improves our interdisciplinary work 6. ERM improves our interdepartmental work 7. ERM improves our understanding of model results 8. ERM improves our problem solving process 9. ERM improves our capacity of mathematical modeling	Dependent. Score index of nine items, each measured on a 5-point scale (Cronbach's alpha 0.93)

Reliability was judged using internal reliability as measured by the Cronbach alpha coefficient, because the variables are built through items without previous scales. The cut-off value considered to be acceptable is 0.7 (Cortina, 1993). All multi-item variables satisfied this (see Table 3).

As well as the testing phase described earlier, validity was assessed using construct validity (Hair, Babin, Money, & Samouel, 2003), in particular convergent validity. This uses the correlation between two constructs to show they are potentially measuring the same concept (Trochim & Donnelly, 2007). The correlations are significant for all pairs of items except two (out of 131 pairs), namely cwc1-cwc5 and qrks1-qrks5, which provides sufficient evidence of convergent validity.

Additionally, a non-response bias test was performed, by comparing the answers of the group of early respondents to the email invitation with those of a second group who only responded after a second invitation to participate in the survey. This second group can be regarded as more representative of non-respondents (Lambert & Harrington, 1990). The results indicate that just two of the 53 items had a significant mean difference between the two groups, thus it is possible to infer the absence of non-response bias.

The ANOVA technique was used to examine whether the variable means and variances differed between categories of the four demographic variables (RM area, RM process, time in position, RM experience). The results of the Levene test indicated that the hypothesis of equal variances was accepted for all variables and all the groups (Table 4). The Tukey test for equal means shows that two variables had a significant mean difference for just one pair of categories in each case. It was thus judged reasonable to aggregate all categories together as a first step.

Each hypothesis was tested in the form of the null hypothesis that there was no association between the variables. The minimum level of significance used was $p=0.05$ throughout.

The first step was to consider the Spearman correlations, as shown in Table 5. The two dependent variables show no significant correlation ($\rho=0.13$). This justifies treating them as two different RM concepts in the analysis.

For the quality of risk control *(qrc)*, all eight correlation coefficients with the independent variables are highly significant and positive. However, only four of the independent variables *(pqc, iis, qrks, cwc)* are significantly correlated with *perm*; it is noticeable that the non-significant correlations are with the four technology variables.

Since all the independent variables show significant positive correlations with each other except the pairs *pqc-isi* and *iis-isi*, it was appropriate to use stepwise regression rather than simple correlation to test the hypotheses. None of the correlations between the independent variables is above 0.7 which is the threshold (Hair et al., 2003) to indicate multicollinearity.

Multivariate Analysis: Regression Diagnostic

The quality of fit of the regression models was judged by the value of R-squared and by power analysis (Cohen, Cohen, West, & Aiken, 2003). The power threshold considered as sufficient is 0.8 (Murphy & Myors, 1998).

The regression models used one dependent and eight independent variables in each case. Two different models were analysed, one for each dependent variable, identified here as the QRC and PERM models. Inspection showed there were no curvilinear patterns that would indicate a non-linear relationship between independent variables and dependent variable for either model.

Table 4. Analysis of variance tests

Anova F-test Tukey -Test Equal Means				
p-value	rmarea	rmprocess	timeposition	rmexperience
isi	0.895	0.321	0.400	0.907
cwc	0.716	0.019	0.206	0.899
qrks	0.490	0.024	0.323	0.356
qrc	0.672	0.576	0.405	0.293
misf	0.622	0.178	0.313	0.643
nccp	0.777	0.084	0.460	0.286
pqc	0.962	0.110	0.052	0.045
iis	0.993	0.083	0.306	0.953
wcf	0.933	0.305	0.259	0.372
perm	0.278	0.585	0.573	0.350

Significance level p=0.05

QRKS differences between groups 1 and 6

PQC difference groups 1 and 4

Anova F-Test Levene test Equal Variances				
p-value	rmarea	rmprocess	timeposit	rmexperience
cwc	0.918	0.517	0.769	0.984
isi	0.908	0.198	0.310	0.442
qrks	0.829	0.934	0.593	0.457
qrc	0.794	0.842	0.799	0.742
misf	0.610	0.899	0.880	0.203
nccp	0.728	0.278	0.426	0.281
pqc	0.828	0.170	0.301	0.175
iis	0.450	0.253	0.677	0.452
wcf	0.888	0.547	0.298	0.829
perm	0.305	0.919	0.667	0.583

Significance level p=0.05

The White test (Pindyck & Rubinfeld, 1998) for the two models was applied and the homoscedasticity assumption is accepted at p=0.05. Neither model has problems with autocorrelation. For the QRC model the Durbin-Watson test statistic is 1.903 and that for the PERM model is 2.112, both in the ranges that indicate independence (Pindyck & Rubinfeld, 1998).

The graphs of the residuals by variable in the QRC model (omitted here for reasons of space) show that the distribution of points is symmetric relative to the mean of the *qrc* variable and so do not suggest the need to transform variables. The residuals and quantile plots show no evidence of non-normality or skewness.

Additionally, the standardized predicted and residuals plot shows 93% of the cases are inside the interval (-2, 2) specified by Hair et al. (2003) suggesting normality as well. The analysis thus confirms that the normality hypothesis for the QRC residuals can be accepted.

Table 5. Spearman correlation coefficients

Spearman Correlation Coefficients, N = 121 Prob > |r| under H0: Rho=0										
	cwc	pqc	iis	qrks	misf	wcf	isi	nccp	qrc	perm
cwc	1	0.53522***	0.44956***	0.64117***	0.49971***	0.47761***	0.37246***	0.68887***	0.61797***	0.18137*
pqc	0.53522***	1	0.53892***	0.57959***	0.41088***	0.28794***	0.10402	0.45073***	0.53377***	0.27389**
iis	0.44956***	0.53892***	1	0.43134***	0.42321***	0.22159*	0.14778	0.39406***	0.40347***	0.25069**
qrks	0.64117***	0.57959***	0.43134***	1	0.56558***	0.36763***	0.23601**	0.55738***	0.64337***	0.20829*
misf	0.49971***	0.41088***	0.42321***	0.56558***	1	0.59586***	0.38091***	0.55309***	0.57992***	0.08741
wcf	0.47761***	0.28794***	0.22159*	0.36763***	0.59586***	1	0.50363***	0.49624***	0.53834***	-0.05781
isi	0.37246***	0.10402	0.14778	0.23601**	0.38091***	0.50363***	1	0.39048***	0.38709***	-0.14987
nccp	0.68887***	0.45073***	0.39406***	0.55738***	0.55309***	0.49624***	0.39048***	1	0.51467***	0.01827
qrc	0.61797***	0.53377***	0.40347***	0.64337***	0.57992***	0.53834***	0.38709***	0.51467***	1	0.1285
perm	0.18137*	0.27389**	0.25069**	0.20829*	0.08741	-0.05781	-0.14987	0.01827	0.1285	1
	*	Significant at 0.05								
	**	Significant at 0.01								
	***	Significant at 0.001								

A similar review of the PERM residuals shows, in the histogram and in the Q-Q and probability plots, one point that is an outlier. Inspection of the original data showed that this record comprises responses of 1 on the Likert scale to all the items comprising the *perm* variable. No other record included a response of 1 for any item for this variable. When including this outlier the tests of normality are not passed. It seems reasonable to conclude this respondent does not believe ERM is of any value at all, irrespective of any KM influence.

After removing this outlier, an acceptable fit to the normal distribution is found. The proportion of values outside the interval (-2, 2) is less than 5% (5 points out of 120). Additionally, the Kolmogorov-Smirnov test indicates that the null hypothesis cannot be rejected at 5%. Thus the normality hypothesis for the PERM residuals can be accepted for the reduced data set.

Collinearity and multicollinearity were tested and there is no indication of collinearity. According to Hair et al. (2003) values of Variance Inflation Factor greater than 10 or condition indices with a value over 30 indicate multicollinearity. Neither model exceeds the stated values.

RESULTS

QRC Model

The application of stepwise regression brought into the model four variables: the perceived quality of risk knowledge sharing *(qrks)*, web channel functionality *(wcf)*, perceived quality of risk communication among people *(pqc)*, and risk management information system functionality *(misf)* (see Table 6). The final stepwise regression model has R-squared of 0.5916 and a power of 1.

Four of the eight hypotheses relating to this model are thus supported. From the people variables group, one variable, the perceived quality of communication among people is positively associated with

Table 6. Stepwise regression results for QRC model, perceived quality of risk control

Stepwise Selection: Step 1					
Variable qrks Entered: R-Square = 0.4177 and C(p) = 50.1031					
Analysis of Variance					
Source	DF	Sum of Squares	Mean Square	F Value	Pr > F
Model	1	802.6062	802.60616	85.38	<.0001
Error	119	1118.698	9.40082		
Corrected Total	120	1921.304			
Variable	Parameter Estimate	Standard Error	Type II SS	F Value	Pr > F
Intercept	-4.20E-16	0.27873	2.14E-29	0	1
qrks	0.70383	0.07617	802.60616	85.38	<.0001

Stepwise Selection: Step 2					
Variable wcf Entered: R-Square = 0.5363 and C(p) = 18.0816					
Analysis of Variance					
Source	DF	Sum of Squares	Mean Square	F Value	Pr > F
Model	2	1030.368	515.18419	68.23	<.0001
Error	118	890.9358	7.5503		
Corrected Total	120	1921.304			
Variable	Parameter Estimate	Standard Error	Type II SS	F Value	Pr > F
Intercept	-1.10E-15	0.2498	1.45E-28	0	1
qrks	0.53864	0.0746	393.65442	52.14	<.0001
wcf	0.29875	0.05439	227.76222	30.17	<.0001

Stepwise Selection: Step 3					
Variable pqc Entered: R-Square = 0.5724 and C(p) = 9.7210					
Analysis of Variance					
Source	DF	Sum of Squares	Mean Square	F Value	Pr > F
Model	3	1099.729	366.57628	52.2	<.0001
Error	117	821.5754	7.02201		
Corrected Total	120	1921.304			
Variable	Parameter Estimate	Standard Error	Type II SS	F Value	Pr > F
Intercept	-1.16E-15	0.2409	1.64E-28	0	1
qrks	0.40652	0.08332	167.14734	23.8	<.0001
pqc	0.22592	0.07188	69.36046	9.88	0.0021
wcf	0.27642	0.05293	191.48362	27.27	<.0001

Stepwise Selection: Step 4					
Variable misf Entered: R-Square = 0.5916 and C(p) = 6.2135					
Analysis of Variance					
Source	DF	Sum of Squares	Mean Square	F Value	Pr > F
Model	4	1136.6	284.14991	42	<.0001
Error	116	784.7045	6.76469		
Corrected Total	120	1921.304			
Variable	Parameter Estimate	Standard Error	Type II SS	F Value	Pr > F
Intercept	-1.34E-15	0.23645	2.18E-28	0	1
qrks	0.31323	0.09102	80.11152	11.84	0.0008
misf	0.18689	0.08005	36.87081	5.45	0.0213
pqc	0.22441	0.07056	68.42698	10.12	0.0019
wcf	0.21834	0.05761	97.1789	14.37	0.0002

Summary of Stepwise Selection									
Step	Variable Entered	Variable Removed	Label	Number Vars In	Partial R-Square	Model R-Square	C(p)	F Value	Pr > F
1	qrks		qrks	1	0.4177	0.4177	50.1031	85.38	<.0001
2	wcf		wcf	2	0.1185	0.5363	18.0816	30.17	<.0001
3	pqc		pqc	3	0.0361	0.5724	9.721	9.88	0.0021
4	misf		misf	4	0.0192	0.5916	6.2135	5.45	0.0213

the perceived quality of risk control (H2a); the process variable perceived quality of risk knowledge sharing is positively associated with perceived quality of risk control (H4a); and from the technology variables, the risk management information system functionality (H5a) and the web channel functionality (H6a) are each positively associated with the perceived quality of risk control. Table 7 summarises the results of the QRC model.

PERM Model

The only variable that entered the stepwise regression model was the perceived quality of risk communication among people *(pqc)* and no other variables were significant (see Table 8a-8c). The model has a low R-squared value of 0.09. Therefore, the results indicate the existence of a weak relationship (Golderberger, 1978; Newman & Newman, 2000). Given the low R-squared value, the power of the model was calculated: the value of 0.65 is smaller than the threshold of 0.8 recommended. Therefore, the model's capacity to describe the relationships of the variables is not sufficient to accept hypothesis H2b, and so none of the hypotheses H1b- H8b are supported.

Table 7. Summary of hypothesis tests for QRC model

Hypotheses	Results
People	
H1a: Organizational capacity for work coordination (*cwc*) is positively associated with the perceived quality of risk control(*qrc*)	Not Supported
H2a: The perceived quality of communication among people (*pqc*) is positively associated with perceived quality of risk control(*qrc*)	Supported
H3a: Perceived quality of people interactions in the ERMIS design (*iis*) is positively associated with the perceived quality of risk control(*qrc*)	Not Supported
Process	
H4a: The perceived quality of risk knowledge sharing (*qrks*)is positively associated with the perceived quality of risk control(*qrc*)	Supported
Technology	
H5a: The risk management information system functionality (*misf*) is positively associated with the perceived quality of risk control(*qrc*)	Supported
H6a: The web channel functionality (*wcf*) is positively associated with the perceived quality of risk control(*qrc*)	Supported
H7a: The perceived integration of the information systems (*isi*) is positively associated with the perceived quality of risk control(*qrc*)	Not Supported
H8a: The quality of the network capacity for connecting people (*nccp*) is positively associated with the perceived quality of risk control(*qrc*)	Not Supported

Table 8a. Stepwise regression results for PERM model, perceived value of enterprise risk management

Root MSE	6.3355	**R-Square**	0.0903
Dependent Mean	0.2568	**Adj. R-Sq.**	0.0826
Coeff. Var	2466.7		

Table 8b. Stepwise regression results for PERM model, perceived value of enterprise risk management

Analysis of Variance					
Source	**DF**	**Sum of Squares**	**Mean Square**	**F Value**	**Pr > F**
Model	1	470.3054	470.31	11.72	0.0009
Error	118	4736.293	40.138		
Corrected Total	119	5206.599			

Table 8c. Stepwise regression results for PERM model, perceived value of enterprise risk management

Parameter Estimates						
Variable	**Label**	**DF**	**Parameter Estimate**	**Standard Error**	**t Value**	**Pr > |t|**
Intercept	Intercept	1	0.24575	0.57835	0.42	0.6717
pqc	pqc	1	0.48032	0.14032	3.42	0.0009

DISCUSSION AND IMPLICATIONS

Four KM variables were found to be significantly associated with the perceived quality of risk control: perceived quality of communication among people (*pqc*, H2a), perceived quality of risk knowledge sharing (*qrks*, H4a), risk management information systems functionality (*misf*, H5a) and web channel functionality (*wcf*, H6a). Risk knowledge sharing and communication among people may be described as social factors, and indeed perceived quality of risk knowledge sharing (*qrks)* is the most influential variable in the perceived quality of risk control model.

Risk knowledge sharing is an activity that financial institutions will always need to deal with. For example, what is done in one area may be an input in another, and there are multiple stakeholders involved. If a "silo organization" obstructs risk knowledge sharing, our findings suggest risk control may be adversely affected.

The support for hypothesis H2a complements the findings of Julibert (2008) and Waldvogel and Whelan (2008) that without communication no KM is possible and that communication and good risk learning support collaboration.

Given the lack of support for H8a, the effect of the quality of the network capacity for connecting people, this suggests that people prefer informal communication mechanisms to formal ones. The result can be considered a change from what Swan et al. (1999) pointed out about knowledge sharing and the web channel. Thus in RM practice, improving social factors, for example by creating spaces for better communication among groups, and improving messages and other means of communication may lead to better risk knowledge sharing, confirming the findings of Bosua and Scheepers (2007) in other sectors. This in turn should contribute positively to the quality of risk control.

As support for risk modelling analysis and work flow (Crouhy et al., 2001; Hormozi & Giles, 2004), better risk management information systems functionality is likely to have a positive effect on risk control.

For the significant association of web channel functionality, several authors (e.g. Desouza & Awazu, 2005; M. Jennex, 2005, 2006; Spies et al., 2005) have indicated that an effective intranet enables an integrated technical infrastructure in order to disseminate knowledge and to support KM processes. The intranet, in financial institutions, provides access to different components of the organization's knowledge and in some cases, to operational tools. RM intranet functionality enhancements, according to the results, are likely to add value to risk control.

Interpreting the PERM Results

The perceived value of ERM implementation is weakly associated with the KM variable perceived quality of communication among people (*pqc*) and is not associated with the other KM variables studied. This means that efforts in improving KM are not perceived as having the same organization-wide effect in implementing ERM as they have in the more local problem of controlling operational risk.

To explain this, consider the demographic characteristics of our sample, especially time in the position and risk management experience: 94% of responses came from people with less than ten years in their current position, and 73% from those with less than ten years of experience. This combination of short time in a position and low experience (many staff spend their entire working careers in RM) indicates a potential lack of an holistic view of the RM function, because there is a reduced exposure to the whole spectrum of RM work; particularly as most junior RM roles are very specialized and lack of seniority potentially restricts access to information and decisions. Francis and Paladino (2008) pointed out that the best practices of ERM require a high degree of participation and involvement in strategic actions. Driscoll (2014) in the description of the seven imperatives or risk management indicates examples where the good practices require the involvements of top management to understand risk in the context of strategy design and execution. Thus it may be that the population would simply not be able to perceive any benefits that existed: indeed, the outlier response removed from this model can be argued to have come from someone who did not perceive ERM to be of any value at all.

CONCLUSION

Our study has examined eight KM constructs associated with knowledge sharing (three relating to people, one relating to process and four relating to technology) and two RM constructs (the perceived quality of risk control and the perceived value of ERM implementation) in the financial services sector. A survey obtained responses from 121 risk management staff in financial institutions. The first step towards our research aim was the identification of these ten variables (eight KM and two RM) and the construction of items and scales for them. These are a contribution to future studies in the field in themselves, given the results of the pilot testing and the Cronbach alpha values for the reliability of the scales.

The overall aim of this research was to study how KM concepts may help improve RM, and help to turn it into true ERM. Specifically, we aimed to identify KM variables associated with knowledge sharing that can influence the perceived quality of risk control and the perceived value of ERM implementation. Eight pairs of hypotheses were tested by stepwise regression. The findings showed that four of the eight variables analysed have a significant positive association overall with the perceived quality of risk control. These are: one people variable (perceived quality of communication among people), the sole process variable (perceived quality of risk knowledge sharing) and two technology variables (risk management information system functionality and web channel functionality).

The perceived quality of risk knowledge sharing accounted for by far the largest part of the variation (31.3%) in the dependent variable perceived quality of risk control. When risk actions, decisions and the experiences that people have in dealing with different risks are shared between different RM groups, they develop awareness and warning signals as inputs to the risk control process. According to our findings, this means that actions to improve risk knowledge sharing are likely to have the most positive effects on the perceived quality of risk control.

The only KM variable associated with the perceived value of ERM implementation (PERM) was the perceived quality of communication among people; however, the low R-squared (0.09) and power (0.65) values make this a tenuous relationship. The use of knowledge based risk management process is clearer individually by risk: operational risk, improvement of risk metrics or credit risk; however, in a general ERM view, it seems that KM has little influence on the value of ERM implementation, or RM workers are not able to perceive that it does for dealing with all types of risks in a holistic view. Two possible explanations to support the latter are: (a) that organizational silos still prevail, and local operational risk control is the main concern, while overall ERM success, for most people, remains someone else's problem; and/or (b) that the majority of relatively junior and inexperienced staff among the respondents are not able to take the holistic view that ERM requires and do not know how valuable or otherwise ERM implementation is in their organization. Padova and Scarso (2012) have recently described similar tensions between local and global knowledge management issues even in an organization as experienced in KM as Ernst & Young. It is also very relevant to note the comments of Zack, McKeen, and Singh

(2009) that "[a] gap exists between KM practices that firms believe to be important and those that were directly related to organizational performance."

The research has thus provided evidence that aligning KM and RM processes should achieve better RM results in risk control. Specifically, recalling the concept of financial institutions as risk and knowledge organizations, the results indicate not only that better interpretation of the meaning of RM information is needed, but also that organizational RM knowledge needs to be managed better (Marshall et al., 1996) through better risk knowledge sharing, better risk management information functionality, better communication among people and better web channel functionality.

There are some limitations to the research process. Our study is based on perceptions of the RM process performance rather than the RM outcome performance. Improvement in the representation of senior staff in the sample could be secured through a stratification according to organizational levels or the number of years of experience. The geographical distribution of the respondents, skewed toward North America, may be a factor influencing results because the degree of maturity in RM might be different in different countries. The period when the data were collected was at the beginning of the global financial crisis, which may have coloured the responses. Finally, the size of the institution in which the RM employee works might affect the level of development and sophistication in RM practice.

This study also raises questions for new research. One is related to the concept of information system integration, which unexpectedly revealed a negative association with the perceived value of ERM implementation, albeit not a statistically significant one. Further work is also needed relating to possible differences between global, top-down business needs for KM and local, bottom-up user perceptions, and how this might affect the management of risk knowledge. Equally, RM needs studies directed to identifying the "soft" part of risk management and its influence in the decision making process and how to grow the capacity for people to develop these activities in a better and more reliable organizational environment. Finally, objective proxy measures of RM performance could be used in studies of specific sectors, such as commercial lending.

REFERENCES

Aaker, D., Day, G., & Kumar, V. (1998). *Marketing Research.* New York: John Wiley and Sons.

Alavi, M., & Leidner, D. (2001). Review: Knowledge management and knowledge management systems: conceptual foundations and research issues. *Management Information Systems Quarterly*, *25*(1), 107–136. doi:10.2307/3250961

Alfares, H., & Duffuaa, S. (2008). Assigning cardinal weights in multi-criteria decision making. *Journal of Multi-Criteria Decision Analysis*, *15*(5-6), 125–133. doi:10.1002/mcda.420

Babbie, E. (1998). *The Practice of Social Research.* Belmont, CA: Wadsworth Publishing Co.

Banham, R. (2004). Enterprising views of risk management. *Journal of Accountancy*, *197*(6), 65–72.

Beasley, M., Bronson, B., & Hancock, B. (2009). ERM: Opportunities for improvement. *Journal of Accountancy*, *208*(3), 175–179.

Bohrnstedt, G., & Knoke, D. (1982). *Statistics for Social Data Analysis*. Itasco, IL: F.E. Peacock Publishers.

Bosua, R., & Scheepers, R. (2007). Towards a model to explain knowledge sharing in complex organizational environments. *Knowledge Management Research and Practice*, *5*(2), 93–109. doi:10.1057/palgrave.kmrp.8500131

Bromiley, P., McShane, M., Nair, A., & Rustambekov, E. (2015). Enterprise Risk Management: Review, Critique, and Research Directions. *Long Range Planning*, *48*(4), 265–276. doi:10.1016/j.lrp.2014.07.005

Brown, B. (2001). Step by step enterprise risk management. *Risk Management*, *48*(9), 43–50.

Callahan, C., & Soileau, J. (2017). Does Enterprise risk management enhance operating performance? *Advances in Accounting*, *37*, 122–139. doi:10.1016/j.adiac.2017.01.001

Champion, D. (2009). Managing risk in the new world. *Harvard Business Review*, *87*(10), 69–75.

Cohen, J., Cohen, P., West, S. G., & Aiken, L. S. (2003). *Applied Multiple Regression/Correlation Analysis for the Behavioral Sciences*. Mahwah, NJ: Lawrence Erlbaum Associates.

Cortina, J. (1993). What Is coefficient alpha? an examination of theory and applications. *The Journal of Applied Psychology*, *78*(1), 98–104. doi:10.1037/0021-9010.78.1.98

Crouhy, M., Galai, D., & Mark, R. (2001). *Risk Management*. New York: McGraw Hill.

Cumming, C., & Hirtle, B. (2001). The challenges of risk management in diversified financial companies. *Federal Reserve Bank of New York Economic Policy Review*, *4*(3), 1–17.

Davenport, T., & Prusak, L. (1998). *Working Knowledge*. Boston, MA: Harvard Business School Press.

Desouza, K., & Awazu, Y. (2005). *Engaged Knowledge Management: Engagement with new realities*. Basingstoke, UK: Palgrave Macmillan. doi:10.1057/9780230006072

Dickinson, G. (2001). Enterprise Risk Management: Its origins and conceptual foundation. *The Geneva Papers on Risk and Insurance*, *26*(3), 360–366. doi:10.1111/1468-0440.00121

Driscoll, M. (2014, November). Seven Imperatives For Process Excellence. *Corporate Finance Review*, 13-19.

Earl, M. (2001). Knowledge management strategies: Toward a taxonomy. *Journal of Management Information Systems*, *18*(1), 215–233. doi:10.1080/07421222.2001.11045670

Edwards, J. S. (2009). Business processes and knowledge management. In M. Khosrow-Pour (Ed.), *Encyclopedia of Information Science and Technology* (2nd ed.; Vol. I, pp. 471–476). Hershey, PA: IGI Global. doi:10.4018/978-1-60566-026-4.ch078

Fourie, L., & Shilawa, J. (2004). The value of concept maps for knowledge management in the banking and insurance industry: A German case study. *Concept Maps: Theory, Methodology, Technology: Proceedings of the First International Conference on Concept Mapping*.

Francis, S., & Paladino, B. (2008). Enterprise Risk Management: A best practice approach. *Journal of Corporate Accounting & Finance*, *19*(3), 19–33. doi:10.1002/jcaf.20382

Fraser, J. R. S., & Simkins, B. J. (2007). Ten Common Misconceptions About Enterprise Risk Management. *Journal of Applied Corporate Finance*, *19*(4), 75–81. doi:10.1111/j.1745-6622.2007.00161.x

Fraser, J. R. S., & Simkins, B. J. (2016). The challenges of and solutions for implementing enterprise risk management. *Business Horizons*, *59*(6), 689–698. doi:10.1016/j.bushor.2016.06.007

Golderberger, A. (1978). *Topics in Regression Analysis*. New York: McMillan.

Hair, J., Babin, B., Money, A., & Samouel, P. (2003). *Essentials of Business Research Methods*. New York: John Wiley and Sons.

Heisig, P. (2009). Harmonisation of knowledge management. *Journal of Knowledge Management*, *13*(4), 4–31. doi:10.1108/13673270910971798

Hobday, M., Davies, A., & Prencipe, A. (2005). Systems integration: A core capability of the modern corporation. *Industrial and Corporate Change*, *14*(6), 1109–1143. doi:10.1093/icc/dth080

Hormozi, A., & Giles, S. (2004). Data mining a competitive weapon for banking and retail industries. *Information Systems Management*, *21*(2), 62–71. doi:10.1201/1078/44118.21.2.20040301/80423.9

Hsu, C., Backhouse, J., & Silva, L. (2013). Institutionalizing operational risk management: An empirical study. *Journal of Information Technology*. doi:10.1057/jit.2013.15

Jennex, M. (2005). Internet Support for Knowledge Management Systems. In M. Khosrow-Pour (Ed.), *Encyclopedia of Information Science and Technology* (pp. 1640–1644). Hershey, PA: Idea Group. doi:10.4018/978-1-59140-553-5.ch289

Jennex, M. (2006). Knowledge Management System Success Factors. In D. Schwartz (Ed.), *Encyclopedia of Knowledge Management* (pp. 436–441). Hershey, PA: Idea Group Reference. doi:10.4018/978-1-59140-573-3.ch057

Jennex, M. E., & Zyngier, S. (2007). Security as a contributor to knowledge management success. *Information Systems Frontiers*, *9*(5), 493–504. doi:10.100710796-007-9053-4

Julibert, S. (2008). Employee attitudes to information sharing: A case study at the European Central Bank. *Records Management Journal*, *18*(3), 194–204. doi:10.1108/09565690810916447

Lam, J. (2003). *Enterprise Risk Management: from Incentives to Controls*. Hoboken, NJ: John Wiley & Sons.

Lambert, D., & Harrington, T. (1990). Measuring nonresponse bias in customer service mail surveys. *Journal of Business Logistics*, *11*(2), 5–25.

Lee, S., & Lam, W. (2007). Application Integration: Pilot Project to Implement a Financial Portfolio System in a Korean Bank. In W. Lam & V. Shankararaman (Eds.), *Enterprise Architecture and Integration: Methods, Implementation and Technologies* (pp. 273–283). Hershey, PA: IGI Global. doi:10.4018/978-1-59140-887-1.ch016

Lehaney, B., Clarke, S., Coakes, E., & Jack, G. (2004). *Beyond Knowledge Management*. Hershey, PA: Idea Group Publishing. doi:10.4018/978-1-59140-180-3

Levine, R. (2004). Risk management systems: Understanding the need. *Information Systems Management*, *21*(2), 31–37. doi:10.1201/1078/44118.21.2.20040301/80419.5

Luburic, R. (2016). Knowledge and Learning in Terms of Operational Risk Management in the Financial and Banking Systems. *International Journal of Qualitative Research, 10*(3), 559–568.

Majchrzak, A., Beath, C., Lim, R., & Chin. (2005). Managing client dialogues during information systems design to facilitate client learning. *Management Information Systems Quarterly, 29*(4), 653–672. doi:10.2307/25148704

Marshall, C., Prusak, L., & Shpilberg, D. (1996). Financial risk and the need for superior knowledge management. *California Management Review, 38*(3), 77–101. doi:10.2307/41165844

Massingham, P. (2010). Knowledge risk management: A framework. *Journal of Knowledge Management, 14*(3), 464–485. doi:10.1108/13673271011050166

McNally, J. S. (2013, June). The 2013 COSO Framework & SOX Compliance. *Strategic Finance*. Retrieved from http://www.coso.org/documents/COSO%20McNallyTransition%20Article-Final%20 COSO%20Version%20Proof_5-31-13.pdf

Meulbroek, L. (2002). The promise and challenge of integrated risk management. *Risk Management & Insurance Review, 5*(1), 55–66. doi:10.1111/1098-1616.00006

Murphy, K., & Myors, B. (1998). *Statistical Power Analysis*. Mahwah, NJ: Lawrence Erlbaum Associates Publishers.

Newman, I., & Newman, C. (2000). A discussion of low r-squares: Concerns and uses. *Educational Research Quarterly, 24*(2), 3–9.

Nonaka, I. (1991). The Knowledge Creating Company. *Harvard Business Review, 69*(6), 96–104.

O'Brien, J. (1996). *Management Information systems*. Chicago: Irwin.

O'Brien, J., & Szerszeń, P. J. (2017). An evaluation of bank measures for market risk before, during and after the financial crisis. *Journal of Banking & Finance, 80*, 215–234. doi:10.1016/j.jbankfin.2017.03.002

Oliveira, M., Curado, C. M. M., Maçada, A., & Nodari, F. (2015). Using alternative scales to measure knowledge sharing behavior: Are there any differences? *Computers in Human Behavior, 44*, 132–140. doi:10.1016/j.chb.2014.11.042

Padova, A., & Scarso, E. (2012). Managing large amounts of knowledge objects: Cognitive and organisational problems. *Knowledge Management Research and Practice, 10*(4), 287–295. doi:10.1057/kmrp.2012.7

Paul, J., Mittal, A., & Srivastav, G. (2016). Impact of service quality on customer satisfaction in private and public sector banks. *International Journal of Bank Marketing, 34*(5), 606–662. doi:10.1108/IJBM-03-2015-0030

Pindyck, R., & Rubinfeld, D. (1998). *Econometric Models and Economic Forecasts*. Boston, MA: Irwin-McGraw Hill.

Ribière, V., & Walter, C. (2013). 10 years of KM theory and practices. *Knowledge Management Research and Practice, 11*(1), 4–9. doi:10.1057/kmrp.2012.64

Rodriguez, E., & Edwards, J. S. (2010). People, technology, processes and risk knowledge sharing. *Electronic Journal of Knowledge Management*, *8*(1), 139–150.

Saha, P., Bose, I., & Mahanti, A. (2016). A knowledge based scheme for risk assessment in loan processing by banks. *Decision Support Systems*, *84*, 78–88. doi:10.1016/j.dss.2016.02.002

Shariq, S., & Vendelo, M. (2006). Tacit Knowledge Sharing. In D. Schwartz (Ed.), *Encyclopedia of Knowledge Management* (pp. 833–839). Hershey, PA: Idea Group Reference. doi:10.4018/978-1-59140-573-3.ch109

Shaw, D., Hall, M., Edwards, J. S., & Baker, B. (2007). Responding to crisis through strategic knowledge management. *Journal of Organizational Change Management*, *20*(4), 559–578. doi:10.1108/09534810710760081

Spies, M., Clayton, A. J., & Noormohammadian, M. (2005). Knowledge management in a decentralized global financial services provider: A case study with Allianz Group. *Knowledge Management Research and Practice*, *3*(1), 24–36. doi:10.1057/palgrave.kmrp.8500046

Swan, J., Newell, S., Scarborough, H., & Hislop, D. (1999). Knowledge management and innovation: Networks and networking. *Journal of Knowledge Management*, *3*(4), 262–275. doi:10.1108/13673279910304014

Taleb, N., Goldstein, D., & Spitznagel, M. (2009). The six mistakes executives make in risk management. *Harvard Business Review*, *87*(10), 78–81. PMID:19968059

Te'eni, D. (2006). Organisational Communication. In D. Schwartz (Ed.), *Encyclopedia of Knowledge Management* (pp. 734–740). Hershey, PA: Idea Group Reference. doi:10.4018/978-1-59140-573-3.ch096

Trkman, P., & Desouza, K. C. (2012). Knowledge risks in organizational networks: An exploratory framework. *The Journal of Strategic Information Systems*, *21*(1), 1–17. doi:10.1016/j.jsis.2011.11.001

Trochim, W., & Donnelly, J. (2007). *The Research Methods Knowledge Base*. Cincinnati, OH: Atomic Dog Publishing.

Uchupalanan, K. (2000). Competition and IT-based innovation in banking services. *International Journal of Innovation Management*, *4*(4), 455–489. doi:10.1142/S1363919600000238

Uzzi, B., & Lancaster, R. (2003). Relational embeddedness and learning: The case of bank loan managers and their clients. *Management Science*, *49*(4), 383–399. doi:10.1287/mnsc.49.4.383.14427

Värlander, S. (2008). The role of situated embodied interaction in the banking customer knowledge creation process. *International Journal of Knowledge Management*, *4*(4), 62–76. doi:10.4018/jkm.2008100104

Waldvogel, A., & Whelan, N. (2008). Towards better financial risk learning. *Journal of Risk Management in Financial Institutions*, *4*(4), 382–393.

Wiig, K. (1997). Knowledge management: Where did it come from and where will it go? *Expert Systems with Applications*, *13*(1), 1–14. doi:10.1016/S0957-4174(97)00018-3

Wiig, K. (2004). *People-Focused Knowledge Management*. Burlington, MA: Elsevier Butterworth Heinemann.

Zack, M., McKeen, J., & Singh, S. (2009). Knowledge management and organizational performance: An exploratory analysis. *Journal of Knowledge Management*, *13*(6), 392–409. doi:10.1108/13673270910997088

Zhao, X., Hwang, B.-G., & Pheng, S. (2016). An enterprise risk management knowledge-based decision support system for construction firms, *Singapore Engineering. Construction and Architectural Management*, *23*(3), 369–384. doi:10.1108/ECAM-03-2015-0042

Zyngier, S. (2008). Risk Management: Strengthening Knowledge Management. *International Journal of Knowledge Management*, *4*(3), 1–32. doi:10.4018/jkm.2008070102

Zyngier, S., & Burstein, F. (2012). Knowledge management governance: The road to continuous benefits realization. *Journal of Information Technology*, *27*(2), 140–155. doi:10.1057/jit.2011.31

Zyngier, S., & Venkitachalam, K. (2011). Knowledge management governance – a strategic driver. *Knowledge Management Research and Practice*, *9*(2), 136–150. doi:10.1057/kmrp.2011.5

Chapter 15
A Business-Driven Process Model for Knowledge Security Risk Management:
Tackling Knowledge Risks While Realizing Business Benefits

Ilona Ilvonen
Tampere University of Technology, Finland

Jari Jussila
University of Jyväskylä, Finland

Hannu Kärkkäinen
Tampere University of Technology, Finland

ABSTRACT

This chapter introduces a model to manage knowledge security risks in organizations. Knowledge security risk management is a process that should always be done in connection with the business benefits assessment and realization. The process model presented helps to identify knowledge security risks and provides a comprehensive approach to evaluating and balancing the costs and benefits of knowledge sharing and knowledge risk management and is discussed in light of the benefits realization process. The process model can be a valuable tool for practitioners aiming to develop knowledge sharing practices in companies, and at the same time need to consider the security of knowledge. It is also a communication tool for managers to identify possible risk sources and sources for business benefits, and as such works as a translation tool in many business contexts.

DOI: 10.4018/978-1-5225-5427-1.ch015

INTRODUCTION

Knowledge and its creation are important sources of competitive advantage and business opportunities for most contemporary organizations (Alavi & Leidner, 2001; Choo, 1996; Grant, 1996; Nonaka & Takeuchi, 1995). Although knowledge creation, sharing and management have been studied extensively (e.g. Bolisani & Scarso, 2014; Matayong & Mahmood, 2013; Tzortzaki & Mihiotis, 2014), there is one viewpoint to knowledge that has received less attention until recent years: knowledge security (Randeree, 2006; Shedden, Scheepers, Smith, & Ahmad, 2011). Despite the importance of knowledge and the need for knowledge protection, there is little literature on knowledge security (Shedden et al. 2010). In terms of knowledge security and risk analysis, most existing risk analysis methods can be regarded as providing a plain technical view on information and technological assets (Ahmad, Bosua, & Scheepers, 2014; Padyab, Päivärinta, & Harnesk, 2014; Shedden et al., 2011; Shedden, Smith, & Ahmad, 2010; Spears, 2006), ignoring that knowledge is bound to people (Shedden et al., 2010, 2011; Ilvonen, 2013; Padyab, Paivarinta, & Harnesk, 2014) and as a consequence people and their communication are significant sources of knowledge security risks.

Since knowledge security is still in the early stage as a research field, it is reasonable to look also for parallel fields in order to understand the principles of security risk management. Information security risk assessment (ISRA) methodologies are means by which organizations aim to manage information security risks (Baskerville, 1991; Siponen, 2005; Whitman & Mattord, 2011). However, typical perspectives on information security risk management, including most ISRA methodologies, largely ignore the business context of information systems (Shedden et al., 2010; Spremic, 2012), and are not framed in terms of competitive advantage (Ahmad et al., 2014). When the business perspective is considered (DeLoach, 2004; Siponen, 2005; Von Solms & Von Solms, 2004), it is mainly limited to the evaluation of individual risk mitigation techniques and their cost reasoning, rather than starting from a broad perspective of reasoning the business benefits of an activity compared to the risks connected to it.

This chapter aims to answer the research question "How can organizations manage knowledge risks and reach business benefits of changes in operation at the same time?" The chapter argues that knowledge security risks should be managed in a systematic process in line with managing business benefits and introduce a conceptually developed process for this purpose.

Several studies point out that increasing the circulation of knowledge also increases the risk of leakage (Desouza, 2006; Desouza & Vanapalli, 2005; Easterby-Smith, Lyles, & Tsang, 2008; Trkman & Desouza, 2012). New forms of organizational operation emphasize opening up of organizational knowledge resources towards customers and other organizational stakeholders. Therefore, when making changes in practices, there is simultaneously a strong need for understanding the potential risks related to open information and knowledge flows, as well as relating these risks to the potential business benefits of the change.

After introducing the theoretical background, the chapter introduces the proposed process model and discusses the relation of the process steps to previous work. After this an analytical case that illustrates the outputs of the process model from a practical perspective is presented. The paper concludes with a brief discussion on avenues for further research.

BACKGROUND

Knowledge and Knowledge Security

In this paper the term knowledge is understood as an intangible asset that is mainly possessed and created by people (Nonaka & Takeuchi, 1995). Knowledge is a result of human thinking and interpretation (Davenport & Prusak, 1998; Thierauf, 2001). Although knowledge in most cases is embedded in people as tacit knowledge, that knowledge can be to some extent shared with other people by externalizing it into a form of explicit knowledge, i.e. written or spoken language (Nonaka & Takeuchi, 1995; Szulanski, 2000). Knowledge generates value to a company through its use and sharing; this requires that the people receiving knowledge are able to interpret the shared knowledge, and use it for learning and reflection (Alavi & Leidner, 2001). Knowledge thus is of value to an organization when the right people have and use the correct knowledge.

When a company creates competitive advantage from the knowledge of its workers, the knowledge becomes the biggest asset of the company (Grant, 1996). This idea is the backbone of the "knowledge-based view" of the firm (Kogut & Zander, 1992; Spender & Grant, 1996). When knowledge is considered a resource of an organization, it is considered bound to the individual people that have it, but also something that is possible to transfer from person to person (Hansen, 1999) on individual and collective (Krogh, 2009) levels. While knowledge is used for making sense of the situation the organization is in, the organization also creates new knowledge and makes decisions based on that knowledge (Choo, 1996). This makes knowledge an elusive object to secure, since new knowledge is created all the time. In fact, the process of knowledge security risk management is in itself a process of new knowledge creation.

The use and sharing of knowledge requires communication between people that possess the knowledge (Hansen, 1999; Nonaka & Takeuchi, 1995). Today there are numerous technical tools to support discussion and knowledge sharing, among them different information systems and social media tools that are designed to mimic face-to-face communication between people that are possibly geographically apart (von Krogh, 2012). From the point of view of knowledge, these kind of tools are repositories for knowledge (Aljafari & Sarnikar, 2009) in addition to being channels for knowledge exchange (Padyab, Paivarinta, et al., 2014). A systematic way to identify and manage risks connected to knowledge would help in establishing a unified level of protection of knowledge, but research shows that this does not exist, at least not widely in organizations (Ahmad et al., 2014). Knowledge security or knowledge protection are not unaddressed in the literature, but individual processes concentrate on narrow areas of knowledge security, such as retaining knowledge of leaving employees (Jennex, 2014) or concentrating on more formal protection mechanisms (Olander, Vanhala, & Hurmelinna-Laukkanen, 2014).

In this paper knowledge security is understood as the managerial process of organizations to identify threats toward important knowledge and secure the knowledge against those threats. As knowledge is bound to people, knowledge security is closely related to managing people and their activities both within and across organizational boundaries.

Risk Management

Risk as a term refers to an event that may have great consequences to an organization, but that is uncertain to happen. In financial terms risk can be understood as either a positive or negative event (DeLoach, 2004). Especially business risks always have the potential of positive financial outcomes. However, this paper concentrates on knowledge risks that, if realized, have negative consequences to an organization. These two, however, are intertwined and cannot be entirely separated from each other, especially when cost-benefit analysis is added as an essential element to risk management.

A risk is constructed of several components: a threat, the consequences of the realization of that threat, and the probability of the realization (Jennex & Zyngier, 2007; Stoneburner, Goguen, & Feringa, 2002). Knowledge risks may harm an organization for example by reducing competitive advantage, by damaging the reputation of a company or by creating distress among employees (Padyab, Päivärinta, & Harnesk, 2014; Väyrynen, Hekkala, & Liias, 2013). Knowledge risks can be seen to originate either from external sources or from internal sources (Ilvonen, 2013; Markus Manhart & Thalmann, 2015). The division to external and internal threats in the context of this paper refers to whether there are external actors involved in the realization of a threat. In case of knowledge risks the vulnerabilities that cause threats and consequently risks can be numerous, and they all have a human element in them, since the knowledge is bound to people.

Generally four basic phases can be identified from risk management processes. These are for example (e.g. Lichtenstein, 1996; Bandyopadhyay, Mykytyn, & Mykytyn, 1999):

- Asset and risk identification
- Risk analysis
- Risk-reducing measures, and
- Risk monitoring

The main steps can be found also from many information security risk management models (Caralli, Stevens, Young, & Wilson, 2007; Shedden et al., 2011). The steps in the process may get different names from different authors, but the essential content of the process remains the same. What is common to the security risk management processes is that many times the trigger to begin a risk management process comes from the reason that risk management requirements and standards need to be met (Webb, Maynard, Ahmad, & Shanks, 2013; Thalmann, Manhart, Ceravolo, & Azzini, 2014; Markus Manhart & Thalmann, 2015).

Project risk management literature considers risk management to be triggered by the need of a development project (e.g. Varnell-Sarjeant, 2008; Jafari, Rezaeenour, Mazdeh, & Hooshmandi, 2011). The focus of project risk management, however, is usually that of successful completion of the project; within deadline and within budget. Security risks get less attention in the project management process models. The idea of the process model presented in this paper is to combine the approaches of project risk management and security risk management to get a comprehensive approach to identify, manage and monitor knowledge security risks.

Benefits Management and Realization

Most organizations focus on implementing information technology, and in doing so, forget to realize the expected business benefits from the IT investment (Peppard, Ward, & Daniel, 2007). Simply put, IT professionals or project managers are assigned the task of implementing the information system, but no one is responsible at the receiving end for making use of the information system and ensuring that benefits that were originally expected and planned are realized in practice. Benefits management approach proposed by Peppard, Ward and Daniel (2007) highlight that business managers and IT professionals must work together and uncover three important aspects of the investment: the target performance improvements (the ends), the ways the business must work differently (the ways), and the enabling IT capabilities (the means). A combination of business changes and new IT capabilities are thus needed for benefits realization, also it must be made clear who needs to become accountable for making the specific changes and delivering the benefits (Peppard et al., 2007).

A benefits management process model has been introduced by Ward and Daniel (2006) as best practice guidelines for benefits management. The process model includes five iteratively following stages: identifying and structuring benefits, planning benefits realization, executing the benefits realization plan, evaluating and reviewing the results and potential for further benefits.

Identifying benefits includes identifying proposed benefits and developing suitable measures (both financial and non-financial) for each proposed benefit (Ward, Taylor, & Bond, 1996). Structuring means understanding the linkages between enabling IT capabilities (technology effects), the ways businesses must work differently (business changes) and the targeted performance improvements (business effects) (Peppard et al., 2007; Ward et al., 1996). Planning benefits realization involves assigning responsibility

Figure 1. A process model of benefits management
Modified from Ward & Daniel 2006.

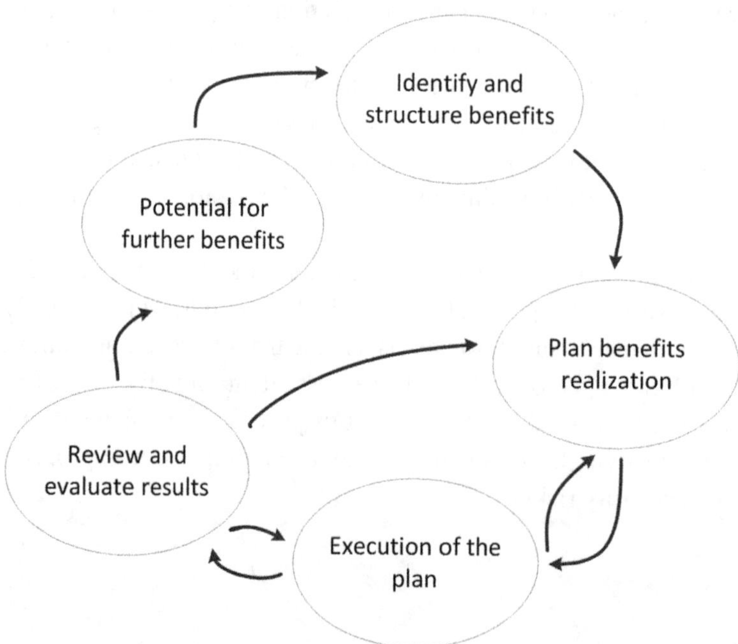

for realizing the identified benefit within the business, and at this stage, the full cost of IT implementation and business changes can be assessed to make an informed judgment of the viability of the proposed project. Executing the benefits realization plan is done alongside the IT implementation, and it ensures that necessary business changes are carried out. Evaluating and reviewing the results follows the IT implementation and business changes and uses the developed measures to evaluate the business effects of the project. Finally, as a result of post-project review, further benefits may become achievable that were not envisaged in the beginning, the last stage also provides opportunity to facilitate the lessons learned to future projects. (Ward et al., 1996) In the following several ways of how the benefits management model can assist in managing knowledge risks, and how knowledge risk management measures can lead to better benefits management are presented.

A PROCESS MODEL FOR KNOWLEDGE SECURITY RISK MANAGEMENT

The knowledge security risk management process model (KSRM) is illustrated in Figure 2. Although the authors of this chapter acknowledge that no risk management model can cover all risks, all activities and all information or knowledge (e.g. Marabelli & Newell, 2012; Trkman & Desouza, 2012) they try to incorporate a wide understanding of the risk management process steps into the KSRM model.

Figure 2. Knowledge security risk management (KSRM) process

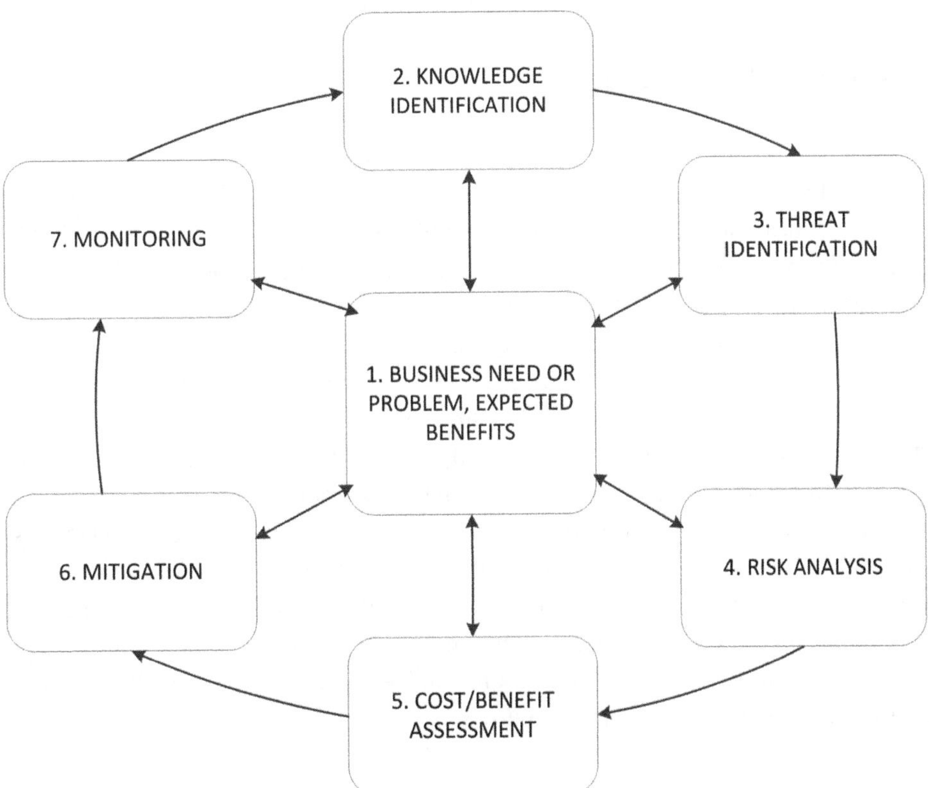

This paper concentrates on discussion of the links between the KSRM model and the benefits management model presented in Figure 1.

Step 1: Business Need

Existing KSRM models do not discuss much what should trigger the process, or their perspective towards business goals. The few links to business include: focusing on processes critical for business (Shedden et al., 2011, 2010), concentrating on stakeholders who are perceived important according to organizational strategic goals and objectives (Padyab, Paivarinta, et al., 2014), identifying strategic assets related to key business processes and involving organization's members and external partners involved in those business processes in the knowledge risk management process (Aljafari & Sarnikar, 2009), supporting reduction of transaction costs in inter-organizational collaboration by making explicit both risks and knowledge transfer benefits (Trkman & Desouza, 2012), and ensuring that knowledge protection meet requirements forced by laws, standards, customers or internal regulations (Manhart & Thalmann, 2013).

The KSRM process illustrates a continuous process that is triggered by a business goal, need or problem that needs to be solved (center of Figure 2). The process may be triggered by changes in the operations or environment of the company. In this step the goals and benefits sought from the change need to be defined, as well as the people that are responsible for the realization of those benefits (Ward & Daniel, 2006) and the evaluating of the risks.

The first step of the process should include an element of sensemaking: understanding the business situation the organization is in, and what is going on around it. The output of the first step of the KSRM process is the business case of the intended change and a description of what kind of benefits are sought by the change, i.e. a first draft of the plan for benefits realization.

Step 2: Knowledge Identification

The first step in the generic risk management process and the second step of the KSRM process is to identify important knowledge that may be at risk in the organization. There are various approaches to knowledge asset identification (A.M. Padyab, Paivarinta, et al., 2014; Shedden et al., 2011; Trkman & Desouza, 2012). Knowledge is bound to people, and there is a lot of knowledge that each individual possesses. However, not all that knowledge is important to the organization.

One way to structure and identify knowledge is to examine different locations, uses, topics and destinations of knowledge to identify knowledge assets, and to prioritize between them (Ilvonen 2013). These dimensions are illustrated in Figure 3. The dimensions are one way to ensure that the organization has a broad enough scope when they begin to identify important knowledge that needs to be secured.

Another way to approach identification of important knowledge is to use the genre-based method for identifying knowledge assets (Padyab, Paivarinta, et al., 2014). In this method knowledge is recognized through identifying actors who communicate within the organization. The different types of communication between different actors form genres of knowledge, that is, in Figure 3, the interaction between the locations and destinations of knowledge. This method is useful in locating knowledge, as well as recognizing the ways and tools that it is communicated with, as it covers all; locations, destinations, topics and uses of knowledge illustrated in Figure 3.

Figure 3. Identification of important knowledge

A method for identification of knowledge assets is also to examine the containers or reservoirs where the knowledge resides (Caralli et al., 2007; Aljafari & Sarnikar, 2009). This can mean either people, as with knowledge is obvious, but also for example social media tools and other information systems that are used for documenting and sharing of knowledge (Aljafari & Sarnikar, 2009). Important knowledge exists both in the individual and collective level (Krogh, 2009) and this should be acknowledged in the identification step. The plan for benefits realization can be updated along the identification of knowledge assets, as new knowledge contributors and possible benefactors are found in the identification process.

The output of the knowledge identification step is an inventory of important knowledge, a list of people that are users or holders of that knowledge, and a list of communication media that are used for sharing that knowledge. The knowledge identification step of risk management can be conducted parallel to the knowledge management process of the organization.

Step 3: Threat Identification

The third step of the KSRM process is threat identification. Many authors (Aljafari and Sarnikar 2009; Padyab et al. 2014; Shedden et al. 2011; Trkman and Desouza 2012) include threat identification either as an individual process phase or as a part of the risk analysis phase in knowledge security risk management process.

A threat is the potential of a negative event (Caralli et al., 2007). It can be further broken down into a vulnerability, an actor or a threat agent that exploits that vulnerability, a motive that the actor has to the exploitation and the outcome of the event (Caralli et al., 2007; Farahmand, Navathe, Enslow, & Sharp, 2003). In order to identify threats to knowledge, an organization needs to recognize vulnerabilities connected to that knowledge, as well as potential actors that threaten the knowledge along with their motives.

One suggestion for performing identification of threats to information is to use standards and ready-made checklists, instead of a deeper down analysis of vulnerabilities and threat agents (e.g. Farahmand et al., 2003; Caralli et al., 2007; Peltier, Peltier, & Blackley, 2005). In addition to just relying on standards and checklists, the step can be performed for example by creating typical and non-typical knowledge use and communication scenarios, and identifying potential threats from these scenarios (Caralli et al., 2007; Peltier et al., 2005). Scenario planning not only helps to identify possible threats, but also highlights the best case scenarios that lead to realization of the expected benefits, and thus a better understanding of the benefits plan.

The scope of the threats is one aspect that needs to be considered when threats are identified: threats may apply only to a limited part of an organization, to the entire organization and in severe situations also to other organizations (Trkman & Desouza, 2012). Especially with knowledge, the target and outcome of threats may be both intangible, so it is not necessary to categorize threats for example to natural, environmental or human threats (Farahmand et al., 2003). Instead, a division to external threats and accidental or deliberate internal threats may be more useful (Trkman & Desouza, 2012) along with the analysis of the result of the threats. The outcome of knowledge threats may, for example, directly benefit a competitor by revealing critical competitive knowledge (Ahmad et al., 2014), or indirectly harm the organization by affecting the trust of customers (Braun & Esswein, 2012).

The outcome of the threat identification step is an analysis of what threatens the important knowledge of the organization, and what are the threat agents. An additional outcome is deeper understanding of the benefits that can be reached in different scenarios, and thus an updated version of the benefits plan.

Step 4: Risk Analysis

The fourth step of the KSRM process is risk analysis. In this step the threats identified in the previous step are individually analyzed to understand what kind of risks they cause and how significant these risks are to the organization. Several models discuss risk analysis as a process phase in knowledge security risk management (Aljafari and Sarnikar 2009; Padyab et al. 2014; Trkman and Desouza 2012). Aljafari & Sarnikar (2009) include sub-phases of making assertions, providing evidence to support them, and calculating risk in the risk analysis process phase and propose the Demspter-Shaefer (Dempster 1967; Shafer 1976) model as an approach for performing the risk analysis. Trkman & Desouza (2012) introduce a framework that categorizes knowledge-sharing risks and propose that managers can use the framework as a guide/sense-making device in identifying the main types of risk facing their organization. The Octave Allegro (Caralli et al. 2007) method introduced in Padyab et al. (2014) provides an approach and a tool for both identifying the threats and vulnerabilities and deciding on the risk mitigation actions (mitigating risks, transferring risks, avoiding risk, or accepting risk).

A risk is the combination of two elements: the consequences of a threat and the probability of it (Stoneburner et al., 2002; Peltier et al., 2005). In this step knowledge of the consequences of a threat is essential, yet difficult to gain (Aljafari & Sarnikar, 2009). In many cases, exact monetary figures to the consequences of a threat are difficult to calculate. In addition to that, the exact probability of a threat is rarely known. Thus, to simplify the risk analysis process a 3x3 or 4x4 matrix with threat consequences categorized from severe to minor and the probability categorized from certain to highly unlikely can be used to assess whether the risks are major or minor (Stoneburner et al., 2002; Caralli et al., 2007). The matrix works as a tool to help prioritize between risks that are too severe to accept as such, and thus should be mitigated, and risks that are small enough to accept. However, since the risks are elusive in terms of exact numbers for probability and consequences, the risk analysis matrix should be used as a tool for collective sense-making of the magnitude of the knowledge risks the organization is facing

The output of the risk analysis phase is a list of identified risks that are associated with the business activities of the organization. The risks are prioritized based on the estimated probability and severity.

Step 5: Cost-Benefit Analysis

The next step, the cost-benefit-analysis, is an essential step of risks management. Elements of this step are performed all along the risk management process, yet it needs to be acknowledged as an individual step in the process in order to systematize the activity. The step gets substantial input from the risk analysis step, but also from the earlier steps of the process, as well as from other sources such as the business case that was built to support the implementation of the change at hand. The cost-benefit analysis gathers together all the benefits and positive elements and all the costs and risks that are connected to the implementation of the IT solution. This step should include the discussion of different managers so that the costs and benefits are thoroughly weighed. Cost-benefit analysis of implementing the solution can be performed in seven steps: 1) calculating cost of the suggested solution, 2) setting an effect of the solution against the problem, 3) estimating the impact the problem has against risk areas, 4) calculating the costs of all estimated impacts of the risk areas and summing it up, 5) calculating the quota to tell if you get "value for money" for the solution compared to other solutions.

The key of this step is to involve both the owners of benefits of changes as well as the owners of the risk management of the change in the sense- making of the costs and benefits. This enables the organisation to weigh the total costs of the implementation against the total expected benefits. The difficulty with risk management is typically that the costs of mitigation are weighed only against the value of individual knowledge assets. Asset valuation is difficult at best, and the value of an asset is tied to the benefits that asset can generate when used. The knowledge risk management cost-benefit assessment tries to address the benefits of a possible change in comparison to the costs of the change, and the possible cost of risks involved with the change. The risk analysis process requires discussion, and the discussion should continue at the cost-benefit analysis stage in order to achieve a common understanding of the sought benefits and risks that are associated with them. In situations where there are alternative solutions to the business problem or change, the cost-benefit-analysis can also compare the benefits, risks and costs of the current situation with the expected benefits, risks and costs brought on by the change.

The output of the cost-benefit analysis is an analysis of what are the costs of implementing the change and mitigating the risks of the change. Depending on the result of the analysis decision can be made whether to go on with the implementation process, or make changes to the approach.

Step 6: Risk Mitigation

Risk mitigation is addressed by several authors also in literature on knowledge risk management (Aljafari & Sarnikar, 2009; Padyab, Paivarinta, et al., 2014; Shedden et al., 2011). Aljafari & Sarnikar (2009) propose developing security policies as the primary means of mitigating risks. Shedden et al. (2011) propose the SECI model (Nonaka & Takeuchi, 1995) of socialization, externalization, combination and internalization as a way to mitigate knowledge risks (Shedden et al., 2011). Thalmann et al. (2014) suggest internal knowledge audits as means of risk mitigation by auditing the performance metrics of knowledge protection controls.

Risk mitigation is begun at the previous stages, when the risks too big to accept, but possible to mitigate are identified. The implementation of risk mitigation controls adds to the costs, but on the other hand diminishes risks, and may thus affect the balance between costs and benefits. Mitigation controls may affect the consequences of a threat, e.g. the knowledge that is shared on a platform is strictly limited to non-strategic knowledge, and thus the consequences of a leak are reduced. Or the controls may reduce the probability of a threat. For example, the number of people that have administrator privileges to a platform is reduced to a minimum to reduce the odds of accidental misconfiguration.

Risk mitigation controls need to be selected after a careful consideration of which risks are worth addressing, and when mitigation is reasonable. The risk analysis and cost-benefit assessment may also result in decisions that render risk mitigation unnecessary, if the change is abandoned because it will not provide enough benefits. On the other hand, the cost-benefit assessment may prove some mitigation choices too costly, and result in selection of other, less costly, means of mitigation.

The output of the mitigation step is a list of risk controls that need to be implemented, and a plan for implementing them. The output also identifies risks that are accepted, i.e. they are not mitigated. However, the mitigation plan may include contingency plans for the potential realization of the accepted risks. In the mitigation stage the benefits plan is also put into the stage or execution, and monitoring of the realization of the benefits is begun.

Step 7: Monitoring

The last, seventh, step in the risk management model is monitoring. Although this is identified as one key step in the general risk management literature (Liechtenstein 1996), most knowledge risk management models do not explicitly discuss monitoring. This step includes the monitoring of the threat environment, as well as the monitoring of the use of knowledge, which translates to monitoring also if the intended benefits are realized. Any changes in the use habits, users, shared knowledge, purpose and technological environment should trigger a re-assessment of the threats and risks that are connected to the activity that initially was processed in the risk management model. Potential new benefits and risks are then processed in the next round of the KSRM process.

The output of the monitoring step of risk management is situational awareness of the risk environment of the organization, and understanding of the status of realization of risks and benefits of a change. This step ties the knowledge security risk management process to other managerial processes of the organization.

ILLUSTRATIVE CASE

In this section the KSRM model is applied to an open innovation, crowdsourcing case, to illustrate how the model works. This case is one example of the kind of contexts this model can be applied to, and this is why it is presented in a general level. The case company is a globally operating manufacturing company, and various internal company functions were involved in knowledge sharing. In addition to this, a crowdsourcing platform provider, an industrial community and individual professionals were involved in the case. The KSRM process is applied to the case analytically, i.e. after the events took place.

The focal company of the case needed to develop a new component to ensure and monitor safe operation of their new core product. The component itself is an independent product that complements the main product. The company decided to gather fresh ideas for this component through crowdsourcing in an open innovation challenge. In addition to this challenge the company decided to schedule development sprints for the component immediately after the open challenge. This way they ensured that the results of the challenge would be taken further without delay. The knowledge security risk management process connected to this open innovation process is described in Table 1.

Table 1. KSRM process phase outputs

No.	Step	Outputs in the Example Case
1	Business need	**Trigger and business need**: Need to develop a new component for monitoring safety and maintenance need of a product Need for more product development resources -> decision to crowdsource ideas from engineering community **Expected benefits**: Faster innovation process of product component through crowdsourcing, novel ideas from outside the organization
2	Knowledge identification	**Important knowledge**: Knowledge about the specifications for the product component Knowledge about the core product, enough in order to fit the component to the core product **Holders of knowledge**: Product designers, crowdsourcing facilitators, crowdsourcing participants **Communication media**: Crowdsourcing platform
3	Threat identification	**Threat agent**: Competitors **Threat**: Exploitation of knowledge about new product development **Threat**: Exploitation of published crowdsourced plans despite copyright **Threat agent**: Crowdsourcing platform **Threat**: Loss of plans submitted to the platform **Threat agent**: Crowdsourcing contributors **Threat**: Exploitation of knowledge about the new product development **Threat**: Cause harm and extra work by contributing intentionally flawed designs
4	Risk analysis	**Exploitation of knowledge about new product development**: potential medium impact, medium probability, medium size risk Loss of plans submitted to the platform: medium impact, low probability **Harm caused by flawed designs**: low probability, medium impact **Reputation damage**: low probability, low impact
5	Cost-benefit analysis	**Business benefits:** faster innovation, faster time-to-market, more ideas for less resource spending. **Business costs:** Cost of crowdsourcing facilitator, time from designers to evaluate the results, time and effort to create specifications **Knowledge risks:** Leakage of knowledge to competitors **Knowledge risk mitigation costs:** Time to monitor the quality of submissions
6	Risk mitigation	**Mitigation:** knowledge that is shared is limited only to the non-strategic component. Knowledge about the core product is not shared to the crowd. Crowdsourcing facilitator enforces the code of conduct in the crowdsourcing platform. Required business changes to achieve the expected benefits involve mainly allocating time for further design sprints immediately after the challenge to beat competitors
7	Monitoring	External crowdsourcing facilitator monitors discussions to ensure quality of submissions Crowdsourcing facilitator and **designers** monitor the incoming submissions and adjust the challenge rules and specifications accordingly

In Table 1 the main benefit that the focal company sought from the crowdsourcing was fresh and innovative ideas with minimal consumed resources. The threat and risk analysis steps emphasize the biggest threat of this open innovation endeavor: the competitors will find out what the company is planning to do. There was significant consideration on what component of the product could be opened up for the challenge. The main reason that tilted the cost-benefit scale toward the benefit-side in this case was the non-strategic nature of the component that was developed through open innovation. The component is vital for the safety of the product, but it is not strategic for the focal company.

DISCUSSION

The KSRM approach presented in this paper differs in several respects from the few existing specifically knowledge- related risk management approaches. First, the identified existing knowledge security risk management models seem to focus strongly on merely the recognition and analysis of knowledge security risks. Some also help to identify the criticality of the various types of knowledge (Aljafari & Sarnikar, 2009; Padyab, Paivarinta, et al., 2014). However, the existing studies do not do this explicitly from the viewpoint of business needs and expected benefits, with the exception of the need to meet the requirements forced by laws, standards, customers, or internal regulations (cf. Thalmann et al., 2014). Second, even if some emphasize the importance of balancing the risk-related costs and benefits, they do not explicitly provide actual concrete approaches to do this or bring this important matter forth as a separate step in the risk management process. From the perspective of contemporary business in general, and especially from the perspective of e.g. relatively novel business approaches such as open innovation, crowdsourcing and social media that emphasize the open sharing of knowledge, it seems evident that all knowledge security risks cannot be eliminated or even controlled. This should not usually even be the priority of business-oriented knowledge security risk management. In this respect, a far more useful approach, due to the potentially huge business benefits received from adopting such business approaches, is trying to balance the benefits to risks and costs.

The illustrative example of a challenging community-based open innovation case demonstrates that risk management is clearly not anymore only a technically-oriented task (cf. Ward & Daniel, 2006; Silvius, 2007; Oza & Abrahamsson, 2009). It also demonstrates the importance of business-orientation (defining expected benefits, and making necessary business changes) as well as the need for specifically knowledge- (not merely information) focused risk management process. It also demonstrates the need for involving various different organizational functions to knowledge risk analysis. The sensemaking nature of knowledge risk analysis, especially in the case of novel types of business approaches such as open innovation, crowdsourcing or social media- and community-based business practices, is demonstrated by the case. These activities often require the sharing of business-critical knowledge in discussions and other formats even outside the company boundaries. The use of traditional information security models, in this case, would have revealed e.g. the technical risks and probably would have strongly emphasized the threats of revealing product development- related information e.g. to competitors, instead of business-related benefits. Thus, there would be a high probability of IT managers to prohibit or limit the use of such approaches, despite their possibly significant potential for business and for the development of new business models.

The model makes knowledge security risk management a business-driven process that balances the costs and benefits of knowledge sharing and protection. The model functions as a communication and sense-making tool for managers across organizational functions and boundaries. This approach is especially useful in situations where a company needs to consider business approaches that require opening up of information and knowledge in novel ways to company outsiders (customers, consumers, suppliers, communities, etc.), as well as when this information and knowledge might be searchable to competitors by novel means of business intelligence, such as data mining and social media analytics approaches. The model also works alongside with the benefits realization model, as both models feed important information to each other, especially in the identification stages of the models.

CONCLUSION

With this paper the authors aimed to answer the research question How can organizations manage knowledge risks and reach business benefits of changes in operations at the same time?" The paper argues that knowledge security risks should be managed as a systematic communication and sensemaking process, instead of risks and business benefits being traditionally evaluated in separate functions of companies. Different functions do not automatically communicate with each other in knowledge security issues, and need a framework in order to successfully do this. The combined use of the KSRM model and a benefits management model would serve the purpose of increased communication and systematic management.

REFERENCES

Ahmad, A., Bosua, R., & Scheepers, R. (2014). Protecting organizational competitive advantage: A knowledge leakage perspective. *Computers & Security*, *42*, 27–39. doi:10.1016/j.cose.2014.01.001

Alavi, M., & Leidner, D. E. (2001). Review: Knowledge Management and Knowledge Management Systems: Conceptual Foundations and Research Issues. *Management Information Systems Quarterly*, *25*(1), 107–136. doi:10.2307/3250961

Aljafari, R., & Sarnikar, S. (2009). A Framework for Assessing Knowledge Sharing Risks in Interorganizational Networks. *AMCIS 2009 Proceedings*.

Bandyopadhyay, K., Mykytyn, P. P., & Mykytyn, K. (1999). A framework for integrated risk management in information technology. *Management Decision*, *37*(5), 437–445. doi:10.1108/00251749910274216

Baskerville, R. (1991). Risk analysis as a source of professional knowledge. *Computers & Security*, *10*(8), 749–764. doi:10.1016/0167-4048(91)90094-T

Bolisani, E., & Scarso, E. (2014). The place of communities of practice in knowledge management studies: A critical review. *Journal of Knowledge Management*, *18*(2), 7–7. doi:10.1108/JKM-07-2013-0277

Braun, R., & Esswein, W. (2012). Corporate Risks in Social Networks–Towards a Risk Management Framework. *Proceedings of the Eighteenth Americas Conference on Information Systems*.

Caralli, R. A., Stevens, J. F., Young, L. R., & Wilson, W. R. (2007). *Introducing octave allegro: Improving the information security risk assessment process*. DTIC Document. doi:10.21236/ADA470450

Choo, C. W. (1996). The knowing organization: How organizations use information to construct meaning, create knowledge and make decisions. *International Journal of Information Management, 16*(5), 329–340. doi:10.1016/0268-4012(96)00020-5

Davenport, T. H., & Prusak, L. (1998). *Working Knowledge: How Organizations Manage What They Know*. Boston: Harvard Business School Press.

DeLoach, J. (2004). The new risk imperative–an enterprise-wide approach. Handbook of Business Strategy, 5(1), 29–34.

Desouza, K. C. (2006). Knowledge Security: An Interesting Research Space. *Journal of Information Science & Technology, 3*(1).

Desouza, K. C., & Vanapalli, G. K. (2005). Securing knowledge in organizations: Lessons from the defense and intelligence sectors. *International Journal of Information Management, 25*(1), 85–98. doi:10.1016/j.ijinfomgt.2004.10.007

Easterby-Smith, M., Lyles, M. A., & Tsang, E. W. (2008). Inter-organizational knowledge transfer: Current themes and future prospects. *Journal of Management Studies, 45*(4), 677–690. doi:10.1111/j.1467-6486.2008.00773.x

Farahmand, F., Navathe, S. B., Enslow, P. H., & Sharp, G. P. (2003). Managing Vulnerabilities of Information Systems to Security Incidents. In *Proceedings of the 5th International Conference on Electronic Commerce* (pp. 348–354). New York: ACM. 10.1145/948005.948050

Grant, R. M. (1996). Toward a knowledge-based theory of the firm. *Strategic Management Journal, 17*(S2), 109–122. doi:10.1002mj.4250171110

Hansen, M. T. (1999). The Search-Transfer Problem: The Role of Weak Ties in Sharing Knowledge across Organization Subunits. *Administrative Science Quarterly, 44*(1), 82–111. doi:10.2307/2667032

Ilvonen, I. (2013). Knowledge Security-A Conceptual Analysis. In *Tampereen Teknillinen Yliopisto.* Julkaisu-Tampere University of Technology.

Jafari, M., Rezaeenour, J., Mazdeh, M. M., & Hooshmandi, A. (2011). Development and evaluation of a knowledge risk management model for project-based organizations: A multi-stage study. *Management Decision, 49*(3), 309–329. doi:10.1108/00251741111120725

Jennex, M. (2014). A proposed method for assessing knowledge loss risk with departing personnel. *Vine, 44*(2), 185–209. doi:10.1108/VINE-07-2012-0028

Jennex, M. E., & Zyngier, S. (2007). Security as a contributor to knowledge management success. *Information Systems Frontiers, 9*(5), 493–504. doi:10.100710796-007-9053-4

Kogut, B., & Zander, U. (1992). Knowledge of the firm, combinative capabilities, and the replication of technology. *Organization Science, 3*(3), 383–397. doi:10.1287/orsc.3.3.383

Lichtenstein, S. (1996). Factors in the selection of a risk assessment method. *Information Management & Computer Security*, *4*(4), 20–25. doi:10.1108/09685229610130503

Manhart, M., & Thalmann, S. (2013). *An Integrated Risk Management Framework: Measuring the Success of Organizational Knowledge Protection*. Academic Press.

Manhart, M., & Thalmann, S. (2015). Protecting organizational knowledge: A structured literature review. *Journal of Knowledge Management*, *19*(2), 190–211. doi:10.1108/JKM-05-2014-0198

Marabelli, M., & Newell, S. (2012). Knowledge risks in organizational networks: The practice perspective. *The Journal of Strategic Information Systems*, *21*(1), 18–30. doi:10.1016/j.jsis.2011.11.002

Matayong, S., & Mahmood, A. K. (2013). The review of approaches to knowledge management system studies. *Journal of Knowledge Management*, *17*(3), 472–490. doi:10.1108/JKM-10-2012-0316

Nonaka, I., & Takeuchi, H. (1995). *The Knowledge-Creating Company: How Japanese Companies Create the Dynamics of Innovation: How Japanese Companies Create the Dynamics of Innovation*. New York: Oxford University Press.

Olander, H., Vanhala, M., & Hurmelinna-Laukkanen, P. (2014). Reasons for choosing mechanisms to protect knowledge and innovationsnull. *Management Decision*, *52*(2), 207–229. doi:10.1108/MD-11-2012-0791

Oza, N., & Abrahamsson, P. (2009). *Building blocks of agile innovation*. Booksurge Llc.

Padyab, A. M., Päivärinta, T., & Harnesk, D. (2014). Genre-Based Approach to Assessing Information and Knowledge Security Risks. *International Journal of Knowledge Management*, *10*(2), 15. doi:10.4018/ijkm.2014040102

Padyab, A. M., Paivarinta, T., & Harnesk, D. (2014). Genre-Based Assessment of Information and Knowledge Security Risks. In *2014 47th Hawaii International Conference on System Sciences (HICSS)* (pp. 3442–3451). Academic Press. 10.1109/HICSS.2014.428

Peltier, T., Peltier, J., & Blackley, J. (2005). *Information Security Fundamentals*. Boca Raton, FL: CRC Press.

Peppard, J., Ward, J., & Daniel, E. (2007). Managing the realization of business benefits from IT investments. *MIS Quarterly Executive*, *6*(1).

Randeree, E. (2006). Knowledge management: Securing the future. *Journal of Knowledge Management*, *10*(4), 145–156. doi:10.1108/13673270610679435

Shedden, P., Scheepers, R., Smith, W., & Ahmad, A. (2011). Incorporating a knowledge perspective into security risk assessments. *Vine*, *41*(2), 152–166. doi:10.1108/03055721111134790

Shedden, P., Smith, W., & Ahmad, A. (2010). Information Security Risk Assessment: Towards a Business Practice Perspective. *Australian Information Security Management Conference*.

Silvius, A. G. (2007). Business & IT Alignment in theory and practice. In *System Sciences, 2007. HICSS 2007. 40th Annual Hawaii International Conference on* (p. 211b–211b). IEEE. 10.1109/HICSS.2007.119

Siponen, M. T. (2005). Analysis of modern IS security development approaches: Towards the next generation of social and adaptable ISS methods. *Information and Organization*, *15*(4), 339–375. doi:10.1016/j.infoandorg.2004.11.001

Spears, J. L. (2006). A holistic risk analysis method for identifying information security risks. In Security Management, Integrity, and Internal Control in Information Systems (pp. 185–202). Springer.

Spender, J.-C., & Grant, R. M. (1996). Knowledge and the firm: Overview. *Strategic Management Journal*, *17*(S2), 5–9. doi:10.1002mj.4250171103

Spremic, M. (2012). Corporate IT Risk Management model: A holistic view at managing information system security risks. In *Information Technology Interfaces (ITI), Proceedings of the ITI 2012 34th International Conference on* (pp. 299–304). IEEE.

Stoneburner, G., Goguen, A., & Feringa, A. (2002). Risk management guide for information technology systems. *Nist Special Publication*, *800*(30), 800–830.

Szulanski, G. (2000). The process of knowledge transfer: A diachronic analysis of stickiness. *Organizational Behavior and Human Decision Processes*, *82*(1), 9–27. doi:10.1006/obhd.2000.2884

Thalmann, S., Manhart, M., Ceravolo, P., & Azzini, A. (2014). An Integrated Risk Management Framework: Measuring the Success of Organizational Knowledge Protection. *International Journal of Knowledge Management*, *10*(2), 28–42. doi:10.4018/ijkm.2014040103

Thierauf, R. J. (2001). *Effective business intelligence systems*. Greenwood Publishing Group.

Trkman, P., & Desouza, K. C. (2012). Knowledge risks in organizational networks: An exploratory framework. *The Journal of Strategic Information Systems*, *21*(1), 1–17. doi:10.1016/j.jsis.2011.11.001

Tzortzaki, A. M., & Mihiotis, A. (2014). A Review of Knowledge Management Theory and Future Directions. *Knowledge and Process Management*, *21*(1), 29–41. doi:10.1002/kpm.1429

Varnell-Sarjeant, J. F. (2008). Managing a Man-rated Software Development Program via Risk Mitigation. *SIGSOFT Softw. Eng. Notes*, *33*(4), 8:1–8:8.

Väyrynen, K., Hekkala, R., & Liias, T. (2013). Knowledge Protection Challenges of Social Media Encountered by Organizations. *Journal of Organizational Computing and Electronic Commerce*, *23*(1–2), 34–55. doi:10.1080/10919392.2013.748607

von Krogh, G. (2009). Individualist and collectivist perspectives on knowledge in organizations: Implications for information systems research. *The Journal of Strategic Information Systems*, *18*(3), 119–129. doi:10.1016/j.jsis.2009.08.001

von Krogh, G. (2012). How does social software change knowledge management? Toward a strategic research agenda. *The Journal of Strategic Information Systems*, *21*(2), 154–164. doi:10.1016/j.jsis.2012.04.003

Von Solms, B., & Von Solms, R. (2004). The 10 deadly sins of information security management. *Computers & Security*, *23*(5), 371–376. doi:10.1016/j.cose.2004.05.002

Ward, J., & Daniel, E. (2006). *Benefits management: delivering value from IS and IT investments*. John Wiley & Sons.

Ward, J., Taylor, P., & Bond, P. (1996). Evaluation and realisation of IS/IT benefits: An empirical study of current practice. *European Journal of Information Systems*, *4*(4), 214–225. doi:10.1057/ejis.1996.3

Webb, J., Maynard, S., Ahmad, A., & Shanks, G. (2013). *Towards an Intelligence-Driven Information Security Risk Management Process for Organisations*. Presented at the *24th Australasian Conference on Information Systems: Intelligence-Driven ISRM for Organisations*, Melbourne, Australia.

Whitman, M., & Mattord, H. (2011). *Principles of information security*. Cengage Learning.

Compilation of References

Aaker, D., Day, G., & Kumar, V. (1998). *Marketing Research*. New York: John Wiley and Sons.

Abereijo, I. O. (2015). Transversing the 'Valley of Death': Understanding the determinants to commercialization of research outputs in Nigeria. *African Journal of Economic and Management Studies*, *6*(1), 90–106. doi:10.1108/AJEMS-10-2012-0066

Abrahamson, E. (1991). Managerial fads and fashions: The diffusion and rejection of innovations. *Academy of Management Review*, *16*(3), 586–612.

Ackoff, R. L. (1989). From Data to Wisdom. *Journal of Applied Systems Analysis*, *16*(1), 3–9.

Ács, Z. J., & Audretsch, D. B. (2006). *Handbook of entrepreneurship research: An interdisciplinary survey and introduction*. Springer Science & Business Media.

Afuah, A. (1998). *Innovation Management: Strategies, Implementation and Profits*. New York: Oxford University Press.

Ahmad, A., Bosua, R., & Scheepers, R. (2014). Protecting organizational competitive advantage: A knowledge leakage perspective. *Computers & Security*, *42*, 27–39. doi:10.1016/j.cose.2014.01.001

Ahmad, S., & Schroeder, R. (2011). Knowledge management through technology strategy: Implications for competitiveness. *Journal of Manufacturing Technology Management*, *22*(1), 6–24. doi:10.1108/17410381111099789

Ajmal, M. M., & Koskinen, U. K. (2008). Knowledge transfer in project-based organizations: An organization culture perspective. *Project Management Journal*, *39*(1), 7–15. doi:10.1002/pmj.20031

Ajzen, I. (1985). From intentions to actions: A theory of planned behavior. In J. Kuhl & J. Beckmann (Eds.), *Action control: From cognition to behavior* (pp. 11–39). Berlin: Springer-Verlag. doi:10.1007/978-3-642-69746-3_2

Akgün, A. E., Keskin, H., Ayar, H., & Okunakol, Z. (2017). Knowledge sharing barriers in software development teams: A multiple case study in Turkey. *Kybernetes*, *46*(4), 603–620. doi:10.1108/K-04-2016-0081

Akhavan, P., & Hosseini, S. M. (2016). Social capital, knowledge sharing, and innovation capability: An empirical study of R&D teams in Iran. *Technology Analysis and Strategic Management*, *28*(1), 96–113. doi:10.1080/09537325.2015.1072622

Aktharsha, U., & Anisa, H. (2011). Knowledge management system and learning organization: An empirical study in an engineering organization. *IUP Journal of Knowledge Management*, *9*(2), 26–43.

Alam, S. S., Abdullah, Z., Ishak, N. A., & Zain, Z. M. (2009). Assessing knowledge sharing behavior among employees in smes: An empirical study. *International Business Research*, *2*(2), 115–122. doi:10.5539/ibr.v2n2p115

Alavi, M., & Leidner, D. E. (1999). Knowledge management systems: Issues, challenges, and benefits. Communications of the AIS, 1(7), 1–37.

Alavi, M., Kayworth, T., & Leidner, D. (2006). An Empirical Examination of the Influence of Organizational Culture on Knowledge Management Practices. *Journal of Management Information Systems*, *22*(3), 191–224. doi:10.2753/MIS0742-1222220307

Alavi, M., & Leidner, D. E. (1999). Knowledge management systems: Issues, challenges, and benefits. *Communications of the AIS*, *1*(2es), 1.

Alavi, M., & Leidner, D. E. (2001). Knowledge management and knowledge management systems: Conceptual foundations and research issues. *Management Information Systems Quarterly*, *25*(1), 107–136. doi:10.2307/3250961

Alazmi, M., & Zairi, M. (2003). Knowledge management critical success factors. *Total Quality Management & Business Excellence*, *14*(2), 199–204. doi:10.1080/1478336032000051386

Alfares, H., & Duffuaa, S. (2008). Assigning cardinal weights in multi-criteria decision making. *Journal of Multi-Criteria Decision Analysis*, *15*(5-6), 125–133. doi:10.1002/mcda.420

Al-Ghamdi, S. M. (1998). Obstacles to successful implementation of strategic decisions: The British experience. European Business Review, 98(6), 322–327. doi:10.1108/09555349810241590 doi:10.1108/09555349810241590

Al-Hakim, L. A. Y., & Hassan, S. (2013). Knowledge management strategies, innovation, and operational performance: *An empirical study of the Iraqi MTS. Journal of Advances in Management Research*, *10*(1), 58–71. doi:10.1108/09727981311327767

Aljafari, R., & Sarnikar, S. (2009). A Framework for Assessing Knowledge Sharing Risks in Interorganizational Networks. *AMCIS 2009 Proceedings*.

Alkhuraiji, A., Liu, S., Oderanti, F. O., Annansingh, F., & Pan, J. (2014). Knowledge network modeling to support decision-making for strategic intervention in IT project-oriented change management. *Journal of Decision Systems*, *23*(3), 285–302. doi:10.1080/12460125.2014.886499

Almond, N. (2001). The transportability and dissemination of knowledge: A surveyor's perspective. *Property Management*, *19*(2), 124–135. doi:10.1108/02637470110387849

Al-Mubaraki, H. M., & Busler, M. (2013). Business Incubation as an Economic Development Strategy: A Literature Review. *International Journal of Management*, *30*(1 Part 2), 362–372.

Alvarez, C. (1998). Making the case for hiring a consultant. *Clinical Nurse Specialist CNS*, *12*(2), 52–65. doi:10.1097/00002800-199803000-00003 PMID:9708106

Al-Zayyat, A. N., Al-Khaldi, F., Tadros, I., & Al-Edwan, G. (2010). The effect of knowledge management processes on project management. *Journal of IBIMA Business Review*, *2010*, 1–6.

Amabile, T. A. (1996). *Creativity and innovation in organizations*. Harvard Business School Press.

Amit, R., & Zott, C. (2010). Business Model Design: An Activity Systems Perspective. *Long Range Planning*, *43*(2-3), 216–226. doi:10.1016/j.lrp.2009.07.004

Åmo, B. (2006). Employee innovation behaviour in health care: The influence from management and colleagues. *International Nursing Review*, *53*(3), 231–237. doi:10.1111/j.1466-7657.2006.00455.x PMID:16879187

Anantatmula, V., & Kanung, S. (2008). Role of IT and KM in improving project management performance. *The Journal of Information and KM Systems*, *38*(3), 357–369. doi:10.1108/03055720810904862

Anaza, N. A., & Nowlin, E. L. (2017). What's mine is mine: A study of salesperson knowledge withholding & hoarding behavior. *Industrial Marketing Management*, *64*, 14–27. doi:10.1016/j.indmarman.2017.03.007

Anbari, F. T., Carayannis, E. G., & Voetsch, R. J. (2008). Post-project reviews as a key project management competence. *Technovation, 28*(10), 633–643. doi:10.1016/j.technovation.2007.12.001

Anderson, R., & Mansingh, G. (2015). KIISA: Towards a Reference Architecture for Integrated Knowledge Services. In Intelligent Systems' 2014 (pp. 247-257). Springer International Publishing.

Anderson, R., & Mansingh, G. (2017). A Workflow Management System for Knowledge Management Initiatives. *Proceedings of the 23rd Americas Conference on Information Systems, AMCIS 2017.*

Anderson, R., Gordon, S., & Mansingh, G. (2014). An assessment of the potential impacts of knowledge-driven decision support in social welfare. In *Global Humanitarian Technology Conference (GHTC)* (pp. 499-505). IEEE. 10.1109/GHTC.2014.6970329

Andersson, T., Formica, P., & Curley, M. G. (2009). *Knowledge-driven entrepreneurship: the key to social and economic transformation.* Springer Science & Business Media.

Andrawina, L., & Govindaraju, R. (2009). Knowledge sharing capability, absorptive capacity, and innovation capability: An empirical study of Indonesia's information and communication technology industry. *Journal of ICT, 8*, 85–102.

Andres, H. P. (2011). Team learning and reflexivity in technology-mediated collaboration. *International Journal of Knowledge Management, 7*(4), 22–36. doi:10.4018/jkm.2011100102

Angel, R. (2006). Putting an innovation culture into practice. *Ivey Business Journal.* Retrieved from http://www.iveybusinessjournal. com/topics/innovation/putting-an-innovation-culture-into-practice

Angele, J., Fensel, D., Landes, D., & Studer, R. (1998). Developing knowledge-based systems with mike. *Automated Software Engineering, 5*(4), 389–418. doi:10.1023/A:1008653328901

Apospori, E., Nikandrou, I., & Panayotopoulou, L. (2006). Mentoring and women's career advancement in Greece. *Human Resource Development International, 9*(4), 509–527. doi:10.1080/13678860601032627

Arnott, D. (2002). *Decision biases and decision support systems development* (Working Paper. No. 2002/04). Monash University, Decision Support Systems Laboratory.

ASAPM. (2012). Please participate in this survey: In this survey learning best practices and project success. *ASAPM February 2012 Newsletter.*

Athar, M. A. Q., & Evans, N. (2015). Deterrents to knowledge-sharing in the pharmaceutical industry: A case study. *Journal of Knowledge Management, 19*(2), 296–314. doi:10.1108/JKM-09-2014-0391

Atzori, L., Iera, A., & Morabito, G. (2010). The Internet of Things: A Survey. *Computer Networks, 54*(15), 2787–2805. doi:10.1016/j.comnet.2010.05.010

Aulawi, H., Sudirman, I., Sudirman, I., Suryadi, K., & Govindarajn, R. (2009). Literature review towards knowledge enablers which is assumed significantly influences ks behavior. *Journal of Applied Sciences Research, 5*(12), 2262–2270.

Aulet & Murray. (2013). *A Tale of Two Entrepreneurs: Understanding differences in the Types of Entrepreneurship in the Economy.* Available at SSRN 2259740.

Austin, R. (2007, January). *The role of IT in ınnovation-based value creation. CIO Digers.*

Babbie, E. (1998). *The Practice of Social Research.* Belmont, CA: Wadsworth Publishing Co.

Badaracco, J. L. (1991). *The Knowledge Link: How Firms Compete through Strategic Alliances.* Boston: Harvard University Press.

Bahrami, M., Salehi, M., & Korkmaz, S. (2014). A Study of the Driving Factors of Knowledge Management and Corporate Entrepreneurship in Iran. *IUP Journal of Knowledge Management, 12*(1), 13.

Baker, D. S., Carson, K. D., & Carson, P. P. (2009). An individual-level examination of the impact of cultural values on organizational identification. *The Journal of Applied Management and Entrepreneurship, 14*(2), 29–43.

Ballou, D. P., Belardo, S., & Pazer, H. L. (2010). A project staffing model to enhance the effectiveness of knowledge transfer in the requirements planning phase for multi-project environments. *International Journal of Knowledge Management, 6*(2), 1–21. doi:10.4018/jkm.2010040101

Bandera, C., Bartolacci, M. R., & Passerini, K. (2016). Knowledge Management and Entrepreneurship: A Contradictory Recipe. *International Journal of Knowledge Management, 12*(3), 1–14. doi:10.4018/IJKM.2016070101

Bandura, A. (1997). *Self-efficacy: The exercise of control*. New York: Freeman.

Bandyopadhyay, K., Mykytyn, P. P., & Mykytyn, K. (1999). A framework for integrated risk management in information technology. *Management Decision, 37*(5), 437–445. doi:10.1108/00251749910274216

Banham, R. (2004). Enterprising views of risk management. *Journal of Accountancy, 197*(6), 65–72.

Banker, R., & Kemerer, C. F. (1992). Performance evaluation metrics for information systems development: A principal-agent model. *Information Systems Research, 3*(4), 379–400. doi:10.1287/isre.3.4.379

Bantel, K. A., & Jackson, S. E. (1989). Top management and innovations in banking: Does the composition of the top team make a difference? *Strategic Management Journal, 10*(S1), 107–124. doi:10.1002mj.4250100709

Barnaghi, P., Wang, W., Henson, C., & Taylor, K. (2012). Semantics for the Internet of Things: Early progress and back to the future. *International Journal on Semantic Web and Information Systems, 8*(1), 1–21. doi:10.4018/jswis.2012010101

Barney, J. (1996). The Resource-Based Theory of the Firm. *Organization Science, 7*(5), 469–476. doi:10.1287/orsc.7.5.469

Baron, R. M., & Kenny, D. A. (1986). The moderator–mediator variable distinction in social psychological research: Conceptual, strategic, and statistical considerations. *Journal of Personality and Social Psychology, 51*(6), 1173–1182. doi:10.1037/0022-3514.51.6.1173 PMID:3806354

Barratt-Pugh, L., Kennett, P., & Bahn, S. (2013). Managing knowledge: The critical role of culture and ownership as mediator of systems. *International Journal of Knowledge Management, 9*(2), 20–37. doi:10.4018/jkm.2013040102

Barson, R. J., Foster, G., Struck, T., Ratchev, S., Pawar, K., Weber, F., & Wunram, M. (2009). Inter-and intra-organisational barriers to sharing knowledge in the extended supply-chain. *Proceedings of the International Conference on e-Business and e-Work*, 367-373.

Bartunek, J. M., & McKenzie, J. (Eds.). (2018). *Academic-practitioner relationships: Developments, complexities and opportunities*. Routledge.

Baskerville, R. L. (1997). Distinguishing action research from participative case studies. Journal of Systems and Information Technology, 1(1), 24–43. doi:10.1108/13287269780000733 doi:10.1108/13287269780000733

Baskerville, R. (1991). Risk analysis as a source of professional knowledge. *Computers & Security, 10*(8), 749–764. doi:10.1016/0167-4048(91)90094-T

Baskerville, R. L., & Wood-Harper, A. T. (1996). A critical perspective on action research as a method for information systems research. *Journal of Information Technology, 11*(3), 235–246. doi:10.1080/026839696345289

Basu, S. (2004). E-government and developing countries: An overview. *International Review of Law Computers & Technology*, *18*(1), 109–132. doi:10.1080/13600860410001674779

Bates, M. J. (2005). Information and Knowledge: An Evolutionary Framework for Information Science. Information Research, 10(4).

Baxter, L. A., & Braithwaite, D. O. (Eds.). (2008). *Engaging theories in interpersonal communication: multiple perspectives*. Los Angeles, CA: Sage. doi:10.4135/9781483329529

Bearden, W. O., Money, R. B., & Nevins, J. L. (2006). A Measure of Long-Term Orientation: Development and Validation. *Academy of Marketing Science Journal*, *34*(3), 456–467. doi:10.1177/0092070306286706

Beasley, M., Bronson, B., & Hancock, B. (2009). ERM: Opportunities for improvement. *Journal of Accountancy*, *208*(3), 175–179.

Becerra-Fernandez, I., Gonzales, A., & Sabherwal, R. (2004). *Knowledge management: Challenges, solutions, and technologies*. Upper Saddle River, NJ: Pearson Education, Inc.

Becerra, M., Lunnan, R., & Huemer, L. (2008). Trustworthiness, risk, and the transfer of tacit and explicit knowledge between alliance partners. *Journal of Management Studies*, *45*(4), 691–713. doi:10.1111/j.1467-6486.2008.00766.x

Befeki, T., Epstein, M., & Yuthas, K. (2009, February). Creating growth from effective opportunity risk management. *CWA Management*, 20-26.

Beijerse, R. P. (2000). Knowledge management in small and medium-sized companies: Knowledge management for entrepreneurs. *Journal of Knowledge Management*, *4*(2), 162–179.

Beijers, R. (1999). Questions in knowledge management: Defining and conceptualizing a phenomenon. *Journal of Knowledge Management*, *3*(2), 94–109. doi:10.1108/13673279910275512

Bell, S. T., Villado, A. J., Lukasik, M. A., Briggs, A. L., & Belau, L. (2011). Getting specific about demographic diversity variable and team performance relationships: A meta analysis. *Journal of Management*, *37*(3), 709–743. doi:10.1177/0149206310365001

Bembenek, B., & Piecuch, T. (2014). Knowledge Management in Industry Clusters As an Indication of Entrepreneurship. *CBU International Conference Proceedings*. 10.12955/cbup.v2.441

Benbasat, I., & Zmud, R. W. (1999). Empirical research in information systems: The practice of relevance. *Management Information Systems Quarterly*, *23*(1), 3–16. doi:10.2307/249403

Benjamin, B., & O'Reilly, C. (2011). Becoming a leader: Early career challenges faced by MBA graduates. *Academy of Management Learning & Education*, *10*(3), 452–472. doi:10.5465/amle.2011.0002

Bennis, W. G., & O'Toole, J. (2005). How business schools lost their way. *Harvard Business Review*, *83*(5), 96–104. PMID:15929407

Berger, P. L., & Luckmann, T. (1966). *The social construction of reality: A treatise on the sociology of knowledge*. New York, NY: Anchor Books.

Berry, M. J., & Linoff, G. (1997). *Data mining techniques: for marketing, sales, and customer support*. John Wiley & Sons, Inc.

Bessant, J., & Venables, T. (2008). *Creating Wealth from Knowledge: Meeting the Innovation Challenge. Cheltenham.* Cheltenham, UK: Edward Elgar Publishing. doi:10.4337/9781848441248

Bettiol, M., Di Maria, E., & Grandinetti, R. (2012). Codification and Creativity: Knowledge Management Strategies in KIBS. Journal of Knowledge Management, 16(4), 550–562. doi:10.1108/13673271211246130doi:10.1108/13673271211246130

Bhagat, R. S., Kedia, B. L., Harveston, P. D., & Triandis, H. C. (2002). Cultural variations in the cross-border transfer of organizational knowledge: An integrative framework. *Academy of Management Review*, *27*(2), 204–221.

Bhirud, S., Rodrigues, L., & Desai, P. (2005). Knowledge sharing practices in km: A case study in Indian software subsidiary. *Journal of Knowledge Management Practice.* Retrieved from http://www. tlainc.com/articl103.htm

Bircham-Connolly, H., Corner, J., & Bowden, S. (2005). An empirical study of the impact of question structure on receipient attitude during knowledge sharing. *Electronic Journal of Knowledge Management*, *32*(1), 1–10.

Birkenshaw, J., Hamel, G., & Mol, M. J. (2008). Management Innovation. Academy of Management Review, 33(4), 825–845. doi:10.5465/AMR.2008.34421969 doi:10.5465/AMR.2008.34421969

Bixler, C. (2002). Codification, personalization, and integration. KMWorld, 11(3), 18.

Blau, P. M. (1977). *Inequality and heterogeneity.* New York: The Free Press.

Bock, G. W., Zmud, R. W., Kim, Y. G., & Lee. (2005). Behavioral intention formation in knowledge sharing: Examining the roles of extrinsic motivators, social-psychological forces, and organizational climate. *Management Information Systems Quarterly*, *29*(1), 87–111. doi:10.2307/25148669

Boella, G., & Van Der Torre, L. (2006). Security policies for sharing knowledge in virtual communities. *IEEE Transactions on Systems, Man, and Cybernetics. Part A, Systems and Humans*, *36*(3), 439–450. doi:10.1109/TSMCA.2006.871793

Bohrnstedt, G., & Knoke, D. (1982). *Statistics for Social Data Analysis.* Itasco, IL: F.E. Peacock Publishers.

Boisot, M., & MacMillan, I. C. (2004). Crossing epistemological boundaries: Managerial and entrepreneurial approaches to knowledge management. *Long Range Planning*, *37*(6), 505–524. doi:10.1016/j.lrp.2004.09.002

Bolisani, E., & Scarso, E. (2014). The place of communities of practice in knowledge management studies: A critical review. *Journal of Knowledge Management*, *18*(2), 7–7. doi:10.1108/JKM-07-2013-0277

Bollinger, A. S., & Smith, R. D. (2001). Managing organizational knowledge as a strategic asset. Journal of Knowledge Management, 5(1), 1–8. doi:10.1108/13673270110384365 doi:10.1108/13673270110384365

Bolognini, M. (2001). *Electronic Democracy. Delphi Method and Public Policy-Making*. Rome: Carocci Editore.

Booker, L., Bontis, N., & Serenko, A. (2008). The relevance of knowledge management and intellectual capital research. *Knowledge and Process Management*, *15*(4), 235–246. doi:10.1002/kpm.314

Booker, L., Bontis, N., & Serenko, A. (2012). Evidence-based management and academic research relevance. *Knowledge and Process Management*, *19*(3), 121–130. doi:10.1002/kpm.1392

Bose, R. (2004). Knowledge management metrics. *Industrial Management & Data Systems*, *104*(6), 457–468. doi:10.1108/02635570410543771

Bosua, R., & Scheepers, R. (2007). Towards a model to explain knowledge sharing in complex organizational environments. *Knowledge Management Research and Practice*, *5*(2), 93–109. doi:10.1057/palgrave.kmrp.8500131

Botero, I. C., & Van Dyne, L. (2009). Employee voice behavior interactive effects of LMX and power distance in the United States and Colombia. *Management Communication Quarterly*, *23*(1), 84–104. doi:10.1177/0893318909335415

Bowman, B. J. (2002). Building knowledge management systems. *Information Systems Management*, *19*(3), 32–40. do i:10.1201/1078/43201.19.3.20020601/37168.5

Boyd, D., & Crawford, K. (2012). Critical Questions for Big Data: Provocations for a Cultural, Technological, and Scholarly Phenomenon. *Information Communication and Society*, *15*(5), 662–679. doi:10.1080/1369118X.2012.678878

Bramwell, A., & Wolfe, D. A. (2008). Universities and regional economic development: The entrepreneurial University of Waterloo. *Research Policy*, *37*(8), 1175–1187. doi:10.1016/j.respol.2008.04.016

Braun, R., & Esswein, W. (2012). Corporate Risks in Social Networks–Towards a Risk Management Framework. *Proceedings of the Eighteenth Americas Conference on Information Systems.*

Brenes, E. R., Mena, M., & Molina, G. E. (2008). Key success factors for strategy implementation in Latin America. Journal of Business Research, 61(6), 590–598. doi:10.1016/j.jbusres.2007.06.033 doi:10.1016/j.jbusres.2007.06.033

Bretschneider, U., & Zogaj, S. (2016). Exploring strategies for capturing customer's tacit knowledge in customer integration methods. *International Journal of Knowledge Management*, *12*(2), 1–19. doi:10.4018/IJKM.2016040101

Brewer, P., & Venaik, S. (2012). On the misuse of national culture dimensions. *International Marketing Review*, *29*(6), 673–683. doi:10.1108/02651331211277991

Bromiley, P., McShane, M., Nair, A., & Rustambekov, E. (2015). Enterprise Risk Management: Review, Critique, and Research Directions. *Long Range Planning*, *48*(4), 265–276. doi:10.1016/j.lrp.2014.07.005

Brown, B. B. (1968). Delphi Process: A Methodology for the Elicitation of Opinions of Experts. Santa Monica, CA: The Rand Corporation. (Possible ASMTE Vectors, February, 1968)

Brown, B. (2001). Step by step enterprise risk management. *Risk Management*, *48*(9), 43–50.

BuBwolder, P., Burgahn, F., Hübner, M., & Werker, M. (2016). *Classification of company-specific influence factors as part of knowledge management system for ramp-up projects.* Paper presented at the 3rd ICRM 2016 International Conference on Ramp-Up Management. doi: 10.1016/j.procir.2016.05.042

Bunderson, J. S., & Sutcliffe, K. M. (2002). Comparing Alternative Conceptualizations of functional diversity in management teams: Process and performance effects. *Academy of Management Journal*, *45*(5), 875–893. doi:10.2307/3069319

Burgelman, R. A. (1991). Intraorganizational ecology of strategy making and organizational adaptation: Theory and field research. Organization Science, 2(3), 239–262. doi:10.1287/orsc.2.3.239 doi:10.1287/orsc.2.3.239

Burns, A. J., Roberts, T. L., Posey, C., Bennett, R. J., & Courtney, J. F. (2015). *Assessing the Role of Security Education, Training, and Awareness on Insiders' Security-Related Behavior: An Expectancy Theory Approach.* Paper presented at the HICSS. 10.1109/HICSS.2015.471

Burns, N., & Grove, S. K. (2005). *The practice of nursing research: Conduct, critique & utilization.* St Louis, MO: Elsevier.

Burstein, F. V., & Gregor, S. (1999). The Systems Development or Engineering Approach to Research in Information Systems: An Action Research Perspective. In *Proceedings of 10th Australasian Conference on Information Systems.* School of Comms & Info Mgmt Vic Uni of Wellington.

Busenitz, L. W., Plummer, L. A., Klotz, A. C., Shahzad, A., & Rhoads, K. (2014). Entrepreneurship research (1985–2009) and the emergence of opportunities. *Entrepreneurship Theory and Practice*, *38*(5), 981–1000. doi:10.1111/etap.12120

Butler, T., & Grace, A. (2010). Building and Maintaining Human Capital with Learning Management Systems. In K. O'Sullivan (Ed.), *Strategic Intellectual Capital Management in Multinational Organizations: Sustainability and Successful Implications.* New York: IGI Global. doi:10.4018/978-1-60566-679-2.ch014

Bygrave, W., & Zacharakis, A. (2014). *Entrepreneurship.* Hoboken, NJ: John Wiley & Sons, Inc.

Bylund, C. L., Peterson, E. B., & Cameron, K. A. (2012). A practitioner's guide to interpersonal communication theory: An overview and exploration of selected theories. *Patient Education and Counseling*, *87*(3), 261–267. doi:10.1016/j.pec.2011.10.006 PMID:22112396

Cabrera, A., & Cabrera, E. F. (2002). Knowledge-Sharing Dilemmas. *Organization Studies*, *23*(5), 687–710. doi:10.1177/0170840602235001

Cabrera, A., Collins, W. C., & Salgado, J. F. (2006). Determinants of individual engagement in knowledge sharing. *International Journal of Human Resource Management*, *17*(2), 245–264. doi:10.1080/09585190500404614

Caimo, A., & Lomi, A. (2015). Knowledge sharing in organizations: A Bayesian analysis of the role of reciprocity and formal structure. *Journal of Management*, *41*(2), 665–691. doi:10.1177/0149206314552192

Callahan, C., & Soileau, J. (2017). Does Enterprise risk management enhance operating performance? *Advances in Accounting*, *37*, 122–139. doi:10.1016/j.adiac.2017.01.001

Caloghirou, Y., Kastelli, I., & Tsakanikas, A. (2004). Internal capabilities and external knowledge sources: Complements or substitutes for innovative performance? *Technovation*, *24*(1), 29–39. doi:10.1016/S0166-4972(02)00051-2

Campatelli, G., Richter, A., & Stocker, A. (2016). Participative knowledge management to empower manufacturing workers. *International Journal of Knowledge Management*, *12*(4), 37–50. doi:10.4018/IJKM.2016100103

Campbell, D. J. (1988). Task Complexity: A Review and Analysis. *Academy of Management Review*, *13*(1), 40–52.

Campbell, D. T., & Stanley, J. C. (1966). *Experimental and Quasi-experimental Designs for Research*. Chicago: Rand McNally.

Cannataro, M., & Talia, D. (2003). The Knowledge Grid. *Communications of the AIS*, *46*(1), 89–93.

Cano, C., & Cano, P. (2006). Human resources management and its impact on innovation performance in companies. *International Journal of Technology Management*, *35*(1-4), 11–28. doi:10.1504/IJTM.2006.009227

Caralli, R. A., Stevens, J. F., Young, L. R., & Wilson, W. R. (2007). *Introducing octave allegro: Improving the information security risk assessment process*. DTIC Document. doi:10.21236/ADA470450

Cárdenas, I. C., Al-Jibouri, S. S. H., Halman, J. I. M., van de Linde, W., & Kaalberg, F. (2014). Using prior risk-related knowledge to support risk management decisions: Lessons learnt from a tunneling project. *Risk Analysis*, *34*(10), 1923–1943. doi:10.1111/risa.12213 PMID:24842516

Carland, J. W., Hoy, F., Boulton, W. R., & Carland, J. A. C. (1984). Differentiating entrepreneurs from small business owners: A conceptualization. *Academy of Management Review*, *9*(2), 354–359.

Carmeli, A., Meitar, R., & Weisberg, J. (2006). Self-leadership skills and innovative behavior at work. *International Journal of Manpower*, *27*(1), 75–90. doi:10.1108/01437720610652853

Carneiro, A. (2000). How does knowledge management influence innovation and competitiveness? *Journal of Knowledge Management*, *4*(2), 87–98. doi:10.1108/13673270010372242

Casadeus-Masanell, R., & Ricart, J. (2010). From Strategy to Business Models and onto Tactics. *Long Range Planning*, *43*(2-3), 195–215. doi:10.1016/j.lrp.2010.01.004

Casson, M. (2010). Entrepreneurship, business culture and the theory of the firm. In Handbook of entrepreneurship research. Springer.

Cater, T., & Pucko, D. (2010). Factors of effective strategy implementation: Empirical evidence from Slovenian business practice. Journal for East European Management Studies, 3, 207–236.

Cavaleri, S. (1994). Soft Systems Thinking: A Pre-Condition for Organizational Learning. *Human Systems Management, 13*(4), 259–267.

Cavusgil, S. T., Calantone, R. J., & Zhao, Y. (2003). Tacit knowledge transfer and firm innovation capability. *Journal of Business and Industrial Marketing, 18*(1), 6–22. doi:10.1108/08858620310458615

Celestine, N. A., & Perryer, C. (2016). The Impact of National Cultural Values on Intrinsic Motivation to Transfer Tacit Knowledge. *International Journal of Knowledge Management, 12*(4), 1–19. doi:10.4018/IJKM.2016100101

Cerchione, R., Esposito, E., & Spadaro, R. M. (2016). A literature review on knowledge management in SMEs. *Knowledge Management Research and Practice, 14*(2), 169–177. doi:10.1057/kmrp.2015.12

Cerpa, N., & Verner, J. M. (2009). Why did your projects fail? *Communications of the ACM, 52*(12), 130–134. doi:10.1145/1610252.1610286

Chakrabarti, A. K. (1974). The role of champion in product innovation. California Management Review, 17(Winter), 58–62. doi:10.2307/41164561 doi:10.2307/41164561

Champion, D. (2009). Managing risk in the new world. *Harvard Business Review, 87*(10), 69–75.

Chang, Y., Hsu, P., Shiau, W., & Tsai, C. (2015). Knowledge sharing intention in the United States and China: A cross-cultural study. *European Journal of Information Systems, 24*(3), 262–277. doi:10.1057/ejis.2014.28

Chaston, I. (2012). Entrepreneurship and knowledge management in small service-sector firms. *Service Industries Journal, 32*(6), 845–860. doi:10.1080/02642069.2011.559224

Chaves, M. S., & Veronese, G. S. (2014). A proposal to manage lessons learned in projects: web 2.0 technologies to promote innovation. International Journal of Innovation, 2(1).

Chell, E. (2008). *The entrepreneurial personality: A social construction.* Routledge. doi:10.4324/9780203938638

Chen, J.S. & Tsou, H.T. (2007). Information technology adoption for service innovation practices and competitive advantage: The case of financial firms. *Information Research, 12*(3).

Chen, C. J., & Hung, S. W. (2010). To give or to receive? Factors influencing members' knowledge sharing and community promotion in professional virtual communities. *Information & Management, 47*(4), 226–236. doi:10.1016/j.im.2010.03.001

Chen, C., & Huang, J. (2009). Strategic human resource practices and innovation performance – the mediating role of knowledge management capacity. *Journal of Business Research, 62*(1), 104–114. doi:10.1016/j.jbusres.2007.11.016

Cheng, M. Y., Ho, J. S. Y., & Lau, P. M. (2009). Knowledge sharing in academic institutions: A study of multimedia university Malaysia. *Electronic Journal of Knowledge Management, 7*(3), 313–324.

Chen, H., Chiang, R. H. L., & Storey, V. C. (2012). Business Intelligence and Analytics: From Big Data to Big Impact. *Management Information Systems Quarterly, 36*(4), 1165–1188.

Chesbrough, H. (2010). Business Model Innovation: Opportunities and Barriers. *Long Range Planning, 43*(2-3), 354–363. doi:10.1016/j.lrp.2009.07.010

Chiang, Y., & Hung, K. (2010). Exploring open search strategies and perceived innovation performance from the perspective of inter-organizational knowledge flows. *R & D Management, 40*(3), 292–306. doi:10.1111/j.1467-9310.2010.00588.x

Chiaroni, D., Chiesa, V., & Frattini, F. (2011). The Open Innovation Journey: How firms dynamically implement the emerging innovation management paradigm. *Technovation*, *31*(1), 34–43. doi:10.1016/j.technovation.2009.08.007

Chilton, M. A., & Bloodgood, J. M. (2008). The Dimensions of Tacit & Explicit knowledge: A Description and Measure. *International Journal of Knowledge Management*, *4*(2), 75–91. doi:10.4018/jkm.2008040106

Chin, W. W., Marcolin, B. L., & Newsted, P. R. (2003). A partial least squares latent variable modeling approach for measuring interaction effects: Results from a Monte Carlo simulation study and an electronic-mail emotion/adoption study. *Information Systems Research*, *14*(2), 189–217. doi:10.1287/isre.14.2.189.16018

Chiu, C.-N., & Chen, H.-H. (2016). The study of knowledge management capability and organizational effectiveness in Taiwanese public utility: The mediator role of organizational commitment. *SpringerPlus*, *5*(1), 1520. doi:10.118640064-016-3173-6 PMID:27652093

Choi, B., Poon, S. K., & Davis, J. G. (2008). Effects of knowledge management strategy on organizational performance: A complementarity theory-based approach. Omega International Journal of Management Science, 36(2), 235–251. doi:10.1016/j.omega.2006.06.007 doi:10.1016/j.omega.2006.06.007

Cho, N., Li, G., & Su, C. (2007). An empirical study on the effect of individual factors on knowledge sharing by knowledge type. *Journal of Global Business and Technology*, *3*(2), 1–15.

Chong, C. W., Yuen, Y. Y., & Gan, G. C. (2014). Knowledge sharing of academic staff: Acomparison between private and public universities in Malaysia. *Library Review*, *63*(3), 203–223. doi:10.1108/LR-08-2013-0109

Choo, C. W. (1996). The knowing organization: How organizations use information to construct meaning, create knowledge and make decisions. *International Journal of Information Management*, *16*(5), 329–340. doi:10.1016/0268-4012(96)00020-5

Choo, C.-W., & de Alvarenga Neto, R.-C.-D. (2010). Beyond the ba: Managing enabling contexts in knowledge organiza5ons. *Journal of Knowledge Management*, *14*(4), 592–610. doi:10.1108/13673271011059545

Chow, C. W., Deng, F. J., & Ho, J. L. (2000). The openness of knowledge sharing within organizations: A Comparative Study in the United States and the People's Republic of China. *Journal of Management Accounting Research*, *12*(1), 65–95. doi:10.2308/jmar.2000.12.1.65

Chowdhury, S. (2005). Demographic diversity for building an effective entrepreneurial team: Is it important? *Journal of Business Venturing*, *3*(6), 727–746. doi:10.1016/j.jbusvent.2004.07.001

Choy, C. S., Yew, K. V., & Lin, B. (2006). Criteria for measuring KM performance outcomes in organisations. Industrial Management & Data Systems, 106(7), 917–936. doi:10.1108/02635570610688850 doi:10.1108/02635570610688850

Christensen, P. H. (2003). *Knowledge sharing - time sensitive and push-pull strategies in a non-hype organization*. Copenhagen: Copenhagen Business School.

Chronéer, D., & Backlund, F. (2015). A holistic view on learning in project-based organizations. *Project Management Journal*, *46*(3), 61–74. doi:10.1002/pmj.21503

Cleveland, S., & Ellis, T. J. (2015). Rethinking knowledge sharing barriers: A content analysis of 103 studies. *International Journal of Knowledge Management*, *11*(1), 28–51. doi:10.4018/IJKM.2015010102

Coakes, E., & Coakes, J. (2009). A meta-analysis of the direction and state of sociotechnical research in a range of disciplines: For practioners and academics. *International Journal of Sociotechnology and Knowledge Development*, *1*(1), 1–52. doi:10.4018/jskd.2009010101

Cody, W. F., Kreulen, J. T., Krishna, V., & Spangler, W. S. (2002). The Integration of Business Intelligence and Knowledge Management. *IBM Systems Journal*, *41*(4), 697–713. doi:10.1147j.414.0697

Cohen, J., Cohen, P., West, S. G., & Aiken, L. S. (2003). *Applied Multiple Regression/Correlation Analysis for the Behavioral Sciences*. Mahwah, NJ: Lawrence Erlbaum Associates.

Cohen, W. M., & Levinthal, D. A. (1990). Absorptive capacity: A new perspective on learning and innovation. *Administrative Science Quarterly*, *35*(1), 128–152. doi:10.2307/2393553

Cole, F. L. (1988). Content analysis: Process and application. *Clinical Nurse Specialist CNS*, *2*(1), 53–57. doi:10.1097/00002800-198800210-00025 PMID:3349413

Constant, D., Sproull, L., & Kiesler, S. (1996). The kindness of strangers: The usefulness of electronic weak ties for technical advice. *Organization Science*, *7*(2), 119–135. doi:10.1287/orsc.7.2.119

Cooke, N. J. (1994). Varieties of knowledge elicitation techniques. *International Journal of Human-Computer Studies*, *41*(6), 801–849. doi:10.1006/ijhc.1994.1083

Cook, T. D., & Campbell, D. T. (1979). *Quasi-Experimentation: Design and Analysis for Field Settings*. Boston, MA: Houghton Mifflin Company.

Cope, J., Jack, S., & Rose, M. B. (2007). Social Capital and Entrepreneurship An Introduction. *International Small Business Journal*, *25*(3), 213–219. doi:10.1177/0266242607076523

Cording, M., Christmann, P., & King, D. R. (2008). Reducing causal ambiguity in acquisition integration: Intermediate goals as mediators of integration decisions and acquisition performance. *Academy of Management Journal*, *51*(4), 744–767.

Cortina, J. (1993). What Is coefficient alpha? an examination of theory and applications. *The Journal of Applied Psychology*, *78*(1), 98–104. doi:10.1037/0021-9010.78.1.98

Costa, E., Soares, A. L., & de Sousa, J. P. (2016). Information, knowledge and collaboration management in the internationalisation of SMEs: A systematic literature review. *International Journal of Information Management*, *36*(4), 557–569. doi:10.1016/j.ijinfomgt.2016.03.007

Cousins, P., Lawson, B., Petersen, K., & Handfield, R. (2011). Breakthrough Scanning, Supplier Knowledge Exchange, and New Product Development Performance. *Journal of Product Innovation Management*, *28*(6), 930–942. doi:10.1111/j.1540-5885.2011.00854.x

Covey, S. R. (1989). *The 7 habits of highly effective people: Powerful lessons in personal change*. New York: Fireside.

Coyle-Shapiro, J., & Conway, N. (2005). Exchange relationships: Examining psychological contracts and perceived organizational support. *The Journal of Applied Psychology*, *90*(4), 774–781. doi:10.1037/0021-9010.90.4.774 PMID:16060794

Creswell, J. W. (2005). Educational Research: Planning, Conducting, and Evaluating Quantitative and Qualitative Research (2nd ed.). Upper Saddle Back River, NJ: Pearson and Merrill Prentice Hall.

Creswell, J. W. (2003). *Research Design: Qualitative, Quantitative, and Mixed Methods Approaches* (2nd ed.). Thousand Oaks, CA: Sage Publications.

Crittenden, V. L., & Crittenden, W. F. (2008). Building a capable organization: The eight levers of strategy implementation. Business Horizons, 51(4), 301–309. doi:10.1016/j.bushor.2008.02.003 doi:10.1016/j.bushor.2008.02.003

Croft, J. (2014c). Rest Assured. *Aviation Week & Space Technology, 176*(18), 38.

Crossnan, M. M., Lane, H. W., & White, R. E. (1999). An organizational learning framework: From intuition to institution. *Academy of Management Review, 24*(3), 522–537.

Crouhy, M., Galai, D., & Mark, R. (2001). *Risk Management*. New York: McGraw Hill.

Crowne, D. P., & Marlowe, D. (1960). A new scale of social desirability independent of psychopathology. *Journal of Consulting Psychology, 24*(4), 349–354. doi:10.1037/h0047358 PMID:13813058

Csikszentmihalyi, M., & Larson, R. (2014). Validity and reliability of the experience-sampling method. In M. Csikszentmihalyi (Ed.), *Flow and the Foundations of Positive Psychology* (pp. 35–54). Dordrecht, The Netherlands: Springer. doi:10.1007/978-94-017-9088-8_3

Cumming, C., & Hirtle, B. (2001). The challenges of risk management in diversified financial companies. *Federal Reserve Bank of New York Economic Policy Review, 4*(3), 1–17.

Dalal, N. (2012). Wisdom Computing: Toward a Framework for Wisdom Research in Information Systems. *AMCIS 2012 Proceedings*. Available at http://aisel.aisnet.org/amcis2012/proceedings/PerspectivesIS/14

Dang & McKelvey. (2016). Knowledge management processes and the formation of entrepreneurial opportunities. *Journal of Innovation Economics & Management*, (1), 31-59.

Darroch, J. (2005). Knowledge management, innovation and firm performance. *Journal of Knowledge Management, 9*(3), 101–115. doi:10.1108/13673270510602809

Das, T. K., & Teng, B.-S. (1997). Time and entrepreneurial risk behavior. *Entrepreneurship Theory and Practice, 22*(2), 69–88. doi:10.1177/104225879802200206

Davel, R., Du Toit, S. A., & Mearns, M. M. (2017). Understanding knowledge networks through social network analysis. *International Journal of Knowledge Management, 13*(2), 1–17. doi:10.4018/IJKM.2017040101

Davenport, T. H., & Prusak, L. (1998). Working Knowledge: How Organizations Manage What They Know. Boston, MA: Harvard Business School Press.

Davenport, T. H., De Long, D. W., & Beers, M. C. (1998, Winter). Successful Knowledge Management Projects. *Sloan Management Review*, 40–57.

Davenport, T. H., & Prusak, L. (1998). *Working Knowledge*. Boston, MA: Harvard Business School Press.

Davenport, T. H., & Prusak, L. (1998). *Working Knowledge: How Organizations Manage What They Know*. Boston: Harvard Business School Press.

Davenport, T. H., & Prusak, L. (1998). *Working knowledge: How organizations manage what they know*. Harvard Business Press.

Davies, M. (2009). *Mixed-use Incubator Handbook: A Start-up Guide for Incubator Developers*. Washington, DC: The World Bank Group.

Davis, F. D. (1989). Perceived usefulness, perceived ease of use, and user acceptance of information technology. *Management Information Systems Quarterly, 13*(3), 319–340. doi:10.2307/249008

Dayan, R., & Evans, S. (2006). KM your way to CMMI. Journal of Knowledge Management, 10(1), 69–80. doi:10.1108/13673270610650111 doi:10.1108/13673270610650111

De Jong, J. P. J., & Den Hartog, D. N. (2007). How leaders influence employees' innovative behavior. *European Journal of Innovation Management, 10*(1), 41-64. 10.1108/14601060710720546

De Jong, J. P. J., & Hartog, D. N. D. (2008). *Innovative work behavior: Measurement and validation*. Scientific Analysis of Entrepreneurship and SMEs. Retrieved from http://www.entrepreneurship-sme.eu/pdf-ez/ h200820.pdf

De Jong, T., & Ferguson-Hessler, M. G. (1996). Types and qualities of knowledge. *Educational Psychologist, 31*(2), 105–113. doi:10.120715326985ep3102_2

De Meyer, A. (2013). The future of doctoral education in business administration. *Journal of Management Development, 32*(5), 477–486. doi:10.1108/02621711311328264

de Vaujany, F.-X., Walsh, I., & Mitev, N. (2011). An historically grounded critical analysis of research articles in IS. *European Journal of Information Systems, 20*(4), 395–417. doi:10.1057/ejis.2011.13

Decarolis, D. M., & Deeds, D. (1999). The impact on stocks and flows of organizational knowledge on firm performance: An empirical investigation of the biotechnology industry. Strategic Management Journal, 20(10), 953–968. doi:10.1002/(SICI)1097-0266(199910)20:10<953::AID-SMJ59>3.0.CO;2-3 doi:10.1002/(SICI)1097-0266(199910)20:10<953::AID-SMJ593.0.CO;2-3

Dee, J., & Leisyte, L. (2017). Knowledge sharing and organizational change in higher education. *The Learning Organization, 24*(5), 355–365. doi:10.1108/TLO-04-2017-0034

DeLoach, J. (2004). The new risk imperative–an enterprise-wide approach. Handbook of Business Strategy, 5(1), 29–34.

DeLong, D. W. (2004). *Lost knowledge: Confronting the threat of an aging workforce*. Oxford University Press. doi:10.1093/acprof:oso/9780195170979.001.0001

Denford, J. S., & Chan, Y. E. (2011). Knowledge strategy typologies: Defining dimensions and relationships. Knowledge Management Research and Practice, 9(2), 102–119. doi:10.1057/kmrp.2011.7 doi:10.1057/kmrp.2011.7

Desouza, K. C., & Awazu, Y. (2005). Maintaining knowledge management systems: A strategic imperative. Journal of the American Society for Information Science and Technology, 56(7), 765–768. doi:10.1002/asi.20149 doi:10.1002/asi.20149

Desouza, K. C. (2006). Knowledge Security: An Interesting Research Space. *Journal of Information Science & Technology, 3*(1).

Desouza, K. C. (2007). *Managing knowledge security: strategies for protecting your company's intellectual assets*. Kogan Page Publishers.

Desouza, K. C., Dingsøyr, T., & Awazu, Y. (2005). Experiences with conducting project postmortems: Reports versus stories. *Software Process Improvement and Practice, 10*(2), 203–215. doi:10.1002pip.224

Desouza, K. C., & Vanapalli, G. K. (2005). Securing knowledge in organizations: Lessons from the defense and intelligence sectors. *International Journal of Information Management, 25*(1), 85–98. doi:10.1016/j.ijinfomgt.2004.10.007

Desouza, K., & Awazu, Y. (2005). *Engaged Knowledge Management: Engagement with new realities*. Basingstoke, UK: Palgrave Macmillan. doi:10.1057/9780230006072

Dessne, K. (2013). Formality and informality: Learning in relationships in an organisation. *International Journal of Knowledge Management, 9*(4), 17–32. doi:10.4018/ijkm.2013100102

Dey, I. (1993). *Qualitative Data Analysis. A user-friendly guide for social scientists*. London: Routledge. doi:10.4324/9780203412497

Diaz-Diaz, N., & De Saa-Perez, P. (2014). The interaction between external and internal knowledge sources: An open innovation view. *Journal of Knowledge Management, 18*(2), 430–446. doi:10.1108/JKM-07-2013-0257

Dickinson, G. (2001). Enterprise Risk Management: Its origins and conceptual foundation. *The Geneva Papers on Risk and Insurance*, *26*(3), 360–366. doi:10.1111/1468-0440.00121

Dictionary.com. (2017, August 30). *Dictionary.com*. Retrieved from http://www.dictionary.com/browse/context?s=t

DiGangi, P., Wasko, M. M., & Tang, X. (2012). Would you share?: Examining knowledge type and communication channel for knowledge sharing within and across the organizational boundary. *International Journal of Knowledge Management*, *8*(1), 1–21. doi:10.4018/jkm.2012010101

Dillard, J. P. (1990). A goal-driven model of interpersonal influence. In J. P. Dillard (Ed.), *Seeking compliance: The production of interpersonal influence messages*. Scottsdale, AZ: Gorsuch Scarisbrick.

Dillard, J. P. (2008). Goals-plans-action theory of message production: Making influence messages. In L. A. Baxter & D. O. Braithwaite (Eds.), *Engaging theories in interpersonal communication: Multiple perspectives* (pp. 65–76). Los Angeles, CA: Sage. doi:10.4135/9781483329529.n5

Disterer, G. (2002). Management of project knowledge and experiences. *Journal of Management Information Systems*, *6*(5), 512–520. doi:10.1108/13673270210450450

Dittes, S., Smolnik, S., Jennex, M. E., & Croasdell, D. T. (2016). Eleven years of the knowledge management track at HICSS: An overview. *International Journal of Knowledge Management*, *12*(4), 51–61. doi:10.4018/IJKM.2016100104

Dodgson, M. (1993). Organizational Learning: A Review of Some Literatures. *Organization Studies*, *14*(3), 375–394. doi:10.1177/017084069301400303

Donate, M., & Guadamillas, F. (2010). The Effect of Organizational Culture on Knowledge Management Practices and Innovation. *Knowledge and Process Management*, *17*(2), 82–94. doi:10.1002/kpm.344

Donate, M., & Guadamillas, F. (2015). An empirical study of the relationships between knowledge management, knowledge-oriented human resource practices and innovation. *Knowledge Management Research and Practice*, *13*(2), 134–148. doi:10.1057/kmrp.2013.36

Dorfman, P. W., & Howell, J. P. (1988). Dimensions of National Culture and Effective Leadership Patterns: Hofstede Revisted. In R. N. Farmer (Ed.), *Advances in International Comparative Management* (Vol. 3, pp. 127–150). Philadelphia, PA: JAI Press.

Dougherty, D. (2004). Organizing Practices in Services: Capturing Practice-Based Knowledge for Innovation. *Strategic Organization*, *2*(1), 35–64. doi:10.1177/1476127004040914

Douglas, E. J. (2013). Reconstructing entrepreneurial intentions to identify predisposition for growth. *Journal of Business Venturing*, *28*(5), 633–651. doi:10.1016/j.jbusvent.2012.07.005

Doz, Y., & Kosonen, M. (2010). Embedding Strategic Agility: A Leadership Model for Accelerating Business Model Renewal. *Long Range Planning*, *43*(2-3), 370–382. doi:10.1016/j.lrp.2009.07.006

Driscoll, M. (2014, November). Seven Imperatives For Process Excellence. C*orporate Finance Review*, 13-19.

Drucker, P. (1954). *The Practice of Management*. New York: Harper & Row.

Drucker, P. (1985). *Innovation and Entrepreneurship*. London: Heinemann.

Du Plessis, M. (2007). The role of knowledge management in innovation. *Journal of Knowledge Management*, *11*(4), 20–29. doi:10.1108/13673270710762684

Duderstadt, J. J. (2000). *The future of higher education in the knowledge-driven, global economy of the 21st century.* Retrieved from http://milproj.ummu. umich.edu/publications/toronto/

Dumas, J. S., & Redish, J. C. (1999). *A Practical Guide to Usability Testing* (revised ed.). Portland, OR: Intellect Books.

Dumrak, J., Baroudi, B., & Hadjinicolaou, N. (2017). Exploring the association between project management knowledge areas and sustainable outcomes. *Journal of Engineering, 182*, 157–164. doi:10.1016/j.proeng.2017.03.152

Dunford, R. (2000). Key challenges in the search for the effective management of knowledge in management consulting firms. Journal of Knowledge Management, 4(4), 295–302. doi:10.1108/13673270010379849 doi:10.1108/13673270010379849

Durst, S., & Runar Edvardsson, I. (2012). Knowledge management in SMEs: A literature review. *Journal of Knowledge Management, 16*(6), 879–903. doi:10.1108/13673271211276173

Earl, M. (2001). Knowledge management strategies: Towards a taxonomy. Journal of Management Information Systems, 18(1), 215–233. doi:10.1080/07421222.2001.11045670 doi:10.1080/07421222.2001.11045670

Easterby-Smith, M. (1997). Disciplines of Organizational Learning: Contributions and Critiques. *Human Relations, 50*(9), 1085–1113. doi:10.1177/001872679705000903

Easterby-Smith, M., Lyles, M. A., & Tsang, E. W. (2008). Inter-organizational knowledge transfer: Current themes and future prospects. *Journal of Management Studies, 45*(4), 677–690. doi:10.1111/j.1467-6486.2008.00773.x

Ebert, C., & De Man, J. (2008). Effectively utilizing project, product, and process knowledge. *Information and Software Technology, 50*(6), 579–594. doi:10.1016/j.infsof.2007.06.007

Edvardsson, I. R., & Durst, S. (2013). Does knowledge management deliver the goods in SMEs? *Business and Management Research, 2*(2), 52. doi:10.5430/bmr.v2n2p52

Edwards, J. S. (2009). Business processes and knowledge management. In M. Khosrow-Pour (Ed.), *Encyclopedia of Information Science and Technology* (2nd ed.; Vol. I, pp. 471–476). Hershey, PA: IGI Global. doi:10.4018/978-1-60566-026-4.ch078

Eidizadeh, R., Salehzadeh, R., & Esfahani, A. C. (2017). Analysing the role of business intelligence, knowledge sharing and organisational innovation on gaining competitive advantage. *Journal of Workplace Learning, 29*(4), 50–267. doi:10.1108/JWL-07-2016-0070

El Sawy, O. A. (2003). The IS Core IX: The 3 Faces of IS Identity: Connection, Immersion, and Fusion. *Communications of the Association for Information Systems, 12*(1), 39.

Elo, S., & Kyngas, H. (2008). The qualitative content analysis process. *Journal of Advanced Nursing, 61*(1), 107–115. doi:10.1111/j.1365-2648.2007.04569.x PMID:18352969

Endres, M. L., Chowdhury, S. K., & Alam, I. (2008). Gender effects on bias in complex financial decisions. *Journal of Managerial Issues, 20*(2), 238–254.

Endres, M. L., Chowdhury, S. K., & Milner, M. (2009). Ambiguity tolerance and accurate assessment of self-efficacy in a complex decision task. *Journal of Management & Organization, 15*(1), 31–46. doi:10.1017/S1833367200002868

Erickson, G. S., & Rothberg, H. N. (2011). Assessing knowledge management needs: A strategic approach to developing knowledge. *International Journal of Knowledge Management, 7*(3), 1–10. doi:10.4018/jkm.2011070101

Eriksson, I., & Raven, A. (1996). *Gaining competitive advantage through shared knowledge creation: in search of a new design theory for strategic information systems*. Academic Press.

Erlandson, D. A., Harris, E. L., Skipper, B. L., & Allen, S. D. (1993). *Doing naturalistic inquiry: A guide to methods.* Newbury Park, CA: Sage Publications.

Ermine, J. L. (2013). Knowledge Management as Knowledge Risk Prevention. *iKNOW, the Magazine for Knowledge Workers, 3*(1), 8-11.

Ettlie, J., & Rosenthal, S. (2011). Services versus Manufacturing Innovation. *Journal of Product Innovation Management, 28*(2), 285–299. doi:10.1111/j.1540-5885.2011.00797.x

European Commission. (2001). *European innobarometer survey.* Luxembourg: European Commission.

Evangelou, C., & Karacapilidis, N. (2005). On the interaction between humans and knowledge management systems: A framework of knowledge sharing catalysts. *Knowledge Management Research and Practice, 3*(4), 253–261. doi:10.1057/palgrave.kmrp.8500076

Fahey, L., & Prusak, L. (1998). The eleven deadliest sins of knowledge management. *California Management Review, 40*(3), 265–276. doi:10.2307/41165954

Farahmand, F., Navathe, S. B., Enslow, P. H., & Sharp, G. P. (2003). Managing Vulnerabilities of Information Systems to Security Incidents. In *Proceedings of the 5th International Conference on Electronic Commerce* (pp. 348–354). New York: ACM. 10.1145/948005.948050

Farzin, M., Kahreh, M., Hesan, M., & Khalouei, A. (2014). A Survey of Critical Success Factors for Strategic Knowledge Management Implementation: Applications for Service Sector. *Procedia: Social and Behavioral Sciences, 109*, 595–599. doi:10.1016/j.sbspro.2013.12.512

Feigenbaum, E. A., & McCorduck, P. (1983). *The Fifth Generation* (1st ed.). Reading, MA: Addison-Wesley.

Feldman, M. P. (1994). The university and economic development: The case of the Johns Hopkins University and Baltimore. *Economic Development Quarterly, 8*(1), 67–76. doi:10.1177/089124249400800105

Ferdousi, Z., & Maeda, A. (2006, September). Anomaly detection using unsupervised profiling method in time series data. ADBIS Research Communications.

Ferrada, X., Núñez, D., Neyem, A., Serpell, A., & Sepúlveda, M. (2016). A lessons-learned system for construction project management: A preliminary application. *Journal of Social and Behavioral Sciences, 226*, 302–309. doi:10.1016/jsbspro.2016.06.192

Ferreira, J. J., Ratten, V., & Dana, L.-P. (2017). Knowledge spillover-based strategic entrepreneurship. *The International Entrepreneurship and Management Journal, 13*(1), 161–167. doi:10.100711365-016-0415-6

Filieri, R. (2010). *Overcoming knowledge sharing barriers through communities ofpractice.* Cambridge Scholars Publishing.

Fisher, R. J. (1993). Social desirability bias and the validity of indirect questioning. *The Journal of Consumer Research, 20*(2), 303–315. doi:10.1086/209351

Flamholtz, E. G., & Randle, Y. (2012). *Growing pains: Transitioning from an entrepreneurship to a professionally managed firm.* John Wiley & Sons.

Fleishman, E. A., & Quaintance, M. K. (2000). *Taxonomies of human performance: The description of human tasks.* Potomac, MD: Management Research Institute, Inc.

Fleming, L., & Frenken, K. (2007). The evolution of inventor networks in the silicon valley and Boston regions. *Advances in Complex Systems, 10*(1), 53–68. doi:10.1142/S0219525907000921

Flick, U. (2002). An introduction to qualitative research (2nd ed.). London: Sage Publications.

Floren, H., Rundquist, J., Schuler, R., Bondarouk, T., & Ruel, H. (2014). HRM and innovation: Themes, contingencies and directions for future research. *European Journal of Innovation Management*, *8*(5), 570–575.

Flores, W. R., Antonsen, E., & Ekstedt, M. (2014). Information security knowledge sharing in organizations: Investigating the effect of behavioral information security governance and national culture. *Computers & Security*, *43*, 90–110. doi:10.1016/j.cose.2014.03.004

Foley, E., & Guillemette, M. G. (2010). What is Business Intelligence? *International Journal of Business Intelligence Research*, *1*(4), 1–28. doi:10.4018/jbir.2010100101

Fong, P. S. W. (2003). Knowledge creation in multidisciplinary project teams: An empirical study of the processes and their dynamic relationships. *International Journal of Project Management*, *21*(7), 479–486. doi:10.1016/S0263-7863(03)00047-4

Fontana, R., Geuna, A., & Matt, M. (2006). Factors affecting university - industry R&D projects: The importance of searching, screening, and signalling. *Research Policy*, *35*(2), 309–323. doi:10.1016/j.respol.2005.12.001

Fornell, C., & Larcker, D. F. (1981). Evaluating structural equation models with unobservable variables and measurement error. *JMR, Journal of Marketing Research*, *18*(1), 39–50. doi:10.2307/3151312

Foss, N. J. (2007). The Emerging Knowledge Governance Approach: Challenges and Characteristics. *Organization*, *14*(1), 29–52. doi:10.1177/1350508407071859

Foss, N. J., & Pedersen, T. (2002). Transferring knowledge in MNCs: The role of sources of subsidiary knowledge and organizational context. *Journal of International Management*, *8*(1), 49–67. doi:10.1016/S1075-4253(01)00054-0

Foss, N., Laursen, K., & Pederson, T. (2011). Linking Customer Interaction and Innovation: The Mediating Role of New Organizational Practices. *Organization Science*, *22*(4), 980–999. doi:10.1287/orsc.1100.0584

Fourie, L., & Shilawa, J. (2004). The value of concept maps for knowledge management in the banking and insurance industry: A German case study. *Concept Maps: Theory, Methodology, Technology: Proceedings of the First International Conference on Concept Mapping*.

Francis, S., & Paladino, B. (2008). Enterprise Risk Management: A best practice approach. *Journal of Corporate Accounting & Finance*, *19*(3), 19–33. doi:10.1002/jcaf.20382

Franklin, C., & Ballan, M. (2001). Reliability and Validity in Qualitative Research. In B. Thyer (Ed.), *The Handbook of Social Work Research Methods*. Thousand Oaks, CA: Sage. doi:10.4135/9781412986182.n15

Fraser, J. R. S., & Simkins, B. J. (2007). Ten Common Misconceptions About Enterprise Risk Management. *Journal of Applied Corporate Finance*, *19*(4), 75–81. doi:10.1111/j.1745-6622.2007.00161.x

Fraser, J. R. S., & Simkins, B. J. (2016). The challenges of and solutions for implementing enterprise risk management. *Business Horizons*, *59*(6), 689–698. doi:10.1016/j.bushor.2016.06.007

Freeze, R. D., & Kulkarni, U. (2007). Knowledge management capability: Defining knowledge assets. *Journal of Knowledge Management*, *11*(6), 94–109. doi:10.1108/13673270710832190

Freeze, R., & Kulkarni, U. (2008). Validating Distinct Knowledge Assets: A Capability Perspective. *International Journal of Knowledge Management*, *4*(4), 40–61. doi:10.4018/jkm.2008100103

Frenkel, S. J., & Sanders, K. (2007). Explaining variations in co-worker assistance in organizations. *Organization Studies*, *28*(6), 797–823. doi:10.1177/0170840607073079

Frese, M., & Gielnik, M. M. (2014). The psychology of entrepreneurship. *Annual Review of Organizational Psychology and Organizational Behavior*, *1*(1), 413–438. doi:10.1146/annurev-orgpsych-031413-091326

Fricke, M. (2007). The Knowledge Pyramid: A Critique of the DIKW Hierarchy. *Journal of Information Science*, *35*(2), 131–142. doi:10.1177/0165551508094050

Friedman, M. (2007). The methodology of positive economics. In D. M. Hausman (Ed.), *The Philosophy of Economics* (Vol. 3, pp. 145–178). Cambridge, UK: Cambridge University Press. doi:10.1017/CBO9780511819025.010

Friedman, T. L. (2005). *The world is flat: A brief history of the twenty-first century*. New York, NY: Macmillan.

Fugate, B., Stank, T., & Mentzer, J. (2009). Linking improved knowledge management to operational and organizational performance. *Journal of Operations Management*, *27*(3), 247–264. doi:10.1016/j.jom.2008.09.003

Fullwood, R., Rowley, J., & Delbridge, R. (2013). Knowledge sharing amongst academics in UK universities. *Journal of Knowledge Management*, *17*(1), 123–136. doi:10.1108/13673271311300831

Gaimon, C., & Bailey, J. (2013). Knowledge Management for the Entrepreneurial Venture. *Production and Operations Management*, *22*(6), 1429–1438. doi:10.1111/j.1937-5956.2012.01337.x

Gallouj, F., & Djellal, F. (Eds.). (2010). *The handbook of innovation and services: a multi-disciplinary perspective*. Cheltenham, UK: Edward Elgar Publishing. doi:10.4337/9781849803304

Gallupe, B. (2001). Knowledge management systems: Surveying the landscape. *International Journal of Management Reviews*, *3*(1), 61–77. doi:10.1111/1468-2370.00054

Ganguly, A., Mostashari, A., & Mansouri, M. (2011). Measuring knowledge management/knowledge sharing efficiency and effectiveness in enterprise networks. *International Journal of Knowledge Management*, *7*(4), 37–54. doi:10.4018/jkm.2011100103

Garbarino-Alberti, H., & Pastorino, S. A. (2014). Knowledge management steps, tools, techniques and influencing factors in SMEs: Systematic literature review. *Computing Conference (CLEI), 2014 XL Latin American*. 10.1109/CLEI.2014.6965108

Garon, S. (2006). Space project management lessons learned: A powerful tool for success. *Journal of Knowledge Management*, *10*(2), 103–112. doi:10.1108/13673270610656665

Gartner Group. (2015). *Gartner Worldwide Spending Forecast*. Retrieved February 2, 2015, from http://www.gartner.com/technology/research/it-spending-forecast

Gauld, R. (2007). Public sector information system project failures: Lessons from a New Zealand hospital organization. *Government Information Quarterly*, *24*(1), 102–114. doi:10.1016/j.giq.2006.02.010

Gavronski, I., Klassen, R., Vachon, S., & Maschado do Nascimento, L. (2012). A learning and knowledge approach to sustainable operations. *International Journal of Production Economics*, *140*(1), 183–192. doi:10.1016/j.ijpe.2012.01.037

Gibb, G. D., & Lofaro, R. J. (1993). Computer-based Assessment of Air Traffic Controller Abilities: Experimental Selection Tasks. *Proceedings of 1993 Aerospace Medical Association Annual Symposium*.

Gibb, G. M., & Lofaro, R. J. (1994). *Airport Security Screener and Checkpoint Security Supervisor Training Workshop. DOT/FAA/CT-94/*109. Springfield, VA: National Technical Information Service.

Gibbons, M., Limoges, C., Nowotny, H., Schwartzman, S., Scott, P., & Trow, M. (1994). *The Production of Knowledge: The Dynamics of Science and Research in Contemporary Societies*. London: Sage.

Gilbert, B. A., McDougall, P. P., & Audretsch, D. B. (2008). Clusters, knowledge spillovers and new venture performance: An empirical examination. *Journal of Business Venturing, 23*(4), 405–422. doi:10.1016/j.jbusvent.2007.04.003

Giles, W. D. (1991). Making strategy work. Long Range Planning, 24(5), 75–91. doi:10.1016/0024-6301(91)90254-L doi:10.1016/0024-6301(91)90254-L

Giles, H. (2008). Communication accommodation theory. In L. A. Baxter & D. O. Braithwaite (Eds.), *Engaging theories in interpersonal communication: Multiple perspectives* (pp. 161–173). Los Angeles, CA: Sage. doi:10.4135/9781483329529.n12

Ginsberg, A., & Abrahamson, E. (1991). Champion of change and strategic shifts: The role of internal and external change advocates. Journal of Management Studies, 28(2), 173–190. doi:10.1111/j.1467-6486.1991.tb00276.x doi:10.1111/j.1467-6486.1991.tb00276.x

Gloet, M., & Berrell, M. (2003). The Dual Paradigm Nature of Knowledge Management: Implications for achieving quality outcomes in human resources. *Journal of Knowledge Management, 7*(1), 78–89. doi:10.1108/13673270310463635

Gloet, M., & Samson, D. (2013). Knowledge Management to Support Systematic Innovation Capability. In *Proceedings of the 46th Hawaii International Conference on the System Sciences (HICSS)*. IEEE. 10.1109/HICSS.2013.374

Goel, A., Rana, G., & Rastogi, R. (2010). Case study: Knowledge management as a process to develop sustainable competitive advantage. *South Asian Journal of Management, 17*(3), 104–116.

Goh, L. A. (2005). Fostering Innovation Through Knowledge-Centered Principles: A Case Study of Singapore Airlines. *International Journal of Knowledge Management, 1*(4), 73–90. doi:10.4018/jkm.2005100104

Gold, A. H., Malhotra, A., & Segars, A. H. (2001). Knowledge management: An organizational capabilities perspective. *Journal of Management Information Systems, 18*(1), 185–214. doi:10.1080/07421222.2001.11045669

Golderberger, A. (1978). *Topics in Regression Analysis*. New York: McMillan.

Gonzales, R. V. D. (2016). Knowledge Retention in the Service Industry. *International Journal of Knowledge Management, 12*(1), 45–59. doi:10.4018/IJKM.2016010104

González-Rojas, O., Pedraza-Garcia, G., Correal, D., & Beltrán, G. (2016). MONO+KM: Knowledge management in collaborative project development, *Lng. University of Bogotá, 20*(2), 267–302. doi:10.11144/Javeriana.iyu20-2.mkkm

Goonatilake, L., Maizza-Neto, O., & Jayawardene, P. (2000). Enhancing enterprise competitiveness in developing countries through the promotion of management information and benchmarking tools. In *Proceedings of the 6th International Working Conference of IFIP WG 9.4: Information Flows, Local Improvisations and Work Practices*. STS Conferences.

Gordon, T. J., & Pease, A. (2006). RT Delphi: An Efficient, "Round-less," Almost Real Time Delphi Method. *Journal of Technological Forecasting and Social Change, 73*(4).

Gordon, M. E., Slade, L. A., & Schmitt, N. (1986). The "science of the sophomore" revisited: From conjecture to empiricism. *Academy of Management Review, 11*(1), 191–208.

Grant, R. M. (1996). Toward a knowledge-based theory of the firm. *Strategic Management Journal, 17*(S2), 109–122. doi:10.1002mj.4250171110

Gray, P. H., & Meister, D. B. (2004). Knowledge sourcing effectiveness. Management Science, 50(6), 821–834. doi:10.1287/mnsc.1030.0192 doi:10.1287/mnsc.1030.0192

Gray, C. D., & Kinnear, P. R. (2012). *IBM SPSS Statistics 19 Made Simple*. New York, NY: Psychology Press.

Greco, A. N., Wharton, R. M., Estelami, H., & Jones, R. F. (2006). The state of Scholarly Journal Publishing: 1981 – 2000. *Journal of Scholarly Publishing*, *37*(3), 155–214. doi:10.3138/jsp.37.3.155

Greenberg, J. (1987). The college sophomore as guinea pig: Setting the record straight. *Academy of Management Review*, *12*, 157–159.

Greenwood, R., Hinings, C. R., & Suddaby, R. (2002). Theorizing change: The role of professional associations in the transformation of institutionalized fields. Academy of Management Journal, 45(1), 58–80. doi:10.2307/3069285 doi:10.2307/3069285

Gregor, S., & Hevner, A. R. (2013). Positioning and presenting design science research for maximum impact. *Management Information Systems Quarterly*, *37*(2), 337–355. doi:10.25300/MISQ/2013/37.2.01

Gressgard, L., Amundsen, O., Aasen, T., & Hansen, K. (2014). Use of information and communication technology to support employee-driven innovation in organizations: A knowledge management perspective. *Journal of Knowledge Management*, *18*(4), 633–650. doi:10.1108/JKM-01-2014-0013

Griffin, A., & Page, A. L. (1993). Introduction to Action Research. Thousand Oaks, CA: Sage.

Guadamillas, F., Donate, M. J., & Pablo, J. D. (2008). Knowledge management for corporate entrepreneurship and growth: A case study. *Knowledge and Process Management*, *15*(1), 32–44. doi:10.1002/kpm.297

Gubbi, J., Buyya, R., Marusic, S., & Palaniswami, M. (2013). Internet of Things (IoT): A vision, architectural elements, and future directions. *Future Generation Computer Systems*, *29*(7), 1645–1660. doi:10.1016/j.future.2013.01.010

Gupta, S., Cadeaux, J., & Dubelaar, C. (2006). Uncovering multiple champion roles in implementing new-technology ventures. Journal of Business Research, 59(5), 549–563. doi:10.1016/j.jbusres.2006.01.003 doi:10.1016/j.jbusres.2006.01.003

Gupta, A. K., & Govindarajan, V. (2000). Knowledge flows within multinational corporations. *Strategic Management Journal*, *21*(4), 473–496. doi:10.1002/(SICI)1097-0266(200004)21:4<473::AID-SMJ843.0.CO;2-I

Haas, M. R., & Hansen, M. T. (2005). When using knowledge can hurt performance: The value of organizational capabilities in a management consulting company. *Strategic Management Journal*, *26*(1), 1–24. doi:10.1002mj.429

Hackett, S., & Dilts, D. (2004). A Systematic Review of Business Incubation Research. *The Journal of Technology Transfer*, *29*(1), 55–82. doi:10.1023/B:JOTT.0000011181.11952.0f

Haeussler, C. (2011). Information-sharing in academia and the industry: A comparative study. *Research Policy*, *40*(1), 105–122. doi:10.1016/j.respol.2010.08.007

Haeussler, C., Jiang, L., Thursby, J., & Thursby, M. (2014). Specific and general information sharing among competing academic researchers. *Research Policy*, *43*(3), 465–475. doi:10.1016/j.respol.2013.08.017

Hafkesbrink, J. (2014). Ambidextrous Organizational and Individual Competencies in Open Innovation: The Dawn of a New Research Agenda. *Journal of Innovation Management*, *2*(1), 9–46.

Haghirian, P. (2010). *Understanding Japanese Management Practices*. New York, NY: Business Expert Press. doi:10.4128/9781606491195

Hair, J. F., Black, W., Babin, B., & Anderson, R. (2014). Multiple Regression Analysis. In J. F. Hair, W. Black, B. Babin, & R. Anderson (Eds.), *Multivariate Data Analysis* (7th ed.; pp. 151–230). Harlow, UK: Pearson Education.

Hair, J. F. Jr, Black, W. C., Babin, B. J., Anderson, R. E., & Tatham, R. L. (2006). *Multivariate data analysis* (6th ed.). Prentice-Hall PTR.

Hair, J. F., Ringle, C. M., & Sarstedt, M. (2011). PLS-SEM: Indeed a silver bullet. *Journal of Marketing Theory and Practice, 19*(2), 139–152. doi:10.2753/MTP1069-6679190202

Hair, J., Babin, B., Money, A., & Samouel, P. (2003). *Essentials of Business Research Methods*. New York: John Wiley and Sons.

Halawi, L., Aronson, J., & McCarthy, R. (2005). Resource-Based View of Knowledge Management for Competitive Advantage. *Electronic Journal of Knowledge Management, 3*(2), 75–86.

Hanberger, A., & Schild, I. (2004). Strategies to evaluate a university-industry knowledge-exchange programme. *Evaluation, 10*(4), 475–492. doi:10.1177/1356389004050217

Handzic, M., & Durmic, N. (2015). Knowledge management, intellectual capital and project management: Connecting the dots. *Electronic Journal of Knowledge Management, 13*(10), 51-61.

Handzic, M. (2004). *Knowledge Management: Through the Technology Glass (Series on Innovation and Knowledge Management)*. Singapore: World Scientific Publishing.

Handzic, M. (2007). *Socio-technical knowledge management: Studies and initiatives*. Hershey, PA: IGI Global Publishing. doi:10.4018/978-1-59904-549-8

Handzic, M. (2011). Integrated Socio-Technical Knowledge Management Model: An Empirical Evaluation. *Journal of Knowledge Management, 15*(2), 198–211. doi:10.1108/13673271111119655

Handzic, M., & Agahari, D. (2004). Knowledge sharing culture: A case study. *Journal of Information &.Knowledge. Management, 3*(2), 135–142.

Handzic, M., & Ozlen, K. (2012). A Descriptive Analysis of KMS Success in Turkish Healthcare Organisations. In J. G. Cegarra (Ed), *Proceedings of the 13th European Conference on Knowledge Management:* Vol 1 (pp. 441-449). Polytechnic University of Cartagena.

Hanisch, B., Lindner, F., Mueller, A., & Wald, A. (2009). KM in project environments. *Journal of Knowledge Management, 13*(4), 148–160. doi:10.1108/13673270910971897 PMID:19622123

Hannabuss, S. (2001). A wider view of knowledge. *Library Management, 22*(8), 357–363. doi:10.1108/01435120110406273

Hansen, T. M., Nohria, N., & Tierney, T. (1999, March). What's your strategy for managing knowledge. Harvard Business Review, 106–116. PubMed PMID:10387767

Hansen, M. T. (1999). The Search-Transfer Problem: The Role of Weak Ties in Sharing Knowledge across Organization Subunits. *Administrative Science Quarterly, 44*(1), 82–111. doi:10.2307/2667032

Hansen, M. T., Nohria, N., & Tierney, T. (1999, March). What's your strategy for managing knowledge? *Harvard Business Review*, 106–116. PMID:10387767

Hansen, R. S. (2006). Benefits and problems with student teams: Suggestions for improving team projects. *Journal of Education for Business, 82*(1), 11–19. doi:10.3200/JOEB.82.1.11-19

Harlow, H. (2008). The effect of tacit knowledge on firm performance. *Journal of Knowledge Management, 12*(1), 148–163. doi:10.1108/13673270810852458

Harwood, T. G., & Garry, T. (2003). An overview of content analysis. *The Marketing Review, 3*(4), 479–498. doi:10.1362/146934703771910080

Harzing, A. W. (2006). Response styles in cross-national survey research a 26-country study. *International Journal of Cross Cultural Management*, *6*(2), 243–266. doi:10.1177/1470595806066332

Hassan, N. R. (2014). Value of IS research: Is there a crisis? *Communications of the Association for Information Systems*, *34*, 801-816.

Hassan, N. R., & Mathiassen, L. (2018). Distilling a body of knowledge for information systems development. *Information Systems Journal*.

Hauschild, S., Licht, T., & Stein, W. (2001). Creating a knowledge culture. *The McKinsey Quarterly*, *1*, 74–81.

Hayton, J. C. (2005). Competing in the new economy: The effect of intellectual capital on corporate entrepreneurship in high-technology new ventures. *R & D Management*, *35*(2), 137–155. doi:10.1111/j.1467-9310.2005.00379.x

Heisig, P. (2009). Harmonisation of knowledge management - comparing 160 KM frameworks around the globe. *Journal of Knowledge Management*, *13*(4), 4–31. doi:10.1108/13673270910971798

Heisig, P., Suraj, O. A., Kianto, A., Kemboi, C., Arrau, G. P., & Easa, N. F. (2016). Knowledge management and business performance: Global experts' views on future research needs. *Journal of Knowledge Management*, *20*(6), 1169–1198. doi:10.1108/JKM-12-2015-0521

Hendry, C., Selim, G., Citron, D., Holtham, C., Brown, J., Holden, J., ... Oehlcke, F. (2008). Facilitating Innovation through the Measurement and Management of Intangibles. In H. Scarbrough (Ed.), *The Evolution of Business Knowledge*. Oxford, UK: Oxford University Press.

Henry, R. M., McCray, G. E., Purvis, R. L., & Roberts, T. L. (2007). Exploiting organizational knowledge in developing IS project cost and schedule estimates: An empirical study. *Information & Management*, *44*(6), 598–612. doi:10.1016/j.im.2007.06.002

Hevner, A. R., March, S. T., Park, J., & Ram, S. (2004). Design science in information systems research. *Management Information Systems Quarterly*, *28*(1), 75–105. doi:10.2307/25148625

Hewes, D. E., & Planalp, S. (1987). The individual's place in communication science. In C. R. Berger & S. H. Chaffer (Eds.), *Handbook of communication science*. Newbury Park, CA: Sage.

Hey, J. (2004). *The Data, Information, Knowledge, Wisdom Chain: The Metaphorical link*. Retrieved June 16, 2008 from: http://www.oceanteacher.org

Higgins, A. (1990). Champions of technological innovation. Administrative Science Quarterly, 35(2), 317–342. doi:10.2307/2393393 doi:10.2307/2393393

Hipp, C., & Grupp, H. (2005). Innovation in the service sector: The demand for service-specific innovation measurement concepts and typologies. *Research Policy*, *34*(4), 517–535. doi:10.1016/j.respol.2005.03.002

Hirai, C., Uchida, Y., & Fujinami, T. (2007). A KM system for dynamic organizational knowledge circulation. *International Journal of IT & Decision Making*, *6*(3), 509–522. doi:10.1142/S0219622007002617

Hitt, M. (1998). Twenty first century organizations? Business firms, business schools, and the academy. *Academy of Management Review*, *23*(2), 218–225.

Hobday, M., Davies, A., & Prencipe, A. (2005). Systems integration: A core capability of the modern corporation. *Industrial and Corporate Change*, *14*(6), 1109–1143. doi:10.1093/icc/dth080

Hodgkinson, G. P., & Starkey, K. (2011). Not simply returning to the same answer over and over again: Reframing relevance. *British Journal of Management*, *22*(3), 355–369. doi:10.1111/j.1467-8551.2011.00757.x

Hoffman, L. R., & Maier, N. R. F. (1961). Quality and acceptance of problem solutions by members of homogeneous and heterogeneous groups. *Journal of Abnormal Psychology*, *62*(2), 401–407. doi:10.1037/h0044025 PMID:13715029

Hofstede, G. (2013). *National Culture - Countries: Turkey*. The Hofstede Centre. Retrieved July 2, 2013, from http://geert-hofstede.com/turkey.html

Hofstede, G. H. (1980). *Culture's consequences: International Differences in Work-Related Values*. Beverly Hills, CA: Sage.

Hofstede, G. H. (2001). *Culture's consequences: Comparing values, behaviors, institutions and organizations across nations*. Beverly Hills, CA: Sage.

Hofstede, G. H., Hofstede, G. J., & Minkov, M. (2010). *Cultures and Organizations: Software of the Mind: Intercultural Cooperation and its Importance for Survival*. New York, NY: McGraw-Hill.

Hofstede, G., & Bond, M. H. (1984). Hofstede's culture dimensions: An independent validation using Rokeach's value survey. *Journal of Cross-Cultural Psychology*, *15*(4), 417–433. doi:10.1177/0022002184015004003

Holsapple, C. W., & Wu, J. (2008). In search of a missing link. *KM Research and Practice, 2008*(6), 31-40.

Holsapple, C. W. (2003). *Knowledge Management Handbook*. Berlin: Springer. doi:10.1007/978-3-540-24748-7

Holsapple, C. W., & Joshi, K. D. (2000). An investigation of factors that influence the management of knowledge in organizations. *The Journal of Strategic Information Systems*, *9*(2), 235–261. doi:10.1016/S0963-8687(00)00046-9

Hong, H., Kim, J.-S., Kim, T., & Leem, B.-H. (2008). The effect of knowledge on system integration project performance. *Industrial Management & Data Systems*, *108*(3), 385–404. doi:10.1108/02635570810858787

Hormozi, A., & Giles, S. (2004). Data mining a competitive weapon for banking and retail industries. *Information Systems Management*, *21*(2), 62–71. doi:10.1201/1078/44118.21.2.20040301/80423.9

Houston, R. D., & Harmon, E. G. (2002). Re-envisioning the information concept: systematic definitions. In H. Bruce, R. Fidel, P. Ingwersen, & P. Vakkari (Eds.), *Emerging frameworks and methods: proceedings of the fourth International Conference on Conceptions of Library and Information Science (CoLIS4)* (pp. 305-308). Greenwood Village, CO: Libraries Unlimited.

Hrebiniak, L. G. (2005). Making strategy work: Leading effective execution and change. Upper Saddle River, NJ: Pearson Education.

Hrebiniak, L. G. (2006). Obstacles to effective strategy implementation. Organizational Dynamics, 35(1), 12–31. doi:10.1016/j.orgdyn.2005.12.001 doi:10.1016/j.orgdyn.2005.12.001

Hsu, C., Backhouse, J., & Silva, L. (2013). Institutionalizing operational risk management: An empirical study. *Journal of Information Technology*. doi:10.1057/jit.2013.15

Hsu, M. L. A., Hou, S., & Fan, H.-L. (2011). Creative self-efficacy and innovative behavior in a service setting: Optimism as a moderator. *The Journal of Creative Behavior*, *45*(4), 258–272. doi:10.1002/j.2162-6057.2011.tb01430.x

Huang, Q., & Davison, R. (2008). Knowledge sharing barriers at the individual level in a Chinese bank. *PACIS 2008 Proceedings*.

Huber, G. P., Davenport, T. H., & King, D. (1998). *Some Perspectives on Organizational Memory*. Presented at the 31st Annual Hawaii International Conference on System Sciences.

Hughes, M., Hughes, P., & Morgan, R. E. (2007). Exploitative learning and entrepreneurial orientation alignment in emerging young firms: Implications for market and response performance. British Journal of Management, 18(4), 359–375. doi:10.1111/j.1467-8551.2007.00519.x doi:10.1111/j.1467-8551.2007.00519.x

Hult, M., & Lennung, S. A. (1980). Towards a definition of action research: A note and bibliography. *Journal of Management Studies*, *17*(2), 241–250. doi:10.1111/j.1467-6486.1980.tb00087.x

Hu, M. M., Horng, J., & Sun, Y. C. (2009). Hospitality teams: Knowledge sharing and service innovation performance. *Tourism Management*, *30*(1), 41–50. doi:10.1016/j.tourman.2008.04.009

Hung, S., Durcikova, A., Lai, H., & Lin, W. (2011). The influence of intrinsic and extrinsic motivation on individuals' knowledge sharing behavior. *International Journal of Human-Computer Studies*, *69*(6), 415–427. doi:10.1016/j.ijhcs.2011.02.004

Hung, S., Lai, H., & Chou, Y. (2015). Knowledge-sharing intention in professional virtual communities: A comparison between posters and lurkers. *Journal of the Association for Information Science and Technology*, *66*(12), 2494–2510. doi:10.1002/asi.23339

Hutchison, J., & Huberman, M. (1994). Knowledge dissemination and use in science and mathematics education: A literature review. *Journal of Science Education and Technology*, *3*(1), 27–47. doi:10.1007/BF01575814

Ibrahim, F., & Reid, V. (2009). What is the Value of Knowledge Management Practices? *Electronic Journal of Knowledge Management*, *7*(5), 567–574.

Ilvonen, I. (2013). Knowledge Security-A Conceptual Analysis. In *Tampereen Teknillinen Yliopisto.* Julkaisu-Tampere University of Technology.

Ilvonen, I., Alanne, A., Helander, N., & Vayrynen, H. (2016). *Knowledge Sharing and Knowledge Security in Finnish Companies.* Paper presented at the HICSS. 10.1109/HICSS.2016.499

Ilvonen, I., Jussila, J., Karkkainen, H., & Paivarinta, T. (2015). *Knowledge Security Risk Management in Contemporary Companies -- Toward a Proactive Approach.* Paper presented at the HICSS.

Ilvonen, I., Jussila, J., Karkkainen, H., & Paivarinta, T. (2016). *Knowledge Security Risk Management in Contemporary Companies -- Toward a Proactive Approach.* Paper presented at the HICSS.

Imran, M. K., Rehman, C. A., Aslam, U., & Bilal, A. R. (2016). What's organization knowledge management strategy for successful change implementation? Journal of Organizational Change Management, 29(7), 1097–1117. doi:10.1108/JOCM-07-2015-0130 doi:10.1108/JOCM-07-2015-0130

Inkinen, H. (2016). Review of empirical research on knowledge management practices and firm performance. *Journal of Knowledge Management*, *20*(2), 230–257. doi:10.1108/JKM-09-2015-0336

Intano, G. P., & Lofaro, R. J. (1989). Developmental Research and Initial Validation of the U.S. *Army Aviator Classification Process in Military Testing Association 1989 National Conference Proceedings.*

Intano, G. P., Howse, W. R., & Lofaro, R. J. (1991). The Selection of an Experimental Test Battery for Aviator Cognitive, Psychomotor Abilities and Personal Traits. U.S. Army Institute for the Behavioral and Social Sciences RN 91-21.

Intano, G. P., & Lofaro, R. J. (1990). Initial Validation of the Army Aviator Classification Process in Human Factors. *Society. Bulletin*, *33*(4).

Intezari, A., & Pauleen, D. J. (2017). The past-present-future conundrum: Extending time-bound knowledge. *International Journal of Knowledge Management*, *13*(1), 1–15. doi:10.4018/IJKM.2017010101

Irma Becerra-Fernandez, R. S. (2001). Organizational knowledge management: A contingency perspective. *Journal of Management Information Systems, 18*(1), 23–55. doi:10.1080/07421222.2001.11045676

Isenberg, D. (2012). Focus Entrepreneurship Policy on Scale-Up, Not Start-Up. *Harvard Business Review*, 21.

Işık, C., & Aydın, E. (2016). Bilgi paylaşımının yenilikçi iş davranışına etkisi: Ayder yaylası konaklama işletmeleri üzerine bir uygulama. *Girişimcilik ve İnovasyon Yönetimi Dergisi, 5*(2), 75–103.

Jacks, T., Wallace, S., & Nemati, H. (2012). Impact of Culture on Knowledge Management: A Meta-Analysis and Framework. *Journal of Global Information Technology Management, 15*(4), 8–42. doi:10.1080/1097198X.2012.10845622

Jacob, M. (2001). Managing the institutionalisation of Mode 2 knowledge production. *Science Studies, 14*(2), 83–100.

Jafari, M., Fathian, M., Jahani, A., & Peyman, A. (2008). Exploring the contextual dimensions of organization from knowledge management perspective. *Vine: The Journal of Information and Knowledge Management Systems, 38*(1), 53 – 71.

Jafari, M., Rezaeenour, J., Mazdeh, M. M., & Hooshmandi, A. (2011). Development and evaluation of a knowledge risk management model for project-based organizations: A multi-stage study. *Management Decision, 49*(3), 309–329. doi:10.1108/00251741111120725

Jagodinski, C. M. (2008). The university press in North America: A brief history. *Journal of Scholarly Publishing, 40*(1), 1–20. doi:10.3138/jsp.40.1.1

Jain, K. K., Sandhu, M. S., & Sidhu, G. K. (2007). Knowledge sharing among academic staff: A case study of business schools in klang valley, Malaysia. *Research Papers* (Vol. 2). Retrieved from http://www. ucsi.edu.my/cervie/ijasa/volume2/pdf/08A.pdf

Janssen, O. (2000). Job demands, perceptions of effort-reward fairness and innovative work behaviour. *Journal of Occupational and Organizational Psychology, 73*(3), 287–302. doi:10.1348/096317900167038

Jaw, C., Lo, J. Y., & Lin, Y. H. (2010). The determinants of new services development: Services characteristics, market orientation, and actualizing innovation effort. *Technovation, 30*(4), 265–277. doi:10.1016/j.technovation.2009.11.003

Jeng, D. J.-F., & Dunk, N. (2013). Knowledge management enablers and knowledge creation in ERP system success. *International Journal of Electronic Business Management, 11*(1), 49–59.

Jennex, M. (2014). Knowledge Management. In Wiley Encyclopedia of Management. Somerset, NJ: Wiley and Sons.

Jennex, M. E. (2001). Research relevance - you get what you reward. *Communications of the Association for Information Systems, 6*, 49-52.

Jennex, M. E. (2008). Impacts from Using Knowledge: A Longitudinal Study from a Nuclear Power Plant. International Journal of Knowledge Management, 4(1), 51–64. doi:10.4018/jkm.2008010105 doi:10.4018/jkm.2008010105

Jennex, M. E. (2009). *Assessing Knowledge Loss Risk.* Paper presented at the AMCIS 2015.

Jennex, M. E. (2009). Revisiting the Pyramid. In *42nd Hawaii International Conference on System Sciences, HICSS42.* IEEE Computer Society.

Jennex, M. E. (2013). Knowledge Management: The Risk of Forgetting. *iKNOW, the Magazine for Knowledge Workers, 3*(1), 4-7.

Jennex, M. E., & Durcikova, A. (2013). *Assessing knowledge loss risk.* Paper presented at the System Sciences (HICSS), 2013 46th Hawaii International Conference on. 10.1109/HICSS.2013.103

Jennex, M. E., & Durcikova, A. (2014). *Integrating IS Security with Knowledge Management: Are We Doing Enough to Thwart the Persistent Threat?* Paper presented at the HICSS. 10.1109/HICSS.2014.429

Jennex, M. E., & Olfman, L. (2006). A Model of Knowledge Management Success. International Journal of Knowledge Management, 2(3), 51–68. doi:10.4018/jkm.2006070104 doi:10.4018/jkm.2006070104

Jennex, M. E., Smolnic, S., & Croasdell, D. T. (n.d.). *Knowledge value, success and performance measurements mini-track*. Retrieved on 8/24/2017 from http://hicss.hawaii.edu/tracks-51/knowledge-innovation-and-entrepreneurial-systems

Jennex, M. E., Smolnik, S., & Croasdell, D. (2012) Where to Look for Knowledge Management Success. In *45th Hawaii International Conference on System Sciences, HICSS45*. IEEE Computer Society. doi:10.1109/HICSS.2012.64110.1109/HICSS.2012.641

Jennex, M. E., Smolnik, S., & Croasdell, D. (2013). The Search for Knowledge Management Success. Global Perspectives on Engineering Management, 2(2), 34–44.

Jennex, M. E., Smolnik, S., & Croasdell, D. T. (2009). Towards a Consensus Knowledge Management Success Definition. VINE. The Journal of Information and Knowledge Management Systems, 39(2), 174–188.

Jennex, M. (2005). Internet Support for Knowledge Management Systems. In M. Khosrow-Pour (Ed.), *Encyclopedia of Information Science and Technology* (pp. 1640–1644). Hershey, PA: Idea Group. doi:10.4018/978-1-59140-553-5.ch289

Jennex, M. (2006). Knowledge Management System Success Factors. In D. Schwartz (Ed.), *Encyclopedia of Knowledge Management* (pp. 436–441). Hershey, PA: Idea Group Reference. doi:10.4018/978-1-59140-573-3.ch057

Jennex, M. (2013). Knowledge Management Success in an Engineering Firm. *Management Review*, *2*(2), •••.

Jennex, M. E. (2005). What is Knowledge Management? *International Journal of Knowledge Management*, *1*(4), i–iv.

Jennex, M. E. (2006). Why We Can't Return to the Moon: The Need for Knowledge Management. *International Journal of Knowledge Management*, *2*(1), i–iv.

Jennex, M. E. (2008). The Impact of Culture and Context on Knowledge Management. In M. E. Jennex (Ed.), *Current Issues in Knowledge Management* (pp. 6–11). IGI. doi:10.4018/978-1-59904-916-8.ch002

Jennex, M. E. (2010). Implementing Social Media in Crisis Response Using Knowledge Management. *International Journal of Information Systems for Crisis Response and Management*, *2*(4), 20–32. doi:10.4018/jiscrm.2010100102

Jennex, M. E. (2012). Identifying the Components of a Knowledge Management Strategy. *18th Americas Conference on Information Systems, AMCIS18.*

Jennex, M. E. (2013). *Where is the Wisdom? In Americas Conference of AIS*. AMCIS.

Jennex, M. E. (2014). A Proposed Method for Assessing Knowledge Loss Risk with Departing Personnel. *VINE: The Journal of Information and Knowledge Management Systems*, *44*(2), 185–209. doi:10.1108/VINE-07-2012-0028

Jennex, M. E. (2017). Big Data, the Internet of Things and the Revised Knowledge Pyramid. *The Data Base for Advances in Information Systems*, *48*(4), 69–79. doi:10.1145/3158421.3158427

Jennex, M. E., & Bartczak, S. E. (2013). A Revised Knowledge Pyramid. *International Journal of Knowledge Management*, *9*(3), 19–30. doi:10.4018/ijkm.2013070102

Jennex, M. E., & Olfman, L. (2005). Assessing Knowledge Management Success. *International Journal of Knowledge Management*, *1*(2), 33–49. doi:10.4018/jkm.2005040104

Jennex, M. E., Smolnik, S., & Croasdell, D. (2014). Knowledge Management Success in Practice. In *47th Hawaii International Conference on System Sciences*. IEEE Computer Society.

Jennex, M. E., Smolnik, S., & Croasdell, D. T. (2009). Towards a Consensus Knowledge Management Success Definition. *VINE. The Journal of Information and Knowledge Management Systems*, *39*(2), 174–188.

Jennex, M. E., & Zyngier, S. (2007). Security as a contributor to knowledge management success. *Information Systems Frontiers*, *9*(5), 493–504. doi:10.100710796-007-9053-4

Jennex, M., Smolinik, S., & Croasdell, D. (2009). Towards a Consensus Knowledge Management Success Definition. *Vine*, *39*(2), 174–188. doi:10.1108/03055720910988878

Jensen, M., Durcikova, A., & Wright, R. (2017). Combating Phishing Attacks: A Knowledge Management Approach. *Proceedings of the 50th Hawaii International Conference on System Sciences*. 10.24251/HICSS.2017.520

Jensen, M., Johnson, B., Lorenz, E., & Lundvall, B. (2007). Forms of knowledge and modes of innovation. *Research Policy*, *36*(5), 680–693. doi:10.1016/j.respol.2007.01.006

Jeon, S., Kim, Y., & Koh, J. (2011). An integrative model for knowledge sharing in communities-of-practice. *Journal of Knowledge Management*, *15*(2), 251–269. doi:10.1108/13673271111119682

Jiao, H., Cui, Y., Zhu, Y., & Chen, J. (2014). Building entrepreneurs' innovativeness through knowledge management: The mediating effect of entrepreneurial alertness. *Technology Analysis and Strategic Management*, *26*(5), 501–516. doi:10.1080/09537325.2013.872774

Jimenez-Jimenez, D., & Sanz-Valle, R. (2011). Innovation, organizational learning, and performance. *Journal of Business Research*, *64*(4), 408–417. doi:10.1016/j.jbusres.2010.09.010

Johansson, T., & Moehler, R. C., & Vahidi. (2013, March). *Knowledge sharing strategies for project knowledge management in the automotive sector.* Paper presented at the 26th IPMA (International Project Management Association), World Congress, Crete, Greece. 10.1016/j.sbspro.2013.03.018

Johns, G. (2017). Incorporating context in organizational research. *Reflections: The SoL Journal.*

Johnson, M. W., Christensen, C. M., & Kagermann, H. (2008). Reinventing Your Business Model. *Harvard Business Review*, *86*, 50–59. PMID:18681297

Johnson, T., Kulesa, P., Cho, Y. I., & Shavitt, S. (2005). The relation between culture and response styles evidence from 19 countries. *Journal of Cross-Cultural Psychology*, *36*(2), 264–277. doi:10.1177/0022022104272905

Jones, G. R. (1983). Psychological orientation and the process of organizational socialization: An interactionist perspective. *Academy of Management Review*, *8*(3), 464–474.

Jorna, R. (2006). *Sustainable Innovation: The Organisational, Human and Knowledge Dimension.* Sheffield, UK: Greenleaf Publishing.

Joshi, A., & Roh, H. (2009). The role of context in work team diversity research: A meta analytic review. *Academy of Management Journal*, *52*(3), 599–627. doi:10.5465/AMJ.2009.41331491

Judd, R. L. (1972). *Use of Delphi Methods in Higher Education in Technological Forecasting and Social Change, 4.* Reed Elsevier.

Jugdev, K. (2007). Closing the circle: The knowledge management spiral of project management. *International Journal of Knowledge Management Studies*, *1*(3/4), 423–441. doi:10.1504/IJKMS.2007.012533

Jugdev, K. (2012). Learning from lessons learned: Project management research program. *American Journal of Economics and Business Administration*, *4*(1), 13–22. doi:10.3844/ajebasp.2012.13.22

Julibert, S. (2008). Employee attitudes to information sharing: A case study at the European Central Bank. *Records Management Journal*, *18*(3), 194–204. doi:10.1108/09565690810916447

Kajiyama, T., Jennex, M. E., Addo, T., & Paolini, C. (2013). *Cloud Computing Security: How Risks And Threats Are Affecting Cloud Adoption Decisions*. Paper presented at the 12th Security Conference.

Kamasak, R., & Bulutlar, F. (2010). The influence of knowledge sharing on innovation. *European Business Review*, *22*(3), 306–317. doi:10.1108/09555341011040994

Kandampully, J. (2002). Innovation as the core competency of a service organization: The role of technology, knowledge and networks. *European Journal of Innovation Management*, *5*(1), 18–26. doi:10.1108/14601060210415144

Kang, M., Kim, Y., & Bock, G. (2010). Identifying different antecedents for closed vs open knowledge transfer. *Journal of Information Science*, *36*(5), 585–602. doi:10.1177/0165551510375667

Kang, M., & Lee, M.-J. (2017). Absorptive capacity, knowledge sharing, and innovative behaviour of R&D employees. *Technology Analysis & Strategic Management Journal*, *29*(2), 219–232. doi:10.1080/09537325.2016.1211265

Kankanhalli, A., Tan, B. C. Y., & Wei, K. K. (2005). Contributing knowledge to electronic knowledge repositories: An empirical investigation. *Management Information Systems Quarterly*, *29*(1), 113–143. doi:10.2307/25148670

Kaplan, R. S., & Norton, D. P. (2006). How to implement a new strategy without disrupting your organization. Harvard Business Review, 84(3), 100–109. PubMed PMID:16515159

Kaps, I. V. (2011). Barriers in intercultural knowledge sharing. learning's from an international plant engineering project. *Journal of Knowledge Management*, 3.

Karlsen, J. T., & Gottschalk, P. (2004). Factors affecting knowledge transfer in IT projects. *Engineering Management Journal*, *16*(1), 3–10. doi:10.1080/10429247.2004.11415233

Keegan, A., & Turner, J. R. (2001). Quantity versus quality in project-based learning practices. *Management Learning*, *32*(1), 77–98. doi:10.1177/1350507601321006

Keen, P. G. W., & Tan, M. (2007). Knowledge Fusion: A framework for extending the rigor and relevance of knowledge management. *International Journal of Knowledge Management*, *3*(4), 1–17. doi:10.4018/ijkm.2007100101

Kenny, D. A., & Judd, C. M. (1984). Estimating the nonlinear and interactive effects of latent variables. *Psychological Bulletin*, *96*(1), 201–210. doi:10.1037/0033-2909.96.1.201

Kenny, D. A., Kashy, D. A., & Cook, W. L. (2006). *Dyadic data analysis*. New York: Guilford Press.

Kessel, M., Hannemann-Weber, H., & Kratzer, J. (2012). Innovative work behavior in healthcare: The benefit of operational guidelines in the treatment of rare diseases. *Health Policy (Amsterdam)*, *105*(2-3), 146–153. doi:10.1016/j.healthpol.2012.02.010

Keupp, M., Palmie, M., & Gassmann, O. (2012). The Strategic Management of Innovation: A Systematic Review and Paths for Future Research. *International Journal of Management Reviews*, *14*(4), 367–390. doi:10.1111/j.1468-2370.2011.00321.x

Keyes, J. (2008). *Identifying the barriers to knowledge sharing in knowledge intensive organizations*. Retrieved from http://www.newarttech. com/KnowledgeSharing.pdf

Khalifa, M., Yan, A., & Shen, K. N. (2008). Knowledge management system success: A contingency perspective. *Journal of Knowledge Management, 1*(12), 119–132. doi:10.1108/13673270810852430

Khalil, O. E. M., & Shea, K. T. (2012). Knowledge sharing barriers and effectiveness at a higher education institution. *International Journal of Knowledge Management, 8*(2), 43–64. doi:10.4018/jkm.2012040103

Khalil, T., & Atieh, K., Mohammed, A. U., & Bagdadlian, F. (2014). Examining the social and technical factors influencing school teachers knowledge sharing intentions in a teachers online professional community. *Electronic Journal of Knowledge Management, 12*(3), 157–165.

Khasawneh, R., & Abu-Shanab, E. (2013). Using KMS as a tool to improve decision making process. In S. Saed & I. Alsmadi (Eds.), Knowledge-Based Processes in Software Development (pp. 1-11). Hershey, PA: Information Science Reference, IGI Global. doi:10.4018/978-1-4666-4229-4.ch001

Khurana, R. (2007). *From higher aims to hired hands: The social transformation of American business schools and the unfulfilled promise of management as a profession.* Princeton, NJ: Princeton University Press.

Kianto, A., Ritala, P., Spender, J.-C., & Vanhala, M. (2014). The interaction of intellectual capital assets and knowledge management practices in organizational value creation. *Journal of Intellectual Capital, 15*(3), 362–375. doi:10.1108/JIC-05-2014-0059

Killingsworth, B., Xue, Y., & Liu, Y. (2016). Factors influencing knowledge sharing among global virtual teams. *Team Performance Management, 22*(5), 284–300. doi:10.1108/TPM-10-2015-0042

Kim, T. W., & Joh, H. W. (2005). *Key barriers and their strategic responses to activate knowledge sharing in construction organizations.* 1st Annual ICCEM (International Conference on Construction Engineering and Management), Seoul, South Korea.

Kim, L. (1997). *Imitation to innovation: The dynamics of Korea's technological learning (management of innovation and change).* Harvard Business School Press.

Kimmerle, J., Cress, U., & Held, C. (2010). The interplay between individual and collective knowledge: Technologies for organizational learning and knowledge building. Knowledge Management Research and Practice, 8(1), 33–44. doi:10.1057/kmrp.2009.36 doi:10.1057/kmrp.2009.36

Kim-Soon, N., & Ahmad, N. (2012). Impact of Motivation to Learn on Perceived Training Transfer: Empirical Evidence from a Bank. In *Proceedings of the International Conference of Technology Management, Business and Entrepreneurship* (pp. 268-383). Melaka: ICTMBE.

Kim, T. Y., Cable, D. M., & Kim, S. P. (2005). Socialization tactics, employee proactivity, and person-organization fit. *The Journal of Applied Psychology, 90*(2), 232–241. doi:10.1037/0021-9010.90.2.232 PMID:15769234

Kim, T., Lee, J., Chun, J., & Benbasat, I. (2014). Understanding the effect of knowledge management strategies on knowledge management performance: A contingency perspective. *Information & Management, 51*(4), 398–410. doi:10.1016/j.im.2014.03.001

Kindstrom, D., Kowalkowski, C., & Sandberg, E. (2013). Enabling service innovation: A dynamic capabilities approach. *Journal of Business Research, 66*(8), 1063–1073. doi:10.1016/j.jbusres.2012.03.003

King, D. A. (2004). The scientific impact of nations: What different countries get for their research spending. *Nature, 430*(6997), 311–316. doi:10.1038/430311a PMID:15254529

Kirkman, B. L., Lowe, K. B., & Gibson, C. B. (2006). A quarter century of culture's consequences: A review of empirical research incorporating Hofstede's cultural values framework. *Journal of International Business Studies*, *37*(3), 285–320. doi:10.1057/palgrave.jibs.8400202

Kirzner, I. M. (1997). Entrepreneurial discovery and the competitive market process: An Austrian approach. *Journal of Economic Literature*, *35*(1), 60–85.

Kleysen, R. F., & Street, C. T. (2001). Toward a multi-dimensional measure of individual innovative behavior. *Journal of Intellectual Capital, 2*(3), 284-296.

Knights, D. (2008). Myopic rhetorics: Reflecting epistemologically and ethically on the demand for relevance in organizational and management research. *Academy of Management Learning & Education*, *7*(4), 537–552. doi:10.5465/AMLE.2008.35882194

Knights, D., & Scarbrough, H. (2010). In search of relevance: Perspectives on the contribution of academic-practitioner networks. *Organization Studies*, *31*(9), 1287–1309. doi:10.1177/0170840610374398

Knouse, S. B., & Chretien, D. (1996). Workforce diversity and TQM. In S. B. Knouse (Ed.), *Human resources management perspectives on TQM: Concepts and practices* (pp. 261–274). Milwaukee, WI: American Society for Quality Control Press.

Koch, T., & Harrington, A. (1998). Reconceptualizing rigour: The case for reexivity. Journal of Advanced Nursing, 28(4), 882–890. doi:10.1046/j.1365-2648.1998.00725.x PubMed doi:10.1046/j.1365-2648.1998.00725.x PMID:9829678

Kock, N. (2015). Common method bias in PLS-SEM: A full collinearity assessment approach. *International Journal of e-Collaboration*, *11*(4), 1–10. doi:10.4018/ijec.2015100101

Kock, N., & Lynn, G. S. (2012). Lateral collinearity and misleading results in variance-based SEM: An illustration and recommendations. *Journal of the Association for Information Systems*, *13*(7), 546–580. doi:10.17705/1jais.00302

Ko, D. G., Kirsch, L. J., & King, W. R. (2005). Antecedents of knowledge transfer from consultants to clients in enterprise system implementations. *Management Information Systems Quarterly*, *29*(1), 59–85. doi:10.2307/25148668

Kogut, B., & Zander, U. (1992). Knowledge of the Firm, Combinative Capabilities, and the Replication of Technology. *Organization Science*, *3*(3), 383–397. doi:10.1287/orsc.3.3.383

Kong, E., Doren, C., & Ramen, R. (2011). The role of human resources practices in developing knowledge and learning capabilities for innovation: a study of service providers in India. *10th International Research Conference on Quality, Innovation and Knowledge Management.*

Kordzadeh, N., Liu, C. Z., Au, Y. A., & Clark, J. G. (2014). A multilevel investigation of participation within virtual health communities. *Communications of the Association for Information Systems*, *34*, 494–512.

Koronios, A., Gao, J., & Selle, S. (2014). Big Data Project Success – A Meta Analysis. *PACIS 2014, Proceedings*. Available at: http://aisel.aisnet.org/pacis2014/376

Kreiser, P. M., Patel, P. C., & Fiet, J. O. (2013). The Influence of Changes in Social Capital on Firm-Founding Activities. *Entrepreneurship Theory and Practice*, *37*(3), 539–568. doi:10.1111/etap.12039

Krenz, P., Basmer, P., Buxbaum-Conradi, S., Redlich, T., & Wulfsberg, J.-P. (2014). Knowledge Management in Value Creation Networks: Establishing a New Business Model through the Role of a Knowledge-Intermediary. *Procedia CIRP*, *16*, 38–43. doi:10.1016/j.procir.2014.01.006

Krueger, N. F. Jr, Reilly, M. D., & Carsrud, A. L. (2000). Competing models of entrepreneurial intentions. *Journal of Business Venturing*, *15*(5/6), 411–432. doi:10.1016/S0883-9026(98)00033-0

Kruglanski, A. W. (1975). The human subject in the psychology experiment: Fact and artifact. In L. Berkowitz (Ed.), Advances in Experimental Social Psychology (Vol. 8, pp. 101–147). New York: Academic Press. doi:10.1016/S0065-2601(08)60249-X

Kulkarni, U. R., Ravindran, S., & Freeze, R. (2007). A Knowledge Management Success Model: Theoretical Development and Empirical Validation. Journal of Management Information Systems, 23(3), 309–347. doi:10.2753/MIS0742-1222230311 doi:10.2753/MIS0742-1222230311

Kuusisto, J., & Meyer, M. (2003). *Insights into services and innovation in the knowledge-intensive economy. Technology Review 134/2003*. Helsinki, Finland: Tekes National Technology Agency.

Kvale, S., & Brinkmann, S. (2008). InterViews – Learning the Craft of Qualitative Research Interviewing (2nd ed.). SAGE Publications.

Lahti, R. K., & Beyerlein, M. M. (2000). Knowledge transfer and management consulting: A look at "the firm". *Business Horizons*, *43*(1), 65–74. doi:10.1016/S0007-6813(00)87389-9

Lai, H., & Chen, T. T. (2014). Knowledge sharing in interest online communities: A comparison of posters and lurkers. *Computers in Human Behavior*, *35*, 295–306. doi:10.1016/j.chb.2014.02.004

Lakshman, C. (2009). Organizational knowledge leadership; An empirical examination of KM by top executive leaders. *Leadership and Organization Development Journal*, *30*(4), 338–364. doi:10.1108/01437730910961676

Lambe, P. (2013). Four Types of Knowledge risk. *iKNOW, the Magazine for Knowledge Workers*, *3*(1), 21-24.

Lambert, D., & Harrington, T. (1990). Measuring nonresponse bias in customer service mail surveys. *Journal of Business Logistics*, *11*(2), 5–25.

Lambooy, J. (2005). Innovation and knowledge: Theory and regional policy. *European Planning Studies*, *13*(8), 1137–1152. doi:10.1080/09654310500336444

Lam, J. (2003). *Enterprise Risk Management: from Incentives to Controls*. Hoboken, NJ: John Wiley & Sons.

Landaeta, R. E. (2008). Evaluating benefits and challenges of knowledge transfer across projects. *Engineering Management Journal*, *20*(1), 29–38. doi:10.1080/10429247.2008.11431753

Laperche, B., & Uzunidis, D. (2008). *The Genesis of Innovation*. Cheltenham, UK: Edward Elgar Publishing. doi:10.4337/9781848442856

Larson, E. W., & Gray, C. F. (2014). *Project management: The managerial process* (6th ed.). New York: McGraw Hill Education.

Laursen, K., & Salter, M. (2006). Open for Innovation: The Role of Openness in Explaining Innovation Performance among U.K. Manufacturing Firms. *Strategic Management Journal*, *27*(2), 131–150. doi:10.1002mj.507

LaValle, S., Lesser, E., Shockley, R., Hopkins, M. S., & Kruschwitz, N. (2013). Big data, analytics and the path from insights to value. *MIT Sloan Management Review*, *52*(2), 21.

Leach, J. C., & Melicher, R. W. (2014). *Entrepreneurial finance*. Cengage Learning.

Lee, H., & Choi, B. (2003). Knowledge management enablers, processes, and organizational performance:An integrative view and empirical examination. Journal of Management Information Systems, 20(1), 179–228. doi:10.1080/07421222.2003.11045756 doi:10.1080/07421222.2003.11045756

Leech, N. L., Barrett, K. C., & Morgan, G. A. (2011). *IBM SPSS for Intermediate Statistics* (4th ed.). New York: Routledge.

Lee, S., & Lam, W. (2007). Application Integration: Pilot Project to Implement a Financial Portfolio System in a Korean Bank. In W. Lam & V. Shankararaman (Eds.), *Enterprise Architecture and Integration: Methods, Implementation and Technologies* (pp. 273–283). Hershey, PA: IGI Global. doi:10.4018/978-1-59140-887-1.ch016

Lee, S., Shin, B., & Lee, H. (2011). Boundary spanning role of the IS development team in consultant-partnered projects: Knowledge management perspective. *International Journal of Knowledge Management*, *7*(2), 22–48. doi:10.4018/jkm.2011040102

Lehaney, B., Clarke, S., Coakes, E., & Jack, G. (2004). *Beyond Knowledge Management.* Hershey, PA: Idea Group Publishing. doi:10.4018/978-1-59140-180-3

Leiponen, A., & Helfat, C. (2010). Innovation Objectives, Knowledge Sources, and the Benefits of Breadth. *Strategic Management Journal*, *31*(2), 224–236. doi:10.1002mj.807

Lengnick-Hall, M., & Lengnick-Hall, C. (2003). *Human Resource Management in the Knowledge Economy: New Challenges, New Roles, New Capabilities*. San Francisco: Berrett-Koehler.

Leonard, D., & Sensiper, S. (1998). The role of tacit knowledge in group innovation. *California Management Review*, *40*(3), 112–132. doi:10.2307/41165946

Leonard, D., & Swap, W. (2004). Deep Smarts. *Harvard Business Review*, *82*(9), 88–97. PMID:15449858

Leonard, D., & Swap, W. (2005). *Deep Smarts: How To Cultivate And Transfer Enduring Business Wisdom.* Boston, MA: Harvard Business School Publishing.

Lesser, E. L., & Storck, J. (2001). Communities of practice and organisational performance. *IBM Systems Journal*, *40*(4), 831–841. doi:10.1147j.404.0831

Levine, R. (2004). Risk management systems: Understanding the need. *Information Systems Management*, *21*(2), 31–37. doi:10.1201/1078/44118.21.2.20040301/80419.5

Lewin, K. (1951). *Problems of research in social psychology. In Field theory in social science: Selected theoretical papers* (pp. 155–169). New York: Harper & Row.

Liao, S., & Wu, C. (2010). System perspective of knowledge management, organizational learning, and organizational innovation. *Expert Systems with Applications*, *37*(2), 1096–1103. doi:10.1016/j.eswa.2009.06.109

Lichtenstein, S. (1996). Factors in the selection of a risk assessment method. *Information Management & Computer Security*, *4*(4), 20–25. doi:10.1108/09685229610130503

Liebowitz, J., & Megbolugbe, I. (2003). A set of frameworks to aid the project manager in conceptualizing and implementing knowledge management initiatives. *International Journal of Project Management*, *21*(3), 189–198. doi:10.1016/S0263-7863(02)00093-5

Lierni, P. C., & Ribière, V. M. (2008). The relationship between improving the management of projects and the use of KM. *The Journal of Information and Knowledge Management Systems*, *38*(1), 133–146.

Lilleoere, A., & Hansen, E. (2011). Knowledge-sharing enablers and barriers in pharmaceutical research and development. *Journal of Knowledge Management*, *15*(1), 53–70. doi:10.1108/13673271111108693

Lin, C. (2007). To share or not to share: Modeling knowledge sharing using exchange ideology as a moderator. *Personnel Review*, *36*(3), 457–475. doi:10.1108/00483480710731374

Lindbergh, L. B. (2009). *The relationship between project manager perceived capability, organizational culture, and project outcome*. Capella University.

Lindner, F., & Wald, A. (2011). Success factors of knowledge management in temporary organizations. *International Journal of Project Management*, *29*(7), 877–888. doi:10.1016/j.ijproman.2010.09.003

Lindsey, K. (2002). Measuring knowledge management effectiveness: A task-contingent organizational capabilities perspective. *AMCIS 2002 Proceedings*, 285.

Lines, R. (2007). Using Power to Install Strategy: The Relationships between Expert Power, Position Power, Influence Tactics and Implementation Success. Journal of Change Management, 7(2), 143–170. doi:10.1080/14697010701531657 doi:10.1080/14697010701531657

Lin, H. (2007). Knowledge sharing and firm innovation capability: An empirical study. *International Journal of Manpower*, *28*(3/4), 315–332. doi:10.1108/01437720710755272

Lin, H. F. (2007). Effects of Extrinsic and Intrinsic Motivation on Employee Knowledge Sharing Intentions. *Journal of Information Science*, *33*(2), 135–149. doi:10.1177/0165551506068174

Lin, H. F., & Lee, G. G. (2004). Perceptions of senior managers toward knowledge-sharing behaviour. *Management Decision*, *42*(1), 108–125. doi:10.1108/00251740410510181

Lin, H., Lee, H., & Wang, D. (2009). Factors influencing knowledge sharing based on a fuzzy AHP approach. *Journal of Information Science*, *35*(1), 25–44. doi:10.1177/0165551508091310

Lin, T. C., & Huang, C. C. (2010). Withholding effort in knowledge contribution: The role of social exchange and social cognitive on project teams. *Journal of International Management*, *47*, 188–196.

Littrell, R. (2012). Cultural Value Dimension Theories: Hofstede – A Work in Progress. *AiB Insights*, *4*(12), 3–6.

Litvaj, I., & Stancekova, D. (2015). Decision - Making, and Their Relation to The Knowledge Management, Use of Knowledge Management in Decision – Making. *Procedia Economics and Finance*, *23*, 467–472. doi:10.1016/S2212-5671(15)00547-X

Liu, S. C., Olfman, L., & Ryan, T. (2005). Knowledge Management System Success: Empirical Assessment of a Theoretical Model. *International Journal of Knowledge Management*, *1*(2), 68–87. doi:10.4018/jkm.2005040106

Li, Y.-H., Huang, J.-W., & Tsai, M.-T. (2009). Entrepreneurial orientation and firm performance: The role of knowledge creation process. *Industrial Marketing Management*, *38*(4), 440–449. doi:10.1016/j.indmarman.2008.02.004

Liyanage, C., Elhag, T., Ballal, T., & Li, Q. (2009). Knowledge communication and translation–a knowledge transfer model. *Journal of Knowledge Management*, *13*(3), 118–131. doi:10.1108/13673270910962914

Llopis, O., & Foss, N. J. (2016). Understanding the climate–knowledge sharing relation: The moderating roles of intrinsic motivation and job autonomy. *European Management Journal*, *34*(2), 135–144. doi:10.1016/j.emj.2015.11.009

Lloria, M. B. (2008). A review of the main approaches to knowledge management. Knowledge Management Research and Practice, 6(1), 77–89. doi:10.1057/palgrave.kmrp.8500164 doi:10.1057/palgrave.kmrp.8500164

Lofaro, R. J. (1992a). A Small Group Delphi Paradigm. *The Human Factors Society Bulletin, 35*(2), 1-4.

Lofaro, R. J. (1998). Identifying and Developing Managerial and Technical Competencies: Differing Methods and Differing Results. Proceedings of 42nd annual Human Factors and Ergonomics Society.

Lofaro, R. J. (1999). Core Competencies and Individual Differences: Chasm or Linkage? Human Factors and Ergonomics Society/HF&ES Individual Differences in Performance News, 12(2).

Lofaro, R. J. (2013). Knowledge Engineering: A Methodology and Examples. In Encyclopedia of Information Science and Technology, 3rd Edition. IGI-Global.

Lofaro, R. J. (2015). Cognitive Engineering: What's Old is New Again. *Proceedings of the 18th International Symposium on Aviation Psychology (ISAP).*

Lofaro, R. J., & Maliko-Abraham, H. (2002). Techniques for Developing High Predictive Validity Selection and Screening Test Batteries. *Proceedings of the 2002 World Aviation Congress.*

Lofaro, R. J., & Smith, K. M. (1999). Operational Decision-Making (ODM) and Risk Management (RM): Rising Risk, the Critical Mission Factors and Training. *Proceedings of Tenth International Symposium on Aviation Psychology.*

Lofaro, R. J., & Smith, K. M. (2001a). A Paradigm for Developing Operational Decision-making (ODM). *Proceedings of 2001 SAE World Aviation Congress (WAC) conference.*

Lofaro, R. J., & Smith, K. M. (2001b). Operational Decision-Making: Integrating New Concepts into the Paradigm. *Proceedings of Eleventh International Symposium on Aviation Psychology.*

Lofaro, R. J., & Smith, K. M. (2003). The Finalized Paradigm for Operational Decision-Making (ODM) Paradigm: Components and Placement. *Proceedings of the 12th International Symposium on Aviation Psychology.*

Lofaro, R. J., & Smith, K. M. (2009). Professionalism in Airline Operations…and Accident Investigations? *Proceedings of the 15th International Symposium on Aviation Psychology.*

Lofaro, R. J., & Smith, K. M. (2011). The Aviation Operational Environment: Integrating a Decision-Making Paradigm, Flight Simulator Training and an Automated Cockpit Display for Aviation Safety. In Technology Engineering and Management in Aviation: Advancements and Discoveries. IGI Global.

Lofaro, R. J. (1992b). *Workshop on Integrated Crew Resource Management (CRM). DOT/FAA/RD-95/5*. Springfield, VA: National Technical Information Service.

Lofaro, R. J., Gibb, G. M., & Garland, D. (1994). *A Protocol for Selecting Airline Passenger Baggage Screeners. DOT/FAA/CT-94/110*. Springfield, VA: National Technical Information Service.

Lofaro, R. J., & Intano, G. P. (1989). Exploratory Research and Development: Army Aviator Candidate Classification By Specific Helicopter *Proceedings of 5th International Symposium of Aviation Psychologists.*

Lofaro, R. J., Intano, G. P., & Howse, W. R. (1990). Final Validation of the Army Aviator Multitrack Algorithm. *Proceedings of the 1990 Annual Military Testing Association Conference.*

Lönnqvist, A. (2017). Embedded knowledge management: Towards improved managerial relevance. *Knowledge Management Research and Practice*, *15*(2), 184–191. doi:10.105741275-017-0053-y

Lopez-Cabrales, A., Perez-Luno, A., & Cabrera, R. (2009). Knowledge as a Mediator Between HRM Practices and Innovative Activity. *Human Resource Management*, *48*(4), 485–503. doi:10.1002/hrm.20295

Luburic, R. (2016). Knowledge and Learning in Terms of Operational Risk Management in the Financial and Banking Systems. *International Journal of Qualitative Research*, *10*(3), 559–568.

Lu, L., Leung, K., & Koch, P. T. (2006). Managerial knowledge sharing: The role of individual, interpersonal, and organizational factors. *Management and Organization Review*, *2*(1), 15–41. doi:10.1111/j.1740-8784.2006.00029.x

Lyons, R., Chatman, J., & Joyce, C. (2007). Innovation in Services: Corporate Culture and Investment Banking. *California Management Review*, *50*(1), 174–191. doi:10.2307/41166422

Maccurtain, S., Flood, P. C., Ramamoorthy, N., West, M. A., & Dawson, J. F. (2008). *Top team trust, knowledge sharing and innovation.* Link Working Paper Series WP 05-08.

Madden, S. (2012). From Databases to Big Data. *IEEE Internet Computing*, *16*(3), 4–6. doi:10.1109/MIC.2012.50

Magnier-Watanabe, R., & Senoo, D. (2010). Shaping knowledge management: Organization and national culture. *Journal of Knowledge Management*, *14*(2), 214–227. doi:10.1108/13673271011032364

Magretta, J. (2002). Why business models matter. *Harvard Business Review*, *6*, 86–92. PMID:12024761

Mahr, D., & Lievens, A. (2012). Virtual lead user communities: Drivers of knowledge creation for innovation. *Research Policy*, *41*(1), 167–177. doi:10.1016/j.respol.2011.08.006

Maier, R. (2002). *Knowledge management systems: Information and communication technologies for knowledge management.* Berlin: Springer. doi:10.1007/978-3-662-04380-6

Majchrzak, A., Beath, C., Lim, R., & Chin. (2005). Managing client dialogues during information systems design to facilitate client learning. *Management Information Systems Quarterly*, *29*(4), 653–672. doi:10.2307/25148704

Makri, M., & Scandura, T. (2010). Exploring the effects of creative CEO leadership on innovation in high technology firms. *The Leadership Quarterly*, *21*(1), 75–88. doi:10.1016/j.leaqua.2009.10.006

Malhotra, N., & Morris, T. (2009). Heterogeneity in Professional Service Firms. *Journal of Management Studies*, *46*(6), 895–922. doi:10.1111/j.1467-6486.2009.00826.x

Malhotra, Y. (2004). *Why knowledge management systems fail: enablers and constraints of knowledge management in human enterprises. In Handbook on Knowledge Management 1* (pp. 577–599). Springer.

Maliko-Abraham, H., & Lofaro, R. J. (2001a). A Screening Test for Military Air Traffic Controllers. Proceedings of the 2001 International Military Testing Association (IMTA) Conference.

Maliko-Abraham, H., & Lofaro, R. J. (2001b). Air Traffic Delays: New ATC Equipment Coupled With New ATC Candidate Selection Methods May Be One Answer. *Proceedings of the 2001 National Air Traffic Controllers Association (NATCA).*

Maliko-Abraham, H., & Lofaro, R. J. (2003). Usability Testing: Lessons Learned and Methodology. *Proceedings of the 2003 International Military Testing Association (IMTA) Conference.*

Manhart, M., & Thalmann, S. (2013). *An Integrated Risk Management Framework: Measuring the Success of Organizational Knowledge Protection.* Academic Press.

Manhart, M., & Thalmann, S. (2015). Protecting organizational knowledge: A structured literature review. *Journal of Knowledge Management*, *19*(2), 190–211. doi:10.1108/JKM-05-2014-0198

Manyika, J., Chui, M., Bughin, J., Dobbs, R., Bisson, P., & Marrs, A. (2013). *Disruptive technologies: Advances that will transform life, business, and the global economy.* Retrieved from http://www.mckinsey.com/business-functions/digital-mckinsey/our-insights/disruptive-technologies

Marabelli, M., & Newell, S. (2012). Knowledge risks in organizational networks: The practice perspective. *The Journal of Strategic Information Systems*, *21*(1), 18–30. doi:10.1016/j.jsis.2011.11.002

Mariscal, G., Marban, O., & Fernandez, C. (2010). A survey of data mining and knowledge discovery process models and methodologies. *The Knowledge Engineering Review*, *25*(02), 137–166. doi:10.1017/S0269888910000032

Markham, S. K. (1998). A longitudinal examination of how champions influence others to support their projects. Journal of Product Innovation Management, 15(6), 490–504. doi:10.1016/S0737-6782(98)00031-9 doi:10.1016/S0737-6782(98)00031-9

Markham, S. K., Ward, S. J., Aiman-Smith, L., & Kingon, A. I. (2010). The Valley of Death as Context for Role Theory in Product Innovation. *Journal of Product Innovation Management*, *27*(3), 402–417. doi:10.1111/j.1540-5885.2010.00724.x

Markus, B., (2005). *Learning Pyramids.* WSVA 1 – Virtual Academy and the Surveying/ Geoinformatics Community –Virtual Academy I.

Marouf, L. N., & Khalil, O. M. (2015). The influence of individual characteristics on knowledge sharing practices, enablers, and barriers in a project management context. *International Journal of Knowledge Management*, *11*(1), 1–27. doi:10.4018/IJKM.2015010101

Marques, C. S., & Ferreira, J. (2009). SME innovative capacity, competitive advantage and performance in a 'traditional' industrial region of Portugal. *Journal of Technology Management & Innovation*, *4*(4). doi:10.4067/S0718-27242009000400005

Marshall, C., Prusak, L., & Shpilberg, D. (1996). Financial risk and the need for superior knowledge management. *California Management Review*, *38*(3), 77–101. doi:10.2307/41165844

Martensson, M. (2000). A critical review of knowledge management as a management tool. *Journal of Knowledge Management*, *4*(3), 204–216. doi:10.1108/13673270010350002

Masa'deh, R., & Shannak, R. O. (2012). Intermediary Effects of Knowledge Management Strategy and Learning Orientation on Strategic Alignment and Firm Performance. *Research Journal of International Studies*, 112-128.

Massingham, P. (2010). Knowledge risk management: A framework. *Journal of Knowledge Management*, *14*(3), 464–485. doi:10.1108/13673271011050166

Matayong, S., & Mahmood, A. K. (2013). The review of approaches to knowledge management system studies. *Journal of Knowledge Management*, *17*(3), 472–490. doi:10.1108/JKM-10-2012-0316

Mathieu, J. E., Tannenbaum, S. I., & Salas, E. (1992). Influences of individual and situational characteristics on measures of training effectiveness. *Academy of Management Journal*, *35*(4), 828–847. doi:10.2307/256317

Matzler, K., Renzl, B., Mooradian, T., von Krogh, G., & Mueller, J. (2011). Personality traits, affective commitment, documentation of knowledge, and knowledge sharing. *International Journal of Human Resource Management*, *22*(2), 296–310. doi:10.1080/09585192.2011.540156

Matzler, K., Renzl, B., Müller, J., Herting, S., & Mooradian, T. A. (2008). Personality traits and knowledge sharing. *Journal of Economic Psychology*, *29*(3), 301–313. doi:10.1016/j.joep.2007.06.004

Mazzucato, M. (2011). *The entrepreneurial state*. London: Demos.

McAdam, R. (2000). Knowledge management as a catalyst for innovation within organizations: A qualitative study. *Knowledge and Process Management*, *7*(4), 233–241. doi:10.1002/1099-1441(200010/12)7:4<233::AID-KPM943.0.CO;2-F

McAfee, A., & Brynjolfsson, E. (2012). Big Data: The Management Revolution. *Harvard Business Review*, *90*(10), 60–68. PMID:23074865

McCabe, J. (2016). *FBI Warns of Dramatic Increase in Business E-Mail Scams*. Retrieved 4/4/2016, 2017, from https://www.fbi.gov/contact-us/field-offices/phoenix/news/press-releases/fbi-warns-of-dramatic-increase-in-business-e-mail-scams

McCafferty, D. (Producer). (2010). *Why IT projects fail*. Retrieved from http://www.cioinsight.com/c/a/IT-Management/Why-IT-Projects-Fail-762340/

McCall, H., Arnold, V., & Sutton, S. (2008). Use of knowledge management systems and the impact on the acquisition of explicit knowledge. *Journal of Information Systems*, *22*(2), 77–101. doi:10.2308/jis.2008.22.2.77

McCumber, J. (2004). *Assessing and managing security risk in IT systems: A structured methodology*. CRC Press. doi:10.1201/9780203490426

McDermott, R., & O'Dell, C. (2001). Overcoming cultural barriers to sharing knowledge. *Journal of Knowledge Management*, *5*(1), 76–85. doi:10.1108/13673270110384428

McKay, D., & Edick, K. (2017, May 23). Global supply chains have changed – Here's how to adapt [Web log post]. Retrieved from https://www.ashford.edu/online-degrees/business/global-supply-chains-have-changed-heres-how-to-adapt

McLaughlin, S., Paton, R. A., & Macbeth, D. K. (2008). Barrier impact on organizational learning within complex organizations. *Journal of Knowledge Management*, *12*(2), 107–123. doi:10.1108/13673270810859550

McLuhan, M. (1962). *The Gutenberg Galaxy: The making of typographic man*. Toronto, Canada: University of Toronto Press.

McNally, J. S. (2013, June). The 2013 COSO Framework & SOX Compliance. *Strategic Finance*. Retrieved from http://www.coso.org/documents/COSO%20McNallyTransition%20Article-Final%20COSO%20Version%20Proof_5-31-13.pdf

McWhorter, D. (2013). *Mandiant Exposes APT1 – One of China's Cyber Espionage Units & Releases 3,000 Indicators*. Retrieved 9/5/2017, 2017, from https://www.fireeye.com/blog/threat-research/2013/02/mandiant-exposes-apt1-chinas-cyber-espionage-units.html

Mehrabani, E. S., & Mohamad, A. N. (2011). Singapore the role of 'training activities' and 'knowledge sharing' in the relationship between leadership development and organizational effectiveness. *International Conference on Sociality and Economics Development IPEDR*, *10*, 172-177.

Meister, D. (1985). *Behavioral Analysis and Measuremen*. New York, NY: Wiley Press.

Merriam, S. (1998). *Qualitative research and case study applications in education*. San Francisco: Jossey-Bass Publishers.

Merton, R. K., & Sztompka, P. (1996). *On social structure and science*. Chicago, IL: University of Chicago Press.

Meulbroek, L. (2002). The promise and challenge of integrated risk management. *Risk Management & Insurance Review*, *5*(1), 55–66. doi:10.1111/1098-1616.00006

Meyer, M. (2010). The rise of the knowledge broker. *Science Communication*, *32*(1), 118–127. doi:10.1177/1075547009359797

Miles, M., & Huberman, A. (1994). *Qualitative data analysis: an expanded sourcebook* (2nd ed.). Thousand Oaks, CA: Sage.

Miles, R. (2007). Innovation and Leadership Values. *California Management Review*, *50*(1), 192–201. doi:10.2307/41166423

Miller, D. (1996). A Preliminary Typology of Organizational Learning: Synthesizing the Literature. *Journal of Management*, *22*(3), 485–505. doi:10.1177/014920639602200305

Milliken, F. J., & Martins, L. L. (1996). Searching for common threads: Understanding the multiple effects of diversity in organizational groups. *Academy of Management Review*, *21*(2), 402–433.

Minbaeva, D. B., Mäkelä, K., & Rabbiosi, L. (2012). Linking HRM and Knowledge Transfer via Individual-Level Mechanisms. *Human Resource Management*, *51*(3), 387–405. doi:10.1002/hrm.21478

Mohsen, Z. A., Ali, M., & Jalal, A. (2011). The Significance of Knowledge Management Systems at Financial Decision Making Process. *International Journal of Business and Management*, *6*(8), 130–142. doi:10.5539/ijbm.v6n8p130

Monavvarian, A., & Kasaei, M. (2007). A KM model for public administration: The case of Labour Ministry. *The Journal of Information and Knowledge Management Systems*, *37*(3), 348–367.

Moreland, R. L., & Myaskovsky, L. (2000). Exploring the performance benefits of group training: Transactive memory or improved communication? *Organizational Behavior and Human Decision Processes*, *82*(1), 117–133. doi:10.1006/obhd.2000.2891

Moss, Q.Z., Alho, J., & Alexander, K. (2007). Performance measurement action research. *Journal of Facilities Management, 5*(4), 290-300.

Mowday, R. T., Steers, R. M., & Porter, L. (1979). The measurement of organizational commitment. *Journal of Vocational Behavior*, *14*(2), 224–247. doi:10.1016/0001-8791(79)90072-1

Mueller, J. (2015). Formal and informal practices and knowledge sharing between project teams and enacted cultural characteristics. *Project Management Journal*, *46*(1), 53–68. doi:10.1002/pmj.21471

Mu, J., Peng, G., & MacLachlan, D. (2009). Effect of risk management strategy on NPD performance. *Technovation*, *29*(3), 170–180. doi:10.1016/j.technovation.2008.07.006

Muller, E., & Doloreux, D. (2009). What we should know about knowledge-intensive business services. *Technology in Society*, *31*(1), 64–72. doi:10.1016/j.techsoc.2008.10.001

Mura, M., Lettieri, E., Radaelli, G., & Spiller, N. (2013). Promoting professionals' innovative behaviour through knowledge sharing: The moderating role of social capital. *Journal of Knowledge Management*, *17*(4), 527–544. doi:10.1108/JKM-03-2013-0105

Murphy, K., & Myors, B. (1998). *Statistical Power Analysis*. Mahwah, NJ: Lawrence Erlbaum Associates Publishers.

Murphy, L. M., & Edwards, P. L. (2003). *Bridging the valley of death: Transitioning from public to private sector financing*. Golden, CO: National Renewable Energy Laboratory.

Myers, P. (1996). *Knowledge Management and Organizational Design*. Boston: Butterworth-Heinemann.

Nahapiet, J., & Ghoshal, S. (1998). Social capital, intellectual capital, and the organizational advantage. *Academy of Management Review*, *23*(2), 242–266.

Negash, S. (2004) Business Intelligence. *Communications of the Association for Information Systems, 13*, 177-195. Available at: http://aisel.aisnet.org/cais/vol13/iss1/15

Nemani, R. (Producer). (2012). *Making your IT project a success*. Retrieved from http://resources.css.edu/academics/cis/graduate/docs/PMWorkshop20120305.pdf

Neville, K., Powell, P., & Panteli, N. (2003). Knowledge and security. *AMCIS 2003 Proceedings*, 262.

Nevo, D. (2003). *Developing effective knowledge management systems*. Academic Press.

Newell, S., Bresnen, M., Edelman, L., Scarbrough, H., & Swan, J. (2006). Sharing knowledge across projects: Limits to ICT-led project review practices. *Management Learning*, *37*(2), 167–185. doi:10.1177/1350507606063441

Newell, S., & Edelman, L. F. (2006). Developing a dynamic project learning and cross-project learning capability: Synthesizing two perspectives. *IT Journal*, *18*, 567–591.

Newman, I., & Newman, C. (2000). A discussion of low r-squares: Concerns and uses. *Educational Research Quarterly*, *24*(2), 3–9.

Noe, R. A., & Schmitt, N. (1986). The Influence of Trainee Attitudes on Training Effectiveness: Test of a Model. *Personnel Psychology*, *39*(3), 497–523. doi:10.1111/j.1744-6570.1986.tb00950.x

Nohria, N., & Eccles, R. G. (1998). Where does management knowledge come from? In J. L. Alvarez (Ed.), *The diffusion and consumption of business knowledge* (pp. 278–304). doi:10.1007/978-1-349-25899-4_13

Nonaka, I., & Takeuchi, H. (1995). The Knowledge-creating Company: How Japanese Companies Create the Dynamics of Innovation. New York: Oxford University Press.

Nonaka, I. (1991). The Knowledge Creating Company. *Harvard Business Review*, *69*(6), 96–104.

Nonaka, I. (1994). A Dynamic Theory of Organizational Knowledge Creation. *Organization Science*, *5*(1), 14–37. doi:10.1287/orsc.5.1.14

Nonaka, I., & Takeuchi, H. (1995). *The Knowledge Creating Company: How Japanese Companies Create the Dynamics of Innovation*. New York: Oxford University Press.

Nonaka, I., & Takeuchi, H. (1995). *The Knowledge-Creating Company – How Japanese Companies Create The Dynamics Of Innovation*. Oxford, UK: Oxford University Press.

Nonaka, I., & Takeuchi, H. (1995). *The knowledge-creating company: How Japanese companies create the dynamics of innovation*. Oxford University Press.

Nonaka, I., & Takeuchi, H. (1995). *The Knowledge-Creating Company: How Japanese Companies Create the Dynamics of Innovation: How Japanese Companies Create the Dynamics of Innovation*. New York: Oxford University Press.

Nonaka, I., Toyama, R., & Konno, N. (2000). SECI, Ba and leadership: A unified model of dynamic knowledge creation. *Long Range Planning*, *33*(1), 5–34. doi:10.1016/S0024-6301(99)00115-6

Nonaka, I., Umemoto, K., & Senoo, D. (1996). From Information Processing to Knowledge Creation: A Paradigm Shift in Business Management. *Technology in Society*, *18*(2), 203–218. doi:10.1016/0160-791X(96)00001-2

Nonaka, I., von Krogh, G., & Voelpel, S. (2006). Organizational knowledge creation theory: Evolution paths and future advances. *Organization Studies*, *27*(8), 1179–1208. doi:10.1177/0170840606066312

Noruzy, A., Dalfard, V., Azhdari, B., Nazari-Shirkouhi, S., & Rezazadeh, A. (2013). Relations between transformational leadership, organizational learning, knowledge management, organizational innovation, and organizational performance: An empirical investigation of manufacturing firms. *International Journal of Advanced Manufacturing Technology*, *64*(5-8), 1073–1085. doi:10.100700170-012-4038-y

Nunnally, J. C. (1978). *Psychometric theory*. New York, NY: McGraw-Hill.

Nunnally, J. C. (1978). *Psychometric Theory*. New York: McGraw-Hill.

O'Brien, J. (1996). *Management Information systems*. Chicago: Irwin.

O'Brien, J., & Szerszeń, P. J. (2017). An evaluation of bank measures for market risk before, during and after the financial crisis. *Journal of Banking & Finance*, *80*, 215–234. doi:10.1016/j.jbankfin.2017.03.002

O'Broin, A., & Palmer, S. (2010). Exploring key aspects in the formation of coaching relationships: Initial indicators from the perspective of the coachee and the coach. *Coaching: An International Journal of Theory. Research and Practice*, *3*(2), 124–143.

O'Dell, C., & Grayson, C. J. (1998). If only we knew what we know: Identification and transfer of internal best practices. *California Management Review*, *40*(3), 154–174. doi:10.2307/41165948

Olander, H., Vanhala, M., & Hurmelinna-Laukkanen, P. (2014). Reasons for choosing mechanisms to protect knowledge and innovationsnull. *Management Decision*, *52*(2), 207–229. doi:10.1108/MD-11-2012-0791

Oliveira, M., Curado, C. M. M., Maçada, A., & Nodari, F. (2015). Using alternative scales to measure knowledge sharing behavior: Are there any differences? *Computers in Human Behavior*, *44*, 132–140. doi:10.1016/j.chb.2014.11.042

Osei-Bryson, K. M., Mansingh, G., & Rao, L. (2014). Understanding and applying knowledge management and knowledge management systems in developing countries: Some conceptual foundations. In Knowledge Management for Development (pp. 1-15). Springer US.

Osterloh, M., & Frey, B. S. (2000). Motivation, Knowledge Transfer, and Organizational Forms. *Organization Science*, *11*(5), 538–550. doi:10.1287/orsc.11.5.538.15204

Owen, J. (2006). Integrating KM with programme management. *International Journal of Knowledge Management*, *2*(1), 41–57. doi:10.4018/jkm.2006010104

Owen, J., Burstein, F., & Mitchell, S. (2004). Knowledge reuse and transfer in a project management environment. *Journal of IT Cases and Applications*, *6*(4), 21–35.

Oyelaran-Oyeyinka, B., & Sampath, P. G. (2009). *Latecomer Development, States, Knowledge and Economic Growth*. London: Routledge.

Oza, N., & Abrahamsson, P. (2009). *Building blocks of agile innovation*. Booksurge Llc.

Ozlen, K., Handzic, M., & Durmic, N. (2014). Empirical Study of KMS Impact on Decision Support. *Eurasian Journal of Business and Economics*, *7*(13), 133–144.

Padova, A., & Scarso, E. (2012). Managing large amounts of knowledge objects: Cognitive and organisational problems. *Knowledge Management Research and Practice*, *10*(4), 287–295. doi:10.1057/kmrp.2012.7

Padyab, A. M., Päivärinta, T., & Harnesk, D. (2014). *Genre-Based Assessment of Information and Knowledge Security Risks*. Paper presented at the HICSS. 10.1109/HICSS.2014.428

Padyab, A. M., Päivärinta, T., & Harnesk, D. (2014). Genre-Based Approach to Assessing Information and Knowledge Security Risks. *International Journal of Knowledge Management*, *10*(2), 15. doi:10.4018/ijkm.2014040102

Palacios, D., Gil, I., & Garrigos, F. (2009). The impact of knowledge management on innovation and entrepreneurship in the biotechnology and telecommunications industries. *Small Business Economics*, *32*(3), 291–301. doi:10.100711187-008-9146-6

Pall, G. A. (1987). *Quality process management*. Englewood Cliffs, NJ: Prentice-Hall.

Pangil, F., & Mohd, N. A. (2009). Assessing the relationship between organisational commitment and knowledge sharing behaviour. *Malaysian Management Journal*, *13*(1 & 2), 35–50.

Parent, R., Roy, M., & St-Jaques, D. (2007). A systems-based dynamic knowledge transfer capacity model. *Journal of Knowledge Management*, *11*(6), 81–93. doi:10.1108/13673270710832181

Parzefall, M. R., Seeck, H., & Leppänen, A. (2008). Employee innovativeness in organizations. A review on the antecedents. *Finnish Journal of Business Economics, 2*, 165–182.

Patterson, F., Kerrin, M., Gatto-Roissard, G., & Crowley, T. (2009). Characteristics & behaviours of innovative people in organisations: from theory to policy and practice. *Proceedings of the European Work and Organizational Psychology Conference*.

Paul, J., Mittal, A., & Srivastav, G. (2016). Impact of service quality on customer satisfaction in private and public sector banks. *International Journal of Bank Marketing, 34*(5), 606–662. doi:10.1108/IJBM-03-2015-0030

Pearson, J. M., Pearson, A., & Shim, J. P. (2005). The relevancy of information systems research: The practitioner's view. *Information Resources Management Journal, 18*(3), 50–67. doi:10.4018/irmj.2005070104

Pelled, L. H., Eisenhardt, K. M., & Xin, K. R. (1999). Exploring the black box: An analysis of work group diversity, conflict, and performance. *Administrative Science Quarterly, 44*(1), 1–28. doi:10.2307/2667029

Pellegrinelli, S., & Bowman, C. (1994). Implementing strategy through projects. Long Range Planning, 27(4), 125–132. doi:10.1016/0024-6301(94)90062-0 doi:10.1016/0024-6301(94)90062-0

Peltier, T., Peltier, J., & Blackley, J. (2005). *Information Security Fundamentals*. Boca Raton, FL: CRC Press.

Peppard, J., Ward, J., & Daniel, E. (2007). Managing the realization of business benefits from IT investments. *MIS Quarterly Executive, 6*(1).

Perez-Soltero, A., Zavala-Guerrero, A. G., Barcelo-Valenzuela, M., Sanchez-Schmitz, G., & Meroño-Cerdan, A. L. (2015). A Methodology for the Development and Implementation of Knowledge Management Strategy in a Mexican SME Trading Company. Journal of Knowledge Management, 13(2), 25–44.

Perlroth, N., & Sanger, D. E. (2017). *In Computer Attacks, Clues Point to Frequent Culprit: North Korea*. Retrieved 9/5/2017, 2017, from https://www.nytimes.com/2017/05/15/us/nsa-hacking-shadow-brokers.html?mcubz=1

Perryer, C., Celestine, N. A., Scott-Ladd, B., & Leighton, C. (2016). Enhancing workplace motivation through gamification: Transferrable lessons from pedagogy. *International Journal of Management Education, 14*(3), 327–335. doi:10.1016/j.ijme.2016.07.001

Petter, S., & Randolph, A. B. (2009). Developing soft skills to manage user expectations in IT projects: Knowledge reuse among IT project managers. *Project Management Journal, 40*(4), 45–59. doi:10.1002/pmj.20130

Pfeffer, J., & Fong, C. T. (2002). The end of business schools? Less success than meets the eye. *Academy of Management Learning & Education, 1*(1), 78–95. doi:10.5465/AMLE.2002.7373679

Phua, C., Lee, V., Smith, K., & Gayler, R. (2010). *A comprehensive survey of data mining-based fraud detection research*. arXiv preprint arXiv:1009.6119

Pill, J. (1970). *The Delphi Method: Substance, Context: Critique and an Annotated Bibliography*. Technical Memorandum 183. Cleveland, OH: Case Western Reserve University.

Pindyck, R., & Rubinfeld, D. (1998). *Econometric Models and Economic Forecasts*. Boston, MA: Irwin-McGraw Hill.

Pineda, R. C., & Lerner, L. D. (2006). Goal attainment, satisfaction and learning from teamwork. *Team Performance Management, 12*(5/6), 182–191. doi:10.1108/13527590610687938

Pirolo, L., & Presutti, M. (2010). The Impact of Social Capital on the Start-ups' Performance Growth. *Journal of Small Business Management, 48*(2), 197–227. doi:10.1111/j.1540-627X.2010.00292.x

Pitt, M., McAulay, L., & Sims, D. (2002). Promoting strategic change: 'playmaker' roles in organizational agenda formation. Strategic Change, 11(3), 155–172. doi:10.1002/jsc.586 doi:10.1002/jsc.586

PMI. (2013). A guide to the project management body of knowledge (PMBOK Guide, 5th ed.). Project Management Institute.

Podsakoff, P. M., MacKenzie, S. B., Lee, J. Y., & Podsakoff, N. P. (2003). Common method biases in behavioral research: A critical review of the literature and recommended remedies. *The Journal of Applied Psychology*, *88*(5), 879–903. doi:10.1037/0021-9010.88.5.879 PMID:14516251

Polanyi, M. (1967). The Tacit Dimension. London: Routledge & K. Paul.

Pomerleau, M. (2015). Cheat sheet for the Internet of Things. *GCN, Technology, Tools, and Tactics for the Public Sector*. Retrieved January 30, 2015 from http://gcn.com/articles/2015/01/30/internet-of-things.aspx?m=1

Ponemon, L. (2013). *2013 Cost of Data Breach: Global Analysis*. Ponemon Institute.

Ponemon, L. (2012). *2012 Cost of Cyber Crime Study*. Ponemon Institute.

Potter, J., Lunati, M., Marchese, M., Martinez-Fernandez, C., & Noya, A. (2010). *SMEs, Entrepreneurship and Innovation*. OECD Publishing.

Pressman, R. S. (2005). *Software engineering: a practitioner's approach*. Palgrave Macmillan.

Pretorius, C. J., & Steyn, H. (2005). KM in project environments. *South African Journal of Business Management*, *36*(3), 41–50.

Provost, F., & Fawcett, T. (2013). Data Science and Its Relationship To Big Data And Data-Driven Decision Making. *Big Data*, *1*(1), 51–59. doi:10.1089/big.2013.1508 PMID:27447038

Prusak, L. (2001). Where did knowledge management come from? *IBM Systems Journal*, *40*(4), 1002–1007. doi:10.1147j.404.01002

Quigley, N., Tesluk, P., Locke, E., & Bartol, K. (2007). A multilevel investigation of the motivational mechanisms underlying knowledge sharing and performance. *Organization Science*, *18*(1), 71–92. doi:10.1287/orsc.1060.0223

Quinn, J. B. (1992). *Intelligent Enterprise: A Knowledge and Service Based Paradigm for Industry*. New York: The Free Press.

Quintas, P., Lefrere, P., & Jones, G. (1997). Knowledge management: A strategic agenda. *Long Range Planning*, *30*(3), 385–391. doi:10.1016/S0024-6301(97)90252-1

Quist, J., & Tukker, A. (2013). Knowledge collaboration and learning for sustainable innovation and consumption: Introduction. *Journal of Cleaner Production*, *48*, 167–175. doi:10.1016/j.jclepro.2013.03.051

Radaelli, G., Lettieri, E., Mura, M., & Spiller, N. (2014). Knowledge sharing and innovative work behaviour in healthcare: A micro-level investigation of direct and indirect effects. *Creativity and Innovation Management*, *23*(4), 400–414. doi:10.1111/caim.12084

Raj, P. (2012). Knowledge management – A road map for winning organization. *International Journal of Research in Economics & Social Sciences*, *2*(2), 363–373.

Ramamoorthy, N., Patrick, C., Flood, S. T., & Sardessai, R. (2005). Determinants of innovative work behavior: Development and test of an integrated model. *Creativity and Innovation Management*, *14*(2), 142–150. doi:10.1111/j.1467-8691.2005.00334.x

Ramjeawon, P. V., & Rowley, J. (2017). Knowledge management in higher education institutions: Enablers and barriers in Mauritius. *The Learning Organization*, *24*(5). doi:10.1108/TLO-03-2017-0030

Ranasinghe, S. B., & Dharmadasa, P. (2013). Intention to knowledge sharing: From planned behavior and psychological needs perspectives. *International Journal of Knowledge Management*, *9*(4), 33–50. doi:10.4018/ijkm.2013100103

Randeree, E. (2005). Secure health knowledge: balancing security, privacy and access. *AMCIS 2005 Proceedings*, 287.

Randeree, E. (2006). Knowledge management: Securing the future. *Journal of Knowledge Management*, *10*(4), 145–156. doi:10.1108/13673270610679435

Rea, L. M., & Parker, R. A. (2005). *Designing and Conducting Survey Research: A Comprehensive Guide* (3rd ed.). San Francisco: Jossey-Bass.

Rebernik, M., & Širec, K. (2007). Fostering innovation by unlearning tacit knowledge. *Kybernetes*, *36*(3/4), 406–419. doi:10.1108/03684920710747039

Recker, J. (2013). *Scientific research in information systems*. Heidelberg, Germany: Springer Berlin. doi:10.1007/978-3-642-30048-6

Reich, B. H. (2007). Managing knowledge and learning in IT projects: A conceptual framework and guidelines for practice. *Project Management Journal*, *38*(2), 5–17. doi:10.1080/1058053080215210

Reich, B., Gemino, A., & Sauer, C. (2014). How knowledge management impacts performance in projects: An empirical study. *International Journal of Project Management*, *32*(4), 590–602. doi:10.1016/j.ijproman.2013.09.004

Reilly, R. C. (2010). Participatory case study. In Encyclopedia of case study research (pp. 658–660). Thousand Oaks, CA: Sage.

Rentz, J. O., Shepherd, C. D., Tashchian, A., Dabholkar, P. A., & Ladd, R. T. (2002). A measure of selling skill: Scale development and validation. *Journal of Personal Selling & Sales Management*, *22*(1), 13–21.

Ribière, V., & Walter, C. (2013). 10 years of KM theory and practices. *Knowledge Management Research and Practice*, *11*(1), 4–9. doi:10.1057/kmrp.2012.64

Richardson, M., Abraham, C., & Bond, R. (2012). Psychological correlates of university students' academic performance: A systematic review and meta-analysis. *Psychological Bulletin*, *138*(2), 353–387. doi:10.1037/a0026838 PMID:22352812

Riecicky, V., & Spichiger, A. (2012). Knowledge asset risk management framework. *Human Resource Management Review*, *2*(4), 38–45.

Riege, A. (2005). Three dozen knowledge sharing barriers managers must consider. *Journal of Knowledge Management*, *9*(3), 18–35. doi:10.1108/13673270510602746

Ries, E. (2011). *The lean startup: How today's entrepreneurs use continuous innovation to create radically successful businesses*. Random House LLC.

Ringle, C. M., Wende, S., & Will, A. (2005). *SmartPLS 2.0 (beta).* University of Hamburg. Retrieved from http://www.smartpls.de/

Rivard, S., Poirier, G., Raymond, L., & Bergeron, F. (1994). Development of a measure to assess the quality of user-developed applications. In J.F. Nunamaker, & R.H. Sprague (Eds.), *Proceedings of the Twenty-Seventh Annual Hawaii International Conference on System Sciences* (pp. 522-531). ICSS. 10.1109/HICSS.1994.323466

Roberts, C. H., & Skeat, T. C. (1983). *The birth of the codex*. London: Oxford University Press.

Rockstuhl, T., Dulebohn, J. H., Ang, S., & Shore, L. M. (2012). Leader–member exchange (LMX) and culture: A meta-analysis of correlates of LMX across 23 countries. *The Journal of Applied Psychology*, *97*(6), 1097–1130. doi:10.1037/a0029978 PMID:22985117

Rodriguez, E., & Edwards, J. S. (2010). People, technology, processes and risk knowledge sharing. *Electronic Journal of Knowledge Management*, *8*(1), 139–150.

Roling, N. G. (1992). The emergence of knowledge systems thinking: A changing perception of relationships among innovation, knowledge process and configuration. *Knowledge and Policy*, *5*(1), 42–64. doi:10.1007/BF02692791

Rolland, C. (1998, June). A comprehensive view of process engineering. In *Advanced Information Systems Engineering* (pp. 1–24). Springer Berlin Heidelberg. doi:10.1007/BFb0054216

Rottman, J. W. (2008). Successful knowledge transfer within offshore supplier networks: A case study exploring social capital in strategic alliances. *Journal of Information Technology*, *23*(1), 31–43. doi:10.1057/palgrave.jit.2000127

Rousseau, D. M., & McCarthy, S. (2007). Educating managers from an evidence-based perspective. *Academy of Management Learning & Education*, *6*(1), 84–101. doi:10.5465/AMLE.2007.24401705

Rousseau, S., & Rousseau, R. (1998). The scientific wealth of European nations: Taking effectiveness into account. *Scientometrics*, *42*(1), 75–87. doi:10.1007/BF02465013

Rowden, R. W. (2002). The relationship between workplace learning and job satisfaction in US small to midsize businesses. *Human Resource Development Quarterly*, *13*(4), 407–425. doi:10.1002/hrdq.1041

Rowley, J. (2007). The wisdom hierarchy: Representations of the DIKW hierarchy. *Journal of Information Science*, *33*(2), 163–180. doi:10.1177/0165551506070706

Rubenstein-Montano, B., Liebowitz, J., Buchwalter, J., McCaw, D., Newman, B., & Rebeck, K. (2001). A system thinking framework for knowledge management. Decision Support Systems, 31(1), 5–16. doi:10.1016/S0167-9236(00)00116-0 doi:10.1016/S0167-9236(00)00116-0

Rubenstein-Montano, B., Liebowitz, J., Buchwalter, J., McCaw, D., Newman, B., & Rebeck, K. (2001). Smartvision: A knowledge-management methodology. *Journal of Knowledge Management*, *5*(4), 300–310. doi:10.1108/13673270110411724

Ruggles, R. (1998). The state of the notion: Knowledge management in practice. California Management Review, 40(3), 80–89. doi:10.2307/41165944 doi:10.2307/41165944

Rumizen. (2002). *The Complete Idiot's Guide to Knowledge Management.* New York: Penguin Group/Alpha Books.

Ryan, S., Harden, G., Ibragimova, B., & Windsor, J. (2012). The business value of knowledge management. *AMCIS 2012 Proceedings*.

Ryan, R. M., & Deci, E. L. (2000). Self-Determination Theory and the Facilitation of Intrinsic Motivation, Social Development, and Well-Being. *The American Psychologist*, *55*(1), 68–78. doi:10.1037/0003-066X.55.1.68 PMID:11392867

Ryan, S. D., Zhang, X., Prybutok, V. R., & Sharp, J. H. (2012). Leadership and knowledge management in an e-government environment. *American Scientist*, *2*(1), 63–81.

Rynes, S. L., Bartunek, J. M., & Daft, R. L. (2001). Across the great divide: Knowledge creation and transfer between practitioners and academics. *Academy of Management Journal*, *44*(2), 350–355.

Sachdev, I., & Bourhis, R. (1987). Status differentials and intergroup behaviour. *European Journal of Social Psychology*, *17*(3), 277–293. doi:10.1002/ejsp.2420170304

Saenger, P. (1975). Colard Mansion and the evolution of the printed book. *The Library Quarterly*, *45*(4), 405–418. doi:10.1086/620426

Safa, N. S., & Von Solms, R. (2016). An information security knowledge sharing model in organizations. *Computers in Human Behavior*, *57*, 442–451. doi:10.1016/j.chb.2015.12.037

Saginova, O., & Belyansky, V. (2008). Facilitating innovations in higher education in transition economies. *International Journal of Educational Management*, *22*(4), 341–351. doi:10.1108/09513540810875671

Saha, P., Parameswaran, N., Chakraborty, B., & Mahanti, A. (2013). *A Formal Analysis of Fraud in Banking Sector.* Paper presented at the System Sciences (HICSS), 2013 46th Hawaii International Conference on. 10.1109/HICSS.2013.36

Saha, P., Bose, I., & Mahanti, A. (2016). A knowledge based scheme for risk assessment in loan processing by banks. *Decision Support Systems*, *84*, 78–88. doi:10.1016/j.dss.2016.02.002

Sajeva, S., & Jucevicius, R. (2010). Determination of essential knowledge management system components and their parameters. *Social Sciences*, *1*(67), 80–90.

Salaman, J. G., & Storey, J. (2002). Managers' theories about the process of innovation. *Journal of Management Studies*, *39*(2), 147–165. doi:10.1111/1467-6486.00286

Salleh, N. A., Jusoh, R., & Isa, C. R. (2010). Relationship between information systems sophistication and performance measurement. *Industrial Management & Data Systems*, *110*(7), 993–1017. doi:10.1108/02635571011069077

Sambamurthy, V., & Subramani, M. (2005). Editorial: Special issue on information technologies and knowledge management. *Management Information Systems Quarterly*, *29*(1), 1–7. doi:10.2307/25148665

Samson, D. (2010). *Innovation for Business Success: achieving a systematic innovation capability.* Report for the Victoria Government Department of Innovation, Industry, Science and Research. Available at http://www.innovation.gov.au/Industry/IndustryInnovationCouncils

San Nicolas-Rocca, T., Schooley, B., & Spears, J. L. (2014). Exploring the effect of knowledge transfer practices on user compliance to is security practices. *International Journal of Knowledge Management*, *10*(2), 62–78. doi:10.4018/ijkm.2014040105

Sánchez, D., Vila, M. A., Cerda, L., & Serrano, J. M. (2009). Association rules applied to credit card fraud detection. *Expert Systems with Applications*, *36*(2), 3630–3640. doi:10.1016/j.eswa.2008.02.001

Santos, V. R., Soares, A. L., & Carvalho, J. Á. (2012). Knowledge sharing barriers in complex research and development projects: An exploratory study on the perceptions of project managers. *Knowledge and Process Management*, *19*(1), 27–38. doi:10.1002/kpm.1379

Sanz-Valle, R., Naranjo-Valencia, J., Jiménez-Jiménez, D., & Perez-Caballero, L. (2011). Linking organizational learning with technical innovation and organizational culture. *Journal of Knowledge Management*, *15*(6), 997–1015. doi:10.1108/13673271111179334

Sareminia, S., Shamizanjani, M., Mousakhani, M., & Manian, A. (2016). Project knowledge management: An ontological view. *Knowledge Management & E-Learning: An International Journal*, *8*(2), 292–316.

Sarigianni, C., Thalmann, S., & Manhart, M. (2016). *Protecting Knowledge in the Financial Sector: An Analysis of Knowledge Risks Arising from Social Media.* Paper presented at the HICSS.

Sarker, S., Sarker, S., Nicholson, D. B., & Joshi, K. D. (2005). Knowledge transfer in virtual systems development teams: An exploratory study of four key enablers. *IEEE Transactions on Professional Communication*, *48*(2), 201–218. doi:10.1109/TPC.2005.849650

Sawhney, M., Verona, G., & Prandell, E. (2005). Collaborating to create: The Internet as a Platform for Customer Engagement in Product Innovation. *Journal of Interactive Marketing*, *19*(4), 1–15. doi:10.1002/dir.20046

Saxton, M. K., Saxton, T., Steen, J., & Verreynne, M. (2010). Failure to advance: Resource logic for early venture failure. *Prometheus*, *28*(1), 15–27. doi:10.1080/08109021003668620

Sazandrishvili, N. (2009). *Contextual and personal antecedents of innovative behavior* (Master's thesis). University of Twente.

Scarbrough, H. (2002). The role of intermediary groups in shaping management fashion. *International Studies of Management & Organization*, *32*(4), 87–103.

Schermerhorn, J. R., & Bond, M. H. (1997). Cross-cultural leadership dynamics in collectivism and high power distance settings. *Leadership and Organization Development Journal*, *18*(4), 187–193. doi:10.1108/01437739710182287

Schinagl, S., Paans, R., & Schoon, K. (2016). *The Revival of Ancient Information Security Models, Insight in Risks and Selection of Measures.* Paper presented at the System Sciences (HICSS), 2016 49th Hawaii International Conference on. 10.1109/HICSS.2016.501

Schiuma, G., Durst, S., & Wilhelm, S. (2012). Knowledge management and succession planning in SMEs. *Journal of Knowledge Management*, *16*(4), 637–649. doi:10.1108/13673271211246194

Schneid, M., Isidor, R., Steinmetz, H., & Kabst, R. (2016). Age diversity and team outcomes: A quantitative review. *Journal of Managerial Psychology*, *31*(1), 2–17. doi:10.1108/JMP-07-2012-0228

Scholl, H. (2005). Innovationas drivers of growth. In *Proceedings of the CBI Conference Synopsis Sharing Knowledge with the Business and Academic Communities*. Carnegie Bosch Institute for Applied Studies in International Management.

Schon, D. A. (1963). Champions for radical new inventions. Harvard Business Review, (March–April): 77–86.

Schoonhoven, C. B., & Eisenhardt, K. M. (2012). Regions as Industrial Incubators of Technology-based Ventures. In Sources of Metropolitan Growth. Transaction Publishers.

Schreiber, G. (2000). *Knowledge engineering and management: the CommonKADS methodology*. MIT Press.

Schreiber, G., Wielinga, B., de Hoog, R., Akkermans, H., & Van de Velde, W. (1994). CommonKADS: A comprehensive methodology for kbs development. *IEEE Expert*, *9*(6), 28–37. doi:10.1109/64.363263

Schumpeter, J. A. (1947). The creative response in economic history. *The Journal of Economic History*, *7*(2), 149–159. doi:10.1017/S0022050700054279

Scott, J. (2000). Rational choice theory. In G. Browning, A. Halcli, & F. Webster (Eds.), *Understanding contemporary society: Theories of the present* (pp. 126–138). London: Sage Publications. doi:10.4135/9781446218310.n9

Scott, S. G., & Bruce, R. A. (1994). Determinants of innovative behavior: A path model of individual innovation in the workplace. *Academy of Management Journal*, *37*(3), 580–607. doi:10.2307/256701

Scovetta, V., & Ellis, T. J. (2015). Leadership social power as a component of KMS success. *International Journal of Knowledge Management*, *11*(2), 1–14. doi:10.4018/IJKM.2015040101

Serenko, A. (2013). Meta-analysis of scientometric research of knowledge management: Discovering the identity of the discipline. *Journal of Knowledge Management, 17*(5), 773–812. doi:10.1108/JKM-05-2013-0166

Serenko, A., Bontis, N., & Hull, E. (2011). Practical relevance of knowledge management and intellectual capital scholarly research: Books as knowledge translation agents. *Knowledge and Process Management, 18*(1), 1–9. doi:10.1002/kpm.363

Serenko, A., Bontis, N., & Moshonsky, M. (2012). Books as a knowledge translation mechanism: Citation analysis and author survey. *Journal of Knowledge Management, 16*(3), 495–511. doi:10.1108/13673271211238797

Serenko, A., & Dumay, J. C. (2015). Citation classics published in knowledge management journals. Part I: Articles and their characteristics. *Journal of Knowledge Management, 19*(2), 401–431. doi:10.1108/JKM-06-2014-0220

Sezgin, O. B., Uçar, Z., & Duygulu, E. (2015). Güven, yenilikçi iş davranişi ilişkisinde bilgi paylaşiminin aracilik rolü. *İşletme Fakültesi Dergisi, 16*(2), 1-20.

Shahid, A., & Alamgir, R. (2011). *ICT enabled knowledge sharing–impact of ict on knowledge sharing barriers: The case of avanade* (Master Thesis).

Shanker, M., & Sayeed, O. B. (2012). Role of Transformational Leaders as Change Agents: Leveraging Effects on Organizational Climate. Indian Journal of Industrial Relations, 47(3), 470–484.

Shanshan, S. (2014). A comprehensive relational model of factors influencing knowledge sharing: An empirical study. *International Journal of Knowledge Management, 10*(1), 1–25. doi:10.4018/ijkm.2014010101

Shariq, S., & Vendelo, M. (2006). Tacit Knowledge Sharing. In D. Schwartz (Ed.), *Encyclopedia of Knowledge Management* (pp. 833–839). Hershey, PA: Idea Group Reference. doi:10.4018/978-1-59140-573-3.ch109

Sharma, N. (2004). *The Origin of the "Data Information Knowledge Wisdom" Hierarchy*. Accessed June 17, 2008: http://wwwpersonal.si.umich.edu/~nsharma/dikw_origin.htm

Shaw, D., Hall, M., Edwards, J. S., & Baker, B. (2007). Responding to crisis through strategic knowledge management. *Journal of Organizational Change Management, 20*(4), 559–578. doi:10.1108/09534810710760081

Shedden, P., Scheepers, R., Smith, W., & Ahmad, A. (2011). Incorporating a knowledge perspective into security risk assessments. *Vine, 41*(2), 152–166. doi:10.1108/03055721111134790

Shedden, P., Smith, W., & Ahmad, A. (2010). Information Security Risk Assessment: Towards a Business Practice Perspective. *Australian Information Security Management Conference.*

Shenhar, A. J., & Dvir, D. (2007a). *Reinventing project management.* Boston: Harvard Business School Press.

Shenhar, A. J., & Dvir, D. (2007b). Project management research-The challenge and opportunity. *Project Management Journal, 38*(2), 93–99.

Shenhar, A. J., Milosevic, D., Dvir, D., & Thamhain, H. (2007). *Linking project management to business strategy.* Newtown Square, PA: Project Management Institute.

Siegel, D. S., Waldman, D. A., Atwater, L. E., & Link, A. N. (2004). Toward a model of effective transfer of scientific knowledge from academicians to practitioners: Qualitative evidence from the commercialization of university technologies. *Journal of Engineering Technology, 21*(1), 115–142.

Silva, L. O. (2002). Outsourcing as an improvisation: A case study in Latin America. *The Information Society, 18*(2), 129–138. doi:10.1080/01972240290075057

Silvius, A. G. (2007). Business & IT Alignment in theory and practice. In *System Sciences, 2007. HICSS 2007. 40th Annual Hawaii International Conference on* (p. 211b–211b). IEEE. 10.1109/HICSS.2007.119

Simmons, P. G., Dawley, D., Richie, W., & Anthony, W. (2001). An exploratory examination of the knowledge transfer of strategic management concepts from the academic environment to practicing managers. *Journal of Managerial Issues, 13*(3), 360–376.

Simpson, B. (2002). The Knowledge Needs of Innovating Organisations. *Singapore Management Review, 24*(30), 51–61.

Siponen, M. T. (2005). Analysis of modern IS security development approaches: Towards the next generation of social and adaptable ISS methods. *Information and Organization, 15*(4), 339–375. doi:10.1016/j.infoandorg.2004.11.001

Sivasubramaniam, N., Liebowitz, S. J., & Lackman, C. L. (2012). Determinants of new product development team performance: A meta-analytic review. *Journal of Product Innovation Management, 29*(5), 803–820. doi:10.1111/j.1540-5885.2012.00940.x

Skulmoski, G. J., Hartman, F. T., & Krahn, J. (2007). The Delphi method for graduate research. *Journal of Information Technology Education, 6*, 1–21. doi:10.28945/199

Smirnov, A. V., & Sandkuhl, K. (2015). Context-Oriented Knowledge Management for Decision Support in Business Networks: Modern Requirements and Challenges. *Joint Proceedings of the BIR 2015 Workshops and Doctoral Consortium co-located with 14th International Conference on Perspectives in Business Informatics Research (BIR 2015), 1420*, 9-23.

Smith, B. R., Matthews, C. H., & Schenkel, M. T. (2009). Differences in Entrepreneurial Opportunities: The Role of Tacitness and Codification in Opportunity Identification*. *Journal of Small Business Management, 47*(1), 38–57. doi:10.1111/j.1540-627X.2008.00261.x

Smith, C. L., Jarrett, M., & Bierer, S. B. (2013). Integrating clinical medicine into biomedical graduate education to promote translational research: Strategies from two new PhD programs. *Academic Medicine, 88*(1), 137–143. doi:10.1097/ACM.0b013e31827644ef PMID:23165264

Smith, T. A., Mills, A. M., & Dion, P. (2010). Linking business strategy and knowledge management capabilities for organizational effectiveness. *International Journal of Knowledge Management, 6*(3), 22–43. doi:10.4018/jkm.2010070102

Snowden, D. (2002). Complex acts of knowing: Paradox and descriptive self-awareness. *Journal of Knowledge Management, 6*(2), 110–111. doi:10.1108/13673270210424639

Solli-Sæther, H., Karlsen, J. T., & van Oorschot, K. (2015). Strategic and cultural knowledge sharing barriers in project networks. *Project Management Journal, 46*(3), 49–60. doi:10.1002/pmj.21501

Soomro, Z. A., Shah, M. H., & Ahmed, J. (2016). Information security management needs more holistic approach: A literature review. *International Journal of Information Management, 36*(2), 215–225. doi:10.1016/j.ijinfomgt.2015.11.009

Sophos. (2013). *Security Threat Report 2013*. Author.

Sözbilir, F., &Yeşil, S. (2016).The Impact of Knowledge Management on Competitive Advantage: An Empirical Study in Turkey. *Bilgi,* (32), 92-116.

Spears, J. L. (2006). A holistic risk analysis method for identifying information security risks. In Security Management, Integrity, and Internal Control in Information Systems (pp. 185–202). Springer.

Spears, J. L., & San Nicolas-Rocca, T. (2016). *Information Security Capacity Building in Community-Based Organizations: Examining the Effects of Knowledge Transfer*. Paper presented at the HICSS. 10.1109/HICSS.2016.498

Spector, P. E. (1987). Method variance as an artifact in self-reported affect and perceptions at work: Myth or significant problem? *The Journal of Applied Psychology*, *72*(3), 438–443. doi:10.1037/0021-9010.72.3.438

Spector, P. E. (2006). Method variance in organizational research: Truth or urban legend? *Organizational Research Methods*, *9*(2), 221–232. doi:10.1177/1094428105284955

Spender, J. (2005). Speaking about management education: Some history of the search for academic legitimacy and the ownership and control of management knowledge. *Management Decision*, *42*(10), 1282–1292. doi:10.1108/00251740510634868

Spender, J.-C., & Grant, R. M. (1996). Knowledge and the firm: Overview. *Strategic Management Journal*, *17*(S2), 5–9. doi:10.1002mj.4250171103

Spies, M., Clayton, A. J., & Noormohammadian, M. (2005). Knowledge management in a decentralized global financial services provider: A case study with Allianz Group. *Knowledge Management Research and Practice*, *3*(1), 24–36. doi:10.1057/palgrave.kmrp.8500046

Spithoven, A., Clarysse, B., & Knockaert, M. (2011). Building absorptive capacity to organize inbound open innovation in traditional industries. *Technovation*, *31*(1), 10–21. doi:10.1016/j.technovation.2010.10.003

Spremic, M. (2012). Corporate IT Risk Management model: A holistic view at managing information system security risks. In *Information Technology Interfaces (ITI), Proceedings of the ITI 2012 34th International Conference on* (pp. 299–304). IEEE.

Stahl, G. K., Maznevski, M. L., Voight, A., & Jonsen, K. (2010). Unraveling the effects of cultural diversity in teams: A meta-analysis of research on multicultural work groups. *Journal of International Business Studies*, *41*(4), 690–709. doi:10.1057/jibs.2009.85

Stam, W., Arzlanian, S., & Elfring, T. (2014). Social capital of entrepreneurs and small firm performance: A meta-analysis of contextual and methodological moderators. *Journal of Business Venturing*, *29*(1), 152–173. doi:10.1016/j.jbusvent.2013.01.002

Standards Australia. (2005). *Australian Standard in Knowledge Management: A Guide AS 5037-2005*. Sydney: SAI Global.

Starkey, K., & Madan, P. (2001). Bridging the relevance gap: Aligning stakeholders in the future of management research. *British Journal of Management*, *12*(S1), S3–S26. doi:10.1111/1467-8551.12.s1.2

Steffens, P. R., Weeks, C. S., Davidsson, P., & Isaak, L. (2014). Shouting from the ivory tower: A marketing approach to improve communication of academic research to entrepreneurs. *Entrepreneurship Theory and Practice*, *38*(2), 399–426. doi:10.1111/etap.12079

Stewart, T. (1997). *Intellectual Capital: The New Wealth of Organizations*. London: Nicholas Brealey.

Stoneburner, G., Gogunen, A., & Feringa, A. (2007). NIST Special Publication 800-30, Risk Management Guide for Information Technology Systems. United States National Institute of Standards and Technology.

Stoneburner, G., Goguen, A., & Feringa, A. (2002). Risk management guide for information technology systems. *Nist Special Publication*, *800*(30), 800–830.

Stonier, T. (2012). *Beyond information: The natural history of intelligence*. Springer Science & Business Media.

Storey, J., & Barnett, E. (2000). Knowledge management initiatives: Learning from failure. *Journal of Knowledge Management*, *4*(2), 145–156. doi:10.1108/13673270010372279

Straub, D., & Ang, S. (2011). Editor's comments: Rigor and relevance in IS research: Redefining the debate and a call for future research. *Management Information Systems Quarterly*, *35*(1), iii–xi. doi:10.2307/23043485

Studer, R., Benjamins, V. R., & Fensel, D. (1998). Knowledge Engineering. *Principles and Methods in Data & Knowledge Engineering, 25*(1-2), 161–197. doi:10.1016/S0169-023X(97)00056-6

Suddaby, R., & Greenwood, R. (2001). Colonizing knowledge: Commodification as a dynamic of jurisdictional expansion in professional service firms. *Human Relations, 54*(7), 933–953. doi:10.1177/0018726701547007

Susanty, A., Puspitasari, D., Puspitasari, N. B., & Ninditarini, M. R. (2011). *Preliminary study of key success factors for effective knowledge transfer in SMEs Batik.* 2011 International Conference on Industrial Engineering and Operations Management, Kuala Lumpur, Malaysia.

Susman, G. I., & Evered, R. D. (1978). An assessment of the scientific merits of action research. *Administrative Science Quarterly, 23*(4), 582–603. doi:10.2307/2392581

Swan, J., Newell, S., & Robertson, M. (2000). Knowledge management when will people management enter the debate? In R.H. Sprague (Ed.), *Proceedings of the 33rd Hawaii International Conference on System Sciences* (pp. 3026–3035). Los Alamitos, CA: IEEE Computer Society Press. doi:10.1109/HICSS.2000.92670910.1109/HICSS.2000.926709

Swan, J., Newell, S., Scarborough, H., & Hislop, D. (1999). Knowledge management and innovation: Networks and networking. *Journal of Knowledge Management, 3*(4), 262–275. doi:10.1108/13673279910304014

Swap, W., Leonard, D., Shields, M., & Abrams, L. (2001). Using Mentoring and Storytelling to Transfer Knowledge in the Workplace. *Journal of Management Information Systems, 18*(1), 95–114. doi:10.1080/07421222.2001.11045668

Syed-Ikhsan, S. O. S., & Rowland, F. (2004). Knowledge management in a public organization: A study on the relationship between organizational elements and the performance of knowledge transfer. *Journal of Knowledge Management, 8*(2), 95–111. doi:10.1108/13673270410529145

Szulanski, G. (1996). Exploring internal stickiness: Impediments to the transfer of best practice within the firm strategic. *Management Journal, 17,* 27–43.

Szulanski, G. (1996). Exploring internal stickiness: Impediments to the transfer of best practice within the firm. *Strategic Management Journal, 17*(1), 27–43. doi:10.1002mj.4250171105

Szulanski, G. (2000). The process of knowledge transfer: A diachronic analysis of stickiness. *Organizational Behavior and Human Decision Processes, 82*(1), 9–27. doi:10.1006/obhd.2000.2884

Taber, T. D., & Deosthali, K. (2014). Analysis of self-reported motives for task-related helping: Implications for an integrated theory of helping. *Journal of Business and Psychology, 29*(3), 343–366. doi:10.100710869-013-9327-4

Tagaris, A., Konnis, G., Benetou, X., Dimakopoulos, T., Kassis, K., Athanasiadis, N., & Koutsouris, D. (2009, November). Integrated web services platform for the facilitation of fraud detection in health care e-government services. In *Information Technology and Applications in Biomedicine, 2009. ITAB 2009. 9th International Conference on* (pp. 1-4). IEEE. 10.1109/ITAB.2009.5394355

Taherparvar, N., Esmaeilpour, R., & Dostar, M. (2014). Customer knowledge management, innovation capability and business performance: A case study of the banking industry. *Journal of Knowledge Management, 18*(3), 430–446. doi:10.1108/JKM-11-2013-0446

Tajfel, H., & Turner, J. C. (1979). An integrative theory of intergroup conflict. In W. G. Austin & S. Worchel (Eds.), *The social psychology of intergroup relations*. Monterey, CA: Brooks.

Taleb, N., Goldstein, D., & Spitznagel, M. (2009). The six mistakes executives make in risk management. *Harvard Business Review, 87*(10), 78–81. PMID:19968059

Tamjidyamcholo, A., Baba, M. S. B., Tamjid, H., & Gholipour, R. (2013). Information security—Professional perceptions of knowledge-sharing intention under self efficacy, trust, reciprocity, and shared-language. *Computers & Education, 68*, 223 232.

Tan, B., Watson, R., & Wei, K. K. (1995). National culture and group support systems: Filtering communication to dampen power differentials. *European Journal of Information Systems, 4*(2), 82–92. doi:10.1057/ejis.1995.10

Tanriverdi, H. (2005). Information technology relatedness, knowledge management capability, and performance of multi-business firms. Management Information Systems Quarterly, 29(2), 311–334. doi:10.2307/25148681 doi:10.2307/25148681

Tanriverdi, H., & Venkatraman, N. (2005). Knowledge relatedness and performance of multibusiness firms. Strategic Management Journal, 26(2), 97–119. doi:10.1002/smj.435 doi:10.1002mj.435

Taormina, R. J., & Bauer, T. N. (2000). Organizational socialization in two cultures: Results from the United States and Hong Kong. *The International Journal of Organizational Analysis, 8*(3), 262–289. doi:10.1108/eb028919

Tapper, J. (2017). *White House Officials Tricked by Email Prankster.* Retrieved 9/5/2017, 2017, from http://www.cnn.com/2017/07/31/politics/white-house-officials-tricked-by-email-prankster/index.html

Tauber, D., & Schwartz, D. G. (2011). *Integrating Knowledge Management with the Systems Analysis Process.* Academic Press.

Tavares, E., & Schneck de Paula Pessoa, M. (2014). Technological innovation projects: Proposal for an integrative model between project management and knowledge management in a customer-supplier perspective. *Journal of Information Systems and Technology Management, 11*(1), 105–130. doi:10.4301/S1807-17752014000100007

Taylor, H., Artman, E., & Woelfer, J. P. (2012). Information technology project risk management: Bridging the gap between research and practice. *Journal of Information Technology, 27*(1), 17–34. doi:10.1057/jit.2011.29

Te'eni, D. (2006). Organisational Communication. In D. Schwartz (Ed.), *Encyclopedia of Knowledge Management* (pp. 734–740). Hershey, PA: Idea Group Reference. doi:10.4018/978-1-59140-573-3.ch096

Teece, D. J. (2000). Strategies for managing knowledge assets: The role of firm structure and industrial context. Long Range Planning, 33(4), 35–54. doi:10.1016/S0024-6301(99)00117-X doi:10.1016/S0024-6301(99)00117-X

Teece, D. (1998). Capturing Value from Knowledge Assets: The new economy, markets for know-how and intangible assets. *California Management Review, 40*(3), 55–79. doi:10.2307/41165943

Teece, D. (2009). *Dynamic Capabilities and Strategic Management: Organizing for Innovation and Growth.* Oxford, UK: Oxford University Press.

Teece, D. (2010). Business Models, Business Strategy and Innovation. *Long Range Planning, 43*(2-3), 172–194. doi:10.1016/j.lrp.2009.07.003

Tenenhaus, M., Vinzi, V. E., Chatelin, Y. M., & Lauro, C. (2005). PLS path modeling. *Computational Statistics & Data Analysis, 48*(1), 159–205. doi:10.1016/j.csda.2004.03.005

Terrell, M. S. (2000, January). Implement a lessons learned process that works. *Hydrocarbon Processing*, 97–100.

Terzieva, M., & Vincenzo, M. (2016). Learning from experience: The project team is the key. *Business Systems Research, 7*(1), 1–15. doi:10.1515/bsrj-2016-0001

Terziovski, M. (2007). *Building innovation capability in organizations - an international cross-case perspective.* Imperial College Press. doi:10.1142/p492

Tesch, R. (1990). *Qualitative research*. New York: The Falmer Press.

Thalmann, S., Manhart, M., Ceravolo, P., & Azzini, A. (2014). An Integrated Risk Management Framework: Measuring the Success of Organizational Knowledge Protection. *International Journal of Knowledge Management*, *10*(2), 28–42. doi:10.4018/ijkm.2014040103

The Hofstede Centre. (2017). National Culture. *The Hofstede Centre Website*. Retrieved August 17, 2017, from, http://geert-hofstede.com/national-culture.html

Thierauf, R. J. (2001). *Effective business intelligence systems*. Greenwood Publishing Group.

Thomas, K. W., & Tymon, W. G. Jr. (1982). Necessary properties of relevant research: Lessons from recent criticisms of the Organizational Sciences. *Academy of Management Review*, *7*(3), 345–352.

Thune, J. (2017). *Senate Committee on Commerce, Science, and Transportation*. Retrieved from https://www.govtrack.us/congress/committees/SSCM

Thuraisingham, V., & Nonmonotonic, A. (1992). *Typed Multilevel Logic for Multilevel Secure Data / Knowledge Base Management Systems - II*. Bedford, MA: The MITRE Corporation.

Tidd, J., & Bessant, J. (2014). *Strategic Innovation Management*. Chichester, UK: John Wiley.

Tobin, T., & Pennington, W. (2015). Organizational networks and the process of corporate entrepreneurship: How the motivation, opportunity, and ability to act affect firm knowledge, learning and innovation. *Small Business Economics*.

Trkman, P., & Desouza, K. C. (2012). Knowledge risks in organizational networks: An exploratory framework. *The Journal of Strategic Information Systems*, *21*(1), 1–17. doi:10.1016/j.jsis.2011.11.001

Trochim, W., & Donnelly, J. (2007). *The Research Methods Knowledge Base*. Cincinnati, OH: Atomic Dog Publishing.

Tsui, E. (2003). Tracking the Role and Evolution of Commercial Knowledge Management Software. In C. W. Holsapple (Ed.), *Handbook on Knowledge Management 2: Knowledge Directions* (pp. 5–27). New York: Springer. doi:10.1007/978-3-540-24748-7_1

Tuomi, I. (2000). Data is More Than Knowledge: Implications of the Reversed Knowledge Hierarchy for Knowledge Management and Organizational Memory. *Journal of Management Information Systems*, *16*(3), 103–117. doi:10.1080/07421222.1999.11518258

Turner, J. R., Zimmerman, T., & Allen, J. M. (2012). Teams as a sub-process for knowledge management. *Journal of Knowledge Management*, *16*(6), 963–977. doi:10.1108/13673271211276227

Turoff, M. & Hiltz, S.R. (1996). Computer based Delphi processes. *Gazing into the oracle: The Delphu method and its application to social policy and public health*, 56-85.

Turoff, M., & Hiltz, R. S. (1996). Computer-based Delphi processes. In Gazing Into the Oracle: The Delphi Method and Its Application to Social Policy and Public Health (pp. 56-88). London: Jessica Kingsley Publishers.

Tushman, M. L., & Nadler, D. (1986). Organizing for innovation. California Management Review, 28(3), 74–92. doi:10.2307/41165203 doi:10.2307/41165203

Tyrrell, S. (2000). The many dimensions of the software process. *Crossroads*, *6*(4), 22–26. doi:10.1145/333424.333435

Tzortzaki, A. M., & Mihiotis, A. (2014). A Review of Knowledge Management Theory and Future Directions. *Knowledge and Process Management*, *21*(1), 29–41. doi:10.1002/kpm.1429

Uchupalanan, K. (2000). Competition and IT-based innovation in banking services. *International Journal of Innovation Management, 4*(4), 455–489. doi:10.1142/S1363919600000238

United States General Accounting Office. (2002). *NASA: Better Mechanisms Needed for Sharing Lessons Learned.* Washington, DC: United States General Accounting Office.

Unsworth, K. L., & Parker, S. K. (2003). Proactivity and innovation: Promoting a new workforce for the new workplace. In D. Holman, T. D. Wall, C. W. Clegg, P. Sparrow, & A. Howard (Eds.), *The new workplace: A guide to the human impact of modern working practices* (pp. 175–196). Chichester, UK: Wiley.

Upadhyava, S., Rao, H. R., & Padmanabhan, G. (2006). Secure Knowledge Management. In D. G. Schwartz (Ed.), *Encyclopedia of Knowledge Management* (pp. 795–801). Idea Group Inc. doi:10.4018/978-1-59140-573-3.ch104

Uzzi, B., & Lancaster, R. (2003). Relational embeddedness and learning: The case of bank loan managers and their clients. *Management Science, 49*(4), 383–399. doi:10.1287/mnsc.49.4.383.14427

Van Aken, J. E. (2005). Management research as a design science: Articulating the research products of Mode 2 knowledge production in management. *British Journal of Management, 16*(1), 19–36. doi:10.1111/j.1467-8551.2005.00437.x

Van de Ven, A. H. (1986). Central problems in the management of innovation. *Management Science, 32*(5), 590–607. doi:10.1287/mnsc.32.5.590

Van de Ven, A. H., & Johnson, P. E. (2006). Knowledge for theory and practice. *Academy of Management Review, 31*(4), 802–821. doi:10.5465/AMR.2006.22527385

Van den Hooff, B., & De Ridder, J. A. (2004). Knowledge sharing in context: The influence of organizational commitment, communication climate and cmc use on knowledge sharing. *Journal of Knowledge Management, 8*(6), 117–130. doi:10.1108/13673270410567675

Van Den Hooff, B., & Van Weenen, F. (2004). Committed to share: Commitment and cmc use as antecedents of knowledge sharing. *Knowledge and Process Management, 11*(1), 13–24. doi:10.1002/kpm.187

Van der Aa, W., & Elfring, T. (2002). Realizing innovation in services. *Scandinavian Journal of Management, 18*(2), 155–171. doi:10.1016/S0956-5221(00)00040-3

van Knippenberg, D. A., De Dreu, C. K. W., & Homan, A. C. (2004). Work group diversity and group performance: An integrative model and research agenda. *The Journal of Applied Psychology, 89*(6), 1008–1022. doi:10.1037/0021-9010.89.6.1008 PMID:15584838

Van Riel, A., Lemmink, J., & Ouwersloot, H. (2004). High-technology service innovation success: A decision-making perspective. *Journal of Product Innovation Management, 21*(5), 348–359. doi:10.1111/j.0737-6782.2004.00087.x

Van Wijk, R., Jansen, J. J., & Lyles, M. A. (2008). Inter-and intra-organizational knowledge transfer: A meta-analytic review and assessment of its antecedents and consequences. *Journal of Management Studies, 45*(4), 830–853. doi:10.1111/j.1467-6486.2008.00771.x

Varadarajan, P. R. (2003). Musing on relevance and rigor of scholarly research in marketing. *Journal of the Academy of Marketing Science, 31*(4), 368–376. doi:10.1177/0092070303258240

Värlander, S. (2008). The role of situated embodied interaction in the banking customer knowledge creation process. *International Journal of Knowledge Management, 4*(4), 62–76. doi:10.4018/jkm.2008100104

Varnell-Sarjeant, J. F. (2008). Managing a Man-rated Software Development Program via Risk Mitigation. *SIGSOFT Softw. Eng. Notes, 33*(4), 8:1–8:8.

Väyrynen, K., Hekkala, R., & Liias, T. (2013). Knowledge Protection Challenges of Social Media Encountered by Organizations. *Journal of Organizational Computing and Electronic Commerce*, *23*(1–2), 34–55. doi:10.1080/10919392.2013.748607

Venkitachalam, K., & Busch, P. (2012). Tacit knowledge: Review and possible research directions. *Journal of Knowledge Management*, *16*(2), 357–372. doi:10.1108/13673271211218915

Vera-Munoz, S. C., Ho, J. L., & Chow, C. W. (2006). Enhancing knowledge sharing in public accounting firms. *Accounting Horizons*, *20*(2), 133–155. doi:10.2308/acch.2006.20.2.133

Verdu, S. (1998). Fifty years of Shannon theory. *IEEE Transactions on Information Theory*, *44*(6), 2057–2078. doi:10.1109/18.720531

Vlacheas, P., Giaffreda, R., Stavroulaki, V., Kelaidonis, D., Foteinos, V., Poulios, G., & Moessner, K. (2013). Enabling smart cities through a cognitive management framework for the internet of things. *Communications Magazine, IEEE*, *51*(6), 102–111. doi:10.1109/MCOM.2013.6525602

Voelpel, S. C., & Han, Z. (2005). Managing knowledge sharing in China: The case of Siemens ShareNet. *Journal of Knowledge Management*, *9*(3), 51–63. doi:10.1108/13673270510602764

von Krogh, G. (2009). Individualist and collectivist perspectives on knowledge in organizations: Implications for information systems research. *The Journal of Strategic Information Systems*, *18*(3), 119–129. doi:10.1016/j.jsis.2009.08.001

von Krogh, G. (2012). How does social software change knowledge management? Toward a strategic research agenda. *The Journal of Strategic Information Systems*, *21*(2), 154–164. doi:10.1016/j.jsis.2012.04.003

Von Solms, B., & Von Solms, R. (2004). The 10 deadly sins of information security management. *Computers & Security*, *23*(5), 371–376. doi:10.1016/j.cose.2004.05.002

Von Zedtwitz, M. (2003). Post-project reviews in R&D. *Research Technology Management*, *46*(5), 43–49.

Waldvogel, A., & Whelan, N. (2008). Towards better financial risk learning. *Journal of Risk Management in Financial Institutions*, *4*(4), 382–393.

Wallace, D. P. (2007). Introduction. In D. Wallace (Ed.), *Knowledge management: historical and cross-disciplinary themes*. London: Libraries Unlimited. doi:10.4135/9781446278581.n1

Wang, Z., Wang, N., Cao, J., & Ye, X. (2016). The impact of intellectual capital – knowledge management strategy fit on firm performance. Management Decision, 54(8), 1861–1885. doi:10.1108/MD-06-2015-0231 doi:10.1108/MD-06-2015-0231

Wang, L., Wang, H., & Wang, K. (2010). The investigation analysis on knowledge-sharing barriers and incentives of university scientific research and innovation teams management and service science (MASS). *Proceedings of the International Conference on*, 1-4.

Wangpipatwong, S. (2009). Factors influencing knowledge sharing among university students. *Proceedings of the 17th International Conference on Computers in Education*, 800-807.

Wang, S., & Noe, R. A. (2010). Knowledge sharing: A review and directions for future research. *Human Resource Management Review*, *20*(2), 115–131. doi:10.1016/j.hrmr.2009.10.001

Wang, S., Noe, R. A., & Wang, Z. M. (2014). Motivating knowledge sharing in knowledge management systems: A quasi–field experiment. *Journal of Management*, *40*(4), 978–1009. doi:10.1177/0149206311412192

Wang, Y., & Zhu, Y. (2012). Organizational socialization, trust, knowledge sharing and innovation behavior: Research on mechanism and path. *R & D Management*.

Wan, Z., Haggerty, N., & Wang, Y. (2015). Individual level knowledge transfer in virtual settings: A review and synthesis. *International Journal of Knowledge Management, 11*(2), 29–61. doi:10.4018/IJKM.2015040103

Ward, J., & Daniel, E. (2006). *Benefits management: delivering value from IS and IT investments*. John Wiley & Sons.

Ward, J., Taylor, P., & Bond, P. (1996). Evaluation and realisation of IS/IT benefits: An empirical study of current practice. *European Journal of Information Systems, 4*(4), 214–225. doi:10.1057/ejis.1996.3

Ward, V., House, A., & Hamer, S. (2009). Knowledge brokering: The missing link in the evidence to action chain? *Evidence & Policy: A Journal of Research. Debate and Practice, 5*(3), 267–279. doi:10.1332/174426409X463811

Wasko, M. M., & Faraj, S. (2000). "It is what one does": Why people participate and help others in electronic communities of practice. *The Journal of Strategic Information Systems, 9*(2), 155–173. doi:10.1016/S0963-8687(00)00045-7

Wasko, M. M., & Faraj, S. (2005). Why should I share? Examining knowledge contribution in networks of practice. *Management Information Systems Quarterly, 21*(9), 35–57. doi:10.2307/25148667

Watson, H. J., & Wixom, B. H. (2007). The Current State of Business Intelligence. *Computer, 40*(9), 96–99. doi:10.1109/MC.2007.331

Watson, R. T. (2008). *Data management, databases and organizations*. John Wiley & Sons.

Weaven, S., Grace, D., Dant, R., & Brown, J. (2014). Value creation through knowledge management in franchising: A multi-level conceptual framework. *Journal of Services Marketing, 28*(2), 97–104. doi:10.1108/JSM-09-2013-0251

Webb, J., Maynard, S., Ahmad, A., & Shanks, G. (2013). *Towards an Intelligence-Driven Information Security Risk Management Process for Organisations*. Presented at the *24th Australasian Conference on Information Systems: Intelligence-Driven ISRM for Organisations*, Melbourne, Australia.

Weber, H. H., Kessel, M., Budych, K., & Schultz, C. (2011). Shared communication processes within healthcare teams for rare diseases and their influence on healthcare professionals' innovative behavior and patient satisfaction. *Implementation Science; IS, 6*(40). Retrieved from http://www.implementationscience.com/content/6/1/40 PMID:21510848

Weber, R. P. (1990). *Basic content analysis*. Newbury Park, CA: Sage Publications. doi:10.4135/9781412983488

Weick, K. E. (2001). Gapping the relevance bridge: Fashion meets fundamentals in management research. *British Journal of Management, 12*(s1), 71–76. doi:10.1111/1467-8551.12.s1.9

Weinberger, D. (2012). *Too Big to Know: Rethinking Knowledge Now That the Facts Aren't the Facts, Experts Are Everywhere, and the Smartest Person in the Room Is the Room*. Basic Books.

Weiss, N. E., & Miller, R. S. (2015). *The Target and Other Financial Data Breaches: Frequently Asked Questions*. Congressional Research Service. Retrieved on September 9, 2017 from https://fas.org/sgp/crs/misc/R43496.pdf

Weitzel, J. R., & Kerschberg, L. (1989). Developing knowledge-based systems: Reorganizing the system development life cycle. *Communications of the ACM, 32*(4), 482–488. doi:10.1145/63334.63340

Wei, X., Liu, Y., & Chen, S. (2015). A meta-analysis of the relationship between team demographic diversity and team performance. *Acta Psychologica Sinica, 47*(9), 1172–1187. doi:10.3724/SP.J.1041.2015.01172

Wenger, E. (1998). *Communities of practice*. Cambridge, UK: Cambridge University Press. doi:10.1017/CBO9780511803932

Wenger, E., McDermott, R., & Snyder, W. (2002). *Cultivating Communities of Practice: A Guide to Managing Knowledge*. Boston, MA: Harvard Business School Press.

Wessner, C. W. (2008). *An Assessment of the SBIR Program*. National Academies Press (US) National Research Council Committee on Capitalizing on Science, Technology, and Innovation.

West, M. A. (1989). Innovation among health care professionals. *Social Behaviour*, *4*, 173–184.

West, M. A., & Farr, J. L. (1989). Innovation at work: Psychological perspectives. *Social Behaviour*, *4*, 15–30.

Westover, J. H. (2010). Managing Organizational Change: Change Agent Strategies and Techniques to Successfully Managing the Dynamics of Stability and Change in Organizations. International Journal of Management and Innovation, 2(1), 45–50.

Wetzels, M., Odekerken-Schroder, G., & Van Oppen, C. (2009). Using PLS path modeling for assessing hierarchical construct models: Guidelines and empirical illustration. *Management Information Systems Quarterly*, *33*(1), 177–196. doi:10.2307/20650284

Wheelen, T. L., & Hunger, D. J. (2006). Concepts in strategic management and business policy. Upper Saddle River, NJ: Pearson – Prentice Hall.

Whitman, M., & Mattord, H. (2011). *Principles of information security*. Cengage Learning.

Wiig, K. (2004). *People-Focused Knowledge Management*. Burlington, MA: Elsevier Butterworth Heinemann.

Wiig, K. M. (1997). Knowledge management: Where did it come from and where will it go? *Expert Systems with Applications*, *13*(1), 1–14. doi:10.1016/S0957-4174(97)00018-3

Wikipedia. (2018). *Big Data*. Retrieved on March 25, 2018 from http://en.wikipedia.org/wiki/Big_data

Wikipedia. (2018). *Internet of Things*. Retrieved on March 25, 2018 from http://en.wikipedia.org/wiki/Internet_of_Things

Wilkesmann, U., Fischer, H., & Wilkesmann, M. (2009). Cultural characteristics of knowledge transfer. *Journal of Knowledge Management*, *13*(6), 464–477. doi:10.1108/13673270910997123

Wilson, S. R. (1997). Developing theories of persuasive message production: The next generation. In J. O. Greene (Ed.), *Advances in communication theory* (pp. 15–43). Hillsdale, NJ: Lawrence Erlbaum.

Winter, R., & Chaves, M. S. (2017). Innovation in the management of lessons learned in an IT project with the adoption of social media. *International Journal of Innovation*, *5*(2), 156–170. doi:10.5585/iji.v5i2.155

Wipawayangkool, K. (2009). Information security compliances and knowledge management capabilities in international diversification. *AMCIS 2009 Proceedings*, 604.

Wood, R. E. (1986). Task Complexity: Definition of the Construct. *Organizational Behavior and Human Decision Processes*, *37*(1), 60–82. doi:10.1016/0749-5978(86)90044-0

World Values Survey. (2017). Data and Documentation. *World Values Survey*. Retrieved August 31, 2017, from, http://www.worldvaluessurvey.org/WVSContents.jsp

Wu, C., Parker, S. K., & De Jong, J. P. J. (2011). Need for cognition as an antecedent of individual innovation behavior. *Journal of Management*, *21*, 1–24.

Wu, I., & Lin, H. (2009). A strategy-based process for implementing knowledge management: An integrative view and empirical study. *Journal of the American Society for Information Science and Technology*, *60*(4), 789–802. doi:10.1002/asi.20999

Wu, J. B., Hom, P. W., Tetrick, L. E., Shore, L. M., Jia, L., Chaoping, L., & Song, L. J. (2006). The norm of reciprocity: Scale development and validation in the Chinese context. *Management and Organization Review*, *2*(3), 377–402. doi:10.1111/j.1740-8784.2006.00047.x

Wu, L.-C., Ong, C.-S., & Hsu, Y.-W. (2008). Active ERP implementation management: A real options perspective. *Journal of Systems and Software*, *81*(6), 1039–1050. doi:10.1016/j.jss.2007.10.004

Xerri, M., & Brunetto, Y. (2011). Fostering the innovative behaviour of sme employees: A social capital perspective. *Research and Practice in Human Resource Management*, *19*(2), 43–59.

Xiaojun, L., & Peng, L. (2010). *The impact of learning culture on individual innovative behaviour*. Institute of Electrical and Electronics Engineers.

Xue, Y., Liang, H., Hauser, R., & O'Hara, M. (2012). An empirical study of knowledge sharing intention within virtual teams. *International Journal of Knowledge Management*, *8*(3), 47–61. doi:10.4018/jkm.2012070103

Xu, J., & Quaddus, M. (2005). Adoption and diffusion of knowledge management systems: An Australian survey. *Journal of Management Development*, *24*(4), 335–361. doi:10.1108/02621710510591343

Yang, J. (2007). The impact of knowledge sharing on organizational learning and effectiveness. *Journal of Knowledge Management*, *11*(2), 83–90. doi:10.1108/13673270710738933

Yang, J. (2010). The knowledge management strategy and its effect on firm performance: A contingency analysis. *International Journal of Production Economics*, *125*(2), 215–223. doi:10.1016/j.ijpe.2010.03.012

Ye, C., Jha, S., & Desouza, K. C. (2015). Communicating the Business Value of Innovation. *International Journal of Innovation Science*, *7*(1), 1–12. doi:10.1260/1757-2223.7.1.1

Yeoh, P., & Roth, K. (1999). An empirical analysis of sustained advantage in the U.S. pharmaceutical industry: Impact of firm resources and capabilities. Strategic Management Journal, 20(7), 637–653. doi:10.1002/(SICI)1097-0266(199907)20:7<637::AID-SMJ42>3.0.CO;2-Z doi:10.1002/(SICI)1097-0266(199907)20:7<637::AID-SMJ423.0.CO;2-Z

Yeşil, S., & Sözbilir, F. (2013). An Empirical Investigation into the Impact of Personality on Individual Innovation Behaviour in the Workplace. *Procedia: Social and Behavioral Sciences*, *81*, 540–551. doi:10.1016/j.sbspro.2013.06.474

Yew Wong, K., & Aspinwall, E. (2005). An empirical study of the important factors for knowledge-management adoption in the SME sector. *Journal of Knowledge Management*, *9*(3), 64–82. doi:10.1108/13673270510602773

Yin, R. K. (2014). *Case Study Research: Design and Methods* (5th ed.). Thousand Oaks, CA: Sage.

Yousuf, M. I. (2007). Using experts' opinions through Delphi technique. *Practical Assessment, Research & Evaluation*, *12*(4), 1–8.

Yukl, G. A. (2002). Leadership in Organizations (5th ed.). Upper Saddle River, NJ: Prentice Hall.

Yusuf, S. (2008). Intermediating knowledge exchange between universities and businesses. *Research Policy*, *37*(8), 1167–1174. doi:10.1016/j.respol.2008.04.011

Yu, X., Shi, Y., Zhang, L., Nie, G., & Huang, A. (2014). Intelligent Knowledge Beyond Data Mining: Influences of Habitual Domains. *Communications of the Association for Information Systems*, *34*, 53. Available at http://aisel.aisnet.org/cais/vol34/iss1/53

Zack, M. H. (1999a). Developing a knowledge strategy. California Management Review, 41(3), 125–145. doi:10.2307/41166000 doi:10.2307/41166000

Zack, M. H. (1999b). Managing codified knowledge. Sloan Management Review, 40(4), 45–58.

Zack, M., McKeen, J., & Singh, S. (2009). Knowledge management and organizational performance: An exploratory analysis. *Journal of Knowledge Management, 13*(6), 392–409. doi:10.1108/13673270910997088

Zaidman, N., & Drory, A. (2001). Upward impression management in the work place cross-cultural analysis. *International Journal of Intercultural Relations, 25*(6), 671–690. doi:10.1016/S0147-1767(01)00031-1

Zhang, X., Chen, Z., Vogel, D., & Guo, C. (2009). Exchange ideology as a moderator of knowledge sharing in virtual teams: A social exchange theory perspective. *International Journal of Internet and Enterprise Management, 6*(2), 143–163. doi:10.1504/IJIEM.2009.023927

Zhao, L., Detlor, B., & Connelly, C. E. (2016). Sharing knowledge in social Q&A sites: The unintended consequences of extrinsic motivation. *Journal of Management Information Systems, 33*(1), 70–100. doi:10.1080/07421222.2016.1172459

Zhao, X., Hwang, B.-G., & Pheng, S. (2016). An enterprise risk management knowledge-based decision support system for construction firms, *Singapore Engineering. Construction and Architectural Management, 23*(3), 369–384. doi:10.1108/ECAM-03-2015-0042

Zheng, W., Yang, B., & McLean, G. (2010). Linking organizational culture, structure, strategy, and organizational effectiveness: Mediating role of knowledge management. *Journal of Business Research, 63*(7), 763–771. doi:10.1016/j.jbusres.2009.06.005

Zhou, H., & Uhlaner, L. (2009). *Knowledge management as a strategic tool to foster innovativeness of SMEs.* ERIM Report Series Reference No. ERS- 2009-025-ORG. Retrieved from http://ssrn.com/abstract=1410468

Zhou, A., & Fink, D. (2003). The intellectual capital Web: A systematic linking of intellectual capital and knowledge management. *Journal of Intellectual Capital, 4*(1), 34–48. doi:10.1108/14691930310455379

Zolnik, E. J. (2010). The role of postdoctoral fellows in technology transfer: Evidence from the National Capital Region of the USA. *International Journal of Knowledge-Based Development, 1*(3), 158–175. doi:10.1504/IJKBD.2010.035657

ZoomInfo. (2010). *Home*. Retrieved September 1, 2010, from http://www.zoominfo.com/search

Zqikael, O., Levin, G., & Rad, P. F. (2008). Top management support - The project friendly organization. *Cost Engineering (Morgantown, W. Va.), 50*(9), 22–30.

Zyngier, S. (2008). Risk Management: Strengthening Knowledge Management. *International Journal of Knowledge Management, 4*(3), 1–32. doi:10.4018/jkm.2008070102

Zyngier, S., & Burstein, F. (2012). Knowledge management governance: The road to continuous benefits realization. *Journal of Information Technology, 27*(2), 140–155. doi:10.1057/jit.2011.31

Zyngier, S., & Venkitachalam, K. (2011). Knowledge management governance – a strategic driver. *Knowledge Management Research and Practice, 9*(2), 136–150. doi:10.1057/kmrp.2011.5

About the Contributors

Murray E. Jennex is a Professor of Management Information Systems at San Diego State University, editor-in-chief of the International Journal of Knowledge Management, co-editor-in-chief of IGI Global book series, co-editor-in chief of the International Journal of Information Systems for Crisis Response and Management, and President of the Foundation for Knowledge Management (LLC). Dr. Jennex specializes in knowledge management, crisis response, system analysis and design, IS security, e-commerce, and organizational effectiveness. Dr. Jennex serves as the Knowledge, Innovation, and Entrepreneurial Systems Track co-chair at the Hawaii International Conference on System Sciences. He is the author of over 150 journal articles, book chapters, and conference proceedings on knowledge management, crisis response, end user computing, international information systems, organizational memory systems, ecommerce, cyber security, and software outsourcing. Dr. Jennex is a former US Navy Nuclear Power Propulsion officer and holds a B.A. in chemistry and physics from William Jewell College, an M.B.A. and an M.S. in software engineering from National University, an M.S. in telecommunications management and a Ph.D. in information systems from the Claremont Graduate University. Dr. Jennex is also a registered professional mechanical engineer in the state of California and a Certified Information Systems Security Professional (CISSP), a Certified Secure Software Lifecycle Professional (CSSLP), and a Project Management Professional (PMP).

* * *

Annabeth Aagaard (Ph.D., MsC) is an Associate Professor at Aarhus University, Department of Business Development and Technology and holds a PhD within Pharmaceutical front-end innovation. She is the Centre director of Centre for Business Development at Aarhus University and has a total of 20 years of experience with innovation management, knowledge management, sustainable business and business model innovation drawing on experience in the academic, public and private sectors, where she was formerly a manager and management specialist working with top 100 Scandinavian companies. Furthermore she has written extensively on management, innovation and sustainability publishing several scientific journal articles in journals like: Journal of Product Innovation Management, Creativity & Innovation management, Scandinavian Journal of Management etc. and has authored nine academic books and management handbook.

Ricardo Anderson is a Lecturer in the Department of Computing at the University of the West Indies Mona, Jamaica. His research interests include Knowledge Management Systems (KMS), Knowledge Discovery in Databases and Data Mining (KDDM). His current work covers the use of KDDM and KMS in several domains including education and social welfare. Ricardo has recent papers published several journals and conferences including IJKM, AMCIS, IJBIR and a chapter in the edited book Knowledge Management for Development (Springer, 2014). Prior to becoming an academic, he worked in IT management and software development for a few years. Ricardo obtained his bachelors and masters degrees in Computer Science and Information Systems respectively from the UWI and is currently a PhD candidate in Computer Science. He is also a Project Management Professional (PMP).

Cesar Bandera is Assistant Professor of Entrepreneurship at the Martin Tuchman School of Management, New Jersey Institute of Technology. His research focus is on the incubation of new ventures, and on entrepreneurship pedagogy. Dr. Bandera is also a serial entrepreneur, and has led multiple technology companies from start-up to successful harvest. He is Senior Member of the IEEE, served as Director of the Academy of the International Council for Small Business, and is an adviser to business incubators in the US, Europe, Africa, and South America. He holds a Ph.D. in Electrical Engineering from SUNY Buffalo.

Michael R. Bartolacci is Professor of Information Sciences and Technology at Penn State University - Berks. He holds a Ph.D. in Industrial Engineering and an MBA from Lehigh University, and a B.A. in Industrial Engineering from Lafayette College. He conducts research in telecommunication systems, supply chain modeling, knowledge management, and cultural aspects of information technologies.

Nicole Celestine received a B.A with first class honours from the University of Western Australia's Business School in 2015, and is currently undertaking a PhD. With a focus in organisational behaviour and facilitating positive social change through management processes, her current research interests include play at work, motivation, and knowledge transfer.

Sanjib Chowdhury is a Professor of Management at Eastern Michigan University. He holds a PhD from the University of North Texas. His current research interests are customer interaction, human capital, and knowledge sharing. His publications have appeared in journals such as Academy of Management Journal, Journal of Business Venturing, and Journal of Business Research.

Alexandra Durcikova is an Assistant Professor at the Price College of Management, the University of Oklahoma. Her research interests include knowledge management and knowledge management systems, the role of organizational climate in the use of knowledge management systems, knowledge management system characteristics, governance mechanisms in the use of knowledge management systems; and human compliance with security policy and characteristics of successful phishing attempts within the area of network security. Her work appeared or will appear in ISR, JMIS, CACM, IJKM, CAIS, ISJ, and JOEUC.

John S. Edwards is Emeritus Professor and Professor of Knowledge Management at Aston Business School, Birmingham, U.K. He holds MA and PhD degrees from Cambridge University. His interest has always been in how people can and do (or do not) use models and systems to help them do things. At present his principal research interests include how knowledge affects risk management; investigating knowledge management strategy and its implementation; and the synergy between knowledge management, analytics and big data. He has written more than 70 peer-reviewed research papers and three books on these topics. He is consulting editor of the journal Knowledge Management Research & Practice.

Tim Ellis' background is an eclectic blend of three career streams: an officer in the United States Navy (Commander, USNR-R), a counseling psychologist (CRC, CAC), and a professor in the informing sciences. His current research and teaching interests focus in two areas: advanced learning technologies, with an emphasis on mobile learning systems, and knowledge management, with an emphasis on knowledge sharing. He is a consistent contributor to both the information systems and educational technology literature, publishing in both highly respected journals (Computers & Education, Informing Science Journal) and conferences (Hawaii International Conference on Systems Sciences, Frontiers in Education). Prior to joining the GSCIS faculty in 1999, was chair of the Business department at Fisher College in Boston, MA where he was instrumental in inaugurating that department and in initiating the college's online learning program.

Megan L. Endres is a Professor of Management at Eastern Michigan University. She holds a PhD from the University of North Texas. Her research focuses on knowledge sharing, self-efficacy, and team processes. Her publications appeared in journals such as Academy of Management Journal, Journal of Managerial Issues, and Theory and Decision.

Marianne Gloet is a Research Fellow in the Department of Management and Marketing at The University of Melbourne, Australia. She holds a PhD from Melbourne University in the field of knowledge management and innovation, and is widely published in the areas of knowledge management, human resource management, innovation and information systems. Marianne has lived and worked in Australia, Canada, the U.S., Malaysia, Hong Kong, Vietnam and the United Arab Emirates, holding senior positions in both the academic and private sectors. Marianne also consults widely to a range of private and public organizations both in Australia and abroad.

Meliha Handzic is Professor of Management and Information Systems at the International Burch University, Sarajevo, Bosnia and Herzegovina. She gained her PhD in Information Systems from the University of New South Wales, Sydney, Australia. Meliha's main research interests lie in the field of knowledge management (KM) with a particular focus on the processes and socio-technological enablers of knowledge development and transfer. She has published widely on these topics in books, journals and conference proceedings. Meliha is also an active member of IAKM, Regional editor of KMRP and serves on editorial boards, executive and program committees for numerous international and national journals and conferences.

Bengü Hirlak is a lecturer at Kilis 7 Aralık University. She graduated from Çukurova University, completed her Master degree at Gaziantep University, and finished her PhD at KSU. Her research lies in the area of HRM and knowledge management.

Ilona Ilvonen is a post-doctoral researcher and lecturer in Tampere University of Technology. Her research focuses on the area of knowledge protection and knowledge management, and she aims to understand the role of people as contributors to security in various contexts.

Jari Jussila, B.Eng., M.Sc., D.Sc., is a Project Researcher at University of Jyväskylä in Finland. He completed his doctoral thesis on social media use in business-to-business companies' innovation. His research is currently focused on social media, big social data analytics, and health informatics.

Hannu Kärkkäinen is Professor of Knowledge Management at the Laboratory of Industrial and Information Management of Tampere University of Technology in Finland. His current research interests include knowledge management and decision making in innovation, organizational learning, customer needs assessment in business-to-business organizations, and co-operation and value networks in innovation. Social media and social data analytics are currently some of his focal research interests, as they bring new important possibilities to the above generic topics. He has published a number of refereed international journal articles in journals like International Journal of Technology Management, R&D Management and International Journal of Production Economics, as well as several books and other publications.

Ronald John Lofaro (PhD, 2 MA's) is an aviation psychologist who has been FAA liaison and visiting professor in Human Factors and Systems at Embry-Riddle Aeronautical University, FL. Prior, he had been the sole FAA liaison to the ten USAF Research Labs and the USAF Aeronautical Systems Center. During that time, he also was a visiting professor, with dual appointments in Aviation and in Systems Engineering at The Ohio State University. He has authored 80 articles and Technical Reports, plus 8 book chapters in reference texts. Topics include knowledge engineering; knowledge management; core competencies; aeronautical decision-making in the operational environment (ODM); operational/usability testing, cognitive ergonomics; crew resource management (CRM); aviator candidate selection and classification; aircrew/pilot training and performance evaluation; aviation security human factors and educational issues. His affiliations include past membership in the Aerospace Medical Association/Human Factors, the International Association of Aviation Psychologists and the Human Factors and Ergonomics Society.

Gunjan Mansingh is a Senior Lecturer at the Department of Computing, the University of the West Indies (U.W.I), Mona, Jamaica. She obtained a PhD. in Information Systems and an M.Phil. in Computer Science from U.W.I. and a B.Sc. from St Xavier's College, Bombay University, India. She worked for a few years in the IT industry in Jamaica before moving to academia. She teaches various courses at the undergraduate and the graduate level in Computer Science and Information Systems. She has conducted workshops in Jamaica on data mining and business intelligence. In 2013 she received a best research publication award and in 2014 she received a teaching award at her university. She is a co-editor of an edited book titled "Knowledge Management for Development: Domains, Strategies and Technologies for Developing Countries", Springer Integrated Series in Information Systems. She also serves on the international editorial review board of International Journal of Knowledge Management. Her research interests are in the following areas; Business Intelligence, Data Mining, Decision Support Systems, Knowledge Management and Knowledge Management Systems, Expert Systems and Technology Adoption. In her research she has worked in different domains in Jamaica such as healthcare, crime, agriculture, e-commerce and financial services. Her research focus has been on harnessing data, information and knowledge in the various sources to assist in the decision-making process. She has over 40 publications in international journals, in the proceedings of several international conferences and as book chapters.

Donald McKay serves as the Associate Dean at the Forbes School of Business and Technology at Ashford University. Don joined Forbes School of Business in 2012. His research interests include knowledge management, project management, and supply chain information systems. Don has presented at the Americas Conference on Information Systems and Hawaii International Conference on Systems Sciences. Prior to joining Forbes School of Business, Don was a consultant and executive in global transportation and logistics. His experience included managing the development of leading edge strategic and tactical decision support systems. Throughout his career, Don also led a number of global business development and operations project teams. He earned his Ph.D. from Nova Southeastern University's Graduate School of Computer and Information Sciences. Don also holds a PMP from the Project Management Institute.

Kurshad Ozlen is Assistant Professor at CENTRUM Católica, Graduate School of Business, Lima, Peru. His research interests include Management Information Systems, Knowledge Management and Production Management. He has substantially published on Knowledge Management and Supply Chain Management.

Katia Passerini is Dean of the College of Professional Studies at St. John's University, where she also holds a Professor appointment in the Division of Computer Science, Mathematics and Science. Prior to joining St. John's, she was Professor and Hurlburt Chair of MIS (2003-2013) and Dean of the Albert Dorman Honors College (2013-16) at NJIT. She holds MBA and PhD degrees in MIS from George Washington University. Prior to academia, she worked as a management consultant in the automotive and telecom industries. Her research focuses on understanding drivers of knowledge management, wireless broadband applications trends; and, computer-supported learning.

Chris Perryer is an Adjunct Lecturer at the University of Western Australia's Business School. He teaches in the areas of international management, international strategy, general management, and business ethics and sustainability. His current research interests encompass a broad range of cross-cultural and cross generational management issues.

Eduardo Rodriguez is the Sentry Endowed Chair in Business Analytics University of Wisconsin-Stevens Point. He has been Analytics Adjunct Professor at Telfer School of Management at Ottawa University, Corporate Faculty of the MSc in Analytics and ISEM MSC at Harrisburg University of Science and Technology Pennsylvania USA, Senior Associate-Faculty of the Center for Dynamic Leadership Models in Global Business at The Leadership Alliance Inc. Toronto Canada, and Principal at IQAnalytics Inc. Research Centre and Consulting Firm in Ottawa Canada. Eduardo has extensive experience in analytics, knowledge and risk management mainly in the insurance and banking industry. He has been Visiting Scholar Chongqing University China, and Risk Monitoring Professor in the Master of Risk Management at EAFIT University in Colombia. He has been Knowledge Management Advisor and Quantitative Analyst at EDC Export Development Canada in Ottawa, Regional Director of PRMIA (Professional Risk Managers International Association) in Ottawa, Vice-President Marketing and Planning for Insurance Companies and Banks in Colombia, professor at Andes University and CESA in Colombia, author of six books in analytics, reviewer of several journals and with publications in peer-reviewed journals and conferences. He created and is Chair of the permanent Think-Tank in Analytics in Ottawa, Chair of the International Conference in Analytics ICAS, member of academic committees for the European and International conferences in Knowledge Management, and international lecturer in the analytics field.

Danny Samson is Professor of Management at the University of Melbourne (since 1988), was Head of the Department of Management in the Faculty of Economics and Commerce for three years and was Director of the Foundation for Sustainable Economic Development there. He has an honors degree in chemical engineering (UNSW) and a PhD in management from the Australian Graduate School of Management. He has published over 100 articles and a dozen books in fields ranging from decision sciences to operations management and business excellence. During his academic career he has been a consultant to senior executives in most manufacturing industries and in numerous service sector organizations in Australia and elsewhere.

Alexander Serenko is a Professor of Management Information Systems & Strategy in the Faculty of Business Administration at Lakehead University, Canada. Dr. Serenko holds a Ph.D. in Management Information Systems from McMaster University. His research interests pertain to scientometrics, knowledge management, and technology addiction. Alexander has published over 70 articles in refereed journals, including MIS Quarterly, European Journal of Information Systems, Information & Management, Communications of the ACM, and Journal of Knowledge Management. He has also won six Best Paper awards at Canadian and international conferences. In 2015, Dr. Serenko received the Distinguished Researcher Award which is the highest honor conferred by Lakehead University for research and scholarly activity.

Salih Yeşil is a full-time lecturer at Kahramanmaraş Sütçü İmam University. He graduated from Hacettepe University in Turkey, completed his MBA at Cleveland State University in the USA, and finished his PhD at Nottingham University in England. His work lies in the area of knowledge management, culture, innovation and related issues. He has published innovation and knowledge management related articles in international journals and conferences.

Index

L

M

N

O

P

R

S

T

U

V

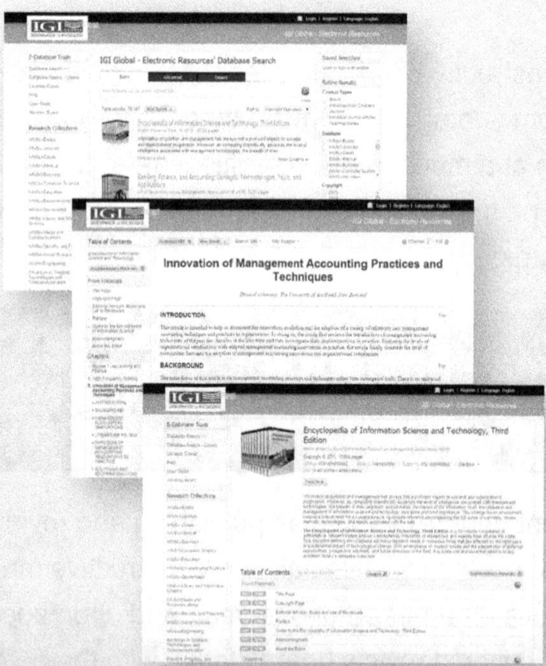
IGI Global - Electronic Resources' Database Search
Innovation of Management Accounting Practices and Techniques
Encyclopedia of Information Science and Technology, Third Edition
Table of Contents

www.ingramcontent.com/pod-product-compliance
Lightning Source LLC
LaVergne TN
LVHW060404300926
849162LV00006B/1340